Mineralogy

Concepts
Descriptions
Determinations

Mineralogy

Second Edition

Concepts
Descriptions
Determinations

L. G. Berry
Late of
Queen's University,
Kingston, Ontario

Brian Mason
United States
National Museum

Second Edition Revised by
R. V. Dietrich
Central Michigan University

 **W. H. Freeman and Company**
San Francisco

Project Editor: Larry Olsen
Copy Editor: Ruth Cottrell
Designer: Christopher Werner
Production Coordinator: Linda Jupiter
Illustration Coordinator: Richard Quiñones
Artists: Roger Hayward and John Foster
Compositor: Interactive Composition Corporation
Printer and Binder: Kingsport Press, Kingsport, Tennessee
Cover: The mineral specimen shown on the cover is the
"Candelabra," sometimes referred to as the "Uncle Sam." It is color-
zoned tourmaline (elbaite) crystals attached to quartz, with albite and
lepidolite. This specimen was removed from the Tourmaline Queen
mine, Pala, San Diego County, California, on New Year's Day, 1972.
It is now on display in the United States National Museum of Natural
History (Smithsonian Institution), Washington, D.C. The overall size
of the specimen is approximately 25 × 22 × 15 centimeters. The
photograph first appeared on the cover of The Mineralogical Record,
Volume 6, Number 2. A black and white photograph of this specimen is reproduced in the text as Figure 15-42. Photograph courtesy
of John S. White, Jr. Cover design by Christopher Werner.

Library of Congress Cataloging in Publication Data

Berry, L. G. (Leonard Gascoigne), 1914–
    Mineralogy—concepts, descriptions, determinations.

    Includes bibliographies and index.
    1. Mineralogy.   I. Mason, Brian Harold, 1917–
II. Dietrich, Richard Vincent, 1924–
III. Title.
QE363.2.B4   1983   549   82-16008
ISBN 0-7167-1424-8

Printed in the United States of America

9 8 7 6 5 4 3 2 1   KP   1 0 8 9 8 7 6 5 4 3

Contents

Part II   Descriptions

Part III   Determinations

viii   Part IV   Appendixes

Preface

This edition of *Mineralogy: Concepts, Descriptions, Determinations* is a major revision of the first edition, published in 1959. As was true for the 1959 edition, this edition is intended primarily to serve as a textbook for a first course in mineralogy. In addition, it should also serve as a useful reference for geologists, petrologists, pedologists, solid-state physicists, and others who use crystallography or mineralogy as a keystone for their professional deliberations.

A fundamental assumption underlying the treatment of mineralogy in this book is that a basic knowledge of minerals provides a key to the interpretation of, for example, geological environments and processes. Consequently, descriptions of both data and investigative procedures are included to acquaint students with the kinds of information that can be obtained from the study and investigation of minerals. Some of the procedures are mentioned only in passing to give beginning students an idea of the range of data that can be determined; other procedures are described in more detail to help prepare students for advanced courses in mineralogy or related disciplines, such as the materials sciences, ceramics, and metallurgy.

As far as possible, the text develops each subject from first principles. It is only assumed that students already understand the basic principles of physics, chemistry, and elementary trigonometry.

The organization of this edition follows that of the first edition. Part I, Concepts, introduces the physical, chemical, and genetic concepts that are relevant to the study of minerals. Part II, Descriptions, gives the properties of nearly 200 minerals and incidentally mentions those of several other minerals. Part III, Determinations, presents a logical procedure and two sets of tables (one based on macroscopically discernable properties, the other based on optical properties) to aid in the examination and identification of "unknown" mineral specimens. Selected Readings appear at the end of each chapter in Part I and in the Introductions to Parts II and III. Appendix A describes two kinds of natural nonmineral substances that commonly occur with minerals—natural glasses and macerals. Appendix B consists of two periodic tables: one including atomic weights and numbers, the other giving the ionic radii of Shannon and Prewitt (as revised by Shannon).

The chapters in Part I have been written so that they can be read and discussed in any order. Some instructors may wish to begin a course by discussing the physical properties of minerals; others may prefer to begin with crystallography; still others may wish to develop two or more topics at the same time. All such schemes should prove successful.

Part I begins with a point by point analysis of the definition of *mineral,* an historical overview of the science of mineralogy, and an introduction to the literature of mineralogy. Since minerals are crystalline solids, central concepts treated include symmetry, crystal systems, and crystal classes as manifested by both external forms and internal arrangement of constituent moieties. Since minerals are chemical entities, the central concepts treated also include the relative sizes of ions, types of bonding, and the significance of both in crystal structures. Brief discussions of isostructuralism, isotypism, isomorphism, polymorphism, pseudomorphism, and "noncrystalline minerals" logically follow. Next, physical properties of minerals are described and correlated with their crystal structures and chemical compositions. Thence, a discussion of mineral genesis emphasizes application of the phase rule and the use and interpretation of phase equilibrium diagrams. Finally, macroscopic, microscopic, and other laboratory methods for identifying and describing minerals are outlined and the rudiments of mineral systematics are treated; these lead directly to the mineral descriptions given in Part II.

Those who have used the first edition of *Mineralogy* may be interested in specific changes in the coverage of topics in this edition. The most notable additions include the sections dealing with imperfections (defects), the phase rule, phase equilibrium diagrams, Pauling's rules, optical mineralogy techniques, and applications of the electron microprobe. Also new are the descriptions for chloritoid, the humites, the melilites, pigeonite, pumpellyite, and stilpnomelane, and the inclusion of space-group and basic optical data for each mineral described in Part II. Significant omissions include the sections dealing with the gnomonic projection and blowpipe analysis, and the descriptions of cerargyrite (chloroargyrite), dumortierite, and tyuyamunite. Also omitted are the blowpipe analysis data and much of the specific locality and production data previously included in the mineral descriptions.

The preparation of this edition of *Mineralogy* has benefited greatly from comments and suggestions of several professional colleagues. Among them are Russell R. Dutcher, Michael Fleischer, Carl A. Francis, Gerald V. Gibbs, David D. Ginsburg, Raymond W. Grant, N. King Huber, Anthony R. Kampf, John F. Karlo, Aphrodite Mamoulides, Wayne E. Moore, Donald R. Peacor, George W. White, John S. White, Hatten S. Yoder, Jr., and Jack Zussman. To each of them I express my appreciation and thanks. Because statements based on comments made by reviewers of the manuscript of the first edition may persist in this edition, it seems only proper also to reacknowledge the late James Gilluly, Arthur Howland, Adolf Pabst, Chalmer Roy, and A. O. Woodford. In addition, I express my gratitude to the late Leonard G. Berry (1914–1982) and to Brian Mason, not only for the foundation laid by their original edition of this book but also for the aid they gave while I was preparing this revision. Brian prepared the Index for this edition. Among other things, Len completed his typically painstaking review of the galley proof for Chapter 14 the day before his fatal heart attack.

In particular, I thank my wife, Frances S. Dietrich, for her persevering and gracious assistance given throughout the time I spent preparing the manuscript and for helping with its conversion into this book.

R. V. Dietrich
September 1982
Mount Pleasant, Michigan

Part I
CONCEPTS

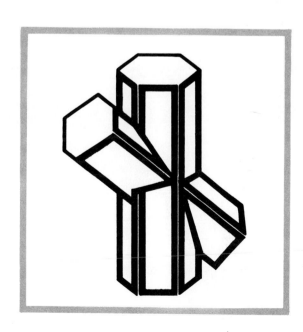

1 Introduction

THE SCIENCE OF MINERALOGY

Mineralogy is the science of minerals—their crystallography, chemical composition, physical properties, and genesis, their identification, and their classification. Mineralogy is closely allied to mathematics (especially geometry), chemistry, and physics. In turn, mineralogy is a fundamental part of geology and also of such diverse subjects as agronomy, astronomy, ceramic engineering, medical science, and metallurgy.

Definitions of the term *mineral* range from the historical (any material that is neither animal nor vegetable) through the legalistic (something valuable that may be extracted from the earth and is subject to depletion) to the scientific (a naturally occurring solid, generally formed by inorganic processes, with an ordered internal arrangement of atoms and a chemical composition and physical properties that either are fixed or that vary within some definite range). A good introduction to the science of mineralogy can be gained by examining each part of this scientific definition.

The qualification *naturally occurring* is considered necessary by some mineralogists and superfluous by others. Gemologists consider it particularly important to subscribe to this aspect of the definition. In fact, they consider it imperative that man-made substances, especially those that are essentially identical to naturally occurring gem minerals, be referred to as synthetic (e.g., synthetic diamond and synthetic ruby).

The requirement that a mineral be *solid* eliminates liquids and gases. This may seem rather arbitrary in that ice is thereby called a mineral whereas water is not. In practice, a few mineralogists ignore this restriction and include water and native mercury as minerals.

The phrase *generally formed by inorganic processes* is probably superfluous. Its purpose is merely to remind us of a former view that even the inorganic substances that are naturally produced by plants and animals (e.g., the aragonite

3

4 that makes up the shells and pearls produced by oysters) should not be called minerals. This former restriction, however, never did eliminate the possibility of an organic compound's being classified as a mineral; a few such substances (certain solid hydrocarbons, calcium oxalates, and similar compounds) have long enjoyed the status of minerals.

An *ordered internal arrangement of atoms* is the criterion of the crystalline state. Another way of expressing this is to say that minerals are crystalline solids. Under favorable conditions of formation, the ordered atomic arrangement may be expressed by an external crystal form. In fact, more than a century before X-rays provided the means of demonstrating the presence of an ordered internal arrangement of "building blocks" within crystalline solids, their existence had already been deduced from the external regularity of crystals. [A few minerals, such as opal, are not crystalline initially but become at least partly crystalline with the passing of geological time; also, a few other minerals, generally referred to as metamict, have had their original crystallinity partially destroyed as a result of irradiation by their radioactive constituents.]

Strictly speaking, the last part of the definition—a *chemical composition and physical properties that either are fixed or vary within some definite range*—merely adds the requirement of a defined chemical composition. As long as the internal atomic arrangement of their constituents is set, minerals with fixed chemical compositions also have fixed physical properties. (The crystal structure restriction is required because of polymorphism; see page 130.) As the definition also indicates, however, the chemical composition and consequently the physical properties of a mineral may vary as long as they vary within a definite range. (This phenomenon is related to solid solution; see page 127.)

THE HISTORY OF MINERALOGY

Certain minerals and rocks were carefully selected for certain purposes long before humans devised any written language. Evidence for such prehistoric uses includes the red and black mineral pigments (hematite and pyrolusite or some of the other manganese oxides, respectively) that were used in cave paintings and the diverse hard or tough minerals and rocks (e.g., jade, flint, and obsidian) that were shaped into tools and weapons. In addition, mining and smelting of metallic minerals to produce gold, silver, iron, copper, lead, and bronze are also known to have predated written records.

Aristotle (384–322 BC) considered it worthwhile to include a section about stones (minerals, metals, and fossils) in his well-known *Meteorologica*. Theophrastus (ca. 372–287 BC), his pupil and successor as head of the Lyceum in Athens, prepared the first known treatise dealing exclusively with substances of the mineral kingdom. The latter, sometimes called the "first mineralogy book," bears the title $\pi\epsilon\rho\iota\,\lambda\iota\theta\omega\nu$ (*On Stones*).

The next widely recognized contribution to mineralogy was by Pliny the Elder (AD 23–79), who recorded a great deal about natural history as it was understood by the Romans. Five of the thirty-seven volumes of his encyclopedic *Historia Naturalis*, which was written about AD 77, dealt directly with minerals that were then being mined for use as gemstones, pigments, or metallic ores.

Advances made during the next 1500 years—apparently, for the most part, during the early and middle Renaissance—were summarized by the German physician and mining expert Georg Bauer (1494–1555). Under his better-known, Latinized name, Georgius Agricola, Bauer prepared two especially important works, *De Natura Fossilium* in 1546 and *De Re Metallica*, published posthumously in 1556. These two volumes include descriptions and exceptionally fine woodcuts that record the state of mineralogy, geology, mining, and metallury of that era. In *De Natura Fossilium*, Agricola enumerated and defined most of the physical properties, such as hardness and cleavage, that are still used in macroscopic identification procedures for minerals. Goethe has compared Agricola to Bacon. Abraham Gottlöb Werner, as well as some science historians, have called Agricola the "Father of Mineralogy."

The next milestone in the development of mineralogy was provided by the Danish scientist Niels Stensen, better known by the Latinized version of his name, Nicolaus Steno. In 1669, Steno showed that the interfacial angles of quartz crystals are constant, no matter what the shape and size of the crystals. This discovery drew attention to the significance of crystal form and ultimately led to the development of the science of crystallography.

Although the term *fossilis*, meaning anything dug from the earth, was used by many people well into the 1700s, there also is good evidence that the term *minerāle* (of Celtic root) had become part of the Latin vocabulary by the end of the twelfth century. For those who used it, *minerāle* had come to include materials previously referred to as *metallum* (metallic ores) and also those called *lapis* (other rocks and minerals). It appears, however, that the word *mineralogy* was not used in print until 1690, when the English natural philosopher Robert Boyle (1627–1691) referred to a group who "love mineralogy much better than they understand it." The late use of this word is somewhat surprising in that the designation *mineralogist* had appeared nearly a half century earlier; in 1646, the English physician Sir Thomas Browne (1605–1682) noted in his *Pseudodoxia Epidemica: Enquiries into the Tenets and Commonly Presumed Truths* that many of the "exactest mineralogists" disagreed with the then rather frequently stated belief that rock crystal represents permanently solidified water.

Throughout the eighteenth century, there was steady, though slow, progress in mineralogy. New minerals were recognized and described, and several attempts were made to achieve a rational classification of minerals. Also, mineralogy became the first of the geological sciences to be recognized to the point of being taught in the universities. The greatest teacher of the time, A. G. Werner (1750–1817), who was Professor at the Mining Academy in Freiberg, made es-pecially noteworthy strides toward standardizing the nomenclature and macroscopic descriptions of minerals.

The period near the turn of the eighteenth century was one of rapid advance. The significance of crystallography in the study of minerals was brought to the forefront, largely through the work of the French scientists Jean Baptiste Louis Romé de l'Isle (1736–1790) and René-Just Haüy (1743–1822). In many people's minds, Haüy's efforts first raised mineralogy to the rank of a true science, and, like Agricola, Haüy has been attributed a paternal role, that of "Father of Mathematical Crystallography." In 1805, the English chemist John Dalton (1766–1844) first published the fundamentals of the atomic theory. Soon thereafter, it was realized that minerals are chemical compounds, each with its own definite composition. The Swedish chemist J. J. Berzelius (1779–1848) and his pupils, especially Eilhardt Mitscherlich (1794–1863), also studied the chemistry of minerals and articulated the principles of a feasible chemical classification of minerals. In 1837, James Dwight Dana (1813–1895) of Yale University completed the first edition of *A System of Mineralogy*; in the fourth edition of this work, which appeared in 1854, Dana introduced the chemically based classification system for minerals that is still used by most mineralogists.

Although the microscope was used to study minerals early in the nineteenth century, it was not until after 1828, when the British physicist William Nicol (1768–1851) invented the polarizer, that optical mineralogy took its place as a major investigative procedure in mineralogy. The first use of the polarizing microscope in the study of minerals and rocks is attributed variously to the French mineralogist A. L. O. LeG. des Cloizeaux (1817–1897) and the English petrographer Henry Clifton Sorby (1826–1908). The latter is rather widely hailed as the "Father of Optical Petrography."

The first great development in the twentieth century came as a result of the experiments made to determine how X-rays might be affected by crystals. The experiments, instigated by Max von

6 Laue (1879–1960), were carried out in 1912 in Munich; the investigators were two students, Walter Freidrich and Paul Knipping. Soon thereafter, W. H. Bragg (1862–1942) and his son W. L. Bragg (1890–1971) of Cambridge (England) investigated and published the results of X-ray studies in which they had determined the atomic arrangements within several minerals and other crystalline solids. In 1916, another far-reaching contribution was made when P. Debye and P. Scherrer in Zurich and A. W. Hull in the United States of America independently devised the technique that is now called the X-ray powder method. Subsequently, several additional techniques that involve X-rays and the coupling of X-ray diffractometers with computers have led to continuing progress in mineralogy and in other sciences and technologies that deal with crystalline materials.

More recent advances include the introduction and widespread use of the electron microprobe and the ever-increasing use of a number of other relatively sophisticated instruments and procedures (e.g., Mössbauer and infrared spectometry), which aid in the determination of certain characteristics of minerals and other crystalline materials. The more frequently used methods are described briefly in Chapter 6.

THE IMPORTANCE OF MINERALOGY

Minerals and consequently mineralogy are extremely important to economics, aesthetics, and science. Economically, utilization of minerals is necessary if we are to maintain anything like our current standard of living. Aesthetically, minerals shine as gems and enrich our lives, especially as we view them in museum displays. Scientifically, minerals comprise the data bank from which we can learn about our physical earth and its constituent materials—how those materials have been formed, where they are likely to be found, and how they can be synthesized in the laboratory. In addition, mineralogy is fundamental to the geological sciences, and its principles are basic to the understanding of a number of diverse aspects of several other disciplines, such as the agricultural

sciences, the materials sciences (ceramic engineering and metallurgy), and even medical science.

One of the greatest stimuli to the development of mineralogy has been and will continue to be the interest in recovering and using mineral resources. Two terms are widely applied to commercially important minerals. The term *ore mineral* embraces minerals from which valuable metallic elements can be extracted. Ore minerals include the metallic elements, the sulfides and several oxides, and a few other minerals that contain relatively high percentages of metallic elements, such as copper, silver, iron, and aluminum. The term *industrial mineral* designates minerals that are used to produce nonmetallic materials, which are used in the manufacture of such things as electrical and thermal insulators, refractories, ceramics, glass, abrasives, cement, fertilizers, and fluxes used in metallurgical processes. Economic mineralogy deals with both kinds of minerals and also with any minerals that are spatially associated with them. The term *gangue mineral* is widely applied to the associated minerals that have no value but nevertheless have to be mined along with the desired minerals from ore deposits.

Unfortunately, there is fairly widespread confusion about the proper use of the term *ore*. Some authorities attach an economic connotation to the term (i.e., they do not consider a mineral to be an ore mineral unless it comes from a deposit that can be worked at a profit). Other experts do not invoke the economic requirement. Still others, apparently a majority, use the term *ore mineral* in the generic sense noted in the preceding paragraph but restrict their use of the noun *ore* and the term *ore deposit* to specify economically workable materials. We subscribe to this last alternative, and thus the pertinent terminology in this book is consonant with a large share of the literature of economic mineralogy and economic geology.

Accordingly, a given ore mineral might be considered to be an ore at some localities but not at others, and an individual occurrence of an ore mineral might change with time from being an ore to being a non-ore (or vice versa). These differences and changes in status relate to changes in demand, changes in technology, and changes in several other economically based contingencies.

In the future, as population and man's needs and desires multiply, ore and industrial minerals will assume an ever-increasing importance in industry and commerce and in international affairs. The political consequences loom especially large because of two main facts: (1) nearly all mineral deposits are nonrenewable and (2) minerals are distributed unequally and erratically within the earth's crust. Thus, it becomes clear that we must continue to increase our knowledge about minerals in general and about economic minerals in particular.

The aesthetic aspects of minerals are manifest and manifold. Gems in jewelry, crown-jewel collections, and other displays attract the attention of millions of people annually. Minerals in national and regional museums and in private collections are also viewed by untold multitudes each year. In addition, minerals, as constituents of building stones, adorn both the interiors and the exteriors of many of man's architectural triumphs (and monstrosities).

Museums do more, however, than just display outstanding gems and mineral specimens. They also have assumed the function of collecting and preserving mineral specimens for posterity. Although a few minerals are common, many occur at only a few localities, and some occur within only a single deposit. Therefore, whenever possible, originally described specimens and other noteworthy specimens need to be preserved. Experience has shown that specimens that are not kept in well-known museum collections are often lost or made inaccessible for further study—study that is often fundamental to the taxonomy of mineralogy (see Chapters 7 through 16). Because this dilemma has been long recognized, most professional mineralogists conscientiously supply original materials to museums. Consequently, museum specimens are standards that are generally available for comparison and other research.

As far as the scientific importance of mineralogy is concerned, attention needs only to be directed to the fact that each individual mineral documents the chemical and physical conditions, and consequently the geological processes, that existed in the specific place at the particular time the mineral was formed. This fact is particularly apparent from the discussion in Chapter 5, "The Genesis of Minerals," and the occurrence data given in the mineral descriptions in Chapters 8 through 16. There you will learn, for example, that the feldspar sanidine crystallizes at high temperatures associated with volcanic activity; that the polymorph of silica called coesite is formed under high-pressure conditions such as those associated with meteorite impact; and that many clay minerals are formed as the result of surface or near-surface weathering. There are, as you might suspect, all sorts of mineralogical indicators for many diverse sets of geological conditions. Thus, the science of mineralogy plays a fundamental role in geological interpretations and, in many cases, both its data and its methods are also applied in several other related fields of scientific and technological endeavor.

THE LITERATURE OF MINERALOGY

The basic literature of mineralogy consists of individual papers that describe original research on minerals. The papers are published in scientific journals all over the world. Important journals, the articles of which are wholly or largely in English, include *The American Mineralogist*, published by the Mineralogical Society of America; *Contributions to Mineralogy and Petrology* and *Physics and Chemistry of Minerals*, both sponsored by the International Mineralogical Association; *The Canadian Mineralogist*, published by the Mineralogical Association of Canada; and *Mineralogical Magazine*, published by the Mineralogical Society of Great Britain. Another English-language periodical that is worthy of note is the privately published *The Mineralogical Record*, which, by its own proclamation, is published for "mineral enthusiasts."

8

There also are a number of publications in which abstracts of papers about minerals appear. Among these are *Chemical Abstracts*, which carries a section entitled "Mineralogical and Geological Chemistry," published by the American Chemical Society; *Mineralogical Abstracts*, published originally as a section in *Mineralogical Magazine* but as a separate publication since 1959, currently under the sponsorship of the Mineralogical Societies of America and Great Britain; *Bulletin Signalétique, Bibliographie des Sciences de la Terre* (Cahier A—Minéralogie et Géochimie), published by the Bureau of Geological and Mineralogical Research (BRGM) in Paris; and *Zentralblatt für Mineralogie*, published by Schweizerbart'sche Verlags Buchhandlung, Stuttgart, West Germany.

In addition, there are several bibliographies, indexes, and computer online capacities for searching in abstract service and bibliography data banks. Especially noteworthy among these are the *Bibliography of North American Geology*, which was formerly published by the U.S. Geological Survey and covers the period from 1732 to 1971; the *Bibliography and Index of Geology Exclusive of North America*, which was published by the Geological Society of America between 1933 and 1968; the *Bibliography and Index of Geology*, which is a continuation of these two bibliographies and is currently published on a monthly basis by the American Geological Institute; *Geotitles Weekly*, published by Geosystems in London, which began its coverage in 1969; and the *Science Citation Index*, published by the Institute of Scientific Information in Philadelphia, Pennsylvania, since 1964. A recently published Data Sheet of the American Geological Institute, A. G. I. Data Sheet #35, Bibliographies, Indexes, and Abstracts, lists several other English-language publications of this type, including those of many disciplines that are supportive of or tangential to mineralogy. Those of you who want to have any of the appropriate computer data banks searched should, at least for your first search, contact the nearest reference librarian.

In America, the standard reference work in mineralogy has long been *A System of Mineral-*

ogy by James Dwight Dana. The first edition of this book, published in 1837, aimed at a complete account of all minerals described up to that time. It was revised and brought up-to-date by five successive editions—dated 1844, 1850, 1854, 1868, and 1892—and by several supplements. The sixth and last complete edition was compiled by Dana's son, Edward Salisbury Dana; the last supplement, dated 1915, was compiled by W. E. Ford. Several years later, the colossal task of preparing a seventh edition was begun. To the present, three volumes of the seventh edition have appeared: Volume I, *The Elements, Sulfides, Sulfosalts,* and *Oxides*; Volume II, *The Halides, Nitrates, Borates, Carbonates, Sulfates, Phosphates, Arsenates, Tungstates, Molybdates, etc.*; and Volume III, *The Silica Minerals*. Volumes I (1944) and II (1951) were by Charles Palache, Harry Berman, and Clifford Frondel; Volume III (1962) was by Clifford Frondel.

Three comprehensive treatises on mineralogy have also been published in German. In some respects, these are even more detailed than Dana's *A System of Mineralogy*. They are *Mineralogische Tabellen* by Hugo Strunz; *Handbuch der Mineral-chemie* by C. Doelter and co-workers, published 1911–1931; and *Handbuch der Mineralogie*, begun by C. A. F. Hintze and carried on after his death by other workers. The first volume of this last mentioned compilation appeared in 1897, the last volume was published in 1939, and supplementary information about previously described minerals and data for new minerals continue to be published on an irregular basis.

Other excellent reference books deal with certain categories of minerals or certain mineral groups. Especially noteworthy are the five-volume work on *Rock-Forming Minerals* by W. A. Deer, R. A. Howie, and J. Zussman, which is currently being revised (Volume 2A, *Single-Chain Silicates*, is already available); the two-volume treatise, *Feldspar Minerals*, by J. V. Smith; and *Reviews in Mineralogy* (formerly short course notes) published by The Mineralogical Society of America.

Also worth mentioning are the *Glossary of*

Mineral Species 1980 by Michael Fleischer and *Klockmanns Lehrbuch der Mineralogie* by Paul Ramdohr and Hugo Strunz. The *Glossary* is an alphabetical list by names that also gives the chemical compositions and symmetry of mineral species. The latest edition of *Klockmanns Lehrbuch der Mineralogie* is a one-volume, 1980 update of the famous text which includes, among other things, at least brief descriptions of all currently recognized minerals.

There are, of course, several other publications dealing primarily with minerals. A few of those that pertain to aspects considered in detail in other chapters in this book are noted in those chapters or in the appropriate list of Selected Readings.

SELECTED READINGS

(Articles and books for which there is sufficient in-text bibliographic data to locate them in library catalogs are not repeated here.)

Bates, R. L., and Jackson, J. A. 1980. *Glossary of Geology*. Falls Church, Va.: American Geological Institute, 749 pp.

Brobst, D. A., and Pratt, W. P. 1973. *United States Mineral Resources*. U.S. Geological Survey Professional Paper 820, 722 pp.

Doelter, C. A., et al. 1911–1931. *Handbuch der Mineralchemie*. Dresden: Steinkopff.

Hintze, C. A. F., et al. 1897–1939. *Handbuch der Mineralogie*. Berlin: DeGruyter. (Early volume, Leipzig: Beit.)

Park, C. F., Jr. 1968. *Affluence in Jeopardy: Minerals and Political Economy*. San Francisco: Freeman, Cooper, 368 pp.

Park, C. F., Jr. 1975. *Earthbound: Minerals, Energy and Man's Future*. San Francisco: Freeman, Cooper, 279 pp.

Ramdohr, Paul, and Strunz, Hugo. 1978 and 1980 (supplement). *Klockmanns Lehrbuch der Mineralogie*. Stuttgart: Ferdinand Enke Verlag, 876 pp. and 56 pp. (supplement).

Sarjeant, W. A. S. 1980. *Geologists and the History of Geology: An International Bibliography from the Origins to 1978*. New York: Arno Press, 5 vols.

Skinner, B. J. 1976. *Earth Resources* (2nd ed.). Englewood Cliffs, N.J.: Prentice-Hall, 152 pp.

Strunz, Hugo. 1970. *Mineralogische Tabellen*. Leipzig: Akad. Verlag. Geest & Portig K.-G., 621 pp.

Theophrastus. 1956. *On Stones* (edited and translated by E. R. Caley and J. F. C. Richards). Columbus: Ohio State University Press, 238 pp.

U.S. Bureau of Mines. 1970. *Mineral Facts and Problems, 1970*. U.S. Bureau of Mines Bulletin 650, 1291 pp.

White, G. W. 1977. *Essays on History of Geology*. New York: Arno Press, 350 pp.

Wood, D. N. (editor). 1973. *Use of Earth Science Literature*. Hamden, Ct.: Achron Books, 459 pp.

2 Crystallography

CRYSTALS, CRYSTALLINE SOLIDS, AND THEIR FORMATION

A *crystal* is a solid body bounded by natural planar surfaces, generally called crystal faces, that are the external expression of a regular internal arrangement of constituent atoms or ions. The term *crystalline*, an adjective, is applied to *any* material having a regular internal arrangement of its constituent atoms or ions; that is, crystalline material may or may not be bounded by crystal faces. Minerals, by definition, are crystalline solids. Some minerals have "grown" as crystals; others exhibit few or no crystal faces.

Natural crystals may grow wherever (1) their constituent atoms and ions are free to come together in the correct proportions, (2) the existing conditions are such that "growth" will take place at a reasonably slow and steady rate, and (3) the external surface of the growing crystal is not constrained physically. Consequently, most well-developed mineral crystals occur lining the walls of open spaces in rocks, such as open fractures, solution cavities, and vesicles. Many such crystals have been deposited from hot aqueous solutions that are widely referred to as hydrothermal solutions; others have been formed by condensation from gaseous fluids. Still others have been formed by other means—for example, as the result of late-stage crystallization of magma within cavities, generally termed *miarolitic*, that were kept open by accumulations of magmatic gases and vapors. The quality of the plane surfaces range from mirrorlike to rough and pitted or deeply striated; the extremes can be observed on some individual crystals as well as from one crystal to another.

Excellent crystals can also be grown in the laboratory. Various processes can be employed —for example, the slow cooling, or evaporating, of a saturated solution of a salt of such compounds as common table salt, an alum, cupric sulfate, or ammonium dihydrogen phosphate. If a small seed crystal is suspended in a saturated solution, and if the solution is then cooled at a slow, steady rate, a large single crystal bounded by smooth plane surfaces will grow as long as the cooling rate and the supply of material can be maintained. Crystals can also be grown at a constant temperature if one uses equipment that permits addition of material to the solution so as to maintain saturated or very slightly supersaturated conditions. In any case, commercial growth of large crystals requires extremely careful control of temperature and other conditions. Numerous methods used for growing crystals are described by Nassau (1980) and in the *Journal of Crystal Growth* (1967—).

Crystalline solids that do not exhibit crystal faces are much more abundant than well-formed crystals. Most of these solids consist of several crystalline grains, the shapes of which reflect mutual interference caused by essentially simultaneous growth. Such solids may be formed whenever a solution is cooled or evaporated quickly, whenever a vapor is cooled quickly, or whenever a molten material is cooled at any rate slower than that termed *quenching*. [Quenching of a magma or other melt results in the formation of noncrystalline glass rather than a crystalline material because sudden cooling does not allow sufficient time for the atoms and ions to attain the regular arrangement characteristic of crystalline materials.]

The first step in the processes termed *crystallization* is the formation of minute nuclei. Under the microscope, many nuclei can be seen to exhibit the geometric shapes of crystals. As crystallization proceeds, however, the nuclei grow until the resulting grains abut one another to form a solid mass of grains, many or all of which do not have crystal shapes. Nonetheless, regardless of their shapes or how they were formed, each individual grain within a crystalline aggregate has the same internal structure and the same physical properties that well-formed crystals of the same mineral, metal, or other crystalline compound have.

Crystalline aggregates that are formed from some solutions, vapors, and melts consist of more than one chemical compound or phase, each having its own composition and orderly arrangement of atoms or ions. The sizes and shapes of the individual grains depend on the sequence of formation of the diverse nuclei, the rate of formation of the nuclei, the composition of the surrounding solutions, and the growth rate of the grains themselves.

In rocks, the shapes of crystalline grains are frequently classified as euhedral, anhedral, or subhedral (Figure 2-1). As shown, euhedral indicates that the grain is bounded by its own crystal faces; anhedral means that the grain exhibits none of its own possible crystal faces; and subhedral indicates that the grain has some, but not all, of its own crystal faces. In accepted practice, even the euhedral grains in rocks are referred to as grains rather than as crystals.

Figure 2-1. The perfection of crystal shape of a mineral—especially where it occurs as a rock constituent—may be indicated by the terms *euhedral* (bounded by its own crystal faces), *subhedral* (bounded by some but not all of its own crystal faces), and *anhedral* (exhibits none of its own possible crystal faces). (From R. V. Dietrich and B. J. Skinner, *Rocks and Rock Minerals,* © 1979 by John Wiley & Sons, New York.)

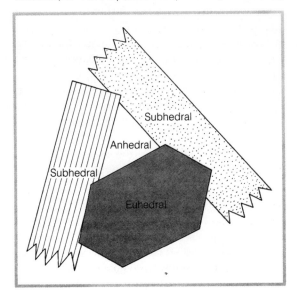

12   The modern science of crystallography embraces the study of both the morphology of crystals and the internal structure of all crystalline substances—minerals, metals, some ceramics, and many other chemical compounds. Different crystallographers refer to themselves as mineralogists, ceramists, chemists, metallurgists, and solid-state physicists. Thus, as might be expected, the growth and development of crystallography have involved scientists and others from several different disciplines.

THE DEVELOPMENT OF CRYSTALLOGRAPHY

The term *crystal* is from the Greek word κρήσταλλοσ, which means ice. The term *ice* was generally employed throughout the Middle Ages to designate rock-crystal (quartz) because such quartz was believed to represent ice that had been frozen so hard that it would no longer melt. Eventually, the term became applied to all solid objects of natural origin that exhibit external crystal forms. [Application of the word *crystal* to any quartz clear enough for vessels or ornaments has persisted; thus, we still hear the word used in reference to crystal ware (now made of specially designed lead-glass)and to fortune tellers' crystal balls, which in the past were fashioned from clear quartz.]

Although a case might be made for tracing crystallographic thinking to Pythagoras (sixth century BC), Plato (427?–347 BC), or Euclid (third century BC), the earliest known written comments about crystals per se and the quality of crystal faces are in Volume XXXVII of Pliny's *Natural History* (AD 77).

Much later, in 1611, the great astronomer Johannes Kepler (1571–1630) wrote a small pamphlet, "A New Year's Gift, or On a Hexagonal Snowflake,"in which he suggested that the regularity of crystal form is probably due to a regular geometric arrangement of minute building units. In 1705, Domenico Guglielmini (1655–1710) suggested an hypothesis of crystal structure that he based on the constancy of cleavage directions in crystals. During the same general era, the French philosopher and mathematician René Descartes (1595–1650), the Dutch physicist–astronomer Christian Huygens (1629–1695), and the English scientist–inventor Robert Hooke (1635–1703) also published information relating to crystallography. It appears, however, that these works were little known and therefore had little, if any, effect on later developments.

As we mentioned in Chapter 1, the first fundamental law of crystallography—The Constancy of Interfacial Angles—was announced by Steno in 1669. Although Steno's work was based on only rough measurements on quartz crystals, in 1783 his pronouncement was duly confirmed as a general law by Romé de l'Isle, who used a simple contact goniometer for measuring interfacial angles (see Figure 2-19). As you will see later, this law is a natural consequence of the regular internal arrangement within crystals.

Haüy published an essay on an hypothesis of crystal structure in 1784 and his well-known *Traité de Mineralogie* in 1801. Apparently, study of the cleavage of calcite led Haüy to suggest that crystals are made up of small polyhedral units, with the unit for each mineral having a characteristic shape. He illustrated his hypothesis with calcite and three cubic crystals (Figure 2-2). He also defined axes of reference for a few crystals and recognized that all faces on crystals of a single substance cut those axes at simple rational multiples of certain lengths. This relationship is known as the Law of Haüy, or the Law of Simple Rational Intercepts.

Between 1815 and 1824, the six crystal systems were distinguished: the isometric, tetragonal, orthorhombic, and hexagonal systems by C. S. Weiss (1780–1856); the monoclinic and triclinic systems independently by Friedrick Mohs (1773–1839) and K. F. Naumann (1797–1873). In 1830, the German mineralogist J. F. C. Hessel (1796–1872) showed that 32 crystal classes are possible. In 1848, the French physicist Auguste Bravais (1817–1863) showed that 14, and only 14, kinds of regular patterns—space lattices—can result from arranging identical points in space such that any point is repeated at regular intervals

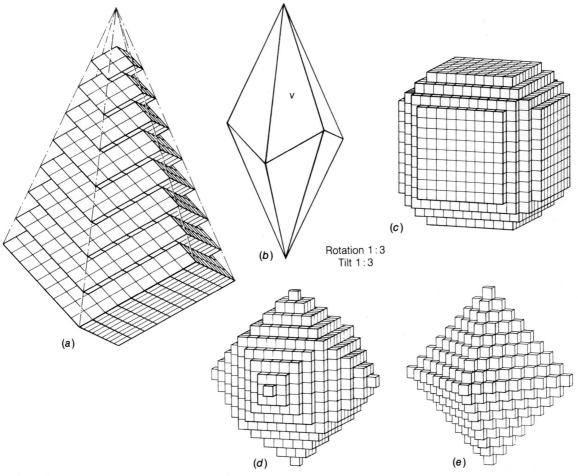

Rotation 1 : 3
Tilt 1 : 3

Figure 2-2. (a) Haüy's conception of structural units, each with the shape of the cleavage rhombohedron, building up a crystal of calcite in the form of the scalenohedron v {21$\bar{3}$1} [shown in (b)]. (c), (d), and (e) Haüy's conception of structural units, each with the form of a cleavage cube of galena or halite, building up a cube modified by dodecahedron faces (c), a dodecahedron (d), and an octahedron (e).

along each row of the pattern. And, to complete the geometric codification of crystallography, between 1885 and 1894, the 230 space groups were derived independently by the Russian crystallographer E. S. Federov (1853–1919), the German mathematician Artur Schönflies (1853–1928), and the semiretired English businessman William Barlow (1845–1934).

During and extending somewhat beyond the same general period, a number of diverse means were devised (or first used) to measure and de-scribe crystals, lattices, and elements of symmetry. The measurement of interfacial angles, previously made with contact goniometers (see Figure 2-19), became a precision operation when, in 1809, the English chemist W. H. Wollaston (1766–1828) described a reflecting goniometer (see Figure 2-20). Two especially noteworthy schemes to measure and describe crystal faces, both based on intercepts of faces on crystallographic axes, were introduced—one in 1818 by C. S. Weiss, the other in 1839 by W. H. Miller

14   (1801–1880); the former are now referred to as Weiss parameters, the latter as Miller indices. In 1851, because Miller indices apply only to three-axis systems, Bravais added a four-integer notation for describing hexagonal crystals; the resulting designations are generally called Miller–Bravais indices. Two sets of nomenclature and symbols for expressing the symmetry of the crystal classes were introduced—one by C. H. Hermann (1898–1961), the other by Schönflies; the former, as modified by Charles Mauguin (1878–1958), is now accepted as the *International Notation*, but many chemists and spectroscopists continue to use the alternative Schönflies notation. In addition, various methods for expressing certain relationships of solid objects on plane surfaces were used for the first time in descriptions of crystals and lattices; for example, F. E. Neumann (1798–1895) introduced the stereographic projection into crystallography in 1823, and Ernest Mallard (1833–1895) used the gnomonic projection to express a number of different crystallographic relationships in a book published in 1879.

As we mentioned in Chapter 1, the first great development of the 1900s was the determination that crystalline substances act as diffraction gratings for X-rays. Now, it is possible, by using X-ray diffraction methods, to measure the absolute dimensions of the so-called unit cell for any crystalline substance, to determine the atomic content of the unit cell, and thus to determine the particular arrangement of individual atoms and ions within any crystal structure.

THE IMPORTANCE OF CRYSTALLOGRAPHY

Crystallography is the key to understanding the structure of crystalline materials—the materials that constitute nearly all of the solids basic to twentieth-century technology. The importance of such an understanding cannot be over-emphasized because crystal structure, along with chemical composition, determines all the properties of all crystalline materials. Furthermore, it is these properties that dictate the uses to which a

material can be put. Examples of properties that can be readily perceived to be influenced by crystal structure include density (specific gravity), hardness, how a material breaks (fracture and cleavage), optical properties, and both thermal and electromagnetic properties.

THE REGULAR ARRANGEMENT OF POINTS IN SPACE

The following sections deal with ordered patterns, nets, and lattices. To appreciate these concepts fully, you must understand the following definitions as they apply to crystalline substances.

Motif: The smallest representative unit of a structure. In essence, it is an atom or group of atoms that, when repeated by translation, give rise to an infinite number of identical regularly organized units.

Unit cell: The fundamental parallelepiped, the content of which is only one or a few formula units, that forms a crystal structure by regular repetition in space. Some unit cells consist of more than one such parallelepiped (i.e., motif) and are termed *multiple.*

Structure: The ordered arrangement (i.e., the actual positions) of the atoms or groups of atoms within a crystalline substance.

Lattice: An imaginary three-dimensional framework that can be referenced to a network of regularly spaced points, each of which represents the position of a motif. Thus, a crystal lattice can be based on lattice points, each imagined to coincide with some particular position within each of several coterminous unit cells.

It follows that any number of unit cells can be arranged so that each has the same three-dimensional orientation *and* so that, as a group, they completely fill a space that is equal to their total volume. In addition, each such total volume would have an identifiable internal symmetry, and, if the total volume constitutes a crystal, it would also exhibit external symmetry. Consequently, the symmetry of crystalline materials is frequently considered from two related, though

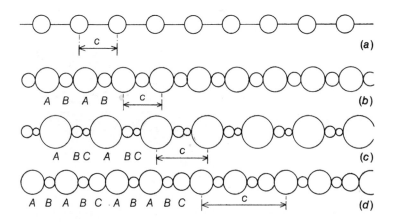

Figure 2-3. Regular arrangement of circles in one dimension with a repeat period c. (a) Circles of one size. (b) Circles of two sizes, composition *AB*, and regular sequence *AB, AB*, ... (c) Composition *ABC*, sequence *ABC, ABC*, ... (d) Composition A_2B_2C, sequence *ABABC, ABABC*, ...

different, viewpoints—either on the basis of structural crystallography or on the basis of morphological crystallography. In essence, structural crystallography involves the description of the internal arrangements of atoms, ions, formula units, and unit cells of crystalline materials, whereas morphological crystallography involves the description of the external forms of crystals. The approaches are, in general, interdependent.

The next three paragraphs may appear to be a digression; they are included because, in our experience, brief considerations of one- and two-dimensional patterns frequently help students to comprehend and master three-dimensional lattices.

One-Dimensional Patterns

The simplest expression of a regular arrangement of objects (which might be atoms) can be found in one-dimensional patterns—e.g., in a straight row of equally spaced, identical circles (Figure 2-3a, Table 2-1). This arrangement possesses a unit length (or *period*) c that is characteristic of the repetition of pattern. This operation of repetition is really a vector **c,** with direction and length c, along which translation will repeat the motif an infinite number of times. Such rows of identical motifs can be seen in most regular patterns of wallpaper, printed draperies, etc. The period in any given row is a constant c. If, for example, the

Table 2-1. *Regular patterns of points in translation rows and nets*

Lattice Type	Dimensions	Symmetry	See Figures	
Linear Row	c	*n/m2/m*	2-3,	2-9a
Planar Nets				
Clino-net	a c	*2/m*	2-6,	2-9b
Ortho-net	a b		2-5a,	2-9c
Rhombo-net	a' γ' (a b)	*2/m2/m2/m*	2-5b,	2-9d
Hexa-net	a	*6/m2/m2/m*	2-5c,	2-9e
Square-net	a	*4/m2/m2/m*	2-5d,	2-9f

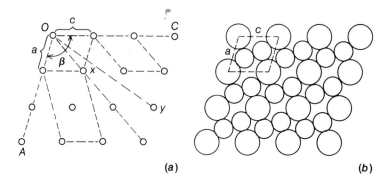

Figure 2-4.   Regular arrangements in two dimensions. (a) Clino-net; smallest least oblique net unit defined by a, c, and β. Two other net units shown, the lower one centered. Any parallelogram may serve as a net unit to build the whole net by translation only. (b) Array of circles with composition AB_2 with unit net of same dimensions as (a).

(a)   (b)

row is made up of alternate circles of different sizes, the true unit length c can be measured from the center of any object to the center of the next identical object (Figure 2-3b). Complex patterns can, of course, be built, but only as long as there is a definite unit that is repeated by translation in a regular fashion along the row (e.g., Figures 2-3c and d).

Two-Dimensional Patterns

Two-dimensional patterns consist of identical rows that are laid out in a plane so that the rows are parallel, equally spaced, have the same directional orientation, and so that a straight line drawn through a series of motifs in each of the rows (e.g., OA in Figure 2-4a) constitutes a "new" row. Each of the original rows (e.g., OC) must contain the same regular repeat arrangement of motifs. The new row is one of a large number of parallel, equally spaced rows; along each row, the motifs are repeated by a translation of unit length a; and, each row is inclined to the original rows (e.g., OC) at an angle β. The overall arrangement can be called a net, which, ideally, is of indefinite extent in one plane. Actually, such a net comprises an unlimited number of non-parallel sets of rows of different unit lengths (e.g., a, c, Ox, and Oy). Furthermore, any two non-parallel unit lengths define two edges of a parallelogram that is usually termed a net unit. A whole net can be developed by translation of a chosen net unit along the two rows that constitute its edges or along both of its diagonal rows. By

convention, the two rows of shortest unit length, with their included angle nearest to 90°, are generally chosen to define the net unit. Nonetheless, in some cases, special conventions are applied, and they commonly result in exceptions to this rule. Also, if a net pattern includes two or more kinds of objects, such as the large and the small circles in Figure 2-4b, the net unit must be defined to include appropriate numbers (or portions) of each object.

Special Types of Planar Nets

In the general case, a planar net can be defined by a unit parallelogram with adjacent sides unequal in length (c and a) and with an included angle not equal to 90°. This type of net can be called a *clino-net* (Figure 2-4a). Four special types of nets can also be recognized (see Figure 2-5 and Table 2-1): an *ortho-net*, in which the net unit has adjoining edges of unequal unit length (a and b) and an included angle of 90°; a *rhombo-net*, in which the simple net unit has edges of equal unit length (a' and a') and an included angle (γ') not equal to either 90° or 60° [in the net shown, the diagonals of the unit rhomb define an ortho-net, with edges a and b and included angle of 90°, plus a lattice point at the center of each ortho-net unit]; a *hexa-net*, in which the net unit has edges of identical unit length (a and a) and an included angle of 120°; and a *square-net* (sometimes called a tetra-net), in which the net unit is a square with edges (a)—that is, it has edges of identical length (a) and included angles of 90°.

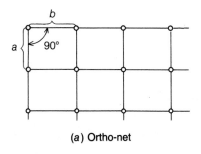

(a) Ortho-net

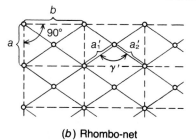

(b) Rhombo-net

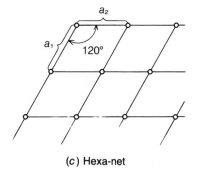

(c) Hexa-net

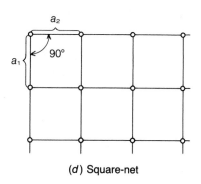

(d) Square-net

Figure 2-5. Four special types of planar nets representing regular two-dimensional translation patterns. (a) Ortho-net, rectangular net unit, edges a, b, included angle 90°. (b) Rhombo-net, centered rectangular net unit, edges a, b, or rhombic net unit with $a_1 = a_2$ and included angle γ'. (c) Hexa-net, rhombic net unit, edges $a_1 = a_2$, included angle 120°. (d) Square- or tetra-net, edges $a_1 = a_2$, included angle 90°.

Each of these four nets differs from the general, the clino-net, only in that its unit parallelogram has special dimensional properties. A complete two-dimensional pattern can be built up by translation of any single net unit (Table 2-1).

The Space Lattice

When a number of identical nets are stacked, the resulting array of lattice points becomes a space lattice if the following four conditions obtain: (1) the nets are parallel to one another; (2) they are equally spaced; (3) they have the edges of a unit net in one net parallel to the corresponding edges of the same unit net in all other nets; and (4) they have one point in each net falling on a straight row that extends through all of the nets (Figure 2-6). That is to say, a space lattice can be generated from a planar net by translation along a noncoplanar vector b. Any three noncoplanar rows (a, b, and c) outline a unit parallelepiped (or *unit*

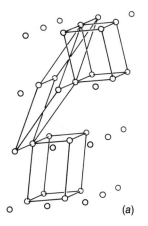

(a)

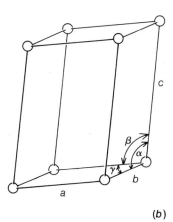

(b)

Figure 2-6. (a) Array of points giving a triclinic lattice; three possible but different unit cells are shown. (b) Enlarged version of the least oblique unit cell, defined by elements a, b, c, and angles α, β, γ.

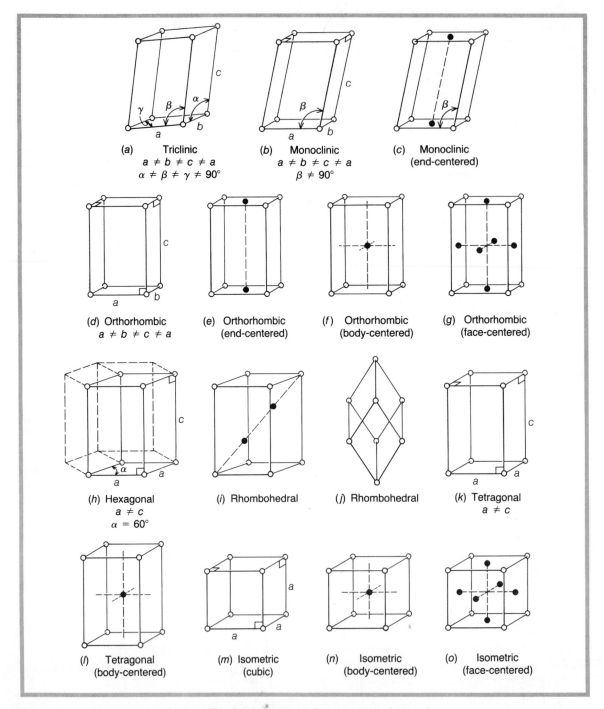

Figure 2-7. The conventional unit cells of the 14 Bravais lattices. Note that each lattice may be described on the basis of points in which each point has identical surroundings; note also that *i* and *j* are different expressions of the same lattice.

cell); lattice points can be considered to correspond with, for example, each of its vertices. Such a unit cell will build up a whole lattice by translation along its edges, along its body diagonals, or along an appropriate set of its face diagonals. Each of these directions is a lattice row. By convention, in most lattices, the three shortest noncoplanar periods are selected to define the unit cell of the lattice. Special conditions, however, require some exceptions to this rule. In general, the angle between any pair of lattice rows will not necessarily be 90°, and the magnitudes of the three unit translations will not be equal.

As just noted, any regular three-dimensional array of points forms a space lattice. With one point per motif, the lattice can be characterized by a unit parallelepiped (unit cell) that is repeated throughout the array by translation along appropriately chosen lattice rows. In crystallography, it is conventional to label the unit cell, and thus the lattices, as follows: edges are *a* if all are equal, *a* and *c* if one edge is not equal to the others, and *a*, *b*, and *c* if all three edges are unequal; included

angles are α $(= b \wedge c)$, β $(= a \wedge c)$, and γ $(= a \wedge b)$ if they are other than right angles (they are not designated if they are right angles). Also, if the unit cell is considered to have lattice points only at its vertices, it is termed *primitive*.

Special Types of Space Lattices

In the general case, the unit cell has three edges of unequal length (*a*, *b*, and *c*), and the angles (α, β, and γ), included by each pair of edges, are not right angles. By convention, a lattice based on a unit cell with this shape is termed *triclinic*, a name also applied to a large group of crystals and crystalline substances. Because it has lattice points only at its vertices, it is primitive.

In addition to this general case, thirteen special types of lattices can be built by stacking one of the five types of planar lattice nets in one or more special ways. These plus the already noted triclinic array are the fourteen Bravais space lattices (Figure 2-7). Their properties, including special designations for the axes, are in Table 2-2.

Table 2-2. *Regular patterns of points in translation lattices*

	Dimensions	Symmetry	See Figures
Triclinic			
1. Primitive *P*	$abc\,\alpha\beta\gamma$	$\bar{1}$	2-6, 2-7a
Monoclinic			
2. Primitive *P*			2-7b, 2-12a
3. End-centered *C* or *A*	$abc\beta$	2/m	2-7c
Orthorhombic			
4. Primitive *P*			2-7d, 2-12b
5. End-centered *C, B,* or *A*	abc	2/m2/m2/m	2-7e
6. Body-centered *I*			2-7f
7. Face-centered *F*			2-7g
Hexagonal and Rhombohedral			
8. Hexagonal *P*	ac	6/m2/m2/m	2-7h, 2-8a, 2-12d
9. Rhombohedral *R*	$ac(a, \alpha)$	$\bar{3}2/m$	2-7j, 2-8b, 2-12e,
Tetragonal			
10. Primitive *P*			2-7k, 2-12c
11. Body-centered *I*	ac	4/m2/m2/m	2-7l
Cubic			
12. Primitive *P*			2-7m, 2-13
13. Body-centered *I*	a	$4/m\bar{3}2/m$	2-7n
14. Face-centered *F*			2-7o

20   The special shapes of the five simplest unit cells of these lattices add five more names—*monoclinic, orthorhombic, hexagonal, tetragonal,* and *isometric.* These, along with *triclinic,* designate the six crystal systems. Each of these six space lattices is primitive; that is, the unit cell of the lattice has identical points, representing identical atoms or an identical array of atoms, at only its corners (Figures 2-7a, b, d, h, k, and m).

Whereas the other eight Bravais lattices also possess primitive unit cells, in practice each is usually described on the basis of a repeat parallelepiped that is not primitive (Figures 2-7c, e, f, g, i, l, n, and o). When a two-, three-, or fourfold unit is chosen, each of these lattices is similar in form to one of the six primitive lattices. But, in each of these multifold units, there also are lattice points at the centers of all of the faces, or of certain opposite faces, of the unit cell, or there are one or two lattice points within the unit cell in addition to the lattice points at the vertices.

In an *end-centered* (i.e., a one face-centered) *lattice,* the edges of the cell are chosen so that a lattice point falls at the center of each unit parallelogram in one set, and in only one set, of parallel planar nets. This choice results in a unit cell with two opposite faces centered. An end-centered lattice can be of monoclinic or orthorhombic type (Figures 2-7c and e). If the centered face is perpendicular to c and includes the a and b axes in orthorhombic lattices, or if it is the plane bounded by a and b in monoclinic lattices, the lattice is designated by the letter C; if the centered face of the lattice includes b and c, the lattice is designated by A; if the face includes a and c, it is designated by B. In a given lattice, an end-centered cell includes twice the number of lattice points of a primitive cell in the same lattice group. Some crystallographers describe this relationship in another way. They note that each of the eight lattice points of a primitive lattice belongs equally to the eight unit cells that come together at the lattice point, so that each primitive lattice contains one ($\frac{8}{8} = 1$) equivalent position, whereas each of the two additional lattice points of end-centered lattices belongs to only two units, so end-centered lattices consist of two ($\frac{8}{8} + \frac{2}{2} = 2$)

lattice points. Hence, they refer to end-centered lattices as *doubly primitive.*

In a *face-centered* (i.e., an all face-centered) *lattice,* each of the three net planes parallel to the faces of the chosen unit cell includes a lattice point at the center of each of its unit parallelograms. This cell has four times the lattice points of a primitive cell of the same lattice group. Such lattices can be orthorhombic or isometric, and they are often designated by the letter F (Figures 2-7g and o). On the alternate description basis outlined in the preceding paragraph, face-centered lattices can be termed *quadruply primitive.* A face-centered isometric lattice defines the type of pattern that results from one method of closest packing of spheres (see page 120), and it represents the atomic arrangement assumed by many metals, such as that of native gold, in which each lattice point represents a single gold atom.

In a *body-centered lattice,* each unit cell has an extra lattice point at its center. This cell, with double the number of lattice points of a primitive cell in the same lattice group, can be orthorhombic, tetragonal, or isometric, and it is designated by the letter I (Figures 2-7f, l, and n). On the alternate description basis outlined in a preceding paragraph, body-centered lattices can be termed *doubly primitive.* A number of metals crystallize with a body-centered isometric structure—for example, iron, in which each lattice point represents a single iron atom.

The remaining type of special lattice is called the *rhombohedral lattice.* It has been described in various ways (Figures 2-7i and j and 2-8b). Perhaps the most commonly used setup is the one shown in Figure 2-8b whereby the array of points gives a primitive cell R that is rhombohedral in form with three equal edges a_r and three equal interedge angles α that are neither 90° nor 60°. As shown, this relationship is established from the regular staggered stacking of a given planar hexa-net so that the fourth net is directly over the first, and the first and fourth hexa-nets form a hexagonal cell with two included points—one each from the second and third nets—that fall on and are equally spaced along one of the long body diagonals of the hexagonal cell. Thence, the primitive

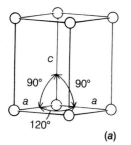

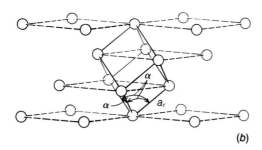

Figure 2-8. Unit cells of lattices: (a) Primitive hexagonal lattice *P*, elements *a*, *c*. (b) Rhombohedral lattice *R*, elements in hexagonal notation, *a*, *c*; rhombohedral unit cell has elements a_r, α.

rhombohedral unit cell assumes as its lattice points single points from the first and fourth nets and three points each from the second and third nets. Another rather frequently used choice of translations is based on the just described right-prism (hexagonal cell) plus the two included points (Figure 2-7i). In line with the alternative point content descriptions, this latter cell can be termed *triply* (or *trebly*) *primitive*.

The character of the rhombohedral lattice is such that crystals with the *R* lattice are usually rather easily distinguished, on the basis of their symmetry, from crystals having a hexagonal lattice. Consequently, in some countries *R* lattice crystals are considered to belong to a seventh crystal system, which is usually referred to as the *trigonal* or *rhombohedral* system. [It is also noteworthy that a rhombohedron with an interedge angle of 90° is a cube, a rhombohedron with an interedge angle of 60° is a primitive unit cell in a face-centered isometric lattice, and a rhombohedron with an interedge angle of 109° 28' is a primitive unit cell of a body-centered isometric lattice.]

SYMMETRY IN TRANSLATION ROWS, NETS, AND LATTICES

The regular arrays of atoms, ions, formula units, and unit cells that result from the crystallization of an element or a compound exhibit symmetry. Such symmetry can be defined as the exact repetition in size, form, and arrangement of motifs on the opposite sides of a plane, a line, or a point. In some cases, the symmetry in the arrangement of the unit cells of crystalline materials is reflected by the development of symmetrical crystals.

To help students to perceive and master the fundamental aspects of both internal and external symmetry, our treatment deals with symmetry in linear rows and planar nets before we describe the symmetry elements and operations as they pertain to three-dimensional lattices. First, however, students must understand the following two definitions.

Symmetry element: The plane, axis, or point (center) around which identical motifs are symmetrically disposed. (There are only 32 possible arrangements of symmetry elements in three-dimensional space that leave a lattice invariant.)

Symmetry operation: A geometric movement of a crystal (or lattice) whereby the shape, position, and orientation of an included motif (or lattice point) is assumed by another identically shaped and oriented motif, etc. (A given operation may involve reflection across a plane, rotation about an axis, inversion, or rotary inversion.)

Symmetry in Linear Rows

The simplest expression of symmetry may be seen in the linear row representing regularly spaced, identical motifs (see, for example, Figure 2-3a). Through each motif (represented by a circle in the diagram), and midway between each two adjacent motifs there is a plane perpendicular to the row, which may be called a *mirror* (or *reflection*) *plane of symmetry*; such planes are designated by the letter *m* in the symmetry symbols used in

22

crystallography. As you can see, every motif along the row, when reflected perpendicularly across the plane, coincides exactly with an identical motif on the opposite side of the plane.

In a similar way, through each motif and midway between any two adjacent motifs there are axes perpendicular to the row, which may be called twofold rotation *axes of symmetry;* such axes are designated by the number 2. When the whole row is rotated by 180° (i.e., 360°/2) about an axis of twofold symmetry, each motif of the row falls into exact coincidence with an equivalent motif. The diagram in Figure 2-9a shows only those axes of symmetry that are perpendicular to and parallel to the page. The latter are indicated by the conventional graphic symbol, a solid ellipse (●).

In this one-dimensional case, each point (motif) and each locus midway between two points acts as a *center of symmetry,* which can be designated by the symbol $\bar{1}$. Because the operation of a center of symmetry repeats every aspect of an array by inversion through the point at which the center of symmetry is located, it is generally referred to as a center of inversion, symbol *i.* [The symbol *1* (without the bar) can be easily dismissed at this point. In the International (Hermann–Mauguin) System, it signifies a onefold axis, which means that repetition occurs only after 360° rotation.]

Additionally, the row of points shown in the diagram can be considered to have infinite rotational symmetry about its length and an infinite number of intersecting mirror planes parallel to its length. This rotational symmetry, which occurs about an axis that is perpendicular to the originally described mirror planes, would be designated *n/m* with *n* indicating that the "-fold" of the rotational symmetry is indeterminate. (In crystals with three-dimensional lattices, *n* can only be 1, 2, 3, 4, or 6.) It follows that, in light of the fact that the previously described axes of twofold symmetry are perpendicular to mirror planes parallel to the length of the row, the twofold rotation axes should be indicated as *2/m,* rather than merely *2.* Thus, the complete symmetry of a row of identical motifs (e.g., identical atoms) can be expressed by the symbol *n/m2/m* (see Table 2-1). You should keep in mind, however, that in linear patterns that contain points of different kinds—as in Figures 2-3c and d—there may be fewer elements of symmetry.

Symmetry in the Five Types of Planar Nets

If a group of four units of a clino-net, with the configuration shown in Figure 2-9b, is rotated 180° about an axis that is perpendicular to the plane of the group at its center point, the new position will be precisely identical to the initial position. If a similar 180° rotation is made about an axis perpendicular to the plane through any net point or any point midway between any two net points, the new position will be like the first although it will be displaced by multiples of the unit lengths. Therefore, if the net is considered to be of infinite extent, rotation of 180° about any of the points indicated by the solid ellipse (●) will result in exact coincidence of the new and the old positions. Thus, there is a twofold rotation axis of symmetry that goes through each net point and is perpendicular to the plane of the net. This fact, along with the fact that the plane of the net is a mirror plane, shows that the symmetry for the clino-net is *2/m.*

Similar twofold axes of symmetry are present at the net points and at any point midway between any two net points in both the ortho-net (Figure 2-9c) and the rhombo-net (Figure 2-9d). In addition, however, each of these nets can be repeated to exact coincidence by folding itself over either of the two linear rows that intersect at right angles (a or b), or over lines parallel to a or b that include the midpoint of a unit net. Therefore, the rows a and the rows b in these nets are both axes of twofold symmetry and the traces of mirror planes of symmetry, and the axes are perpendicular to the planes, irrespectively. Thus, the complete symmetry for each of these nets is *2/m2/m2/m.*

As indicated by the conventional symbols (⬟ and ▲), the hexa-net has both sixfold and

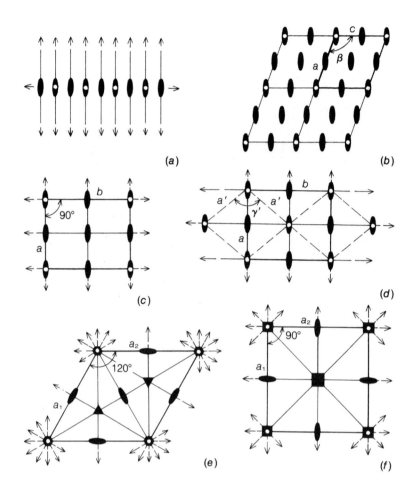

Figure 2-9. Symmetry in a translation row and in the five types of translation net. Net nodes are shown as small open circles. (a) Linear row of points, symmetry $n/m2/m$. (b) Clino-net, symmetry $2/m$. (c) Ortho-net, symmetry $2/m2/m2/m$. (d) Rhombonet, symmetry $2/m2/m2/m$. (e) Hexanet, symmetry $6/m2/m2/m$. (f) Square-net (tetra-net), symmetry $4/m2/m2/m$.

threefold rotation axes of symmetry (Figure 2-9e). These axes require repetition six times and three times, respectively, in a single 360° rotation. Twofold axes are also present. The short diagonal of the simplest unit parallelogram in this net is identical to, and thus indistinguishable from, the edges of the net unit (a_1 and a_2). Each of these three equivalent rows—usually designated by the translation magnitudes a_1, a_2, and a_3—may be the trace of a mirror plane of symmetry. In addition, the rows that are the long diagonals of the unit parallelogram, and thus at 30° to the rows generated by translations along a_1 and a_2, may also be the traces of mirror planes of symmetry perpendicular to the net. If the net points are joined using all three a directions, the resulting net unit has the form of a hexagon with one point at its

center. This unit, which better displays the symmetry of the whole net, will also build up the whole net pattern by translations. In any case, the complete symmetry for the basic hexa-net is $6/m2/m2/m$. This type of net, by the way, represents the closest possible packing of spheres in two dimensions.

The square-net shows fourfold axes of symmetry (symbol ■) that require exact repetition four times during a full rotation of 360° (Figure 2-9f). As is indicated, twofold axes are also present through certain points on the net. The net displays mirror planes parallel to the edges (a_1 and a_2) and also parallel to the diagonals of the square. Diagonals of four unit squares with a common node (corner point) define a square with a center point and twice the area of the basic net

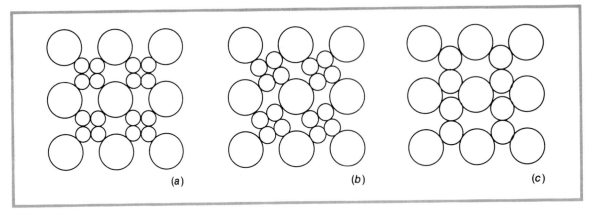

Figure 2-10. (a) Array of circles, AB_4, with a square-net unit, symmetry $4/m2/m2/m$. (b) Array of circles, AB_4, with a square-net unit, symmetry $4/m$. (c) Array of circles, AB_2, with a square net under the conditions shown, symmetry $2/m2/m2/m$.

unit. The symmetry of the unit square-net is $4/m2/m2/m$.

In each of these five nets, which represent all of the possible planar arrays: (1) each net point, as well as each point midway between any pair of net points, is a center of symmetry; (2) each net row that is perpendicular to the trace of a symmetry plane is a twofold axis of symmetry; and (3) each net plane is a mirror plane of symmetry to which the two-, three-, four-, or sixfold axes are perpendicular. (Axes of five- and sevenfold symmetry are impossible in crystals and crystal lattices.)

All regular planar arrangements of points can be described in terms of one of the five nets. Some of the arrangements will have the same symmetry as its type net; others may have less symmetry than that of its type. In Figure 2-4b, the pattern with two sizes of circles has the same symmetry as the clino-net. In Figure 2-10, the first array (a) has all of the elements of symmetry of the type square-net, whereas the second array (b) possesses the fourfold axis of symmetry but lacks mirror planes of symmetry perpendicular to the plane of the page. Therefore, a can be described as $4/m2/m2/m$ whereas b is merely $4/m$.

Figure 2-10c shows an hypothetical array of circles that appears to constitute a square-net, but, on the basis of its symmetry, it is an ortho-net; that is, it displays only twofold axes of symmetry along with its mirror planes of symmetry. The equality of edge lengths is, therefore, strictly fortuitous and, under changes of temperature that would cause changes in the sizes of the circles (i.e., if they were atoms or ions), the two edges might be expected to change at different rates and thus become unequal.

The Elements of Symmetry

We have noted the seven symmetry elements that are possible in regular two-dimensional arrays. These are the one-, two-, three-, four-, and sixfold rotation axes of symmetry (1, 2, 3, 4, and 6); mirror reflection planes of symmetry (m); and the center of symmetry (i). Taken as a whole, three-dimensional lattices may possess these symmetry elements and, in addition, axes of symmetry that involve rotatory inversion. The rotoinversion axes, which like the rotation axes may be one-, two-, three-, four-, or sixfold, are given the symbols $\bar{1}$, $\bar{2}$, $\bar{3}$, $\bar{4}$, and $\bar{6}$ (generally stated as bar-one, bar-two, etc.). [To generalize, we can say that the operation involving rotoinversion axes of symmetry includes a rotation of 360° divided by the "-fold" of the axis followed by an inversion across the center, for example, $\bar{3}$ rotoinversion involves rotation of 120° (360°/3) and then inversion through the center (Figures 2-11a and b).] As previously mentioned, the $\bar{1}$ rotoinversion axis is the same as the center of symmetry (i.e., $\bar{1} = 1 + i$). In addition, the $\bar{2}$ rotoinversion symmetry dupli-

cates $1/m$, $\bar{3} = 3 + i$, and $\bar{6} = 3/m$.

Two other kinds of symmetry, each of which combines two operations, also warrant mention here even though it is beyond the scope of this book to describe them in any detail. They are

Figure 2-11.   Other kinds of symmetry operations: (a) rotoinversion axis, $\bar{2}$ ($= m$); (b) rotoinversion axis, $\bar{3}$ ($= 3 + i$); (c) glide plane; (d) screw axis. Numbers indicate a stepwise method of considering each operation. See text for description.

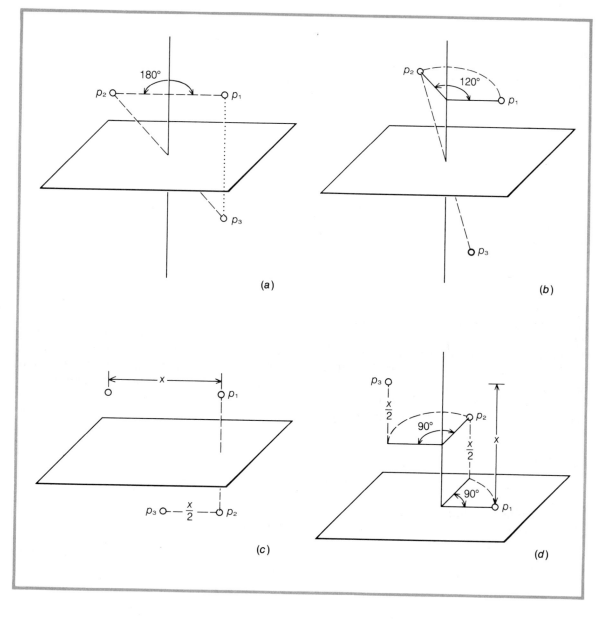

26

generally referred to as glide planes of symmetry and screw axes of symmetry. Glide planes of symmetry may be looked upon as combining reflection across the plane followed by a certain translation that is termed the glide component (Figure 2-11c). Screw axes of symmetry may be considered as rotation followed by translation that is parallel to the axis of rotation but along only a given fraction of the length of the row period of that axis (Figure 2-11d). Three-dimensional arrays of atoms in some crystal structures may possess two-, three-, four-, or sixfold screw axes of symmetry. Bloss (1971) presents a well-illustrated, fairly concise discussion of these kinds of symmetry.

Symmetry in the Space Lattices

The classification of the 14 translation lattices (Bravais' 14 space lattices) into the 6 crystal systems is based primarily on symmetry, even though the names of the systems are derived from the special dimensional properties of the lattices. The fundamental symmetry of each crystal system can be described as follows.

Triclinic Lattice

In the triclinic lattice (Figure 2-7a) with three principal lattice periods unequal in length $(a \neq b \neq c \neq a)$* and interaxial angles (α, β, γ) not 90°, the lattice must be rotated 360° about any lattice row before it comes into coincidence with the former position. Each lattice point and each point midway between any pair of lattice points is, however, a center of symmetry; that is, the whole array of points in any triclinic lattice comes into coincidence when inverted through a center of symmetry. Thus, the symmetry is designated $\bar{1}$, which becomes the symmetry symbol for the lattice. The presence of a center of symmetry does not require any pair of nonparallel lattice rows to be equal in length or to be at any special

*The conventional expression of this relationship, which is merely $a \neq b \neq c$, does not preclude $a = c$.

angle to each other. Nonetheless, in a given crystal structure, the lengths a, b, and c and the angles α, β, and γ do have specific values, and, in some crystals, two lattice translations may (albeit fortuitously) be equal within the limits of measurement and/or an angle between two lattice rows may be essentially indistinguishable from 90°.

Monoclinic Lattices

The primitive monoclinic lattice P shown in Figure 2-7b is derived readily by stacking identical clino-nets directly over one another and parallel to each other with a regular spacing; that is, identical clino-nets are stacked so that the lattice period b is perpendicular to the plane of ac. The resulting lattice has three unequal lattice periods a, b, and c, with $\alpha \equiv \gamma \equiv 90°$ and $\beta \neq 90°$. According to a widely accepted convention, the lattice row that is perpendicular to the set of clino-nets, and is a twofold rotation axis of symmetry, is labeled b. Consequently, the plane through the lattice that is either coincident with the ac clino-net or midway between two ac clino-nets, and is perpendicular to b, is a mirror plane of symmetry. As in the triclinic lattice, each lattice point and each point midway between any two lattice points is a center of symmetry. Since the twofold rotation axis is perpendicular to the mirror plane, the symmetry symbol becomes $2/m$. The symmetry arrangement is shown in Figure 2-12a.

Thus, by definition, monoclinic crystals have a lattice in which the symmetry of the array of points is $2/m$—a symmetry that requires one lattice row, here designated b, to be perpendicular to one set of lattice nets, in this case clino-nets that are parallel to the plane of symmetry.

A second method of stacking clino-nets, one in which alternate nets are directly over one another, results in either an end-centered lattice (Figure 2-7c) or a body-centered lattice. In either, the conventionally defined unit cell of the lattice is monoclinic in both symmetry ($2/m$) and shape ($a \neq b \neq c \neq a$, $\alpha \equiv \gamma \equiv 90°$, and $\beta \neq 90°$).

In monoclinic lattices, a and c are usually chosen arbitrarily as the two shortest lattice periods in the clino-net, which is perpendicular to b,

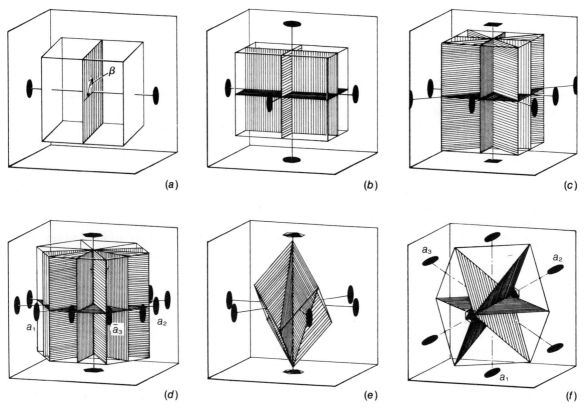

Figure 2-12. Symmetry in lattices. (a) Monoclinic lattice, symmetry $2/m$. (b) Orthorhombic lattice, symmetry $2/m2/m2/m$. (c) Tetragonal lattice, symmetry $4/m2/m2/m$. (d) Hexagonal lattice, symmetry $6/m2/m2/m$. (e) Rhombohedral lattice, symmetry $\bar{3}2/m$, c axis vertical. (f) Same as (e), with c axis horizontal.

often with $a > c$. In the centered lattice, if a and c are chosen so that the centered lattice point is in the ab plane, the lattice is designated C; if the centered point is in the bc plane, the lattice is designated A; if the point is body-centered, it is designated I; and if the point is face-centered, it is designated F. As previously noted, A, C, and I lattices are doubly primitive and F lattices are quadruply primitive.

Orthorhombic Lattices

The primitive lattice P arises from the regular stacking of identical ortho-nets to form a unit cell with $a \neq b \neq c \neq a$ and $\alpha \equiv \beta \equiv \gamma \equiv 90°$. In an array of points with these restrictions, each of the principal lattice rows with periods a, b, and c is perpendicular to an ortho-net (bc, ac, and ab, respectively). Each row is also a twofold rotation axis of symmetry, and there is a mirror plane coincident with or midway between each ortho-net (Figure 2-12b). This results in the symmetry symbol $2/m2/m2/m$. All lattice points and points midway between each pair of lattice points are centers of symmetry.

Ortho-nets can also be stacked so that alternate nets are directly over one another, which again results in an orthorhombic lattice but with one ortho-net centered (an end-centered lattice, Figure 2-7e) or with the unit cell body-centered (Figure 2-7f). In both of these lattices, the three edges of the unit cell are again twofold rotation axes of symmetry and each is perpendicular to a

28

mirror plane; in addition, the unit is of the same general form with $a \neq b \neq c \neq a$ and all interaxial angles 90°. Both lattices are doubly primitive.

Orthorhombic lattices also result from the regular stacking of rhombo-nets. [As we have already shown, a rhombo-net can be described alternatively as a centered ortho-net.] If rhombo-nets are stacked directly over one another, an end-centered orthorhombic lattice results (Figure 2-7e). This lattice is of the same type as that described in the preceding paragraph. If, however, the rhombo-nets are stacked in a staggered fashion so that alternate nets are directly over one another, a lattice with a face-centered (F) orthorhombic unit cell results (Figure 2-7g). This lattice is quadruply primitive.

Note that all four types of orthorhombic lattices possess three mutually perpendicular twofold rotation axes of symmetry, each of which is perpendicular to a mirror reflection plane of symmetry that is parallel to the net (bc, ac, or ab) that includes the other two axes. This requirement is fundamental to any orthorhombic lattice, 2/m2/m2/m.

The three axes are always chosen to coincide with the a, b, and c principal periods of the lattice. They are characteristically but not necessarily of unequal length. The three centered lattices are described in terms of a centered orthorhombic unit cell rather than in terms of simple cells of diverse form because each possesses the same rotation axes and reflection planes as the orthorhombic primitive lattice.

In labeling the twofold axes a, b, or c, most crystallographers choose an orientation with $c < a < b$; this choice usually corresponds with habitual elongation of crystals along c. The end-centered lattices are described as A, B, or C, depending on the choice of labels for the axes.

Hexagonal and Rhombohedral Lattices

Some crystallographers and mineralogists consider hexagonal and rhombohedral lattices to constitute two different subdivisions of the hexagonal crystal system. Others consider them to constitute two crystal systems, generally termed hexagonal and trigonal. Under the latter scheme, there are seven, rather than the conventional six, crystal systems.

The primitive hexagonal lattice results from the regular stacking of hexa-nets such that the shortest translation that joins points in successive nets is perpendicular to the hexa-net (Figure 2-7h). This lattice possesses one sixfold rotation axis of symmetry that is perpendicular to the hexa-net, three twofold rotation axes that are at 60° to one another in the plane of the hexa-net and coincident with the three shortest rows in the net (generally designated a_1, a_2, and a_3), and three twofold axes that are at 30° to the first set in the plane of the hexa-net. Each of these seven symmetry axes is perpendicular to a mirror plane of symmetry. Thus, the symmetry of the hexagonal lattice requires three identical lattice rows (a_1, a_2, and a_3) at 60° in one plane (the plane of the hexa-net) and a row with a principal lattice period c at right angles to the plane (and thus perpendicular to a_1, a_2, and a_3). Although c is generally shorter or longer than a, it may appear to be the same length within the limits of observation. The symmetry is 6/m2/m2/m.

The second type of lattice results from the regular stacking of hexa-nets, staggered so that every third net is connected by a lattice row perpendicular to the net. This lattice is hexagonal in form, but each unit cell, with $a_1 \equiv a_2 \neq c$ and $a_1 \wedge a_2 = 120°$, contains two extra lattice points that are located on one long diagonal of the cell at distances of one-third and two-thirds along the length of the diagonal as measured from either end (Figure 2-8a). The fact that the diagonals chosen are parallel in adjacent hexagonal unit cells is consistent with the symmetry, which is noted at the end of this paragraph. A primitive cell of this lattice can be readily outlined in three adjoining hexagonal cells (Figure 2-8b). The primitive unit, so defined, is a rhombohedron (R lattice). This case of centering is unique, compared with centered lattices in other systems, in that the centered lattice (generally termed rhombohedral) does not have the same symmetry as the primitive hexagonal lattice. In the rhombohedral lattice, the unique body diagonal of the rhombohedron (i.e., the c lattice period of the hexagonal cell, perpendicular to the hexa-net) is a threefold rotoinversion axis of symmetry. The three planes, each of which is perpendicular to one of the a translation directions and includes c, are mirror planes of symmetry (Figure 2-12e); the three planes, each of which is parallel to one of the a translations and to c, are not planes of symmetry (Figure 2-12f). Also, whereas the a translation directions coincide with twofold rotation axes of symmetry, the long diagonals of the 60° rhomb in the hexa-net are not symmetry axes. Thus the symmetry of the rhombohedral lattice is designated $\bar{3}2/m$.

Both the hexagonal and rhombohedral lattices have centers of symmetry at all lattice points and at each point midway between each pair of lattice points.

Tetragonal Lattices

If square nets are stacked one directly over another so that successive nets are parallel, so that like edges are parallel and equally spaced, so that the spacing is not equal to the edge length of the square, and so that the shortest translation connecting successive nets is perpendicular to the square net, a primitive tetragonal lattice results (Figure 2-7k). The rows perpendicular to the square net are fourfold axes of symmetry. The two edge rows, which may be designated a_1 and a_2, and the two diagonal rows of the square nets are twofold rotation axes of symmetry. Each of these axes of symmetry is perpendicular to a mirror plane of symmetry (Figure 2-12c). Therefore, the symmetry is described by the symbol $4/m2/m2/m$. Again, each lattice point is also a center of symmetry.

The lattice dimensions are defined by the edge length of the square-net and the principal period of the row normal to the square-net so $a_1 \equiv a_2 \neq c$. These three directions are mutually perpendicular, and c can be either longer or shorter than a.

If identical square-nets are stacked in a regular staggered fashion such that the row joining alternate nets is perpendicular to the square and points of the intermediate nets fall midway along the body diagonal of the lattice, a body-centered tetragonal lattice (I) results (Figure 2-7l). [Some crystallographers refer to C and F tetragonal lattices. Prove to yourself that C = P and F = I in this system.]

In all tetragonal crystals, the symmetry $4/m2/m2/m$ requires that the lattice be square in any section perpendicular to the fourfold axis and that the interaxial angles be right angles. Therefore, in tetragonal crystals the a and c translations are characteristically unequal in length, but they may appear equal within the limits of observation in some cases.

Isometric Lattices

The most specialized type of lattice is in the isometric system where the unit parallelepiped is a cube with three lattice rows of equal period intersecting at 90° (Figures 2-7m, n, and o). The array of points of each of the three types of isometric lattice has the same symmetry as the cube. There are three mirror planes of symmetry, one parallel to each of the three cube edges (generally designated a_1, a_2, and a_3) and thus perpendicular to each cube face; each has a perpendicular fourfold axis of symmetry $(4/m)$. Each of the four body-diagonal axes of the cube is a threefold rotoinversion axis of symmetry $(\bar{3})$. Each lattice plane that includes two opposite, parallel, cube-face diagonals is also a mirror plane of symmetry (there are six of these), and each of these also has a perpendicular axis of twofold symmetry $(2/m)$. Therefore, the full symmetry of the isometric lattice is expressed in the symbol $4/m\bar{3}2/m$ (Figure 2-13).

RELATION OF THE CRYSTAL LATTICE TO THE CRYSTAL

A crystal lattice is a three-dimensional array of identical lattice points, each point representing a single motif of structure. This array of points has an arrangement such that all points fall on straight

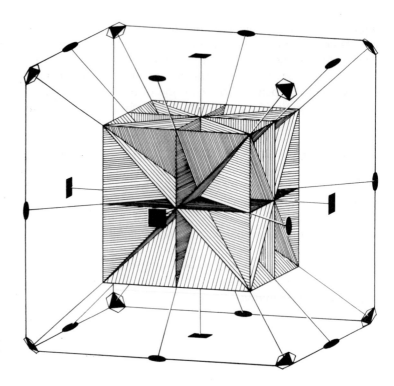

Figure 2-13. Symmetry in the isometric lattice, $4/m\bar{3}2/m$.

rows, and each point is repeated at a regular interval along the row. Every pair of nonparallel rows defines a planar net in which a unit parallelogram is repeated by translation throughout the net. Any three noncoplanar rows define a unit parallelepiped, which is repeated by translation throughout the array. Special dimensional and angular relations between rows result in thirteen special types of lattice, making fourteen lattices in all. These can be classified into six systems on the basis of axes of reference, or into seven systems on the basis of lattice symmetry.

The plane surfaces that bound natural crystals (i.e., the crystal faces) develop parallel to certain sets of net-planes in the crystal lattice of any specific substance or mineral. Each edge between any pair of nonparallel faces is parallel to a lattice row. If the lattice for a substance has certain linear and angular dimensions, the angles between corresponding planes in each lattice domain for the given substance will be identical as long as they are measured under conditions of constant temperature and pressure. This condition is in agreement with the aforementioned Law of Constancy of Angles:

The angles between corresponding faces on different crystals of a substance are constant.

The symmetry displayed by some lattices requires that certain nonparallel sets of lattice planes be identical in spacing and bear identical relations to the crystal structure. Consequently, crystal faces developed parallel to these equivalent sets of lattice planes will be symmetrically equivalent and may be expected to appear in equal development on a crystal. (Together, such symmetrically equivalent faces constitute a *crystal form*—see page 56.) Therefore, the identification of certain crystal forms (i.e., the determination of the symmetric relations between external faces) provides information about the symmetry of both the lattice and the structure and, thus, about the relative orientations and magnitudes of lattice translations.

CRYSTALLOGRAPHIC NOTATION FOR PLANES AND AXES

In the following section, we describe how to determine and express Miller indices, the notation that is currently used in crystallography for referring to both the planes and rows in lattices and the planes and axes of reference for crystals. As introduced, the system applied to three axes. The adaptation of the system of Miller indices to the four-axis setup often used for describing hexagonal crystals—a system widely referred to as *Miller–Bravais indices*—is outlined in the section in which we describe the hexagonal crystal class (page 70).

Miller Indices

In the general case, the three shortest noncoplanar translations and their inter-row angles are used to define the unit cell. By convention, the translations (*a*, *b*, and *c*) are generally termed *a*, *b*, and *c* axes. Planes of lattice points are present in many different orientations (not just in those parallel to the faces of the unit cell); in fact, there are such planes parallel to two of the reference axes and intersecting the third, other planes parallel to one axis and intersecting the other two, and still others intersecting all three axes.

One plane of a crystal lattice, a clino-net, is shown in Figure 2-14. The shortest lattice periods in this net are *a* and *b*; therefore, the lattice rows *OA* and *OB* are taken as reference axes. For simplicity, let the third axis *OC* be vertical (i.e., perpendicular to the *AOB* plane). The crystal can then be bounded by vertical faces, all parallel to *OC*, that for convenience will be labeled as *RDE*, *EF*, *FP*, *PI*, *IHJ*, *JK*, etc. Face *RDE* cuts axis *OA* at *D* and is parallel to axes *OB* and *OC*; therefore, it can be denoted by 4, ∞, ∞. Similarly, face *FGP* can be denoted by intercepts ∞, 4, ∞, face *EF* can be denoted 6, 6, ∞, etc. (These, of course, can be reduced to their smallest equivalent intercepts—

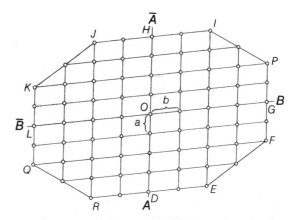

Figure 2-14. A planar net with shortest rows *a, b* representing one plane of a lattice with a third axis *c* emerging perpendicularly from the plane of the drawing. The lines *RDE, EF*, etc. are the traces of lattice planes which are taken as parallel to *c* in the text.

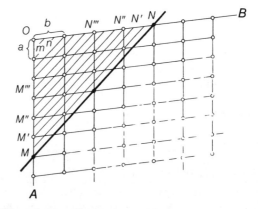

Figure 2-15. The same net as that in Figure 2-14, showing another set of planes, *mn, MN*, et cetera.

1, ∞, ∞; ∞, 1, ∞; and 1, 1, ∞; respectively. In Figure 2-15, however, the vertical plane *MN*, which is shown in the same net, has its smallest integral intercepts equal to 3, 2, ∞. In the three-dimensional general case shown in Figure 2-16, plane *XYZ* has its smallest integral intercepts equal to 2, 3, 6. The intercepts, however, are not the Miller indices. Rather the Miller indices are prime integers proportional to the reciprocals of the intercepts; that is,

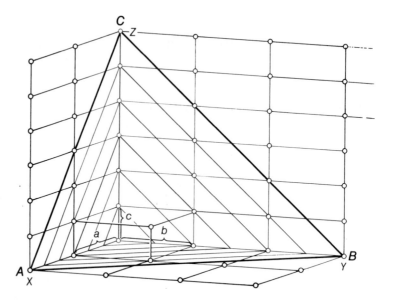

Figure 2-16.   The cystal lattice with a unit cell defined by the cell edges a, b, c, and by the interedge angles (not labeled here). The set of planes XYZ has Miller indices (321).

Miller indices are the coprime integers that are proportional to the reciprocals of the intercepts of crystallographic axes by the crystallographic plane being described.

Consequently, the indices are obtained by taking the reciprocals of the intercepts (where $\frac{1}{\infty} = 0$) and clearing fractions so that the indices are coprime integers. Thus, the crystal faces in Figures 2-14, 2-15, and 2-16 receive the following notations:

Face	Intercepts	Reciprocals	Indices
RDE	4, ∞, ∞	$\frac{1}{4}, \frac{1}{\infty}, \frac{1}{\infty}$	100
FGP	∞, 4, ∞	$\frac{1}{\infty}, \frac{1}{4}, \frac{1}{\infty}$	010
EF	6, 6, ∞	$\frac{1}{6}, \frac{1}{6}, \frac{1}{\infty}$	110
MN	6, 4, ∞	$\frac{1}{6}, \frac{1}{4}, \frac{1}{\infty}$	230
XYZ	2, 3, 6	$\frac{1}{2}, \frac{1}{3}, \frac{1}{6}$	321

In Figures 2-14, 2-15, and 2-16, it can also be noted that each crystal face—*EF, MN, XYZ,* etc.—is the outermost lattice plane of a set of

lattice planes that are parallel, equidistant, and geometrically equivalent. Each plane of a set will have different but proportional intercepts, the values of which depend on the physical size of the crystal and the location of the origin O. It can be seen, for example, that for the set parallel to *MN* in Figure 2-15, each plane has the same Miller indices when the fractions in the reciprocals are cleared to the simplest integers. For the planes noted in the figure, the intercepts and indices are as follows:

Face	Intercepts	Reciprocals	Miller Indices
MN	6, 4, ∞	$\frac{1}{6}, \frac{1}{4}, \frac{1}{\infty} \times 12$	230
M'N'	5, $\frac{10}{3}$, ∞	$\frac{1}{5}, \frac{3}{10}, \frac{1}{\infty} \times 10$	230
M''N''	4, $\frac{8}{3}$, ∞	$\frac{1}{4}, \frac{3}{8}, \frac{1}{\infty} \times 8$	230
M'''N'''	$\frac{5}{2}, \frac{5}{3}$, ∞	$\frac{2}{5}, \frac{3}{5}, \frac{1}{\infty} \times 5$	230
mn	$\frac{1}{2}, \frac{1}{3}$, ∞	$\frac{2}{1}, \frac{3}{1}, \frac{1}{\infty} \times 1$	230

Thus, we see two of the advantages of Miller indices: (1) they apply to a whole set of parallel planes and (2) they indicate the orientation of a face in terms of its axes of reference, irrespective of the size of a crystal and the position of the origin.

From the example just given, and from Figures 2-15 and 2-16, it can be observed that each set of lattice planes subdivides the edges or axial lengths of the unit lattice cell a, b, and c into h, k, and l, equal parts, respectively, where h, k, and l are whole numbers. Therefore, the Miller indices of a set of lattice planes or for the corresponding crystal face can be usefully defined as the number of parts into which the set of planes subdivides the axial periods (a, b, and c) of the lattice. It then follows that any crystal face with Miller indices (hkl) intersects a, b, and c at lengths of a/h, b/k, and c/l from any lattice point as origin (Figure 2-16). [Indices that are enclosed in parentheses, such as those in the preceding sentence, refer to a single plane or face. This should be kept in mind when you deal with crystallographic literature because the indication is accepted internationally. Also, as you will see, indices enclosed in brackets and braces have other meanings. To anticipate, those enclosed in square brackets designate lattice rows or axes, and those enclosed in braces denote one face of a crystal form of which the indices are for the face found in the front right, or in rare cases the front left octant.]

Thus far, we have referred only to those planes that cut the positive ends of axes with origin O. For a full description of the planes in a lattice, or for the faces on a crystal, it is necessary to recognize the directions $O\bar{A}$, $O\bar{B}$, and $O\bar{C}$ as negative directions. Thence, the intercepts and indices of the appropriate crystal faces have the following values, with negative values being indicated by a bar (Figure 2-14):

Face	Intercepts	Miller Indices
JHI	$\bar{4}$, ∞, ∞	$\bar{1}00$
KLQ	∞, $\bar{4}$, ∞	$0\bar{1}0$
JK	$\bar{6}$, $\bar{6}$, ∞	$\bar{1}\bar{1}0$
IP	$\bar{6}$, 6, ∞	$\bar{1}10$
QR	6, $\bar{6}$, ∞	$1\bar{1}0$

With regard to Miller indices, for proper designation of the position of a crystal face, the origin O must be taken inside the crystal. Consequently, parallel but opposite crystal faces—both members of the same set of lattice planes—will have the same Miller indices, except that corresponding indices will be opposite in sign; examples are (100) versus ($\bar{1}00$) and ($1\bar{1}0$) versus ($\bar{1}10$).

From Figures 2-14 and 2-15 it is further apparent that each set of lattice planes has a specific spacing and point density. The absolute values of the spacings (i.e., the perpendicular distances between any two adjacent planes of a set of lattice planes) are proportional and can be calculated, or obtained graphically, from the lattice parameters a, b, c, α, β, γ, and the Miller indices. Therefore, it can be seen, for example, that the spacings and point densities decrease with an increase in the numerical value of the indices of the lattice planes.

By further examination of any regular array of points (e.g., Figure 2-16), such as those in a lattice, it is evident that all planes that contain three or more points not in a straight line are lattice planes that can be designated with reference to the three noncoplanar axes a, b, and c, by intercepts that are whole numbers or infinity and thus by Miller indices that are whole numbers or zero.

It was first observed by Haüy, and it has subsequently been recognized as a law of crystallography, that

crystal faces make simple rational intercepts on suitable crystal axes.

This statement is now widely known as the Law of Haüy. Since crystal faces are parallel to lattice planes, it becomes axiomatic that the intercepts are whole multiples or submultiples of the lattice periods a, b, and c, and thus these lattice periods can be considered to be axes of reference for the crystal. Note that the Law of Haüy requires that the intercepts be simple; thus, it is to be expected that the Miller indices of observed crystal faces should be simple whole numbers. In fact, most crystals of most substances do have faces that will index with small single-digit numbers as Miller indices when the faces are referenced to the con-

34   ventionally chosen set of axes *a*, *b*, and *c*. Only rarely are Miller indices greater than five for crystals of simple development. Crystals of very complex development, however, are occasionally found (e.g., rare crystals of calcite), and some of those faces do have two digit indices.

It follows, and is generally supported empirically, that the faces most likely to be found on crystals are those that are parallel to lattice planes of greatest point density. This characteristic has been enunciated as the Law of Bravais:

> *The relative importance of crystal forms is proportional to the point densities or spacings of the respective lattice planes.*

This statement is nearly equivalent to Haüy's law, but it requires that the order of importance of crystal faces correspond with the decreasing order of interplanar spacing within the lattice. Bravais' law, however, is the more specific because point density or spacing can be expressed exactly. It is unfortunate that for many individual crystals the "law" is only an approximation. This is so because numerous factors, such as the particular atomic arrangement, the physical conditions of formation, and the presence of minor chemical elements in the parent solution influence the general development of crystals, which is widely known as their crystal *habit*. In general, therefore, it is only when crystals of a single mineral from many different localities are studied and the order of decreasing size of their faces is determined statistically that Bravais' law is more or less satisfied.

Crystal Edges and Zone Axes

Two nonparallel crystal faces meet in a line that is called an *edge;* edges are parallel to lattice rows that are common to the intersecting planes. The indices for an edge, which are similar in form to Miller indices, are the coordinates of the point that is nearest the origin and on the lattice row that both passes through the origin and is parallel to the given edge. As is true for *all indices that*

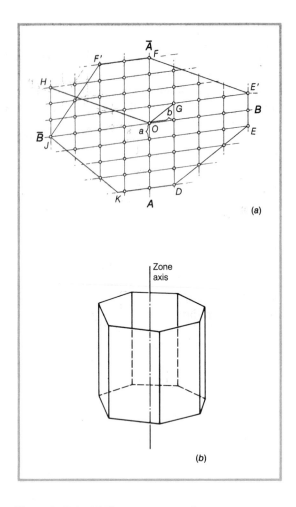

Figure 2-17.   (a) The same net as that in Figure 2-14, showing the edge *DE* with indices [$\bar{1}$10] and *E'F* with indices [$\bar{3}\bar{4}$0]. (b) Three-dimensional sketch of the net, extended by repetition along its perpendicular axis, shown as the zone axis.

designate lattice rows or axes, these three digits are enclosed in square brackets—for example, [*uvw*], where *u, v,* and *w* are integers. Consider, for example, Figure 2-17a to represent the cross section of a crystal that is bounded by the faces (001)—parallel to the drawing, along with (100)—*DK,* (110)—*DE,* (010)—*EE',* ($\bar{5}$40)—*E'F,* ($\bar{1}$00)—*FF',* ($\bar{1}\bar{2}$0)—*F'J,* and (2$\bar{3}$0)—*JK.* The edge *DE* between (001) and (110) is equivalent to the lattice row joining the origin (O) to the lattice point G, the coordinates of which are $\bar{1}10$. Therefore, the indices of crystal edge DE are [$\bar{1}10$]. Similarly, crystal edge E'F receives the coordinates of H on lattice row OH, namely [$\bar{3}40$]. In like manner, a crystal edge inclined to all three axes is denoted by the coordinates of the first lattice point uvw that is both on its parallel lattice row and passes through the lattice origin.

On some crystals, several faces intersect in parallel edges that are, as a consequence, all parallel to the same lattice row. Such a group of faces constitutes a *zone*, and the lattice row that is common to two or more such nonparallel crystal faces may be called a *zone axis* (e.g., the vertical axes in Figures 2-14 and 2-17b).

From Figures 2-14 and 2-15, we can see that each set of lattice rows has a specific period and consequently a specific point density. The period is the distance between successive points in the row; the point density is equal to the reciprocal of the period. These can be calculated from a, b, c, α, β, γ, and u, v, and w, or they can be obtained graphically. In any case, it is apparent that increases in the complexity of [uvw] indices are accompanied by an increase in period and a decrease in linear lattice point density. This fact suggests a useful alternative statement of the Law of Bravais:

The rate of crystal growth in any lattice direction is proportional to the point density in that direction.

In Figure 2-17a, for example, suppose that the plane (001) has the greatest areal point density in the lattice, which would be true if c were longer than a and b. According to the usual statement of Bravais' law, (001) would be the largest face. Also, if the (001) plane has the greatest point density, the two densest lattice rows will be parallel to [100] and [010]. Therefore, according to the alternative statement of Bravais' law, crystal growth will be most rapid in these directions so the (001) plane will again be the largest. In either form, this law of crystal growth is clearly displayed by habitually elongated crystals in which the shortest lattice period coincides with the directions of elongation, and by habitually tabular crystals in which the plane of tabular development contains the two shortest lattice periods. It also follows that the nearer the three shortest lattice periods approach one another in length, the more nearly equidimensional the crystal should be. As stated previously, however, Bravais' law is only approximately true as far as the development of any particular crystal is concerned because it is well established that atomic arrangement and environmental conditions that exist at the time of formation affect crystal growth. Therefore, it should be kept in mind that the Law of Bravais, no matter how stated, is a generalization, not a rigorous law.

Zone Relationships

Certain zone relationships that apply to the Miller indices and zone axis indices are very useful to the crystallographer. These relations can be derived from the equations of planes and lines that pass through the origin. A brief derivation of these relations follows.

Consider, as an example, the three planes $(h_1k_1l_1)$, $(h_2k_2l_2)$, $(h_3k_3l_3)$ that pass through the origin and intersect along the same straight line, which is equivalent to saying that the following three equations for the planes have one system of solutions. (Donnay, 1934):

$$h_1\frac{x}{a} + k_1\frac{y}{b} + l_1\frac{z}{c} = 0$$

$$h_2\frac{x}{a} + k_2\frac{y}{b} + l_2\frac{z}{c} = 0$$

$$h_3\frac{x}{a} + k_3\frac{y}{b} + l_3\frac{z}{c} = 0$$

These relations obtain if the determinant of the coefficients is equal to zero,

$$\begin{vmatrix} h_1 & k_1 & l_1 \\ h_2 & k_2 & l_2 \\ h_3 & k_3 & l_3 \end{vmatrix} = 0$$

for when all the elements of a row are multiplied by the same factor, the determinant is multiplied by that factor. If this determinant is multiplied by cofactors, the condition that the planes belong to the same zone can be written

$$\begin{vmatrix} k_2 & l_2 \\ k_3 & l_3 \end{vmatrix} h_1 + \begin{vmatrix} l_2 & h_2 \\ l_3 & h_3 \end{vmatrix} k_1 + \begin{vmatrix} h_2 & k_2 \\ k_3 & k_3 \end{vmatrix} l_1 = 0$$

which, by expansion and substitution, gives the formula

$$h_1 u + k_1 v + l_1 w = 0$$

where $u = k_2 l_3 - k_3 l_2$, $v = l_2 h_3 - l_3 h_2$, $w = h_2 k_3 - h_3 k_2$.
Therefore:

1. The plane (hkl) belongs to the zone $[uvw]$; that is, it includes the lattice row $[uvw]$ if

$$hu + kv + lw = 0$$

2. The plane (hkl) belongs to two zones $[u_1v_1w_1]$ and $[u_2v_2w_2]$ if

$h:k:l = (v_1w_2 - v_2w_1):(w_1u_2 - w_2u_1):(u_1v_2 - u_2v_1)$.

This relation can be remembered by the following operation of cross-multiplication.

$$\begin{matrix} u_1 & v_1 & w_1 & u_1 & v_1 & w_1 \\ u_2 & v_2 & w_2 & u_2 & v_2 & w_2 \end{matrix}$$

3. The zone $[uvw]$ is common to the two planes $(h_1k_1l_1)$ and $(h_2k_2l_2)$ if $u:v:w = (k_1l_2 - k_2l_1):(l_1h_2 - l_2h_1):(h_1k_2 - h_2k_1)$. This is analogous to the preceding formula and may be remembered in the same way. From Figure 2-17a the zone axis common to (001) and $(\bar{4}30)$ can be determined as follows:

$$\begin{matrix} 0 & 0 & 1 & 0 & 0 & 1 \\ \bar{4} & 3 & 0 & \bar{4} & 3 & 0 \end{matrix}$$

from which $[uvw] = [0 - 3, \bar{4} - 0, 0 - 0] = [\bar{3}40]$.

4. The plane $(h_3k_3l_3)$ will belong to the same zone as $(h_1k_1l_1)$ and $(h_2k_2l_2)$ if

$$h_3 = h_1 + h_2, k_3 = k_1 + k_2, \text{ and } l_3 = l_1 + l_2$$

or if

$$h_3 = h_1 - h_2, k_3 = k_1 - k_2, \text{ and } l_3 = l_1 - l_2$$

or, in general, if

$$h_3 = mh_1 \pm nh_2, k_3 = mk_1 \pm nk_2 \text{ and } l_3 = ml_1 \pm nl_2$$

where m and n are integers.

STUDY AND MEASUREMENT OF CRYSTALS

In the macroscopic and/or microscopic examination of crystals of minerals or of other chemical substances, measurement of interfacial angles and the study of the symmetry can provide diagnostic information. Symmetry can sometimes be deduced from the overall appearance of the geometric development of the crystal, especially if it is supplemented by the measurement of interfacial angles. Such angular measurements enable the plotting of a stereographic or gnomonic projection of the crystal, and the resulting array of face poles will display the symmetry of the crystal.

In an atomic arrangement possessing a certain symmetry, it is required that certain inter-row and interplanar angles be equal in value and that certain nonparallel lattice planes intersect certain rows at equal angles. This fact is illustrated in a simple manner in Figure 2-18 in which cross sections of two crystals, each with a principal row perpendicular to the plane of the drawing, are compared. In Figure 2-18a, the lattice net perpendicular to the principal row is a clino-net; in Figure 2-18b, it is an ortho-net. In Figure 2-18b, lattice row OA is the trace of a mirror plane of symmetry perpendicular to the plane of the drawing. [The presence of this mirror plane of symmetry requires the following: $OG = OH$; spacing of plane (110) equal to that of $(1\bar{1}0)$ (both also

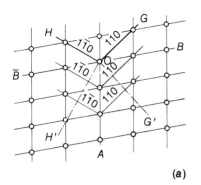 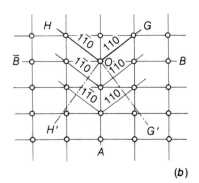

(a)   (b)

Figure 2-18. One lattice plane in two different crystals: (a) a clino-net, (b) an ortho-net. The plane of symmetry in (b) parallel to OA requires the equivalence of OG and OH in period, the spacing of (110) and (1$\bar{1}$0), the angles AOB, AO$\bar{B}$, and the angles BOG, HO$\bar{B}$.

perpendicular to the drawing); and $\angle AOB = \angle AO\bar{B} = 90°$.] Therefore, the planes (110) and (1$\bar{1}$0) are equivalent and, from Bravais' law, we can expect the two faces (110) and (1$\bar{1}$0) to be equal in size. As we have implied, however, even though symmetry is a feature of the atomic arrangement and the equivalence of certain lattice planes is a consequence of the symmetry, the growth of a crystal can be different in different equivalent directions in space. Therefore, we may find that the planes OH and OG make equal angles with OA and OB within a given crystal, but that the resulting crystal faces (110) and (1$\bar{1}$0) are different in size because of, for example, the addition of more material in the direction of OG' than in that of OH' or vice versa. This fact points out an important aspect of crystal symmetry: the symmetry of the structure is present whether the crystal does or does not grow exactly the same amount in each equivalent direction.

In any case, it is rare to find a crystal as perfect in development as the ideally shaped wooden models commonly used for study. Rather the infinite variety of individual aspects that prevails in nearly all natural objects is also evident in crystals. In fact, it is rare to find two identical crystals of an individual mineral, even among many hundreds of crystals formed under essentially the same conditions at a particular locality. There are far too many continually changing influences that promote diversity.

Occasionally, influences cause two faces that are parallel to nonequivalent lattice planes to develop to about the same extent. Thus, mineral crystals from certain localities can display an unusual, but characteristic, development of faces. Fortunately, most crystals of "average development" for a given substance show the geometric symmetry that correctly reflects the symmetry of the substance's internal atomic arrangement.

Before we continue, here are two reminders: Although geometric solid models can be used to illustrate the features of crystal symmetry, much crystal symmetry does not have either the internal or the external perfection of geometric symmetry. Furthermore, symmetrically equivalent crystal faces—no matter how different their appearances—have formed with identical internal interplanar spacings and therefore have identical angles with respect to their symmetry elements.

Measurement of Interfacial Angles

One of the best ways to comprehend the distinction between crystal symmetry and geometric symmetry, and at the same time to observe the relationships that led to formulation of the Law of Constancy of Interfacial Angles, is to study a number of hexagonal crystals of two or more substances such as those that are readily available in most collections.

When a number of single crystals of quartz, apatite, beryl, or corundum are examined, we

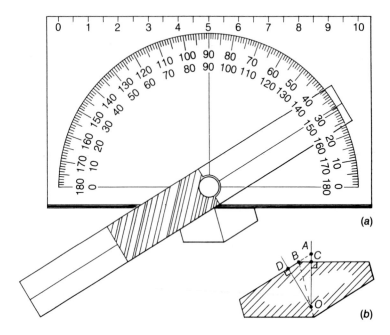

Figure 2-19. (a) A contact goniometer on which the interfacial angle $CBD = 148\frac{1}{2}°$ (or the polar angle $COD = ABC = 31\frac{1}{2}°$) can be read directly for the example shown in (b).

find that a cross section of the crystal in a plane perpendicular to the typical axis of elongation is essentially six-sided or hexagonal. The six faces that outline the crystal and are parallel to the elongation constitute what is generally termed a hexagonal prism. Examination of the cross-sectional outlines of several crystals, measurement of the lengths of the sides and the angles between the faces in the plane of each section, and taking into account possible errors of measurement show that the interfacial angles are equal even though the lengths of the sides are not. On some crystals, all of the sides are nearly identical and the outline approaches a geometrically perfect hexagon; on others, there is great diversity in the lengths of the sides. On all of these crystals, however, the angle between any two prism faces that meet at an edge is 120° (i.e., a 60° polar angle) within the errors of measurement. Therefore, crystals of this sort are described as being hexagonal in outline, and the angles between corresponding pairs of faces on different crystals of the same substance are identical (constant). As already mentioned, this prop-

erty of crystals so interested Nicholaus Steno that, in 1669, he sawed through several quartz crystals so he could measure the angles directly and thus convince himself of the Law of Constancy of Interfacial Angles.

It is imperative that interfacial angles be measured in a plane perpendicular to both of the crystal faces concerned. This measurement can be done with a simple contact goniometer, which generally consists of a printed protractor with a straight strip of transparent plastic pivoted at the center and with a hairline mark that can be read against the scale (Figure 2-19). The goniometer is held with the straight edge of the protractor in contact with one face, the straight edge of the plastic strip in contact with the other face and, as just noted, with the plane surface of the protractor and the strip perpendicular to both crystal faces. Two values of the interfacial angle, which total 180°, can be read from the protractor. One is the internal angle DBC, the other is the external angle ABC between one face and the other face extended. This latter angle, which is equal to the angle COD between the perpendiculars to the two faces (since $ODA = OCB = 90°$), is generally called the polar angle. It is this polar angle (COD) that is quoted in this book, and in most of the literature of crystallography, in descriptions of crystals.

With small crystals, interfacial angles can be measured more conveniently, as well as more accurately, with a reflecting goniometer. This instrument has wider application than the contact goniometer because, for most minerals, small crystals occur more commonly than large ones. In fact, many minerals are rarely, if ever, found in crystals large enough for contact goniometry.

In its simplest form, a reflecting goniometer consists of a rotating spindle, a collimator, and a telescope. The spindle is located at the center of a divided circular scale; the collimator and telescope are in a plane perpendicular to the spindle

Figure 2-20.   A two-circle reflecting goniometer.

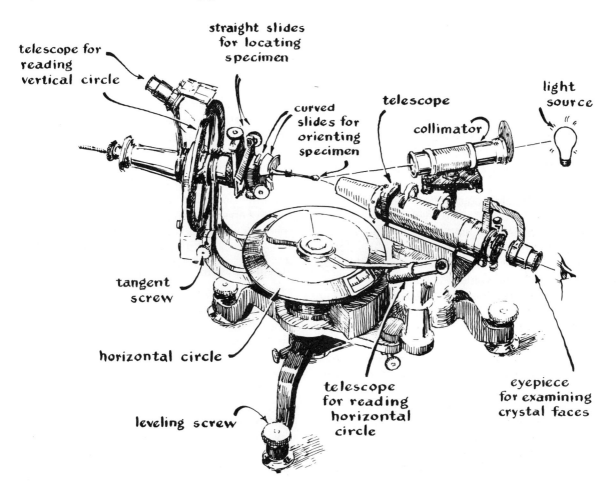

40

and have their axes intersecting the axis of the spindle. The crystal is mounted at the point of intersection. The collimator and telescope are separate, and the angle between them is usually set at about 60°. The crystal is mounted so that a prominent zone axis is parallel to the spindle axis. The angular position at which each face of the zone reflects the collimated beam into the receiving telescope is easily read on the divided circle. The difference between any pair of readings from adjacent faces is the interfacial (polar) angle.

A much more convenient method of crystal measurement is possible with a two-circle reflecting goniometer such as that shown in Figure 2-20. With this instrument, the crystal is adjusted to rotate about the axis of a vertically divided circle that intersects the axes of the collimator and telescope at the axis of the horizontal divided circle. By the use of two curved slides, each operated by a rack and pinion, the crystal can be adjusted such that the axis of a prominent zone of faces coincides with the axis of the vertical circle. By the use of straight rack and pinion slides, the crystal can be moved parallel to or at right angles to the axis, so that the face to be observed can fall on the axis of the horizontal circle in reflecting position. The arm supporting the vertical circle is connected to the moving portion of the horizontal circle. With this instrument, all faces in the zone used for adjustment and all faces intersecting one end of this zone axis can be brought into reflecting position. For each face, two angles are recorded. The angles V, read on the vertical circle, become the azimuth angles in preparing a projection. The angles H, read on the horizontal circle and corrected so that the reading for a face perpendicular to the vertical circle axis is zero degrees, are equal to the angle ρ between the axis of adjustment and the perpendicular to each face. From these measurements, all face poles can be plotted quickly on a projection. [The relations of the measured angles to the axial ratio of a mineral, as established on a morphological basis, can also be calculated or determined graphically (see Table 2-3, p. 52).]

Axes in Crystals

Crystallographic axes within crystals should be designated so they are parallel to the edges of the conventionally defined unit cell. Before the advent of X-ray crystallographic analysis, the orientation of crystals and the choosing and measuring of their axes were accomplished according to conventions established and adhered to by morphological crystallographers. Today, since the directions of unit cell translations can be defined by X-ray diffraction, axes can be chosen parallel to those directions and thence correlated to directions in actual crystals.

Considering the way a crystal's external morphology reflects its internal lattice geometry, it should come as no surprise to learn that the data gained from the two approaches generally agree. Perhaps the most common discrepancy relates to the fact that morphology reflects only the directions of axes and the ratios of their magnitudes, not the true magnitudes. Thus, we can predict which refinements and modifications of the morphologically based data are most likely to be required; for example, we know that for certain monoclinic minerals there is likely to be a reversal of the positions of a and c.

The conventions ("rules") that have been established for orienting and choosing the directions of crystallographic axes on the basis of morphological considerations follow.

Triclinic System

As we noted earlier, the three unit cell translations that are used as the crystal axes of reference in the triclinic system are referred to as a, b, and c, and the interaxial angles α, β, and γ are not 90° (Figure 2-6b). In addition, it has always been customary to think of c as being vertical, positive end up; b as trending right and left, positive end toward the right; and a as trending front and back, positive end toward the observer. The axes a, b, and c are usually chosen to coincide with the three shortest noncoplanar periods in the lattice, and a commonly adopted convention results in a unique setting in which $c < a < b$; α and β are obtuse; and γ can be either acute or obtuse in a given lattice. In this setting, c, as the shortest translation of the lattice, generally corresponds to the axis of habitual elongation in a crystal.

The conventions for orientation of crystals in the other systems are similar but are modified by the fact that one or more of the lattice translations is parallel to one or more directions of rotation symmetry and/or perpendicular to one or more planes of symmetry.

Monoclinic System

In the monoclinic system, it has been a nearly universal convention to choose the twofold symmetry axis (i.e., the direction perpendicular to the mirror plane) as the b axis, and c and a as the two shortest lattice translations in the plane of symmetry perpendicular to b. This choice results in $b \wedge c \equiv b \wedge a \equiv 90°$. Under this convention, there is no arbitrary choice, and b can be longer or shorter than either or both a and c. [It warrants mention, however, that in recent years, a few crystallographers have suggested that it is more elegant to make the unique twofold axis c, as in each of the other systems that possesses a unique axis.] In the monoclinic system, as in all systems except the isometric, c is set vertical, positive end up. The angle β is obtuse between the positive ends of a and c, with a sloping downward toward the observer. In the end-centered monoclinic lattice, the choice of $c < a$ can result in the lattice's being centered either in the ab plane (and thus called a C lattice) or in the bc plane (and termed an A lattice); and, for the same lattice, alternative choices may lead to a body-centered (I) lattice.

Orthorhombic System

In the orthorhombic system with $a \neq b \neq c$ ($\neq a$) and $\alpha \equiv \beta \equiv \gamma \equiv 90°$, all three axes of the unit parallelepiped are twofold rotation axes and thus unique within the lattice (Figure 2-12b). The international symmetry symbol for the lattice is $2/m2/m2/m$. Therefore, the twofold axes become the axes of reference, often with $c < a < b$. This setting usually corresponds to habitual elongation of crystals parallel to c and flattening normal to b. The end-centered orthorhombic lattice can be set with $c < a < b$, in which case the centering can be on the ab, bc, or ac planes (termed C, A, or B lattices, respectively). Some workers, however, ignore the $c < a < b$ rule and always set the lattice so that the ab plane is centered (i.e., the C lattice). For many minerals, it is impossible to follow both conventions.

Hexagonal System

In the primitive hexagonal lattice, the unique axis, which is normal to the hexa-net, is sixfold in symmetry. It becomes c and is always set vertical, positive end up. The two edges of the primitive hexagonal unit cell in the plane of the hexa-net have twofold symmetry and become the a (a_1 and a_2) with an angle of 120° between their positive ends. For many purposes, it is convenient to consider a third a (a_3), the positive end of which is at 120° to a_1 and a_2. The third, a_3, is not required for geometric reasons, but its use does help display the hexagonal symmetry (Figure 2-7h). Strictly speaking, a_3 is parallel to the short diagonal of the 60° rhomb of the a_1 and a_2 hexa-net and is indistinguishable from a_1 and a_2. There is a second set of twofold symmetry axes parallel to the long diagonals of the 60° rhombs in the hexa-net. Although these axes can be distinguished from the a axes by X-ray diffraction, their distinction is not possible from observation of only the geometric forms of crystals. In the symmetry symbol for the lattice ($6/m2/m2/m$), the $6/m$ applies to c and the plane perpendicular to it, the first $2/m$ applies to the three a translations and the planes normal to them, and the second $2/m$ applies to the long diagonals of the 60° rhomb in the hexa-net (at 30° to a_1, a_2, and a_3) and the planes normal to them.

In the hexagonal (and also in the tetragonal) system, c is chosen on the basis of symmetry, not on the basis of length as it is in the triclinic, mono-

42   clinic, and orthorhombic systems. Nonetheless, c is either longer or shorter than a, although in rare cases it may closely approach a in length. If c is shorter than a, the Law of Bravais would lead us to expect crystals to be habitually elongated along c, as in apatite. On the other hand, if c is longer than a, we should expect the crystals to be tabular normal to c, as in molybdenite.

Tetragonal System

In the tetragonal system, the unique lattice period c, which is normal to the square net of the lattice and is either longer or shorter than the two identical translations (a_1 and a_2) of that net, is an axis of fourfold symmetry. This axis, termed c, is always set vertical, positive end up. The other two translations, identical in length and at 90° to each other (Figure 2-7k) and thus indistinguishable, are both referred to as a (or as a_1 and a_2 when it is necessary to refer to them separately). In tetragonal lattices, there also are two pairs of twofold symmetry axes—one pair parallel to the edges of the square-net and the other parallel to the diagonals of the square-net.

In our discussion of the tetragonal lattice, we gave the symmetry symbol as $4/m2/m2/m$; thus the $4/m$ applies to c, the first $2/m$ to the two a axes, and the second $2/m$ to the diagonal axes that are at 45° to a_1 and a_2 in the a_1a_2 plane. When the dimensions of the square-net are measured by absolute methods, such as by X-ray diffraction, there is no doubt about the choice of a. When only the geometric form is observed, it is not possible to decide which pair of twofold axes is parallel to the diagonals of the square-net.

Isometric System

In the isometric system, the lattices have four axes of threefold symmetry and three equal axes of either fourfold or twofold symmetry that are at right angles to each other. These fourfold or twofold axes are used as the crystallographic axes of reference. The translations along these axes, being identical in length, are designated a_1 (front and back, positive forward), a_2 (right and left,

positive to the right), and a_3 (vertical, positive upward) (see Figure 2-7m). They are also the three densest rows of the primitive lattice. In the symmetry symbol for the lattice ($4/m\bar{3}2/m$), $4/m$ applies to the crystallographic axes and the planes perpendicular to them; $\bar{3}$ applies to the cube body diagonals; and $2/m$ applies to the cube face diagonals (i.e., the two diagonals in each square-net) and the planes perpendicular to them (Figure 2-13).

CRYSTAL PROJECTIONS

In crystallography, as in many other disciplines, we are continually confronted with the problem of representing three-dimensional objects on a two-dimensional surface. To achieve this goal, we use different kinds of projections. In this book, for example, we use the orthographic and clinographic views in the pictoral crystal drawings and stereograms to describe the symmetry of the crystal classes.

The orthographic view is a plan or elevation drawn perpendicular to a major axis of a crystal (e.g., see Figures 2-68 and 15-30). The clinographic view is a trimetric drawing using a standard rotation and tilt of the principal directions while retaining a parallel perspective. (Most of the other crystal drawings in this book are of this type.) Although drawings of this kind are useful for showing the shapes and relative sizes of crystal faces, they are not convenient for displaying the more significant angular relationships between crystal faces or for deriving the dimensional properties of crystal lattices. It is fortunate, therefore, that these angular relations can be readily shown by other kinds of diagrams—those more generally referred to as projections.

The *spherical projection*, although of little use in itself, is fundamental to three projections that have found wide use in crystallography: the orthographic, the gnomonic, and the stereographic projections. In constructing the spherical projection, the crystal (or lattice) is imagined to be located at the center of a sphere. Thence, as is shown in Figure 2-21, by extending a line out-

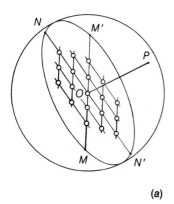

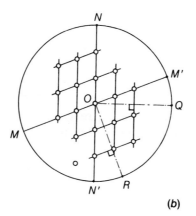

Figure 2-21. (a) A lattice plane *MNM'N'* passing through the center of a sphere at *O*. *OP* is normal to the plane, *P* is the pole of the plane on the sphere. (b) The same lattice plane parallel to the page. *OR* is normal to *MM'*, and *OQ* is normal to *NN'*.

(a)

(b)

ward from the center of the sphere perpendicularly through a face (or plane) to the surface of the sphere, the face (or plane) becomes represented on the surface of the sphere by a point, usually referred to as a *pole*. Also, crystal axes that are not perpendicular to crystal faces can be plotted; they are merely extended directly to the surface of the sphere and the resulting points are marked appropriately.

Relationships between the spherical projection and the other three projections are shown in Figure 2-22. As you can see, each is constructed by projecting poles of a spherical projection onto a plane that is perpendicular to a diameter of the reference sphere. Each type of projection thereby accomplishes the desired effect of expressing three-dimensional attributes on a plane

surface, such as a sheet of paper. As is also apparent, the differences among these three projections are based on different positions of the eye when the sphere is projected onto the plane surface.

In an *orthographic projection*, the eye is at infinity along the reference sphere diameter that is perpendicular to the plane of projection (i.e., along diameter *XOY* extended); the pole *P* would be projected onto a plane at either *R* or *R'*. In the *gnomonic projection*, the eye is at the center of the reference sphere; the pole *P* would be projected to *G*. In the *stereographic projection*, the eye is at the distal end of the reference diameter (i.e., at *Y*), which by convention is the south pole; the pole *P* would be projected backward, so to speak, onto the equatorial plane at *Q*.

The orthographic and gnomonic projections

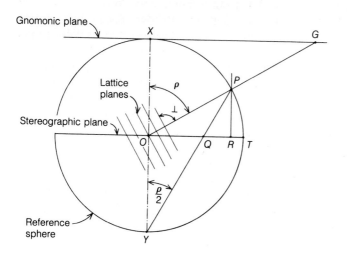

Figure 2-22. The relation between the stereographic, orthographic, gnomonic, and spherical projections in a section including an axis of the sphere, *XOY*, and the perpendicular *OP* to a set of lattice planes; *P* is the pole point, the intersection of the perpendicular *OP* with the sphere; *Q* is the corresponding stereographic projection pole; *R* is the orthographic projection pole; and *G* is the gnomonic projection pole representing the lattice planes. In this drawing, the plane of the orthographic projection is shown coinciding with the stereographic plane.

44   are largely of historical interest. The orthographic and gnomonic projections have been used chiefly for crystal drawings; the gnomonic projection provides a convenient method for treating measurements obtained from the two-circle goniometer, especially to index faces and derive elements for crystals with numerous faces. In addition, the gnomonic projection is still used by some crystallographers in conjunction with calculations involving *reciprocal lattices*. [In essence, a reciprocal lattice is a lattice that can be derived from a direct lattice as follows: the axes of the reciprocal lattice are perpendicular to the planes (100), (010), and (001) of the direct lattice (or unit cell), and the repeat distance along each of the axes is the reciprocal of the appropriate interplanar spacing of the direct lattice. Consequently, the reciprocal lattice is frequently used in interpreting X-ray and electron and neutron diffraction patterns of single crystals.] Unlike the orthographic and gnomonic projections, the stereographic projection finds both broad-scale and widespread use in crystallography (and also in structural geology and structural petrology). Therefore, the remainder of our considerations in this section deal with stereographic projections.

THE STEREOGRAPHIC PROJECTION

As we have already mentioned, in this method of projection, the center of the crystal (or lattice unit) is imagined to coincide with the center of a sphere that completely surrounds the crystal. In the first step, lines from the center of the sphere and perpendicular to each face are extended to the surface of the sphere (Figure 2-22). The points where these lines intersect the spherical surface are the spherical poles of the faces and the assembly of spherical poles thus established is called a spherical projection. Thence, the stereographic projection can be derived from the spherical projection by projecting the poles of the spherical projection to the south pole of the sphere. The points where these latter lines of projection intersect the equatorial plane of the sphere become the stereographic poles of the faces and the complete pattern is the stereographic projection.

Stereographic projections of most crystals can be plotted readily by simple constructions that are outlined in many books (e.g., see Amorós et al. and Wolfe, both of which are cited in the Selected Readings at the end of this chapter). In most cases, however, manual construction of stereographic projections is accomplished by using the Wulff stereographic net (Figure 2-23).

The Wulff net is simply the stereographic projection of a sphere marked off at 2° intervals of both latitude (small circles) and longitude (great circles). As you can see, the radii of this net from the center to N, E, W, and S, are graduated in a scale of stereographically projected degrees and the circumference is graduated in true degrees (azimuth angles).

In general, it is most convenient to plot data on a transparent sheet of paper laid over the net so that the tracing paper is free to move about a pin that is placed through it and the center of a printed net. The methods used in plotting data can differ greatly with the data available. For example, poles of planes are readily plotted if the azimuth angle and the angle (ρ) between some reference axis and the perpendicular to the plane are known. A zero azimuth reading is marked on the circle; the tracing paper is turned counterclockwise to the required azimuth angle from E (Figure 2-23); then the ρ angle is read from the center along the radius to E. (The procedure may, of course, be repeated for other poles.)

With but a few simple constructions, it becomes evident that stereographic projections possess several interesting and, for crystallographic problem solving, advantageous relationships. Among other things, circles on spherical projections plot as circles on stereographic projections, and angular relations on the surfaces of spherical projections plot as identical angular relationships on stereographic projections. This latter relationship is particularly useful because angles between planes, between poles, and between planes and poles can be plotted accurately and thereby become subject to additional analyses through relatively simple measurements. As another example of its utility, the plotted data can be rotated into diverse orientations.

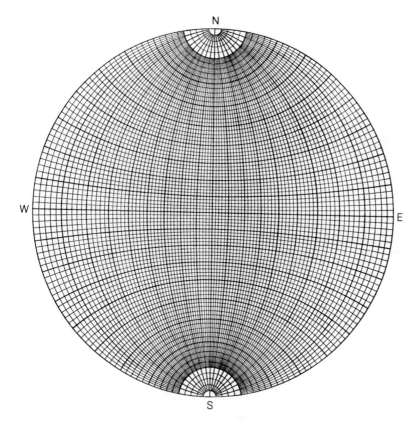

Figure 2-23. A Wulff net. Stereographically projected great circles appear as lines of equal longitude, and small circles centered about the north and south poles appear as lines of equal latitude at two-degree intervals.

(If you want to rotate data into a new position on a stereographic projection, you can do it as follows: first, set the desired axis of rotation parallel to NS; then, each pole can be moved through the desired angle, along the latitude line that passes through the pole, by counting off the angle in degrees of longitude.)

Crystals in Stereographic Projection

In preparing a stereographic projection, the standard orientation of a crystal is with the vertical crystallographic axis c (or a_3) taken as the north-south axis of the sphere (XOY in Figure 2-22), positive end north and negative end south, and the b (or a_2) crystallographic axis east-west, positive end to the east [except for triclinic crystals for which the pole of the face (010) is to the east]. With this orientation, the positive end of c (or a_3) emerges at the center of the projection, and the

poles of all faces in the vertical zone, parallel to c, fall on the equatorial circle, which is generally referred to as the primitive. Also, the pole for a face perpendicular to c falls at the center of the projection.

The projection of the poles to faces of the lower half of a crystal will be outside the primitive if the faces are projected to the south pole. Thus, for convenience, such faces are generally projected by joining them to the north pole. Thence they can be distinguished from faces projected from the northern hemisphere by representing them by circles instead of the small crosses generally used for northern hemisphere poles (see, for example, Figure 2-36).

To provide an example, the relations of the stereographic projection to the crystal faces of an orthorhombic crystal are shown in Figures 2-24 through 2-26. The drawings are intended to indicate the relations between different sections of a crystal lattice and the stereographic projection;

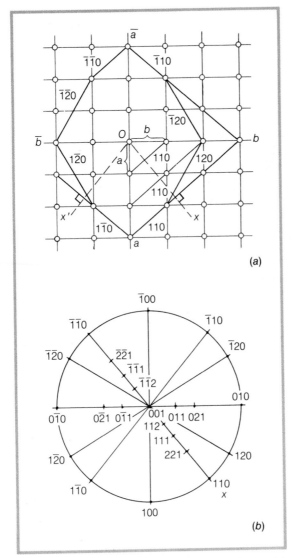

Figure 2-24. Section of an orthorhombic lattice (a) perpendicular to c; the planes (110), (120), (100), (010), . . . are represented on the primitive circle of the stereographic projection (b).

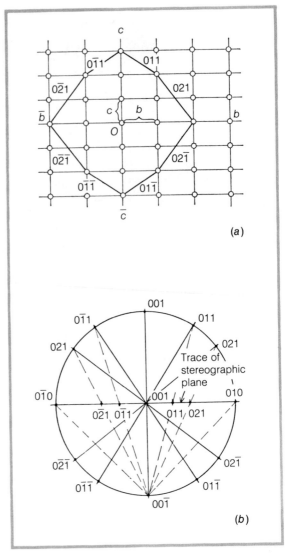

Figure 2-25. Vertical sections of an orthorhombic lattice (a) and the sphere of projection (b) perpendicular to the a axis. The poles of the planes (001), (011), (021), (010), . . . fall on the circumference of the sphere. The stereographic projection points are found on the horizontal line by joining each pole to the pole of $(00\bar{1})$. These points will fall on the east-west diameter of Figure 2-24b.

they do *not* depict a method often used for constructing projections from angular crystal measurements.

An ortho-net with unit translations a and b is shown in Figure 2-24a. The planes (hk0) of the zone [001] are shown as lines that are traces of

planes perpendicular to the drawing. Therefore, normals to these planes can be drawn radially from an arbitrary center and transferred to the

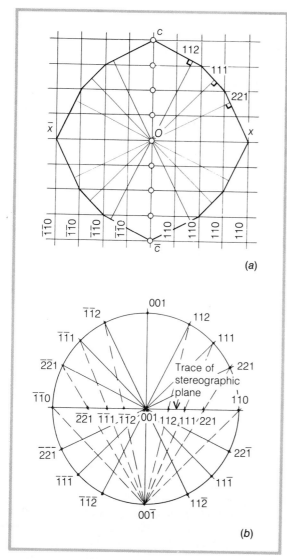

Figure 2-26. Vertical sections of an orthorhombic lattice (a) and the sphere of projection (b) perpendicular to (110), ($\bar{1}\bar{1}$0), and (001). The poles of the planes (001), (112), (111), (221), and (110) fall on the circumference of the sphere. The stereographic projection points fall at the intersection of the horizontal diameter and the line joining each pole to (00$\bar{1}$). These points will fall on the diameter (110)–($\bar{1}\bar{1}$0) in Figure 2-24b.

the poles at the intersection of their normals with the primitive (Figure 2-24b).

In Figure 2-25a, a vertical section is drawn parallel to b and c; in this case, planes (0kl) in the zone [100] can be indicated by straight lines that are the traces of planes normal to the paper. The perpendiculars to these planes radiating from a common center can then be transferred to a vertical section of the projection sphere, and their poles fall on a vertical great circle (Figure 2-25b). These poles can then be transferred to the trace of the stereographic plane by joining each to the south pole (the pole of 00$\bar{1}$) and noting the intersection on the primitive. This zone of faces then appears on the stereographic plane (Figure 2-24a) on an east-west great circle through the center.

A similar construction of the ac plane results in the plotting of planes (h0l) of the zone [010].

All of these relations can be combined in a full projection, as is shown in Figure 2-27.

If we now look back to the net drawing, Figure 2-24a, and visualize c emerging perpendicular to the paper at the center, with planes cutting both a and b at one unit but tilted to cut c at one, two, three, etc. units, or cutting both a and b at two units, and c at one, two, three, etc. units, we have the planes (hhl); that is, we have (111), (221), etc., in the zone [1$\bar{1}$0]. (The planes (001) and (110) are also in this zone.) A cross section of this zone is shown in Figure 2-26a; it can be transferred to the sphere with the normals to the planes intersecting the sphere on a great circle. On the stereographic projection, this great circle becomes a diameter of the primitive joining (110) and ($\bar{1}\bar{1}$0) and including all planes whose Miller indices have the first two indices alike.

Several other relationships will become apparent as you use the projection. For example, you will see that all planes in which two of the indices have the same ratio throughout belong to the same zone, for example, on Figure 2-27, see (201), (211), (221), (241), etc.

Lattice Symmetry Displayed in Stereographic Projections

As is indicated on the stereograms for the crystal

equatorial section of the spherical projection, which is also the plane of the stereographic projection. Thence, the planes are represented by

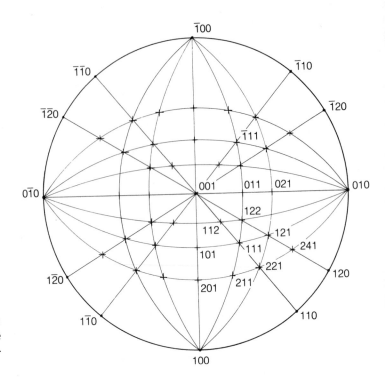

Figure 2-27. Stereographic projection of an orthorhombic crystal, including the points located in the constructions of Figures 2-24 and 2-25.

classes, the stereographic projection of a crystal reflects the symmetry of the crystal lattice. [When so used, it is customary to indicate a plane of symmetry by a solid line. Therefore, a horizontal (ab) plane of symmetry appears as a solid line for the primitive circle, which otherwise is represented by a broken line; vertical planes of symmetry appear as solid diameters; etc. In addition, the positions of symmetry axes can be shown, with their type being indicated by an appropriate symbol, as follows: a solid ellipse for a twofold axis, a solid triangle for a threefold axis, a solid square for a fourfold axis, a solid hexagon for a sixfold axis, a solid triangle within a hexagon for a threefold inversion axis, and a solid ellipse within a square for a fourfold rotoinversion axis (see Figure 2-38).]

The distribution of the lattice points displayed in Figures 2-24 and 2-25 clearly shows that the lattice has twofold axes of symmetry parallel to c and to a and mirror planes of symmetry parallel to OA, OB, and OC. A third drawing of

the lattice in the ac plane would show, in addition, a twofold axis of symmetry parallel to b. This is the full symmetry $2/m2/m2/m$ of the orthorhombic lattice. Since the distribution of points as shown in these drawings is required by this symmetry, it can be noted further that the planes (110), (1$\bar{1}$0), and ($\bar{1}$10), and ($\bar{1}\bar{1}$0) are equivalent in spacing and point density and are therefore equivalent in this crystal lattice. Such a combination of symmetrically equivalent planes is known as a crystal form, and enclosing the Miller indices in braces—{110}—implies all of the faces of the form in a given symmetry. On Figures 2-24 and 2-25, it is evident or readily implied that for this symmetry there are other groups of equivalent planes that can be denoted by form symbols, for example, {120}, {011}, and {021}, etc.

The equivalence of planes in this symmetry is also evident in the stereographic projection (Figure 2-27). The poles of (110), (1$\bar{1}$0), ($\bar{1}$10), and $\bar{1}\bar{1}$0) fall on the corners of a rectangle with sides parallel to the east-west and north-south directions of the projection; therefore, these directions are the traces of vertical symmetry planes, parallel to the ac and bc planes of the lattice. Further, a line perpendicular to the center of the rectangle has twofold rotational symmetry that coincides in direction with the vertical symmetry axis, c, in the lattice. And, the poles of the forms {120}, etc. can also be shown to correspond with this symmetry.

The poles of the zone $(0kl)$ show, on projection, a distribution that fits in with the vertical twofold axis and two vertical mirror planes, and, in addition, a horizontal mirror plane when each form {$0kl$} includes (011), $(0\bar{1}\bar{1})$, $(0\bar{1}1)$, $(01\bar{1})$ or (021), $(02\bar{1})$, $(0\bar{2}1)$, $(0\bar{2}\bar{1})$, etc. (Figure 2-25). Planes with l plus are shown by $+$ and planes with l minus by ò in later projections (see Figures 2-39, 2-41, etc.). If we examine the poles of faces (hkl), we find that, for example, (111), $(\bar{1}11)$, $(1\bar{1}1)$, and $(\bar{1}\bar{1}1)$ are equivalent in position just as in (110), but each is represented in the projection by both $+$ and ○ (Figure 2-43); therefore, $(11\bar{1})$, $(\bar{1}11)$, $(1\bar{1}\bar{1})$, and $(\bar{1}\bar{1}\bar{1})$, with l minus, are equivalent to those with l plus, making eight equivalent planes. This fact is evident from the lattice drawings of Figures 2-25a and 2-26a if we realize that the section of the lattice parallel to OX (Figure 2-24) has its exact counterpart in the section parallel to OX'. In fact, they are indistinguishable except by our arbitrary labeling of the axes. Furthermore, it becomes clear that all forms of the type (hkl) include eight equivalent planes. Also, as can be seen in Figure 2-24a, the stereographic projection can be used to illustrate crystal symmetry no matter what the relative sizes of a crystal's faces. This is true because the positions of the face poles are determined by the angular relations of their faces, not by their sizes.

Determination of Axial Ratios in Stereographic Projections

In stereographic projections, the ratios of the lengths of any two crystallographic axes can be determined graphically if the position of any face

cutting those two axes and parallel to the third crystallographic axis is also known on the projection: that is, a face $(hk0)$ will give the ratio $a:b$; $(h0l)$, the ratio $a:c$; and $(0kl)$, the ratio $b:c$. [These ratios are, of course, morphologically based; today, X-ray diffraction data can be used to establish cell dimensions. Measurements of axes made on the basis of morphological studies can be compared with cell dimensions, such as those given in the mineral descriptions in this book, by making simple calculations; for example, $a = 4.76$, $b = 10.20$, $c = 5.98$ gives $0.467:1:0.586$ (the values for forsterite).]

In any case, in morphological crystallography, any two ratios will suffice to determine the axial ratio for an orthorhombic crystal, which is always stated as $a:b:c = x:1:z$; the location of one face cutting c and either a axis will give the ratio for tetragonal and hexagonal crystals, both stated as $a:c = 1:z$; but, for monoclinic and triclinic crystals, more information is required.

The following paragraphs illustrate the graphic derivation of the axial ratio for orthorhombic crystals.

On a circle of any diameter, lay off the position of the poles of (100), $(hk0)$, (110, 120, etc.), and (010) (Figures 2-28a and b). Draw a tangent to the circle at $(hk0)$ [this tangent is parallel to the trace of a lattice plane $(hk0)$ in the plane ab]. This tangent cuts the radius of the circle through (100) (the a axis) at e, and the radius through (010) (the b axis) at f. The axial ratio $a:b = h(oe):k(of) = (h \cot \phi)/k:1$, where ϕ is the polar angle (010) $\wedge$ $(hk0)$.

The $a:c$ and $b:c$ ratios can be obtained in a similar way from faces $(h0l)$ or $(0kl)$ (Figures 2-28c, d, e, and f) where $b:c = k(of):l(og) = 1:(l \tan \rho)/k$ and $a:c = h(oe):l(og) = 1:(l \tan \rho)/h$.

The ratio can also be obtained from the position of a point hkl (Figures 2-29a and b). Draw a radius of the primitive circle through the point (hkl) to the circle; its intersection on the primitive is $(hk0)$, which may or may not be present in the particular case. Draw a tangent to the primitive at $(hk0)$ to cut the radii through (100) and (010) at e

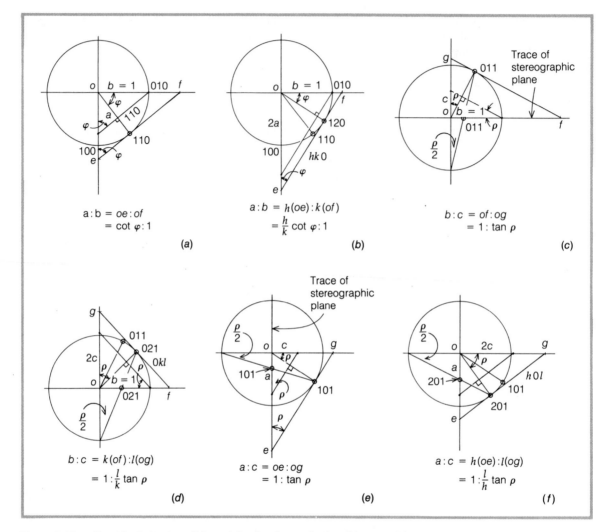

Figure 2-28.   Graphic derivation of the axial ratio of an orthorhombic crystal in stereographic projection. (a) a:b from the pole (110). (b) a:b from the pole (120). (c) b:c from the pole (011). (d) b:c from the pole (021). (e) a:c from the pole (101). (f) a:c from the pole (201).

and f, respectively. The a:b ratio is then h(oe):k(of). Draw a diameter parallel to the tangent at hk0; this diameter will represent the c axis folded down onto the paper with the positive end in the northeast quadrant. Draw a line mn from the negative end of c at the primitive through the pole of hkl to the primitive beyond at n. The radius on is normal to (hkl) in the new orientation. Draw the trace of (hkl), through the pole

of (hk0), normal to on, and extend it to cut the c axis at g. The complete axial ratio is now a:b:c = h(oe):k(of):l(og).

Plotting Faces When the Axial Ratio Is Known

If you know the axial ratio, you can plot faces simply by reversing the foregoing procedure and

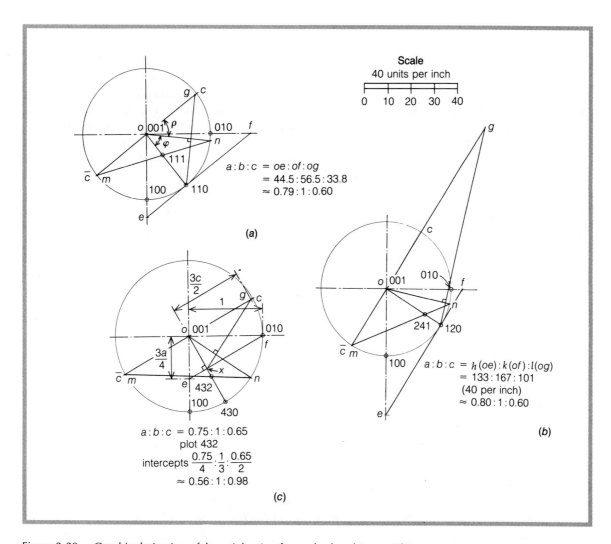

Figure 2-29. Graphic derivation of the axial ratio of an orthorhombic crystal in stereographic projection, and locating any face pole for a given axial ratio. (a) a:b:c from the pole (111). (b) a:b:c from the pole (241). (c) Locating the pole (432) for the axial ratio a:b:c = 0.75:1:0.65.

constructions. First plot the position of a face (hk0) for a crystal with axial ratio x:1:z (Figure 2-29c). On a primitive circle of any suitable radius, with the north-south axis as the pole of (100) and the east-west axis as (010), take the radius as unity to represent b, and measure the length (ka)/h from (001) toward (100) at e. Join ef, and draw a radius of the circle perpendicular to this line; the place where this perpendicular cuts the primitive circle is the position of the pole of (hk0). On Figure 2-29c, the line ef is the trace of the (hk0) plane. Draw c as a line $\overline{C}OC$ parallel to ef, and lay off og equal to (kc)/l. The trace of (hkl) is the line joining g with x, the intersection of ef, and the radius to (hk0). Draw the radius on perpendicular to gx; join mn; the correct projection point for (*hkl*) is the intersection of *mn* with the radius to (*hk*0). And, similar constructions can be used to plot (0*kl*) and (*h*0*l*) faces.

Axial Ratios from Interfacial Angles

In the case of crystals of simple development, axial ratios can be calculated directly from interfacial angles. The formulas relating interfacial angles to axial elements in the general case of a triclinic crystal have been given by, for example, G. Tunell and J. Murdoch (complete reference is given at end of this chapter). These formulas can be used as a basis for deriving simpler formulas for obtaining the axial ratio in the other systems. For crystals other than those of the triclinic and isometric systems, some of the simpler relations are given in Table 2-3 and are illustrated in Figures 2-30 through 2-33.

Tetragonal and Hexagonal Crystals

For tetragonal or hexagonal crystals, the axial ratio can be calculated from the interfacial angle between a face perpendicular to the *c* axis and a face intersecting the *c* axis and one or more *a* axes; it can also be calculated from an interfacial angle between a face parallel to *c* and a face intersecting *c*, providing both cut the same *a* axes (Figures 2-30, 2-31).

Orthorhombic Crystals

To derive the two values of the axial ratio for orthorhombic crystals it is necessary to have two interfacial angles. If one of these angles is between two faces that are parallel to the same axis of reference, this angle will yield the ratio between the other two axes. A second angle between one of these faces and a face intersecting

Table 2-3.   Some formulas relating interfacial angles to axial elements

Tetragonal Crystals (Figure 2-30) (010) $\wedge$ (110) = 45°; $a_1 = a_2$ for all crystals

(001) $\wedge$ (*h*0*l*) or (0*hl*)	$a:c = 1:[l \tan (001) \wedge (h0l)]/h$
(001) $\wedge$ (*hhl*)	$a:c = 1:[l \tan (001) \wedge (hhl) \sin 45°]/h$
(100) $\wedge$ (111)	$a:c = 1: \dfrac{\cot (100) \wedge (111)}{\cos [\sin^{-1} \cot (100) \wedge (111)]}$
(001) $\wedge$ (*hkl*)	$a:c = 1:[l \sin (\tan^{-1} h/k) \tan (001) \wedge (hkl)]/h$

Hexagonal Crystals (Figure 2-31) $(10\bar{1}0) \wedge (01\bar{1}0) = 60°$ and $(10\bar{1}0) \wedge (11\bar{2}0) = 30°$

(0001) $\wedge$ ($h\bar{2}hl$)	$a:c = 1:[l \tan (0001) \wedge (hh\bar{2}hl)]/2h$
(0001) $\wedge$ ($h0\bar{h}l$)	$a:c = 1:[l \tan (0001) \wedge (h0\bar{h}l) \sin 60°]/h$

Orthorhombic Crystals (Figure 2-32)

(010) $\wedge$ (*hk*0)	$a:b = [h \cot (010) \wedge (hk0)]/k:1$
(010) $\wedge$ (0*kl*)	$b:c = 1:[l \cot (010) \wedge (0kl)]/k$
(001) $\wedge$ (*h*0*l*)	$a:c = 1:[l \tan (001) \wedge (h0l)]/h$
(001) $\wedge$ (*hkl*)	$a:b:c = [h \cot (010) \wedge (hk0)]/k:1:[l \cos (010) \wedge (hk0) \tan (001) \wedge (hkl)]/k$

Monoclinic Crystals (Figure 2-33) (100) $\wedge$ (001) = μ = 180° $-$ β

(100) $\wedge$ (110)	$a:b = [\tan (100) \wedge (110)]/\sin \beta:1$
(001) $\wedge$ (011)	$b:c = 1:[\tan (001) \wedge (011)]/\sin \beta$

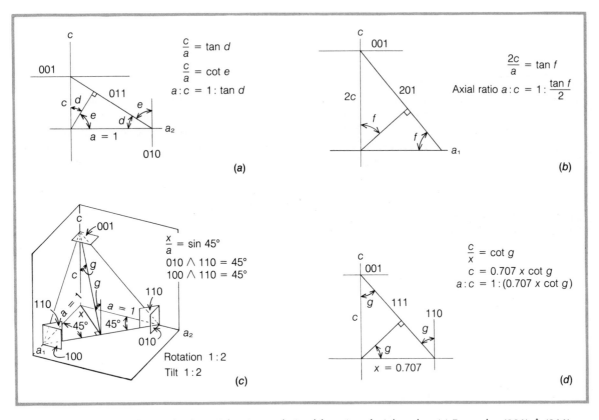

Figure 2-30.   Tetragonal crystals, the axial ratio a:c derived from interfacial angles. (a) From $d = (001) \wedge (011)$ or $e = (010) \wedge (011)$. (b) From $f = (001) \wedge (201)$. (c) and (d) From $g = (110) \wedge (111)$.

the third axis will readily give the relative length of the third axis (Figure 2-32).

Monoclinic Crystals

For monoclinic crystals, calculation of the axial ratio is more involved and requires at least three measured angles. The angle β—the obtuse angle between the a and c axes—is most easily obtained from the polar angle $(100) \wedge (001)$. An angle between two faces parallel to the c axis will give the ratio between b and the projected length of a in a direction perpendicular to c. Since this projected length of a is equal to $a \sin \beta$, a can be determined. Similarly, an angle between two faces parallel to a will give the ratio of b to the projected length of c in a direction normal to a. This projected length is equal to $c \sin \beta$. The

simpler angular relations leading to an axial ratio are given in the diagrams of Figure 2-33.

THE CRYSTAL CLASSES

As already mentioned, Hessel predicted the existence of 32 exclusive crystal classes in 1830. Since then, the limitation to 32 classes has been confirmed by both graphic and mathematical proofs (see Boisen and Gibbs, 1976).

The 32 classes are distributed among the 6 crystal systems as shown in Table 2-4. Crystals of each of the classes within a system have one of the lattices characteristic of that system. In addition, crystals of all of the classes in the ortho-rhombic, tetragonal, and isometric systems also have one or more symmetry elements in com-

54

mon. The hexagonal classes fall into two groups: those with a vertical axis of sixfold symmetry and those with a vertical axis of threefold symmetry.

In each system, the crystal class in which member crystals have the same symmetry as the lattice is generally called the *holohedral* class. Other classes, which have somewhat fewer symmetry elements, are referred to as *merohedral* classes. The symmetry of a merohedral class can, for example, represent an array of lattice points,

each of which is surrounded by an array of atoms with less symmetry than the array of the lattice points themselves.

Examples of holohedral versus merohedral classes that are perhaps easiest to visualize are some of the structures in the isometric system. Native copper, as an example, crystallizes with a face-centered cubic lattice in which each lattice point can be considered to be at the center of a copper atom (see Figures 8-1a and 8-2a) and each

Figure 2-31.　Hexagonal crystals, the axial ratio a:c derived from interfacial angles. (a) and (b) From $d = (0001) \wedge (11\bar{2}1)$ or $e = (11\bar{2}0) \wedge (11\bar{2}1)$. (c) and (d) From $f = (0001) \wedge (10\bar{1}1)$ or $g = (10\bar{1}0) \wedge (10\bar{1}1)$.

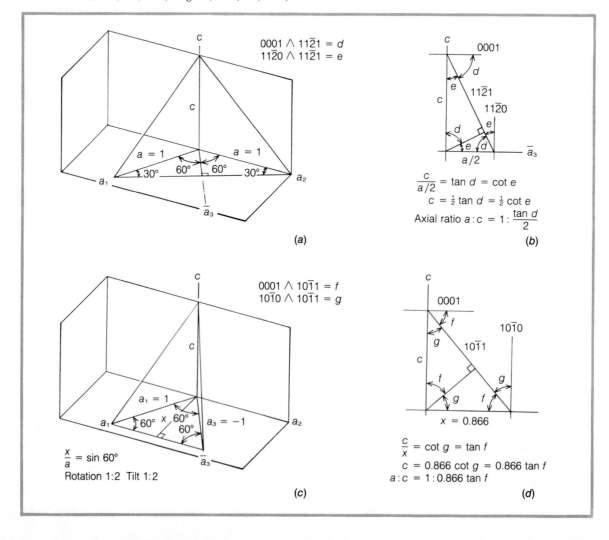

copper atom on a lattice point. Also, if the atoms are considered as spheres, the symmetry of the atomic arrangement is the same as that of the lattice; that is, it is holohedral. On the other hand, in sphalerite (ZnS), atoms of one kind fall at the points of a face-centered cubic lattice while atoms of the second kind are at the points of a second interpenetrating face-centered lattice that is displaced from the first by one-fourth

the length of the body diagonal of the cube of the first (see Figure 9-7). Consequently, this structure possesses no mirror planes of symmetry parallel to cube faces, no twofold axes parallel to face diagonals of the cube, and no center of symmetry. Therefore, the atomic arrangement in sphalerite does not have the same symmetry as the lattice; that is, it is merohedral.

It seems important here to emphasize the

Figure 2-32. Orthorhombic crystals, the axial ratio $a:b:c$ derived from interfacial angles. (a) $a:b$ from $d = (100) \wedge (110)$ or $e = (010) \wedge (110)$. (b) $a:b$ from $f = (100) \wedge (120)$ or $g = (010) \wedge (120)$. (c) and (d) $a:b:c$ from $a:b$ and $i = (001) \wedge (111)$ or $h = (110) \wedge (111)$.

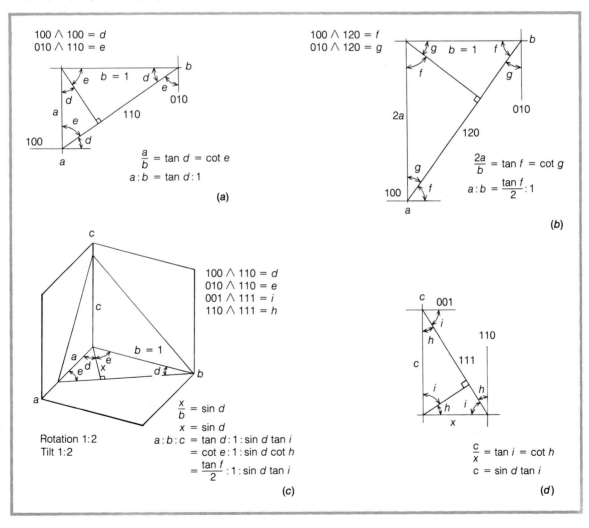

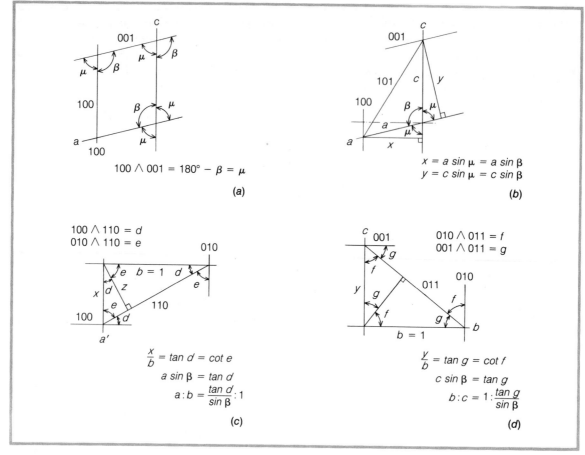

Figure 2-33.   Monoclinic crystals, the axial elements a:b:c and β derived from interfacial angles. (a) $\beta = (001) \wedge (\overline{1}00)$ or $(00\overline{1}) \wedge (100)$, $\mu = 180° - \beta = (100 \wedge (001)$ or $(\overline{1}00) \wedge (00\overline{1})$. (b) x = projected length of a normal to c, and y = projected length of c normal to a. (c) a:b from d = (100) $\wedge$ (110) or e = (010) $\wedge$ (110), (d) b:c from g = (001) $\wedge$ (011) or f = (010) $\wedge$ (011).

meaning of the term *crystal form:*

> *Those crystal faces that are equivalent in a particular symmetry class constitute a crystal form.*

Two examples are: (1) in a triclinic crystal with a center of symmetry, each form will comprise two faces that are parallel to each other and on opposite sides of the crystal; and (2) in an isometric crystal of any class, if we start with the face (010), operation of the symmetry elements develops the six-sided form called a cube. Some crystallographers refer to the former as an *open form*

because it does not enclose space and to the latter (the cube) as a *closed form* because it does enclose space. Under ideal conditions, if one face of a certain crystal form develops on a growing crystal, all of the equivalent faces in the form will also develop. As we mentioned earlier, however, many factors influence the growth of a crystal, so it is rather common to find the different faces of a given form different in size and in distance from the center of the crystal. And, in the case of forms represented by relatively small faces, one or more of the faces may even be absent.

In the following discussion, *each crystal form*

is denoted by the Miller indices of one face, the one with the "simplest" indices, *enclosed in braces*: {*hkl*}. In classical crystallography, it has also been customary in describing the morphology of crystals of a mineral to designate each observed form by a letter in addition to its Miller indices—for example, *m*{110}. Lower case italic letters have usually been used, but for minerals exhibiting many forms, upper case italic and Greek letters have also been used. These letters

Table 2-4. *The thirty-two crystal classes*

#	Axes				Planes	Center	Hermann–Maugin Symbols	Class Name	System
	2-Fold	3-Fold	4-Fold	6-Fold					
1	—	—	—	—	—	—	1	Pedial	Triclinic
2	—	—	—	—	—	yes	$\bar{1}$	Pinacoidal	
3	—	—	—	—	1	—	m	Domatic	
4	1	—	—	—	—	—	2	Sphenoidal	Monoclinic
5	1	—	—	—	1	yes	2/m	Prismatic	
6	1	—	—	—	2	—	mm2	Orthorhombic pyramidal	
7	3	—	—	—	—	—	222	Orthorhombic disphenoidal	Orthorhombic
8	3	—	—	—	3	yes	2/m2/m2/m	Orthorhombic dipyramidal	
9	—	1	—	—	—	—	3	Trigonal pyramidal	
10	—	1	—	—	—	yes	$\bar{3}$	Rhombohedral	
11	—	1	—	—	3	—	3m	Ditrigonal pyramidal	(Trigonal)
12	3	1	—	—	—	—	32	Trigonal trapezohedral	
13	3	1	—	—	3	yes	$\bar{3}$2/m	Trigonal scalenohedral	
14	—	1	—	—	1	—	$\bar{6}$	Trigonal dipyramidal	
15	—	—	—	1	—	—	6	Hexagonal pyramidal	
16	—	—	—	1	1	yes	6/m	Hexagonal dipyramidal	
17	3	1	—	—	4	—	$\bar{6}$m2	Ditrigonal dipyramidal	Hexagonal
18	—	—	—	1	6	—	6mm	Dihexagonal pyramidal	
19	6	—	—	1	—	—	622	Hexagonal trapezohedral	
20	6	—	—	1	7	yes	6/m2/m2/m	Dihexagonal dipyramidal	
21	1	—	—	—	—	—	$\bar{4}$	Tetragonal disphenoidal	
22	—	—	1	—	—	—	4	Tetragonal pyramidal	
23	—	—	1	—	1	yes	4/m	Tetragonal dipyramidal	
24	3	—	—	—	2	—	$\bar{4}$2m	Tetragonal scalenohedral	Tetragonal
25	—	—	1	—	4	—	4mm	Ditetragonal pyramidal	
26	4	—	1	—	—	—	422	Tetragonal trapezohedral	
27	4	—	1	—	5	yes	4/m2/m2/m	Ditetragonal dipyramidal	
28	3	4	—	—	—	—	23	Tetartoidal	
29	3	4	—	—	3	yes	2/m$\bar{3}$	Diploidal	
30	3	4	—	—	6	—	$\bar{4}$3m	Hextetrahedral	Isometric
31	6	4	3	—	—	—	432	Gyroidal	
32	6	4	3	—	9	yes	4/m$\bar{3}$2/m	Hexoctahedral	

58 are especially convenient in designating the faces of forms on crystal drawings. Such usage, however, may differ with authors as well as with crystal system. Perhaps the closest to a standard of reference is Victor Goldschmidt's nine-volume *Atlas der Kristallformen*, the complete citation of which is in the Selected References at the end of this chapter.

In all crystal classes, any crystal form with faces that intersect each of the three axes of reference at different multiples of the unit lengths—for example, {321} or {432}—is known as a general form. In all classes, a general form has the largest number of equivalent faces possible in that class or, to state it conversely, in nearly all crystal classes, crystal forms comprising faces such as {101}, {010}, {011}, etc. have fewer equivalent faces than general forms do. In each crystal class, then, the general form is diagnostic of the symmetry of that class, whereas other forms may be identical in two or more classes of a given system. Hence, the name of a general form in the class is also used as the name of the class (see Table 2-4).

Crystals belonging to certain of the merohedral classes display special features of both crystal form and physical properties. Crystals on which different crystal forms can be found terminating a unique axis (i.e., crystals on which the faces at the two ends of c differ) are described as *hemimorphic*. Such crystals lack a center of symmetry, a mirror plane perpendicular to the unique axis, and symmetry axes perpendicular to the unique axis. Such crystals also commonly display such directional properties as pyroelectricity and piezoelectricity (see page 158). Examples of hemimorphic crystals are hemimorphite, as the name implies, tourmaline, and zincite.

A few mineral species have crystals that can be described as *enantiomorphic*. This term indicates that a given pair of their crystals may be mirror images of each other instead of equivalent. The relationship is expressed by the general form and, in most instances, by other forms as well. Enantiomorphic crystals are found in those crystal classes that lack mirror planes and a center of symmetry. Because they are related to each other in the same manner as a person's right and left hands, the crystals are widely referred to as right-versus left-handed. Perhaps the best example of this property in minerals is found in quartz (Figure 15-8). There are a number of physical properties that give evidence of the right- or left-handed character of enantiomorphic crystals. For example, if a beam of plane polarized light (light vibrating in only one plane parallel to the beam) is passed parallel to the c axis through a plate of quartz, the plane of polarization of the light is rotated to the right or to the left during transmission, with the angle of rotation proportional to the thickness of the quartz and inversely proportional to the wavelength of light. (This property is also possessed by many crystalline organic compounds, such as many of the sugars that crystallize in the monoclinic sphenoidal class, symmetry 2. With the organic compounds, the property persists even in solutions of the compounds and can be detected by an instrument called a polarimeter or saccharimeter. In fact, it is possible to determine the concentration of sugar in a solution by measuring the degree of rotation of the plane of polarization of a known pathlength of light.) Like hemimorphic crystals, crystals of enantiomorphic substances can also display pyroelectricity and piezoelectricity.

In 1913, George Friedel noted that hemimorphic and enantiomorphic crystal classes cannot be distinguished on the basis of their X-ray diffraction patterns. His statement, which became known as *Friedel's Law*, was based on the observation that most X-ray patterns do not clearly exhibit the presence versus the absence of a center of symmetry. This "law" and its consequences have been repeated in several books on crystallography, including some written within the last decade. It is also frequently noted that supplementary tests, such as those involving pyro- and piezoelectric effects, can serve to determine noncentrosymmetry. Actually, in 1962, A. J. C. Wilson pointed out that the presence versus absence of a center of symmetry can be deduced reliably on the basis of a statistical survey of the intensities of the weak versus the strong X-ray reflections, and this relationship is now used routinely by many X-ray crystallographers.

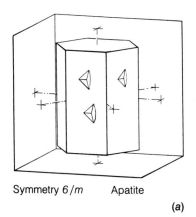 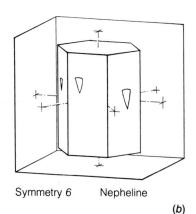

Symmetry 6/m   Apatite   Symmetry 6   Nepheline

(a)   (b)

Figure 2-34. Etch figures on crystals. (a) Pits formed on {10$\bar{1}$0} prism faces of apatite, showing a horizontal symmetry plane but no vertical symmetry planes, consistent with the symmetry 6/m. (From data given by Honess, 1927.) (b) Pits formed on {10$\bar{1}$0} prism faces of nepheline, showing no symmetry planes, consistent with the symmetry 6. (Redrawn from Rogers, *Introduction to the Study of Minerals*.)

Etch Figures and Solution Pits

Crystals, when treated with suitable reagents, can develop pits or etch figures on some or all of their crystal faces. Generally the pits enlarge with accumulated time of exposure to the corrosive reagent, in some cases to the point that several pits merge leaving only residual "hills." The faces of different crystal forms are usually attacked at different rates and, with some reagents, the faces of some crystal forms are greatly affected while others remain quite unaffected. Crystals of some minerals rather commonly exhibit solution pits etched by natural solutions. Many such solution pits have geometric form, and the symmetry apparent in the pits of many minerals is consistent with the orientation of symmetry elements throughout the entire crystal. Therefore, the shape of etch figures on crystal faces of known orientation can be used tentatively to deduce the correct symmetry of the crystal. Hence, etch figures can be of great value in the megascopic study of any mineral that only rarely occurs as crystals exhibiting the general crystal forms that are diagnostic of the mineral's specific class (see Figure 2-34). There are, however, numerous technical difficulties in applying this method and, in some cases, anomalous results have been obtained. Despite the possible problems, in several instances, careful observations have led to deductions of symmetry that have been confirmed later by detailed crystal structure analyses. A

good discussion of the nature of etch figures is given by Honess (see Selected Readings at end of this chapter).

Symmetry Operations

Operations of the basic symmetry elements have been described and/or illustrated. The effect of each of the individual elements of symmetry in producing equivalent faces of a crystal form plus the stereogram for each are shown in Figures 2-9 through 2-13. (The term stereogram is often applied to a stereographic projection that is diagramatic rather than specific.) As previously noted, a cross indicates the pole of a crystal face that intersects the positive end of a vertical axis, and a circle indicates the pole of a face that cuts the negative end of a vertical axis. We begin with a pole in the front right portion of each stereogram, and then show the equivalent poles required by the symmetry. Below each stereogram, there is a drawing of the resulting crystal form. The orthographic (plan) view is shown in Figures 2-35 and 2-36; the clinographic view is shown in Figures 2-35 and 2-37.

Figure 2-35 shows the operation of each type of rotation axis when it is parallel to the vertical axis. Figure 2-36 shows each type of rotoinversion (rotatory inversion) axis when it is parallel to the vertical axis. Figure 2-37 shows the effect of a mirror reflection plane parallel to the

Figure 2-35. Rotation axes of symmetry perpendicular to the plane of a stereographic projection, and the crystal form developed from one face not parallel to the axis by the operation of the symmetry. Form names: (a) pedion, (b) dome, (c) trigonal pyramid, (d) tetragonal pyramid, (e) hexagonal pyramid.

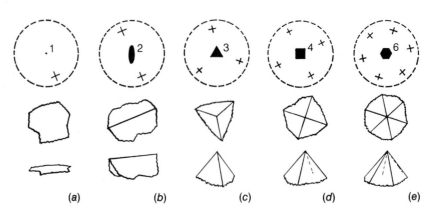

(a) (b) (c) (d) (e)

vertical axis; a twofold rotation axis perpendicular to the vertical axis; and the combinations two-, four-, and sixfold axes perpendicular to mirror planes. Figure 2-38 shows the operation of the major elements of symmetry in a different way.

The Crystal Systems and Classes

In the following descriptions, which are given according to crystal system, all of the classes are tabulated and shown in stereograms. The only classes that are dealt with in any detail in the text, however, are those represented by relatively common, well-known minerals (see Tables 2-5 through 2-15).

Triclinic System

Triclinic crystals have a lattice with $a \neq b \neq c \neq a$ and $\alpha \neq \beta \neq \gamma \neq 90°$; the axes are usually chosen to give a primitive cell. Triclinic crystals commonly have several zones of faces, no two of which intersect at right angles. One habitually elongated zone-axis is usually chosen as [001], defining the c axis. If crystals are habitually tabular, the largest face is usually made (010), and the two most prominent edges of the tabular form define the a and c directions. These choices are usually consistent with $c < a < b$, α and β obtuse, and γ either acute or obtuse. In a stereographic projection with c vertical and the pole of (010) to the east, the pole of (100) will be either to the right or left of south and (001) will be

Figure 2-36. Rotary inversion axes of symmetry perpendicular to the plane of a stereographic projection, and the crystal form developed from one face not parallel to the axis by the operation of the symmetry. Form names: (a) pinacoid, (b) dome, (c) rhombohedron, (d) tetragonal disphenoid, (e) trigonal dipyramid.

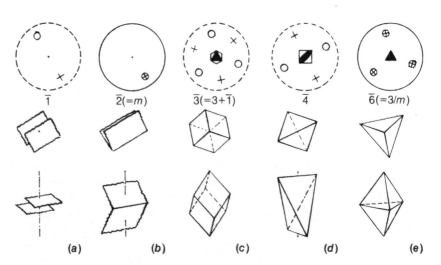

$\overline{1}$ $\overline{2}(=m)$ $\overline{3}(=3+\overline{1})$ $\overline{4}$ $\overline{6}(=3/m)$

(a) (b) (c) (d) (e)

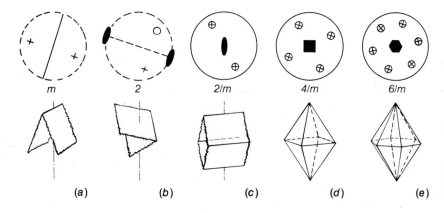

Figure 2-37. The crystal forms developed by (a) a mirror plane of symmetry parallel to the vertical axis, (b) a twofold rotation axis perpendicular to the vertical axis, (c) 2/m, (d) 4/m, (e) 6/m. Each is shown with the symmetry axis vertical. Form names: (a) dome, (b) sphenoid, (c) prism, (d) tetragonal dipyramid, (e) hexagonal dipyramid.

in the southeast or southwest quadrant, respectively.

In the *holohedral class*, with symmetry $\bar{1}$, a center of symmetry requires that each form consist of two parallel and opposite faces; each form may therefore be properly called a pinacoid—a form consisting of two parallel and opposite faces. The relation of any form to the crystal axes, in any particular case, is best indicated by the Miller indices. In the stereographic projection (Figure 2-39), several forms are plotted, each consisting of two poles. The indices for any form depend on the choice of axial elements, which is purely arbitrary, but an author of mineral descriptions is usually guided by the rules given in the preceding paragraph. Many authors, however, have not followed the same rules for orientation,

especially in the past. As an illustration it might be noted that the mineral axinite has been described in twenty different settings, six of which are different orientations of the same unit cell. Figure 15-39 shows two of these orientations. Table 2-5 lists the different types of forms in terms of Miller indices, together with the names of the forms. It is evident that there is no distinction in number of faces between any form—they are all pinacoids. Figure 2-40 shows a simple triclinic crystal in ideal development.

For a few substances, the atomic arrangement with triclinic symmetry may have a lattice in which two or three axes are equal in length, or in which one or more interaxial angles are 90° within the limits of measurement. Because crystals of these substances commonly present the

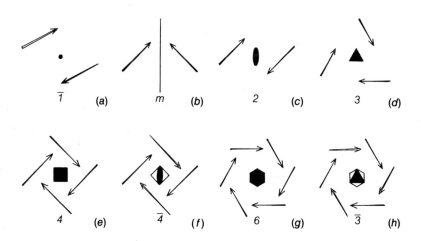

Figure 2-38. The operation of the symmetry elements $\bar{1}$, m, 2, 3, 4, $\bar{4}$, 6, $\bar{3}$, perpendicular to the page, in relating arrows.

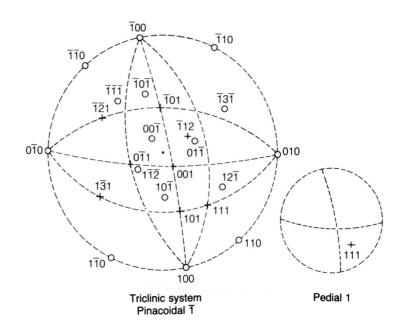

Figure 2-39. Stereographic projection of some forms in the pinacoidal class $\bar{1}$ (Table 2-5). The small stereogram shows the pedion {111} in the pedial class 1. Here and in later figures a face with l positive is shown by + , one with l negative by O, and one with l zero by O on the primitive.

Triclinic system
Pinacoidal $\bar{1}$

Pedial 1

appearance of higher symmetry, the proof of triclinic symmetry requires careful study by other methods.

For some minerals, such as the feldspars, a nonprimitive unit cell is chosen because it shows the similarity of the lattice to the lattice of the monoclinic feldspars. In other minerals, a nonprimitive cell, which shows a pronounced similarity to the cell of a more symmetric lattice, may be chosen for convenience. In such a case, the most important crystal faces will not always be those of simplest indices.

Some crystals with a triclinic lattice have no center of symmetry. In this case, no crystal face has an equivalent face, and a crystal may be bounded by several single faces, often without parallel opposite faces. Each such face is a separate crystal form, called a pedion, and the class is called the pedial class. Such crystals may be pyroelectric and piezoelectric (see Chapter 4). In fact, the identification of directional properties such as these may be taken as evidence, albeit not absolute, for the lack of a center of symmetry.

Figure 2-40. A hypothetical triclinic crystal with the three pinacoids {100}, {010}, {001}. (b) A crystal of rhodonite showing the forms a {100}, c {001}, m {110}, M {1$\bar{1}$0}, n {11$\bar{1}$}, and k {1$\bar{1}$$\bar{1}$}, all pinacoids.

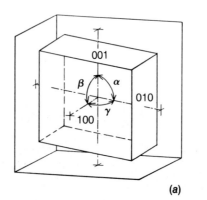

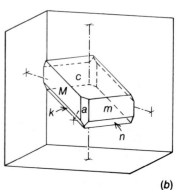

(a)

(b)

Monoclinic System

All monoclinic crystals possess a lattice with $a \neq b \neq c\ (\neq a)$, and with $c \wedge a\ (= \beta)$ greater than 90°; the lattice is one for which the conventionally chosen cell can be primitive or end-centered. Monoclinic crystals, however, can belong to any one of three symmetry classes; the *prismatic, 2/m*; the *domatic, m*; or the *sphenoidal, 2*. Most monoclinic substances crystallize in the prismatic class; the other classes are represented in few minerals. It is possible for an arrangement of atoms to have monoclinic symmetry and a lattice in which two axes are of equal length, within the limits of measurement, and/or in which the β angle is apparently 90°. Such crystals may appear in their geometric form to have higher symmetry, which is known as pseudosymmetry. The true symmetry can, in most cases, be determined only by careful optical or X-ray diffraction studies. We note this here to re-emphasize the fact that, in general, the symmetry is the defining property of a mineral, and the lattice geometry needs only to be compatible with the appropriate symmetry, not vice versa.

The twofold axis of the lattice is chosen as the

Table 2-5. Triclinic system*

Name	Lattice P; Axes $a \neq b \neq c \neq a$; Angles $\alpha \neq \beta \neq \gamma \neq 90°$ Pinacoidal			Pedial	
International symmetry symbol		$\bar{1}$		1	
{001}		pinacoid	2	pedion	1
{00$\bar{1}$}				pedion	1
{010}		pinacoid	2	pedion	1
{0$\bar{1}$0}				pedion	1
{100}		pinacoid	2	pedion	1
{$\bar{1}$00}				pedion	1
{0kl}	{011}	pinacoid	2	pedion	1
{0$\bar{k}l$}	{0$\bar{1}\bar{1}$}			pedion	1
{0$\bar{k}l$}	{0$\bar{1}$1}	pinacoid	2	pedion	1
{0$k\bar{l}$}	{0$1\bar{1}$}			pedion	1
{$h0l$}	{101}	pinacoid as for {0kl}	2	pedion as for {0kl}	1
{$hk0$}	{110}	pinacoid as for {0kl}	2	pedion as for {0kl}	1
{hkl}	{111}	pinacoid	2	pedion	1
{$\bar{h}kl$}	{$\bar{1}$11}	pinacoid	2	pedion	1
{$h\bar{k}l$}	{1$\bar{1}$1}	pinacoid	2	pedion	1
{$hk\bar{l}$}	{$\bar{1}\bar{1}$1}	pinacoid	2	pedion	1
				four others with $\bar{1}$	
Minerals		plagioclase, kyanite, axinite, turquoise		(calcium thiosulphate) bustamite	

* In the accompanying stereographic projections for each class (Figure 2-39), + represents a face with *l* positive, and ○ a face with *l* negative; the large projections of the seven holohedral classes show all the type crystal forms; the small projections show only one form of the general type {hkl}, except in isometric classes where {100}, {110}, and {111} are also shown.

64

b axis and is normal to the mirror plane of symmetry. In minerals with habitual elongation, the *b* axis is typically perpendicular to the axis of elongation; in a few minerals, such as epidote, *b* is parallel to the elongation. In tabular minerals, the *b* axis may lie in the plane of tabular development, as in most micas, or perpendicular to the tabular plane, as in gypsum. If the elongation lies in the symmetry plane, the axis of elongation is usually chosen as the *c* axis. Most common monoclinic minerals—including orthoclase, the amphiboles and pyroxenes, gypsum, and malachite—are described in terms of the *c* axis as the axis of typical elongation. Some crystals of orthoclase, however, are elongated on the *a* axis in terms of the usual description; crystals of some varieties of pyroxene fail to show any marked elongation; etc.

*Table 2-6. Monoclinic system**

Lattices: $P(= B)$, $C(= A$ or $I)$; Axes $a \neq b \neq c \neq a$;
Axial angles $a{:}b = b{:}c = 90°$, $a{:}c = \beta \neq 90°$

Name	Prismatic		Sphenoidal (Enantiomorphic)		Domatic	
International symmetry symbol	2/m		2		m	
{001}	pinacoid (basal)	2	pinacoid (basal)	2	pedion	1
{00$\bar{1}$}					pedion	1
{010}	pinacoid (clino)	2	pedion	1	pinacoid	2
{0$\bar{1}$0}			pedion	1		
{100}	pinacoid (ortho)	2	pinacoid	2	pedion	1
{$\bar{1}$00}					pedion	1
{h0l} {101}	pinacoid (ortho)	2	pinacoid	2	pedion	1
{$\bar{h}$0$\bar{l}$} {$\bar{1}$0$\bar{1}$}					pedion	1
{$\bar{h}$0l} {$\bar{1}$01}	pinacoid (ortho)	2	pinacoid	2	pedion	1
{h0$\bar{l}$} {10$\bar{1}$}					pedion	1
{0kl} {011}	prism	4	sphenoid	2	dome	2
{0$\bar{k}$l} {0$\bar{1}$1}			sphenoid	2		
{0k$\bar{l}$} {01$\bar{1}$}					dome	2
{hk0} {110}	prism	4	sphenoid	2	dome	2
{h$\bar{k}$0} {1$\bar{1}$0}			sphenoid	2		
{$\bar{h}$k0} {$\bar{1}$10}					dome	2
{hkl} {111}	prism	4	sphenoid	2	dome	2
{h$\bar{k}$l} {1$\bar{1}$1}			sphenoid	2		
{$\bar{h}$k$\bar{l}$} {$\bar{1}$1$\bar{1}$}					dome	2
{$\bar{h}$kl} {$\bar{1}$11}	prism	4	sphenoid	2	dome	2
{$\bar{h}$$\bar{k}$l} {$\bar{1}$$\bar{1}$1}			sphenoid	2		
{hk$\bar{l}$} {11$\bar{1}$}					dome	2
Minerals	gypsum, hornblende, augite, orthoclase		(sugar, tartaric acid), afwillite		clinohedrite	

* Stereographic projections are shown in Figure 2-41.

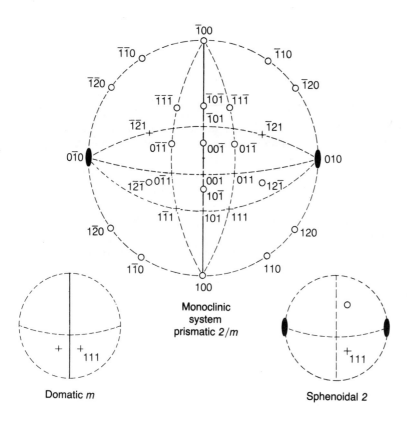

Monoclinic
system
prismatic 2/m

Domatic m

Sphenoidal 2

Figure 2-41. Stereographic projection of some forms in the prismatic class 2/m (Table 2-6). The small stereograms show a dome {111} in the class m and a sphenoid {111} in the class 2.

In monoclinic crystals, any two zone axes parallel to the symmetry plane will not, in general, be at right angles to each other. However, any zone parallel to the symmetry plane is perpendicular to the zone parallel to b. It is usual to choose the axes of the two most prominent zones in the symmetry plane as the a and c axes and to give the largest faces in those zones the simplest indices—(001), (011), and (010) for the zone parallel to a, and (100), (110), and (010) for the zone parallel to c. The crystal forms developed on crystals of the class 2/m are listed in Table 2-6 and are shown on the stereogram of Figure 2-41. As in the triclinic system, c\{001\}, b\{010\}, and a\{100\} are pinacoids. In a widely used nomenclature scheme: \{001\} is the basal pinacoid since it cuts the c axis; \{010\} is the clinopinacoid since it is parallel to a clino-net and perpendicular to the twofold axis; and \{100\} is an orthopinacoid since it is parallel to an ortho-net and perpendicular to the mirror plane.

The shape of an individual face depends largely on its interrelations with the combination of all the other faces present on a crystal. On well-developed crystals, the shapes also reflect certain relations to the symmetry axes. In the monoclinic system, any plane (101) or ($h0l$) is repeated by the symmetry as two faces. Each form of \{$hk0$\} and \{$0kl$\}, in two principal zones parallel to the mirror plane, consists of four equivalent faces, all parallel to one axis; each form is a prism that is rhombic in cross section. If a monoclinic crystal is incompletely developed, showing only the faces of the prisms (\{$hk0$\} or \{$0kl$\}) of the zone of elongation, it cannot be distinguished from an orthorhombic crystal.

Each form \{hkl\} and \{$\bar{h}kl$\} is also a prism with four faces all parallel to one lattice row. These prisms constitute zones, the axes of which lie in the mirror plane and are diagonals, such as [101] or [$\bar{1}01$], in the clino-net with edges of a and c. Thus (101), (212), (111), and (010) make up one

66

such zone, [101]; and ($\bar{2}01$), ($\bar{2}11$), ($\bar{2}21$), and (010) make up another zone, [102]. The type crystal forms, together with a few typical crystals, are shown in Figure 2-42.

In most amphibole and pyroxene crystals, there are two zones that are at nearly the same angle, about 75°, to the prominent zone of habitual elongation (i.e., parallel to c). One of these zones is made the ($0k\bar{l}$) zone, thus defining the a axis; the other becomes the ($\bar{1}kl$) zone. It is usual to choose the angle that is closer to 90° as β, but very careful measurements are required to distinguish the two angles in these minerals because the difference is usually less than 1°. It has been established by X-ray study that the lattice is centered in these minerals and, with one choice of

the a axis, the lattice is C-centered, whereas, with the other choice, the lattice is I-centered. The Law of Bravais indicates that {001}, {$\bar{1}11$}, and {111} are to be expected as important forms in a C lattice but not in an I lattice, since in the latter the spacing of {001} and {111} are halved by the centering. Thus, a setting in which {001}, {111}, and/or {$\bar{1}11$} are important should be that with a C lattice.

More minerals—more than 300 in all—crystallize in the monoclinic prismatic class than in any other single class. Some of the more common ones are gypsum, orthoclase, hornblende, augite, chlorite, the micas, the clay minerals, epidote, heulandite, talc, pyrophyllite, malachite, azurite, borax, sphene, realgar, and orpiment.

Figure 2-42.   The type crystal forms of the prismatic class: prisms {hkl}, {$0kl$}, {$hk0$}; pinacoids {$h0l$}, {100}, {010}, {001}. (a) A combination of three pinacoids: {100}, {010}, {001}. (b) A combination of a prism {110} and a pinacoid {001}. (c) A combination of three pinacoids: {101}, {10$\bar{1}$}, {010}. (d) A crystal of pyroxene with the forms: pinacoids a{100}, b{010}, c{001}, p{$\bar{1}01$}; prisms u{111}, v{221}. (e) A crystal of orthoclase elongated along a with the forms: pinacoids b{010}, c{001}, y{$20\bar{1}$}; prisms m{110}, z{130}, n{021}, o{11$\bar{1}$}.

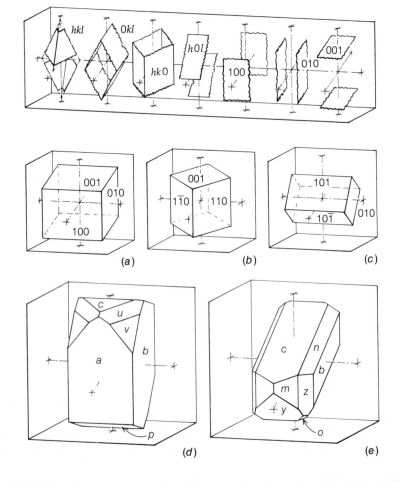

Table 2-7. Orthorhombic system*

Lattices: *P, C* (= *B* or *A*), *I* or *F*; Axes *a* ≠ *b* ≠ *c* ≠ *a*; Axial angles all 90°

Name	Rhombic Dipyramidal		Rhombic Disphenoidal (Enantiomorphic)		Rhombic Pyramidal (Hemimorphic)	
International symmetry symbol	*2/m2/m2/m*		*222*		*mm2* [a]	
{001}	pinacoid (basal)	2	pinacoid (basal)	2	pedion	1
{00$\bar{1}$}					pedion	1
{010}	pinacoid (side)[b]	2	pinacoid	2	pinacoid	2
{100}	pinacoid (front)[c]	2	pinacoid	2	pinacoid	2
{0*kl*} {011}	rhombic prism[b]	4	rhombic prism	4	dome	2
{0*k$\bar{l}$*} {01$\bar{1}$}					dome	2
{*h*0*l*} {101}	rhombic prism[c]	4	rhombic prism	4	dome	2
{*h*0*$\bar{l}$*} {10$\bar{1}$}					dome	2
{*hk*0} {110}	rhombic prism	4	rhombic prism	4	rhombic prism	4
{*hkl*} {111}	rhombic dipyramid	8	+rhombic disphenoid	4	rhombic pyramid	4
{*h$\bar{k}$l*} {1$\bar{1}$1}			−rhombic disphenoid	4		
{*hk$\bar{l}$*} {11$\bar{1}$}					rhombic pyramid	4
Minerals	barite, stibnite, marcasite, aragonite, etc.		epsomite		hemimorphite, enargite	

* Stereographic projections are shown in Figure 2-43.
[a] This nomenclature of forms applies when the single twofold axis is made the *c* axis. If it is set as the *a* or *b* axis, the names of the forms are transposed.
[b] In many older textbooks, these forms are known as brachypinacoid and brachydome; that nomenclature is appropriate only when *a* is the short or brachyaxis.
[c] These forms have been called macropinacoid and macrodome since they are parallel to the *b* axis, on the basis of the assumption that the *b* axis is always the longer or macroaxis; this is not the case in some minerals.

The domatic class *m*, as already mentioned, is represented by only a few rare minerals, such as clinohedrite. Therefore, in this book, we describe it only with the stereogram (Figure 2-41) and the tabulation of forms (Table 2-6).

The monoclinic sphenoidal class, *2*, is not represented to any extent in minerals. It does, however, include a great many organic compounds such as sugars, tartaric acid, various tartrates, and rochelle salts. Many of these substances are of interest from the standpoint of crystallography since they rotate the plane of polarized light and are pyroelectric and piezoelectric. Plates of a certain orientation, cut from single crystals of rochelle salt, are used extensively in electronic equipment because of their piezoelectric properties. The *b* axis is the polar axis, and the class may, at the same time, be thought of as hemimorphic with opposite ends of the twofold axis having different crystal forms.

Orthorhombic System

In the orthorhombic system, there are three symmetry classes: *2/m2/m2/m*, *mm2* (or *2mm*), and *222*. In the first and third, the three twofold axes of symmetry become the crystal axes *a*, *b*, and *c*.

The habitual elongation of crystals of many species often leads to the choice of the axis parallel to the elongation as the c crystallographic axis. Thus, c usually corresponds to the shortest of the three edges of the unit cell. If the crystals tend to be tabular, the next longest edge of the tabular crystal is designated the a axis, and the normal to the tabular development then becomes the b axis. These criteria for labeling the three axes of reference in orthorhombic crystals often correspond to a unit cell of the lattice in which c < a < b. There are, however, numerous exceptions to these conventions; for example, in the barite group of minerals, there is a long standing tradition whereby the crystals are oriented so that the prominent cleavages become {001} and {110}. Actually, in the unit cell of barite, as determined by X-ray methods, the a axis is doubled. Thus, the orientation of the axes is retained but the cleavages become {001} and {210}, and the unit cell has b < c < a.

The crystal forms are listed in Table 2-7. The symmetry of the rhombic dipyramidal class (Figure 2-43) requires that the forms c{001}, b{010}, and a{100} (Figure 2-44) consist of two parallel, opposite faces that are therefore pinacoids. It is often convenient to refer to them as basal-, side-, and front-pinacoids (or as c-, b-, or a-pinacoids), respectively; it is, however, more concise and less subject to misinterpretation to use the appropriate Miller indices. The forms {011}, {101}, and {110} consist of four equivalent faces each; these forms are best called rhombic prisms, with the Miller index added to indicate their orientations. The forms {0kl}, {h0l}, and {hk0} with any other integral values of h, k, or l are also rhombic prisms. They differ only in the angles of the rhombic cross section, which depend on the values of a, b, and c, and h, k, and l, that apply in any particular case.

The form {111}, {321}, or {hkl}—the general form—consists of eight faces and is called a

Figure 2-43. Stereographic projection of some forms in the rhombic dipyramidal class 2/m2/m2/m (Table 2-7). The small stereograms show the rhombic pyramid {111} in class mm2, and the rhombic disphenoid {111} in class 222.

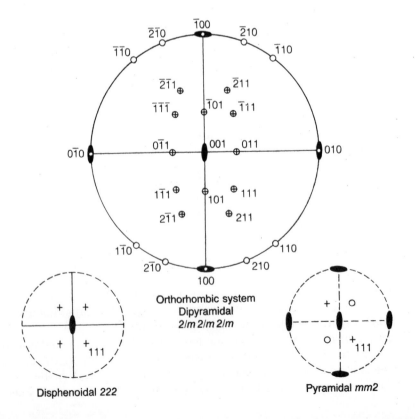

Orthorhombic system
Dipyramidal
2/m 2/m 2/m

Disphenoidal 222

Pyramidal mm2

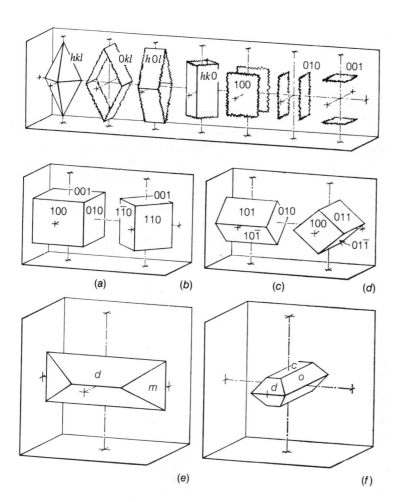

Figure 2-44. Type crystal forms: rhombic dipyramid {*hkl*}; rhombic prisms {*0kl*}, {*h0l*}, {*hk0*}; pinacoids {100}, {010}, {001}. (a) A combination of three pinacoids. (b), (c), (d) Combinations of a prism, with one pinacoid. (e), (f) Typical crystals of barite with rhombic prisms *m*{210}, *d*{101}, *o*{011}, and basal pinacoid *c*{001}. (See also Figures 13-1 and 13-2.)

rhombic dipyramid since the cross section normal to any axis (*a*, *b*, or *c*) is rhombic. With one axis vertical, the upper end is terminated by a four-faced pyramid that is repeated exactly below, thus terminating the negative end of the *c* axis in the same fashion to give the *di*pyramid.

As previously stated, the Laws of Haüy and Bravais lead us to expect that the lattice planes with the highest density of points per unit area will be those that are most likely to occur parallel to crystal faces. Conversely, when you handle a crystal for the first time, after you ascertain the axial directions from the symmetry and choose which axes will be *a*, *b*, and *c* from the elongation or tabular development of the crystal, you should assign the simplest indices to the largest faces in

each of the zones of faces (*0kl*), (*h0l*), (*hk0*), and (*hkl*). If the Law of Bravais were strictly true, the angular relations of the faces chosen as {011}, {101}, and {111} would lead to the axial ratio *a* : *b* : *c*, which would agree with the ratio of the lengths of the edges of the unit cell as determined by some absolute method such as X-ray diffraction. You should recall, however, that an orthorhombic lattice may be primitive, end-centered, body-centered, or face-centered, and that factors such as lattice centering thus affect the order of form importance in one or more zones. Also, as you have learned, certain features of the particular atomic arrangement in the lattice, or the physical conditions present during formation of the crystal, or some combination of these controls,

70   may affect the order of form importance within a zone. Consequently, it may be prudent, especially when dealing with only a single crystal of a mineral, to compromise and adjust the indices in one or more zones to fit the axial ratio determined from the well-developed zones. Along this line, it seems noteworthy that several such determinations of axial ratios have been found to be incorrect by a simple factor of two or three for only one or two of the axes, when compared with the cell dimensions determined by X-ray methods.

The other two classes of the orthorhombic system are represented by relatively few minerals; enargite, hemimorphite, natrolite, and prehnite crystallize in the pyramidal class, epsomite in the disphenoidal class.

The rhombic pyramidal class $mm2$ has one unique axis of symmetry that is usually labeled c regardless of its length relative to a or b. The class is hemimorphic so, with the unique axis as the c axis, all forms intersecting one end of this axis differ from those intersecting the other end. Forms $\{100\}$, $\{010\}$, and $\{hk0\}$ are indistinguishable from similar forms in the class $2/m2/m2/m$, whereas forms $\{001\}$ and $\{00\bar{1}\}$ are pedions, forms $\{h0l\}$ and $\{0kl\}$ are domes (each with two nonparallel faces), and forms $\{hkl\}$ are rhombic pyramids (Figure 2-45a).

The rhombic disphenoidal class 222 is enantiomorphic and the general form, $\{111\}$ or $\{hkl\}$, which is required to identify the class, is a rhombic disphenoid (Figure 2-45b). The dis-

phenoid may be of either right- or left-handed character.

Hexagonal System

Hexagonal crystals are distinct from those of other systems in that they possess one unique axis that is either sixfold or threefold in symmetry. This unique axis is perpendicular to a hexa-net in which the three shortest periods are identical and are labeled a_1, a_2, and a_3. The relation between the a and c axial lengths is expressed in an axial ratio in which $a:c = 1:x$.

The primitive unit cell of the hexagonal lattice (Figure 2-7h) is completely described by $a_1 \equiv a_2 \neq c$, with $a_1 \wedge a_2 = 120°$ and $a_1 \wedge c = a_2 \wedge c = 90°$. It may be easily seen (Figure 2-46) that the short diagonal of a 60° rhomb with edges a_1, a_2 is exactly equivalent to a_1 and a_2; this row, which is indistinguishable from a_1 and a_2, is usually designated a_3. Also, this third axis has led to the development of a four-digit system for indexing crystal faces and lattice planes in this system. Although the index referring to the a_3 axis is superfluous and many crystallographers do not use it, its use should be learned because it is referred to rather frequently. The four-index symbols $(hk\bar{i}l)$, which refer to the four coordinate axes $(a_1, a_2, a_3,$ and $c)$, constitute the aforementioned Miller–Bravais indices. They are derived from intercepts on the axes in the same manner that the three-number Miller indices are. The derivation is illustrated in Figure 2-46, and the intercepts and

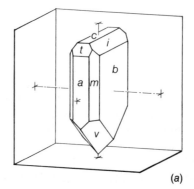
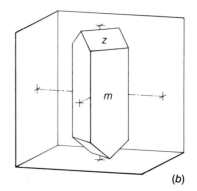

Figure 2-45.   (a) Typical crystal of hemimorphite with symmetry $mm2$ and forms rhombic pyramid $v\{12\bar{1}\}$, rhombic prism $m\{110\}$, domes $i\{031\}$, $t\{301\}$, pinacoids $a\{100\}$, $b\{010\}$, pedion $c\{001\}$. (b) Typical crystal of epsomite with symmetry 222 and forms rhombic disphenoid $z\{111\}$ and rhombic prism $m\{110\}$.

(a)

(b)

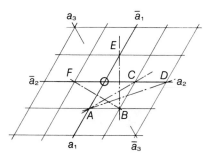

Figure 2-46. A plane of a hexgonal lattice normal to c, showing the derivation of four index Bravais indices: AB is the trace of $(10\bar{1}0)$, AC of $(11\bar{2}0)$, AD of $(21\bar{3}0)$, BE of $(\bar{1}2\bar{1}0)$, and BF of $(2\bar{1}\bar{1}0)$.

indices can be tabulated as follows for planes parallel to the c axis.

Face	Intercepts	Reciprocals	Indices
AB	$1, \infty, \bar{1}, \infty$	$\dfrac{1}{1}, \dfrac{1}{\infty}, -\dfrac{1}{1}, \dfrac{1}{\infty}$	$10\bar{1}0$
BC	$\infty, 1, \bar{1}, \infty$	$\dfrac{1}{\infty}, \dfrac{1}{1}, -\dfrac{1}{1}, \dfrac{1}{\infty}$	$01\bar{1}0$
AC	$1, 1, \dfrac{\bar{1}}{2}, \infty$	$\dfrac{1}{1}, \dfrac{1}{1}, -\dfrac{2}{1}, \dfrac{1}{\infty}$	$11\bar{2}0$
AD	$1, 2, \dfrac{\bar{2}}{3}, \infty$	$\dfrac{1}{1}, \dfrac{1}{2}, -\dfrac{3}{2}, \dfrac{1}{\infty}$	$21\bar{3}0$
BE	$\bar{1}, \dfrac{1}{2}, \bar{1}, \infty$	$-\dfrac{1}{1}, \dfrac{2}{1}, -\dfrac{1}{1}, \dfrac{1}{\infty}$	$\bar{1}2\bar{1}0$
BF	$\dfrac{1}{2}, \bar{1}, \bar{1}, \infty$	$\dfrac{2}{1}, -\dfrac{1}{1}, -\dfrac{1}{1}, \dfrac{1}{\infty}$	$2\bar{1}\bar{1}0$

It is important to remember that the negative end of any a axis falls between the positive ends of the other two, since the positive ends are 120° apart. The introduction of an intercept on the c axis results in whole numbers in the fourth position so, for example, the intercepts $1, 1, \frac{\bar{1}}{2}, 2$ become the indices $22\bar{4}1$. Note that the third index (i) referring to the a_3 axis is the algebraic sum of the first two indices (h and k), but with opposite sign. This is always the case. Therefore, as al-

ready noted, the third index (i) is, in essence, superfluous; consequently, in practice, it is frequently symbolized by a dot (e.g., hk.l).

The symmetry of the lattice and of crystals in the holohedral class of the hexagonal system is expressed by the symbol $6/m2/m2/m$, where $6/m$ refers to the plane of the hexa-net and the unique axis, c, normal to that plane. The first $2/m$ refers to the a axes (the shortest lattice rows in the hexa-net) and the lattice planes $(11\bar{2}0)$, $(\bar{1}2\bar{1}0)$, and $(2\bar{1}\bar{1}0)$, to which the axes are perpendicular. The second $2/m$ refers to the axes at 30° to the a axes (i.e., the long diagonals in the 60° rhomb in the hexa-net) and the ac planes of the lattice [$(10\bar{1}0)$, $(01\bar{1}0)$, and $(1\bar{1}00)$] to which these axes are perpendicular. This symmetry is displayed on the stereogram (Figure 2-47), and the nature of the crystal forms characteristic of this symmetry can be predicted (see Table 2-8).

The face (0001) has one other equivalent face, $(000\bar{1})$, resulting in a form referred to as the basal pinacoid. The faces $(10\bar{1}0)$ and $(11\bar{2}0)$ give rise to hexagonal prisms with 6 faces each: these are commonly distinguished as I-order $\{10\bar{1}0\}$ and II-order $\{11\bar{2}0\}$. The faces $(10\bar{1}1)$ or any face $(h0\bar{h}l)$ will generate a I-order hexagonal dipyramid with 12 faces; the faces $(11\bar{2}1)$ or $(hh\overline{2h}l)$ will generate a II-order hexagonal dipyramid. Each of these prisms and pyramids has a 6-sided cross section. The faces $(21\bar{3}0)$ or $(hk\bar{i}0)$ and $(21\bar{3}1)$ or $(hk\bar{i}l)$ will generate a dihexagonal prism with 12 faces and a dihexagonal dipyramid with 24 faces, respectively (Figure 2-48). The outline of a dihexagonal prism in a plane perpendicular to c is 12-sided. No combination of whole numbers for h and k, however, will give a regular dodecagon with interfacial angles of 30° between every pair of faces; instead, alternate interfacial angles are equal but not equal to the other set of alternate angles. A beryl crystal typical of this class is shown in Figure 2-49d.

Because Cartesian coordinates are not applicable to more than three axes of reference, the third axis, a_3, is disregarded in dealing with zones in the hexagonal system. The face indices (hk.l) and (h'k'.l') are used in obtaining the zone sym-

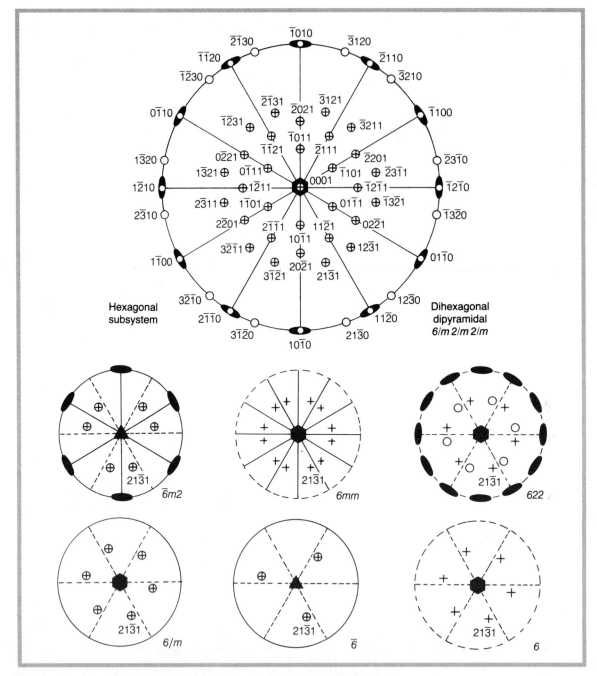

Figure 2-47. Stereographic projection of some forms on the dihexagonal dipyramidal class $6/m2/m2/m$ (Table 2-8). Small stereograms show one general form $\{21\bar{3}1\}$ for each merohedral class $\bar{6}m2$, $6mm$, 622, $6/m$, $\bar{6}$, and 6 (Tables 2-8 and 2-9).

Table 2-8. *Hexagonal system (hexagonal subsystem)*

Lattice P; Axes $a_1 = a_2 = a_3 \neq c$;

Axial angles $+a_1 \wedge +a_2 = +a_2 \wedge +a_3 = 120°$, $a_1 \wedge c = a_2 \wedge c = a_3 \wedge c = 90°$

Name		Dihexagonal Dipyramidal	Hexagonal Trapezohedral (Enantiomorphic)	Dihexagonal Pyramidal (Hemimorphic)	Ditrigonal Dipyramidal
International symmetry symbol		$6/m2/m2/m$	622	$6mm$	$\bar{6}m2$ (or $\bar{6}2m$[a])
$\{0001\}$ $\{000\bar{1}\}$		pinacoid (basal) 2	pinacoid (basal) 2	pedion 1 pedion 1	pinacoid (basal) 2
$\{10\bar{1}0\}$ $\{01\bar{1}0\}$		hexagonal prism 6	hexagonal prism 6	hexagonal prism 6	trigonal prism 3 trigonal prism 3
$\{11\bar{2}0\}$		hexagonal prism 6	hexagonal prism 6	hexagonal prism 6	hexagonal prism 6
$\{hk\bar{i}0\}$	$\{2\bar{1}\bar{3}0\}$	dihexagonal prism 12	dihexagonal prism 12	dihexagonal prism 12	ditrigonal prism 6
$\{kh\bar{i}0\}$	$\{1\bar{2}\bar{3}0\}$				ditrigonal prism 6
$\{h0\bar{h}l\}$	$\{10\bar{1}1\}$	hexagonal dipyramid 12	hexagonal dipyramid 12	hexagonal pyramid 6	trigonal dipyramid 6
$\{h0\bar{h}\bar{l}\}$	$\{10\bar{1}\bar{1}\}$			hexagonal pyramid 6	
$\{0h\bar{h}\bar{l}\}$	$\{01\bar{1}\bar{1}\}$				trigonal dipyramid 6
$\{hh\bar{2}hl\}$	$\{11\bar{2}1\}$	hexagonal dipyramid 12	hexagonal dipyramid 12	hexagonal pyramid 6	hexagonal dipyramid 12
$\{hh\bar{2}h\bar{l}\}$	$\{11\bar{2}\bar{1}\}$			hexagonal pyramid 6	
$\{hkil\}$	$\{2\bar{1}\bar{3}1\}$	dihexagonal dipyramid 24	right hexagonal trapezohedron 12	dihexagonal pyramid 12	ditrigonal dipyramid 12
$\{i\bar{k}\bar{h}l\}$	$\{\bar{3}\bar{1}21\}$		left hexagonal trapezohedron 12		
$\{hki\bar{l}\}$	$\{2\bar{1}\bar{3}\bar{1}\}$			dihexagonal pyramid 12	
$\{khi\bar{l}\}$	$\{1\bar{2}\bar{3}\bar{1}\}$				ditrigonal dipyramid 12
Minerals		beryl, pyrrhotite, molybdenite, graphite	β-quartz (high)	zincite, iodyrite,	benitoite wurtzite

[a] This symbol applies where [[10.0]] axes are twofold and $\{10\bar{1}0\}$ planes are mirror reflection planes; in this case $\{10\bar{1}0\}$ and $\{h0\bar{h}l\}$ are hexagonal forms, and $\{11\bar{2}0\}$ and $\{hh\bar{2}hl\}$ are trigonal forms. Stereographic projections are shown in Figure 2-47.

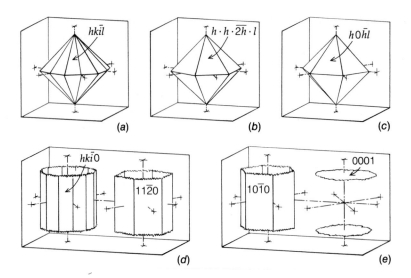

Figure 2-48. Type forms in the dihexagonal dipyramidal class. (a) Dihexagonal dipyramid, (b) II-order hexagonal dipyramid, (c) I-order hexagonal dipyramid, (d) dihexagonal prism and II-order hexagonal prism, (e) I-order hexagonal prism and basal pinacoid.

bol [uv.w], and the equation of zone control $(hu + kv + lw = 0)$ is valid as in the other systems.

There are six classes within the hexagonal subsystem with less symmetry than $6/m2/m2/m$. Each has a general form $\{hk\bar{i}l\}$ that characterizes the symmetry, and each has one-half or one-quarter the number of faces found in a dihexagonal dipyramid (i.e., 12 or 6 faces, as the case may be). In some classes, however, forms other than $\{hk\bar{i}l\}$ also differ from corresponding forms in class $6/m\,2/m\,2/m$. These six classes fall naturally into two groups: those with symmetry axes perpendicular to c or with mirror planes par-

Figure 2-49. Diagnostic forms in the hexagonal dipyramidal class. (a) III-Order hexagonal prism, (b) III-order hexagonal dipyramid. (c) Relation of III-order and I-order hexagonal prisms. (d) Crystal of beryl, dihexagonal dipyramidal, showing forms hexagonal prism $m\{10\bar{1}0\}$; basal pinacoid $c\{0001\}$; hexagonal dipyramids $p\{10\bar{1}2\}$, $u\{10\bar{1}1\}$, $s\{11\bar{2}2\}$; dihexagonal dipyramid $v\{21\bar{3}2\}$. (e) Crystal of apatite, hexagonal dipyramidal, showing forms hexagonal prisms $m\{10\bar{1}0\}$, $a\{11\bar{2}0\}$; basal pinacoid $c\{0001\}$; hexagonal dipyramids $r\{10\bar{1}2\}$, $x\{10\bar{1}1\}$, $z\{30\bar{3}1\}$, $v\{11\bar{2}2\}$, $s\{11\bar{2}1\}$, $\mu\{21\bar{3}1\}$, $n\{31\bar{4}1\}$.

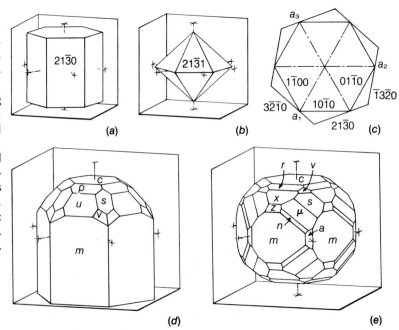

allel to c (Table 2-8) and those with no symmetry of this sort (Table 2-9). In the first group, there are the three classes $\bar{6}m2$, $6mm$, and 622; in the second group, there are the three classes $6/m$, 6, and $\bar{6}$ ($= 3/m$).

The first three classes are represented by a few minerals, most of which rarely exhibit crystal forms. Class $\bar{6}m2$ is represented by the mineral benitoite. The hexagonal forms of zinc sulfide (wurtzite), zinc oxide (zincite), and several other substances (including ice) crystallize with symmetry $6mm$, which is a hemimorphic class. Crystals of β-quartz, which is formed at temperatures above 573°C, have the symmetry 622. Many crystals of this form of silica show only the hexagonal dipyramid. The hexagonal trapezohedron, typical of this class, has not been observed. Therefore, unless structural studies are made of the mineral in question, the correct identification of the class usually depends on determining cer-

Table 2-9. Hexagonal system* (hexagonal subsystem)

Name		Hexagonal Dipyramidal		Hexagonal Pyramidal (Hemimorphic)		Trigonal Dipyramidal	
International symmetry symbol		$6/m$		6		$\bar{6}(3/m)$	
{0001}		pinacoid (basal)	2	pedion	1	pinacoid (basal)	2
{000$\bar{1}$}				pedion	1		
{10$\bar{1}$0}		hexagonal prism	6	hexagonal prism	6	trigonal prism	3
{01$\bar{1}$0}						trigonal prism	3
{11$\bar{2}$0}		hexagonal prism	6	hexagonal prism	6	trigonal prism	3
{2$\bar{1}\bar{1}$0}						trigonal prism	3
{hk$\bar{i}$0}	{21$\bar{3}$0}	hexagonal prism	6	hexagonal prism	6	trigonal prism	3
{kh$\bar{i}$0}	{12$\bar{3}$0}	hexagonal prism	6	hexagonal prism	6	trigonal prism	3
{i$\bar{k}\bar{h}$0}	{3$\bar{1}$20}					trigonal prism	3
{i$\bar{h}\bar{k}$0}	{3$\bar{2}\bar{1}$0}					trigonal prism	3
{h0$\bar{h}$l}	{10$\bar{1}$1}	hexagonal dipyramid	12	hexagonal pyramid	6	trigonal dipyramid	6
{h0$\bar{h}\bar{l}$}	{10$\bar{1}\bar{1}$}			hexagonal pyramid	6		
{0h$\bar{h}$l}	{01$\bar{1}$1}					trigonal dipyramid	6
{hh$\overline{2h}$l}	{11$\bar{2}$1}	hexagonal dipyramid	12	hexagonal pyramid	6	trigonal dipyramid	6
{hh$\overline{2h}\bar{l}$}	{11$\bar{2}\bar{1}$}			hexagonal pyramid	6		
{2$\bar{h}\bar{h}$l}	{2$\bar{1}\bar{1}$1}					trigonal dipyramid	6
{hk$\bar{i}$l}	{21$\bar{3}$1}	hexagonal dipyramid	12	hexagonal pyramid	6	trigonal dipyramid	6
{kh$\bar{i}$l}	{12$\bar{3}$1}	hexagonal dipyramid	12	hexagonal pyramid	6	trigonal dipyramid	6
{hk$\bar{i}\bar{l}$}	{21$\bar{3}\bar{1}$}			hexagonal pyramid	6		
{kh$\bar{i}\bar{l}$}	{12$\bar{3}\bar{1}$}			hexagonal pyramid	6		
{i$\bar{k}\bar{h}$l}	{3$\bar{1}\bar{2}$1}					trigonal dipyramid	6
{i$\bar{h}\bar{k}$l}	{3$\bar{2}\bar{1}$1}					trigonal dipyramid	6
Minerals		apatite, vanadinite		nepheline		disilver orthophosphate (artificial)	

* Stereographic projections are shown in Figure 2-47.

76 tain physical properties, such as the rotation of the plane of polarized light, that indicate enantiomorphism. This enantiomorphism and the structural relations with α-quartz, for example, establish *622* as the correct class for β-quartz.

The first class of the second group has the symmetry *6/m*. It is represented by crystals of apatite (Figure 2-49e). The forms $hk\bar{i}0$ and $hk\bar{i}l$ are III-order hexagonal prisms and dipyramids (Table 2-9). The recognition of a III-order form is absolutely necessary to establish the symmetry of crystals in this class on the basis of morphology. Unfortunately, these forms have more complex indices and smaller spacings in the lattice than I-

Table 2-10. Hexagonal system* (trigonal subsystem)

		Lattice *P* or *R*; Axes and angles as in Table 2-8		
Name		**Trigonal Scalenohedral**	**Trigonal Trapezohedral (Enantiomorphic)**	**Ditrigonal Pyramidal (Hemimorphic)**
International symbol		$\bar{3}2/m$ [a]	32 [a]	$3m$ [a]
$\{0001\}$		pinacoid (basal) 2	pinacoid (basal) 2	pedion 1
$\{000\bar{1}\}$				pedion 1
$\{10\bar{1}0\}$		hexagonal prism 6	hexagonal prism 6	trigonal prism 3
$\{01\bar{1}0\}$				trigonal prism 3
$\{11\bar{2}0\}$		hexagonal prism 6	trigonal prism 3	hexagonal prism 6
$\{2\bar{1}\bar{1}0\}$			trigonal prism 3	
$\{hk\bar{i}0\}$	$\{21\bar{3}0\}$	dihexagonal prism 12	ditrigonal prism 6	ditrigonal prism 6
$\{i\bar{k}\bar{h}0\}$	$\{3\bar{1}\bar{2}0\}$		ditrigonal prism 6	
$\{khi0\}$	$\{12\bar{3}0\}$			ditrigonal prism 6
$\{h0\bar{h}l\}$	$\{10\bar{1}1\}$	+rhombohedron 6	+rhombohedron 6	trigonal pyramid 3
$\{0h\bar{h}l\}$	$\{01\bar{1}1\}$	−rhombohedron 6	−rhombohedron 6	trigonal pyramid 3
$\{h0\bar{h}\bar{l}\}$	$\{10\bar{1}\bar{1}\}$			trigonal pyramid 3
$\{0h\bar{h}\bar{l}\}$	$\{01\bar{1}\bar{1}\}$			trigonal pyramid 3
$\{hh\overline{2h}l\}$	$\{11\bar{2}1\}$	hexagonal dipyramid 12	trigonal dipyramid 6	hexagonal pyramid 6
$\{2\bar{h}\bar{h}\bar{h}l\}$	$\{2\bar{1}\bar{1}1\}$		trigonal dipyramid 6	
$\{hh\overline{2k}\bar{l}\}$	$\{11\bar{2}\bar{1}\}$			hexagonal pyramid 6
$\{hk\bar{i}l\}$	$\{21\bar{3}1\}$	+trigonal scalenohedron 12	right + trigonal trapezohedron 6	ditrigonal pyramid 6
$\{k\bar{h}il\}$	$\{12\bar{3}1\}$	−trigonal scalenohedron 12	left − trigonal trapezohedron 6	ditrigonal pyramid 6
$\{i\bar{h}\bar{k}l\}$	$\{3\bar{2}\bar{1}1\}$		right − trigonal trapezohedron 6	
$\{i\bar{k}\bar{h}l\}$	$\{3\bar{1}\bar{2}1\}$		left + trigonal trapezohedron 6	
$\{hk\bar{i}\bar{l}\}$	$\{21\bar{3}\bar{1}\}$			ditrigonal pyramid 6
$\{k\bar{h}i\bar{l}\}$	$\{12\bar{3}\bar{1}\}$			ditrigonal pyramid 6
Minerals		calcite, hematite, corundum, siderite	α-quartz (low), cinnabar	tourmaline, pyrargyrite

* Stereographic projections are shown in Figure 2-50.

[a] All forms except c{0001} are consistent with the setting in which the *a* axes have twofold symmetry and/or the planes {11$\bar{2}$0} are reflection planes; if the symmetry axes are the lattice rows [[21.0]] and/or the planes {10$\bar{1}$0} are reflection planes, the symmetry symbols become $\bar{3}12/m, 312, 31m$, and the nature of the {h0--} and {hh--} forms are transposed.

and II-order forms do; therefore, they are not likely to be developed. For example, the forms $\{10\bar{1}0\}$ and $\{10\bar{1}1\}$ are common on most apatite crystals, whereas II- and III-order forms occur on only exceptionally well-developed crystals of the mineral. The crystal class with symmetry 6 is represented by nepheline, but crystals unambiguously displaying this symmetry are unknown. The correctness of this class for nepheline was established by etching tests on simple crystals and confirmed by a detailed analysis of the arrangement of the atoms based on X-ray diffraction investigations. No minerals are known to belong to the crystal class $\bar{6}$.

Trigonal Subsystem (of the Hexagonal System)

There are five crystal classes in which the vertical axis c is either a threefold rotation or threefold rotoinversion axis; the lattice for each can be either rhombohedral or hexagonal (Tables 2-10 and

Table 2-11. *Hexagonal system* (trigonal subsystem)*

Name		Lattice P or R; Axes and angles as in Table 2-8			Trigonal Pyramidal (Hemimorphic)	
		Rhombohedral				
International symmetry symbol		$\bar{3}$			3	
$\{0001\}$		pinacoid	2		pedion	1
$\{000\bar{1}\}$					pedion	1
$\{10\bar{1}0\}$		hexagonal prism	6		trigonal prism	3
$\{01\bar{1}0\}$					trigonal prism	3
$\{11\bar{2}0\}$		hexagonal prism	6		trigonal prism	3
$\{2\bar{1}\bar{1}0\}$					trigonal prism	3
$\{hk\bar{i}0\}$	$\{21\bar{3}0\}$	hexagonal prism	6		trigonal prism	3
$\{kh\bar{i}0\}$	$\{12\bar{3}0\}$	hexagonal prism	6		trigonal prism	3
$\{i\bar{k}\bar{h}0\}$	$\{3\bar{1}\bar{2}0\}$				trigonal prism	3
$\{\bar{i}\bar{h}k0\}$	$\{3\bar{2}\bar{1}0\}$				trigonal prism	3
$\{h0\bar{h}l\}$	$\{10\bar{1}1\}$	+rhombohedron	6		trigonal pyramid	3
$\{0h\bar{h}l\}$	$\{01\bar{1}1\}$	−rhombohedron	6		trigonal pyramid	3
$\{h0\bar{h}\bar{l}\}$	$\{10\bar{1}\bar{1}\}$				trigonal pyramid	3
$\{0h\bar{h}\bar{l}\}$	$\{01\bar{1}\bar{1}\}$				trigonal pyramid	3
$\{hh\overline{2h}l\}$	$\{11\bar{2}1\}$	+rhombohedron	6		trigonal pyramid	3
$\{2\bar{h}\bar{h}\bar{h}l\}$	$\{2\bar{1}\bar{1}1\}$	−rhombohedron	6		trigonal pyramid	3
$\{hh\overline{2h}\bar{l}\}$	$\{11\bar{2}\bar{1}\}$				trigonal pyramid	3
$\{2\bar{h}\bar{h}\bar{h}\bar{l}\}$	$\{2\bar{1}\bar{1}\bar{1}\}$				trigonal pyramid	3
$\{hk\bar{i}l\}$	$\{21\bar{3}1\}$	+rhombohedron	6		trigonal pyramid	3
$\{kh\bar{i}l\}$	$\{12\bar{3}1\}$	+rhombohedron	6		trigonal pyramid	3
$\{i\bar{k}\bar{h}l\}$	$\{3\bar{1}\bar{2}1\}$	−rhombohedron	6		trigonal pyramid	3
$\{\bar{i}\bar{h}kl\}$	$\{3\bar{2}\bar{1}1\}$	−rhombohedron	6		trigonal pyramid	3
					four others with $\bar{l}$	
Minerals		dolomite, ilmenite, willemite, dioptase			parisite, roentgenite	

* Stereographic projections are shown in Figure 2-50.

78

2-11). The lattice is rhombohedral for many minerals in these classes but hexagonal for minerals such as α quartz, the stable polymorph of SiO_2 from room temperature up to 573°C.

Class $\bar{3}2/m$ A large number of well-known substances—including calcite and hematite—crystallize in this class, which may be looked upon as the holohedral class of the rhombohedral lattice. The unique axis c, which is normal to the hexa-net, is $\bar{3}$; the three a axes of the net have twofold symmetry perpendicular to the three mirror planes [($11\bar{2}0$), ($\bar{1}2\bar{1}0$), and ($2\bar{1}\bar{1}0$)] and thus are $2/m$. The distribution of symmetry is shown

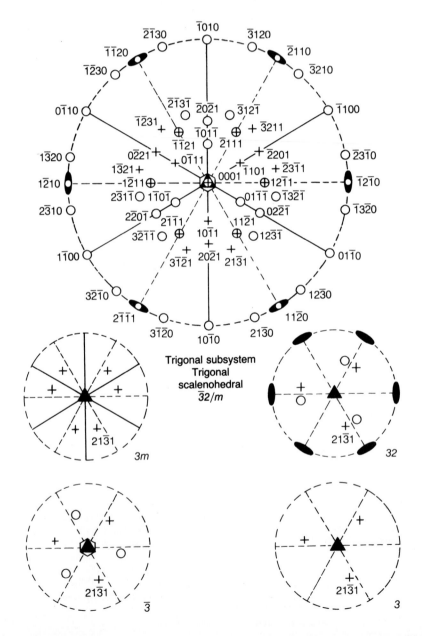

Figure 2-50. Stereographic projection of some forms in the trigonal scalenohedral class $\bar{3}2/m$ (Table 2-10). The small stereograms show the symmetry, and a general form $\{21\bar{3}1\}$ in each class, $3m$, 32, $\bar{3}$, 3 (Tables 2-10 and 2-11).

on the stereogram (Figure 2-50). As noted in Table 2-8, the lattice rows [21.0], instead of the a axes, can have twofold symmetry; and/or the planes $\{10\bar{1}0\}$, instead of $\{11\bar{2}0\}$, can be mirror planes. When this is the case, the symmetry symbols become $\bar{3}12/m$, 312, and $31m$, and the character of the $\{h0—\}$ and $\{hh—\}$ forms are transposed. The two types of forms, $\{h0\bar{h}l\}$ and $\{hh\bar{i}l\}$ (Figure 2-53), distinguish this class from the class $6/m2/m2/m$. The form $\{h0\bar{h}l\}$ is a positive rhombohedron whereas the form $\{0hh\bar{l}\}$ is a negative rhombohedron; both have six faces. Positive and negative rhombohedra with the same indices, $\{10\bar{1}1\}$, $\{01\bar{1}1\}$ (Figures 2-51a and b), are geometrically indistinguishable; that is, when one is turned 60°, it is precisely like the other. The forms

$\{hk\bar{i}l\}$ and $\{kh\bar{i}l\}$ are positive and negative trigonal scalenohedra with twelve faces each. In cross section, the scalenohedra are ditrigonal, not hexagonal; that is, alternate angles are equal but not equal to 60°.

For the rhombohedral lattice, the rhombohedral centering of the lattice affects the expected order of form importance per the Law of Bravais. Those lattice planes whose spacings (between identical planes of a set) are reduced by one-third, due to the centering, are moved down the list of expected crystal faces to the position indicated by the new spacing. Therefore, $\{0001\}$ for which the lattice spacing is divided by three is much less likely to occur on a crystal, and is, in fact, rarely observed (e.g., on calcite). The planes

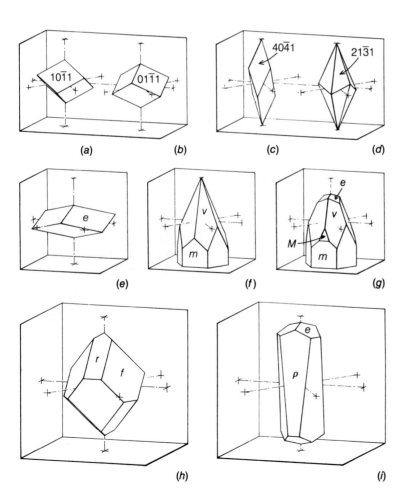

(a)   (b)   (c)   (d)

(e)   (f)   (g)

(h)   (i)

Figure 2-51. Some crystal forms in calcite. (a) Positive rhombohedron $\{10\bar{1}1\}$; (b) negative rhombohedron $\{01\bar{1}1\}$; (c) positive rhombohedron $\{40\bar{4}1\}$; (d) trigonal scalenohedron $\{21\bar{3}1\}$; (e) negative rhombohedron e$\{01\bar{1}2\}$; (f) combination of hexagonal prism m$\{10\bar{1}0\}$ and trigonal scalenohedron v$\{21\bar{3}1\}$; (g) combination of m$\{10\bar{1}0\}$, M$\{40\bar{4}1\}$, v$\{21\bar{3}1\}$, e$\{01\bar{1}2\}$; (h) combination of r$\{10\bar{1}1\}$ with negative rhombohedron f$\{20\bar{2}1\}$; (i) combination of e$\{01\bar{1}2\}$ and positive rhombohedron $\rho\{16\cdot0\cdot\bar{16}\cdot1\}$.

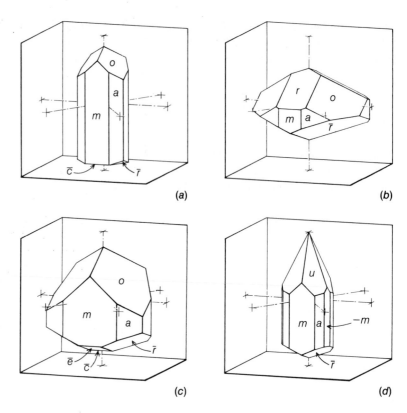

Figure 2-52.   Some typical crystals of tourmaline, showing the hemimorphic character of class $3m$ and the forms $m\{10\bar{1}0\}$, $a\{11\bar{2}0\}$, $-m\{01\bar{1}0\}$, $r\{10\bar{1}1\}$, $o\{02\bar{2}1\}$, $u\{32\bar{5}1\}$, $\bar{c}\{000\bar{1}\}$, $\bar{e}\{10\bar{1}\bar{2}\}$, $\bar{r}\{01\bar{1}\bar{1}\}$. Figure 15-41 shows another habit in which $-m(M')$ is the dominant prism.

$(10\bar{1}1)$, $(\bar{1}101)$, and $(0\bar{1}11)$ (the upper faces of a positive rhombohedron) do not have their spacing so reduced, whereas $(01\bar{1}1)$, $(\bar{1}011)$, and $(1\bar{1}01)$ (the upper faces of a negative rhombohedron) do have their lattice spacing divided by three. As one would expect from this, the positive rhombohedron $r\{10\bar{1}1\}$, which is parallel to the principal planes of the rhombohedral cell, is an important crystal form, but the negative rhombohedron $\{01\bar{1}1\}$ is rarely observed. On the other hand, the negative rhombohedra $e\{01\bar{1}2\}$ and $f\{02\bar{2}1\}$, whose lattice plane spacing is not reduced by centering, occur prominently on calcite crystals, whereas the corresponding positive rhombohedra are only rarely observed. As a general rule, if $h - i + l = 3n$ or zero, the lattice plane spacing is not divided by three, due to rhombohedral centering. Thus, $\{21\bar{3}1\}$, but not $\{12\bar{3}1\}$, would be expected to occur on rhombohedral crystals; some calcite crystals are examples. Also, the positive rhombohedron $\{40\bar{4}1\}$ is

more likely to be found than $\{20\bar{2}1\}$ or $\{30\bar{3}1\}$; crystals of calcite are also fine examples of this development (Figure 2-50).

Class 3m   The mineral tourmaline crystallizes in the ditrigonal pyramidal class $3m$. The symmetry is illustrated on the stereogram (Figure 2-50) with the threefold axis vertical and the m planes normal to each a axis, the same orientation as that given for class $\bar{3}2/m$. This class is hemimorphic in character since it lacks any axes or planes of symmetry perpendicular to c, the threefold axis. Crystals of tourmaline, when sufficiently developed, display different crystal forms at opposite ends of the c axis (Figures 2-52, 15-41). The form $c\{0001\}$ is a pedion, and all forms $\{h0\bar{h}l\}$, $\{hh\overline{2h}l\}$, $\{hki\,\bar{l}\}$ are pyramids, not dipyramids. The nature of the prisms also leads to a diagnostic character for tourmaline crystals. The most commonly observed prism to appear on crystals, $m\{10\bar{1}0\}$, is a trigonal prism with three faces; thus many tourmaline crystals are needles or thin prisms, commonly showing a nearly equilateral triangular outline (Figure 15-42). When the hexagonal prism a{11$\bar{2}$0} and ditrigonal prism {hki$\bar{}$0} are present, they modify these trigonal prisms, commonly producing irregular prisms that are deeply striated parallel to their c axes but still generally trigonal in outline.

In tourmaline, the c axis is the polar axis, and, under changes of heat or pressure, opposite electric charges develop at opposite ends of the c axis. (Because of its piezoelectric character, tourmaline is often used in pressure gauges.) The hemimorphic character of tourmaline is also sometimes manifested by a change of color along the c axes of some crystals; this change is generally interpreted as an indication that the crystal grew largely in one direction along that axis. The fact that authigenic growth on detrital grains of tourmaline in sediments tends to favor one end of the c axis over the other may corroborate this interpretation.

Class 32. Quartz crystallizes in crystal class 32 when formed below 573°C. The symmetry distribution is readily shown on a stereogram (Figure 2-50); it consists of a threefold axis, which is vertical, plus three twofold axes normal to the threefold axis and parallel to the a axes. In this class, the basal pinacoid {0001} and the I-order hexagonal prism m{10$\bar{1}$0} are normal hexagonal forms. Thus, quartz crystals displaying m{10$\bar{1}$0} as the prism almost exclusively are hexagonal in

cross section. The two prisms {11$\bar{2}$0} trigonal and {hki$\bar{}$0} ditrigonal are rare on quartz.

The forms {h0$\bar{h}$l} and {0h$\bar{h}$l} are positive and negative rhombohedra as in calcite, but since the lattice is not rhombohedral in quartz, the lattice plane spacings for the negative rhombohedron {01$\bar{1}$1} are not reduced by one-third. Although quartz crystals may display both rhombohedra r{10$\bar{1}$1} and z{01$\bar{1}$1} in about equal development (Figure 15-8), typically the positive rhombohedron faces are distinctly larger than those of the negative form and, not uncommonly, the negative form is not developed. The faces (10$\bar{1}$1) and (01$\bar{1}$1) will be inclined to the prism faces (10$\bar{1}$0) and (01$\bar{1}$0) at the same angle. The two forms r{10$\bar{1}$1} and z{01$\bar{1}$1}, when equally developed, will together simulate a I-order hexagonal dipyramid.

The forms {hh$\,$2hl} comprise six faces forming a trigonal dipyramid that may be positive or negative. None of the special forms so far mentioned serves to distinguish unambiguously the correct symmetry of quartz, although the combination of a rhombohedron {10$\bar{1}$1} with a trigonal dipyramid {11$\bar{2}$1} is diagnostic of this class. It should be noted that the occurrence of striations on the faces of s{11$\bar{2}$1} or 's{2$\bar{1}\bar{1}$1} parallel to the edge with r{10$\bar{1}$1} serves to distinguish the positive and negative trigonal dipyramids on quartz crystals.

The face (hki$\bar{}$l) develops into a general form with only six faces; this is called a trigonal trapezohedron (Figure 2-53). Since there are 24 possi-

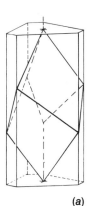 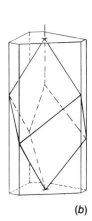 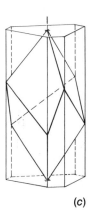 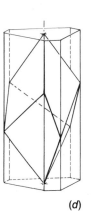

(a) (b) (c) (d)

Figure 2-53. Trigonal trapezohedra in symmetry class 32. (a) Left negative {1$\bar{2}$31}, (b) right positive {2$\bar{1}$31}, (c) left positive {3$\bar{1}\bar{2}$1}, (d) right negative {3$\bar{2}\bar{1}$1}.

82

ble crystal faces parallel to lattice planes $\{21\bar{3}1\}$, there must be four distinct forms in this class: $\{21\bar{3}1\}$ and $\{3\bar{1}\bar{2}1\}$ are right- and left-handed enantiomorphic pairs; $\{3\bar{2}\bar{1}1\}$ and $\{12\bar{3}1\}$ are also right- and left-handed pairs. The two right-handed forms are not separately distinguishable, one turned 60° being identical with the other; the same is true of the two left-handed forms. No rotation, however, will bring a right-handed form into coincidence with a left-handed form. The two right trigonal trapezohedra combined would simulate the right hexagonal trapezohedron.

The right- and left-handed character (enantiomorphism) can be traced back to the structure of quartz where it is seen that the SiO_4 tetrahedra are grouped about the vertical axis like the steps on a spiral staircase (Figure 2-54). The spiral arrangement may form as either a right-or left-handed screw thread, and therefore the quartz structure has either a right- or left-handed spiral in the arrangement of its SiO_4 tetrahedra. The presence of right- or left-handed trigonal tra-

pezohedra, or the striations on $s\{11\bar{2}1\}$ or $'s\{2\bar{1}\bar{1}1\}$, give visual evidence as to the right- or left-handed character of a given crystal of quartz (Figure 15-8). Physical properties such as rotation of the plane of polarization of a beam of light are characteristic of enantiomorphic substances (p. 58). Crystals of quartz, and of other nonconducting cyrstals of enantiomorphic character, are also pyroelectric and piezoelectric. In quartz, the a axes are polar in character, and the opposite ends become oppositely charged; right- and left-handed crystals give oppositely directed results under the same conditions.

The remaining two classes of the trigonal subsystem have the symmetry $\bar{3}$ and 3. The first is

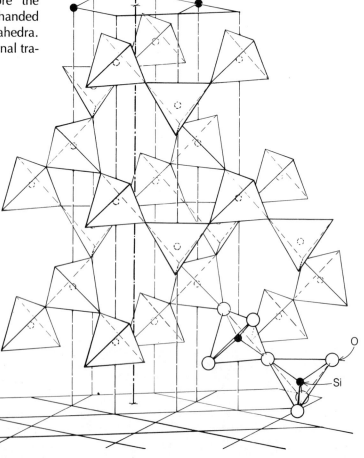

Figure 2-54. The structure of low quartz, SiO_2, shown as SiO_4 tetrahedra with a small silicon atom at the center of a group of four oxygens in tetrahedral coordination. Each tetrahedron shares a corner (oxygen) with an adjoining tetrahedron. For simplicity, most of the tetrahedra are shown without the circles representing oxygen. The screw axis symmetry (threefold) of the c axis is shown by the dot-dash line, repeating each tetrahedron with a rotation of 120° and a translation of c/3. Two units of structure along c are shown.

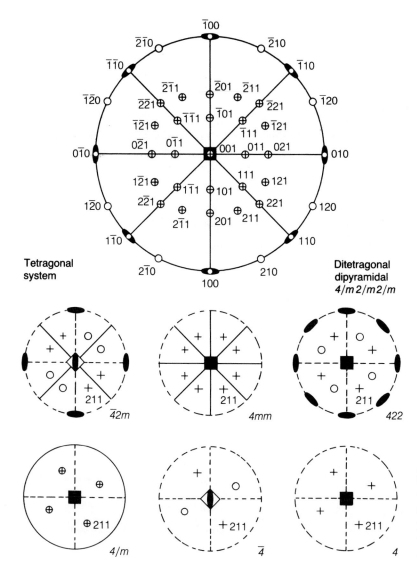

Figure 2-55. Stereographic projection of some forms in the ditetragonal dipyramidal class $4/m2/m2/m$ (Table 2-12). The small stereograms show one general form {211} in each merohedral class, $\bar{4}2m$, $4mm$, 422, $4/m$, $\bar{4}$, 4 (Tables 2-12 and 2-13).

important because it includes the minerals dolomite and ilmenite. In crystal form and cleavage, dolomite appears to be much like calcite. In rare crystals, however, it is clear that the general form $(hk\bar{i}l)$ for dolomite is rhombohedral in character instead of scalenodral as for calcite. Actually, this symmetry for dolomite was established by etch figures and confirmed by the determination of the atomic structure of the mineral. Ilmenite shows a similar relation to hematite. The other class, 3, does not warrant discussion here.

Tetragonal System

All crystals that are classified in the tetragonal system have either a primitive or a body-centered tetragonal lattice with $a_1 \equiv a_2 \neq c$ and interaxial angles equal to 90°. On the basis of symmetry, a tetragonal crystal can belong to any one of seven different crystal classes. Each of these classes has a fourfold rotation or fourfold rotoinversion axis of symmetry perpendicular to the square-net of the lattice, and that axis is chosen as c. In the

84

holohedral class, the symmetry is identical to that of the lattice $4/m2/m2/m$ (Figure 2-7k), where 4 is parallel to c and perpendicular to a mirror plane parallel to the square-net; the first 2 applies to the a_1 and a_2 axes, each of which is perpendicular to a mirror plane—one parallel to the a_2c ortho-net and the other parallel to the a_1c ortho-net; and the second 2 applies to the two diagonals of the square-net, each perpendicular to a mirror plane

that is parallel to both the other diagonal and to c. In crystals of the six classes with symmetry less than $4/m2/m2/m$, the atomic arrangement lacks symmetry about one or more of the axes or planes just noted. The c axis, however, is 4 or $\bar{4}$ in all the classes.

Ditetragonal Dipyramidal Class. This class is known as the holohedral class since it has the full

Table 2-12.   *Tetragonal system*

	Ditetragonal Dipyramidal (Holohedral)	Tetragonal Trapezohedral (Enantiomorphic)	Ditetragonal Pyramidal (Hemimorphic)	Tetragonal Scalenohedral
Name		Lattices P ($= C$) or I ($= F$); Axes $a_1 = a_2 \neq c$; Axial angles all 90°		
International symmetry symbol	$4/m2/m2/m$	422	$4mm$	$\bar{4}2m$ [a]
{001} {00$\bar{1}$}	pinacoid (basal) 2	pinacoid (basal) 2	pedion 1 pedion 1	pinacoid (basal) 2
{100}	tetragonal prism 4	tetragonal prism 4	tetragonal prism 4	tetragonal prism 4
{110}	tetragonal prism 4	tetragonal prism 4	tetragonal prism 4	tetragonal prism 4
{hk0} {210}	ditetragonal prism 8	ditetragonal prism 8	ditetragonal prism 8	ditetragonal prism 8
{h0l} {101} {h0$\bar{l}$} {10$\bar{1}$}	tetragonal dipyramid 8	tetragonal dipyramid 8	tetragonal pyramid 4 tetragonal pyramid 4	tetragonal dipyramid 8
{hhl} {111} {hh$\bar{l}$} {11$\bar{1}$} {h$\bar{h}$l} {1$\bar{1}$1}	tetragonal dipyramid 8	tetragonal dipyramid 8	tetragonal pyramid 4 tetragonal pyramid 4	+tetragonal disphenoid 4 −tetragonal disphenoid 4
{hkl} {211} {h$\bar{k}$l} {2$\bar{1}$1} {hk$\bar{l}$} {21$\bar{1}$}	ditetragonal dipyramid 16	right tetragonal trapezohedron 8 left tetragonal trapezohedron 8	ditetragonal pyramid 8 ditetragonal pyramid 8	+tetragonal scalenohedron 8 −tetragonal scalenohedron 8
Minerals	zircon, rutile	retgersite	diaboleite	chalcopyrite

[a] If the crystal is oriented such that the symmetry reflection planes are perpendicular to a and the symmetry axes are parallel to [110] and [1$\bar{1}$0], the symmetry symbol becomes $\bar{4}m2$. Stereographic projections are shown in Figure 2-61.

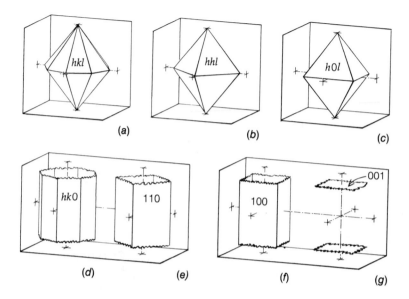

Figure 2-56. Type forms in the ditetragonal dipyramidal class. (a) Ditetragonal dipyramid, (b) I-order tetragonal dipyramid, (c) II-order tetragonal dipyramid, (d) ditetragonal prism, (e) I-order tetragonal prism, (f) II-order tetragonal prism, (g) basal pinacoid.

symmetry of the lattice 4/m2/m2/m. The distribution of symmetry is shown in the stereogram (Figure 2-55) with the c fourfold axis vertical and intersecting the center of the projection. The pole of (010) intersects the primitive on the east and coincides with the a_2 axis (twofold), and the pole of (100) intersects the primitive on the south and coincides with the a_1 axis (also twofold).

The crystal forms developed by the operation of this symmetry are shown in Table 2-12. The face (001) is repeated by the symmetry in the face (00$\bar{1}$); thus the form {001} is a basal pinacoid. The faces (100) and (110) both result in forms with four symmetrically equivalent faces. These forms are called tetragonal prisms because of their square cross sections and the fact that all faces are parallel to the c axis. Since these two tetragonal prisms, {100} and {110}, are identical in form, it is usual to distinguish them by referring to them as II-order or I-order, respectively. Alternatively, the Miller indices clearly define the two prisms.

The face (210) requires seven other equivalent faces, resulting in a ditetragonal prism {210} or {hk0}. Although the ditetragonal prism has eight faces, it is not a regular octagon in cross section. No combination of simple whole numbers for h and k will result in a regular octagon with exactly 45° as the polar angle between adjacent faces. The four alternate angles are equal but never identical with the other four alternate angles.

The face (111), by the operation of this symmetry, results in a form {111} with eight faces, called a tetragonal dipyramid. This form is square in any cross section perpendicular to the c axis. Any face with indices of the type (hhl) will also result in an eight-faced tetragonal dipyramid form differing only in the slope of the face; the faces are always isosceles triangles in ideal development. All poles of indices (hhl) lie on the vertical great circles that coincide with the vertical diagonal mirror planes; therefore, these dipyramids are of I-order type.

The face (101), or any face (h0l), also results in an eight-faced tetragonal dipyramid, but of II-order type. Again the faces are ideally isosceles triangles whose slopes depend on the a:c ratio and the indices (h0l). Poles of faces (h0l) are always located on the north-south vertical great circle of the projection parallel to the a_1c mirror plane. It is important to note that the symmetry requires these faces to be exactly equivalent to faces (0hl) whose poles lie on the east-west vertical great circle parallel to the a_2c mirror plane.

Any face (hkl) that might include (211) as well as (321) will fall in the open triangular area of the

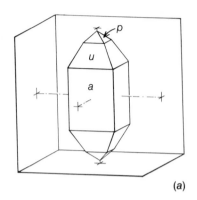

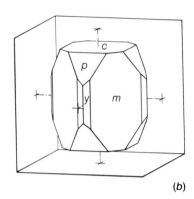

Figure 2-57. (a) Typical zircon crystal with a{100}, p{101}, u{301}. (b) Typical crystal of apophyllite with c{001}, p{101}, m{110}, y{210}.

(a)

(b)

projection between the trace of the mirror planes. The form {hkl} developed by a face of this type is a ditetragonal dipyramid with sixteen faces. Any cross section perpendicular to the c axis is ditetragonal, as is characteristic of the prisms (hk0) (Figure 2-56). Crystals of zircon, apophyllite (Figure 2-57), rutile (Figure 10-12), and cassiterite (Figure 10-15) display the symmetry of this class.

Tetragonal Scalenohedral Class. In this class with symmetry $\bar{4}2m$, the basal pinacoid {001} and the three types of prism {110}, {100}, and {hk0} are identical to the corresponding forms in the normal class. If the a_1 and a_2 axes are the symmetry axes as shown by the symbol, the forms {h0l}, are also like the normal class forms. The forms {hhl}, however, have only four faces; this form is called a tetragonal disphenoid and may

occur as either a positive form {hhl} or a negative form {h$\bar{h}$l}. If the a:c ratio of the crystal is close to 1:1, the tetragonal disphenoid {111} will approximate an isometric tetrahedron (Figure 2-63a) in appearance, but the faces of an ideal form {hhl} (Figure 2-58a) are always isosceles triangles, whereas those on a tetrahedron are equilateral triangles. The most commonly observed form on chalcopyrite crystals is a tetragonal disphenoid {112}, which approximates a tetrahedron in appearance because the axial ratio of chalcopyrite is a:c = 1:1.97. The general form {hkl} consists of eight faces (Figure 2-58b) and is known as a tetragonal scalenohedron (the faces are scalene triangles in outline). This form is uncommon on chalcopyrite crystals.

   An alternative setting of crystals of this symmetry is possible and results in the symmetry sym-

Figure 2-58. (a) Tetragonal disphenoid {111} with a:c = 1:1.645. (b) Tetragonal scalenohedron {322} with a:c = 1:0.833.

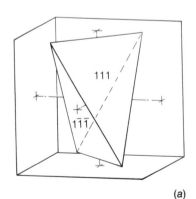

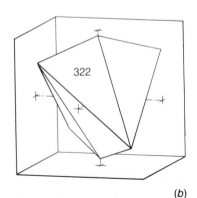

(a)

(b)

bol $\bar{4}m2$. In this orientation, the diagonals of the square-net, rather than the edges, are twofold axes. It is equivalent to turning the symmetry stereogram 45°. In this case, the form {101} or {h0l} is a tetragonal disphenoid, and {111} or {hhl} is a I-order tetragonal dipyramid. There is no change in the other forms.

Tetragonal Dipyramidal Class. In this class, the a axes and the diagonal axes in the square mesh are not symmetry axes and the planes (100), (010), (110), and (1$\bar{1}$0) are not symmetry planes. The crystal forms {001}, {100}, {110}, {h0l}, {hhl} are identical to those of the same indices in the holohedral class (Table 2-13).

In this class, the forms {hk0} and {hkl} have only four and eight faces, respectively, and are called tetragonal prisms and tetragonal dipyramids (Figures 2-59a and b). The fourfold symmetry axis requires a square cross section for both. For this reason, they are indistinguishable individually in form from the I- and II-order tetragonal prisms (Figure 2-59c) and dipyramids; they are commonly designated III-order forms.

To prove by crystal morphology that crystals of any mineral belong in this class, it is necessary to identify at least one of the III-order forms. This identification is generally difficult, if not impossible, because forms with the simple indices [{001}, {100}, {110}, {101}, and {111}], none of

Table 2-13. Tetragonal system*

Lattices P ($= C$), or I($= F$); Axes $a_1 = a_2 \neq c$; Axial angles all 90°

Name	Tetragonal Dipyramidal		Tetragonal Pyramidal (Hemimorphic)		Tetragonal Sphenoidal	
International symmetry symbol	$4/m$		4		$\bar{4}$	
{001}	pinacoid	2	pedion	1	pinacoid	2
{00$\bar{1}$}			pedion	1		
{100}	tetragonal prism	4	tetragonal prism	4	tetragonal prism	4
{110}	tetragonal prism	4	tetragonal prism	4	tetragonal prism	4
{hk0} {210}	tetragonal prism	4	tetragonal prism	4	tetragonal prism	4
{kh0} {120}	tetragonal prism	4	tetragonal prism	4	tetragonal prism	4
{h0l} {101}	tetragonal dipyramid	8	tetragonal pyramid	4	+tetragonal disphenoid	4
{h0$\bar{l}$} {10$\bar{1}$}			tetragonal pyramid	4		
{0hl} {011}					−tetragonal disphenoid	4
{hhl} {111}	tetragonal dipyramid	8	tetragonal pyramid	4	+tetragonal disphenoid	4
{hh$\bar{l}$} {11$\bar{1}$}			tetragonal pyramid	4		
{h$\bar{h}$l} {1$\bar{1}$1}					−tetragonal disphenoid	4
{hkl} {211}	tetragonal dipyramid	8	tetragonal pyramid	4	tetragonal disphenoid	4
{h$\bar{k}$l} {2$\bar{1}$1}	tetragonal dipyramid	8	tetragonal pyramid	4	tetragonal disphenoid	4
{hk$\bar{l}$} {21$\bar{1}$}			tetragonal pyramid	4	tetragonal disphenoid	4
{h$\bar{k}$$\bar{l}$} {2$\bar{1}$$\bar{1}$}			tetragonal pyramid	4	tetragonal disphenoid	4
Minerals	scapolite					

* Stereographic projections are shown in Figure 2-55.

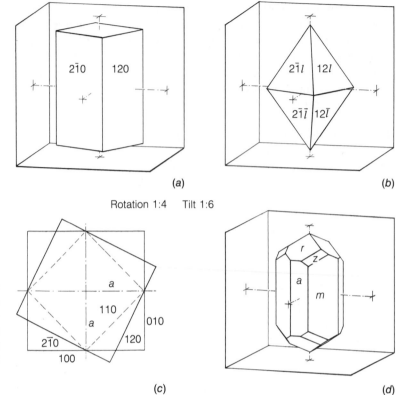

Rotation 1:4   Tilt 1:6

Figure 2-59.   (a) III-Order tetragonal prism {120}. (b) III-Order tetragonal dipyramid {12l}. (c) Relation of III-order prism {120} to
I-order {110} and II-order {100}. (d)
Typical crystal of scapolite with
a{100}, m{110}, r{101}, z{211}.

which are diagnostic of this class, are much more
likely to appear on the crystals than the diagnostic
forms are. For example, most crystals of scapolite
show the I- and II-order forms (Figure 2-59),
whereas very few crystals of scapolite show any
faces of a III-order prism or dipyramid.

The other classes of the tetragonal system are
of rare occurrence in minerals. Therefore, they
are not discussed in this book.

Isometric System

Hexoctahedral Class.   The arrangement of symmetry elements in the holohedral class of the isometric system is shown in the stereographic
projection (Figure 2-60). If one starts with any
possible crystal face located on the stereogram,
the location and number of equivalent faces can
be found; this operation, then, will give the number of faces in a particular form. Thus, starting

with the face (010), we find five other poles
[(100), ($\bar{1}$00), (0$\bar{1}$0), (001), and (00$\bar{1}$)], each of
which coincides with a fourfold symmetry axis
and which together constitute a crystal form
called the cube {100} (Figure 2-61a). The form
{111} with eight faces, for which the poles coincide with the threefold axes, is called an octahedron (Figure 2-67b). The form {110} with 12
faces, for which the poles coincide with twofold
axes, is called a dodecahedron (Figure 2-67c).
These forms—the cube, octahedron, and
dodecahedron—are called fixed forms because,
ideally, they are invariable in shape. The cube
alone will exhibit the full symmetry of this class.
The face (210) is repeated 23 times to produce a
pattern of poles on the projection in accordance
with the symmetry. This form, {210}, is called a
tetrahexahedron (Figure 2-62a) because of the
distribution of four poles near each cube-face
pole; it results in a form with a four-faced pyramid

in place of each cube face. This type of form also results from any other face symbol of the type $(hk0)$—for example, (310), (320), (410), (430); therefore, it can be said that this form can occur with different interfacial angles in agreement with the Law of Rational Intercepts. For each of these form symbols, the general shape of the faces remains the same and the poles fall on the same arcs of great circles, which coincide with the principal mirror planes, between the poles of the "fixed forms" {100} and {110}. In a similar fashion, the faces (221) and (211) result in 24-faced forms called the trisoctahedron and trapezohedron (Figures 2-62b and c), respectively; these forms (Table 2-14) may vary within the limits indicated by the general symbols (hhk) and (hkk), re-

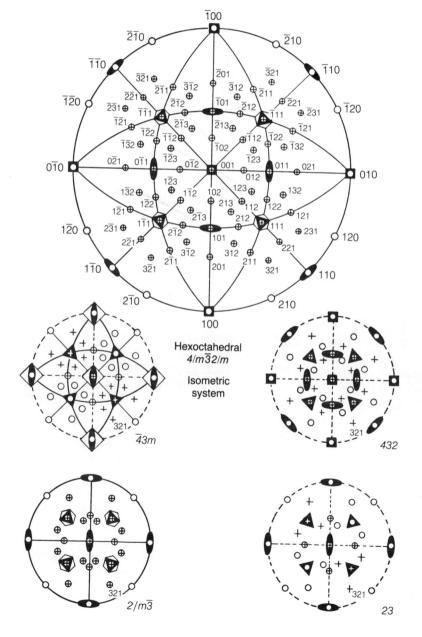

Hexoctahedral
$4/m\overline{3}2/m$

Isometric
system

Figure 2-60. Stereographic projection of seven type forms in the hexoctahedral class $4/m\overline{3}2/m$ (Table 2-14). The small stereograms show one general form {321} and the forms {100}, {110}, {111} in each merohedral class, $\overline{4}3m$, 432, $2/m\overline{3}$, and 23 (Tables 2-14 and 2-15).

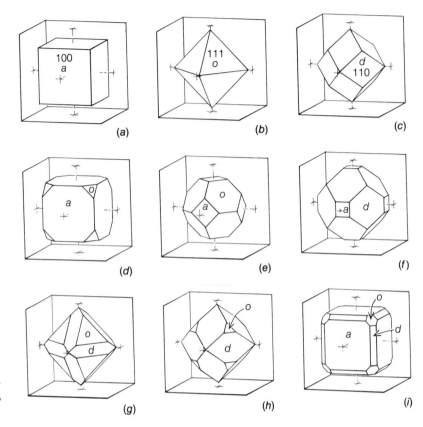

Figure 2-61. The fixed iso-metric forms cube a{100}, do-decahedron d{110}, octahedron o{111}, and various combina-tions of two or three of them, as found in the hexoctahedral class.

spectively, where $h > k$ and the face poles re-main on the arcs of the great circles that coincide with the diagonal mirror planes between the fixed forms {100}, {110}, and {111}. One other type of form, the general form with indices {321}—that is, {hkl} in which all indices are different—is also found on isometric crystals. This form, with 48 equivalent planes in this symmetry class, is called a hexoctahedron because there are six faces in each octant (Figure 2-62d). This form may differ, but the indices of any plane must be three differ-ent whole numbers (hkl). The pole, however, will always fall in the open triangular area on the projection, between a principal mirror plane and two diagonal mirror planes. This form is called the general form because it has the largest num-ber of equivalent planes, and the poles of those planes fall in general positions on the projection rather than in special positions as the poles of the

faces for the other six forms do. It is the only form that requires all of the symmetry, $4/m\bar{3}2/m$, to generate it from one of its faces. Therefore, it is completely diagnostic of this crystal class and has led to the distinctive name for the class.

It can be seen in Tables 2-14 and 2-15 that the six special forms may occur in one or more of the other four classes of the isometric system. Also, it should be recalled that the "laws" of crys-tallography indicate that, as a general rule, the form {321} will be less common than {211} or {221}, and that the forms {100}, {110}, and {111} will be most common.

Many important minerals, and hundreds of inorganic compounds, crystallize in the hex-octahedral class. Among the minerals are the gar-net group, analcime, the spinel group, copper, silver, gold, diamond, galena, halite, sylvite, pentlandite, fluorite, and uraninite.

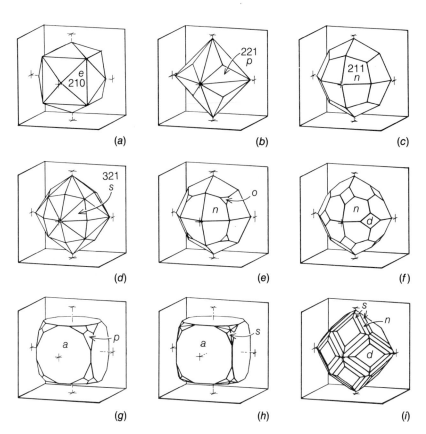

Figure 2-62. Crystal forms of the hexoctahedral class. (a) Tetrahexahedron e{210}, (b) trisoctahedron p{221}, (c) trapezohedron n{211}, (d) hexoctahedron s{321}, (e) octahedron and trapezohedron, (f) dodecahedron and trapezohedron, (g) cube and trisoctahedron, (h) cube and hexoctahedron, (i) dodecahedron, trapezohedron, and hexoctahedron.

Hextetrahedral or Tetrahedral Class. In this class, the symmetry symbol $\bar{4}3m$ indicates a fourfold rotoinversion axis parallel to each a axis, a threefold rotation axis parallel to each cube-body diagonal, and a mirror plane parallel to each lattice plane {110}. The result is the symmetry shown in the stereogram, Figure 2-60. It is readily apparent that the forms {100}, {110}, {hk0} are required by the symmetry to have the same number of faces as in the hexoctahedral class (Table 2-14). The face (111), however, is repeated by the $\bar{4}$ axes as only four, instead of eight, equivalent faces, as in the hexoctahedral class, and the form is called a tetrahedron (Figure 2-63). If, on the other hand, one starts with ($1\bar{1}1$) as the initial pole, the symmetry requires addition of ($\bar{1}11$), ($11\bar{1}$), and ($\bar{1}\bar{1}\bar{1}$), and these four faces also constitute a tetrahedron. The two tetrahedra, although geometrically identical, are not equivalent to

each other in the symmetry of this class. To distinguish between them, we arbitrarily designate {111} as the positive tetrahedron and {$1\bar{1}1$} as the negative tetrahedron. The fundamental difference between the two forms can be appreciated by examining the internal structures of galena and of sphalerite (see pages 254 and 257).

From the projection of Figure 2-60, we can also see that the forms {hkk}, {hhk}, and {hkl} are present as either positive or negative forms, with half the number of faces as represented by the same Miller indices in the hexoctahedral class. The names of these forms are given in Table 2-14. With the largest number of faces, and being completely diagnostic of this class, the hextetrahedron is the general form and consequently gives the class name. The tetrahedron, however, is more often recognized and also serves to distinguish this class from the other commonly oc-

Table 2-14.   Isometric system*

Lattices *P, I,* or *F;* Axes $a_1 = a_2 = a_3$; Axial angles all 90°

Name	Hexoctahedral		Gyroidal (Enantiomorphic)		Hextetrahedral (Tetrahedral)	
International symmetry symbol	$4/m\bar{3}2/m$		*432*		$\bar{4}3m$	
{100}	cube	6	cube	6	cube	6
{110}	dodecahedron (rhombic)	12	dodecahedron (rhombic)	12	dodecahedron (rhombic)	12
{111} {1$\bar{1}$1}	octahedron	8	octahedron	8	+tetrahedron −tetrahedron	4 4
{hk0} {210}	tetrahexahedron	24	tetrahexahedron	24	tetrahexahedron	24
{hhk} {221} {$h\bar{h}k$} {2$\bar{2}$1}	trisoctahedron	24	trisoctahedron	24	+deltohedron −deltohedron	12 12
{hkk} {211} {$h\bar{k}k$} {2$\bar{1}$1}	trapezohedron	24	trapezohedron	24	+tristetrahedron −tristetrahedron	12 12
{hkl} {321} {$h\bar{k}l$} {3$\bar{2}$1}	hexoctahedron	48	right gyroid left gyroid	24 24	+hextetrahedron −hextetrahedron	24 24
Minerals	galena, fluorite				sphalerite, tetrahedrite	

* Stereographic projections are shown in Figure 2-60.

curring classes (but not from the rare class with the symmetry *23*); consequently, the class name is frequently given as tetrahedral rather than as hextetrahedral.

Some crystal forms in this class—{100}, {110}, and {210}—are indistinguishable individually from the same forms on crystals of symmetry $4/m\bar{3}2/m$. Thus, for example, crystals of sphalerite exhibiting only the cube or dodecahedron cannot be identified as belonging to the hextetrahedral class. If, however, both hemihedral forms (the positive tetrahedron and the negative tetrahedron) occur on a crystal, one is commonly represented by larger faces than the other, or the two are represented by faces of different surface quality and thus are distinctive. If, however, a crystal displays both forms with about equal development, it will very likely be indistinguishable

from a crystal of higher symmetry; hence the crystal could be said to display pseudosymmetry. Therefore, as we have noted previously, it is desirable—and, in some cases, imperative—to examine a large number of crystals of a mineral from a wide variety of occurrences before making a conclusion as to its symmetry class. And, again, we emphasize the fact that in morphological studies, the positive evidence of the presence of a merohedral form is absolutely necessary to establish the correct symmetry for most crystals. Nonetheless, it also is true that lacking such features, etch tests or certain physical tests may permit macroscopic determination of the correct symmetry class in some cases.

Isometric Diploidal (Pyritohedral) Class. This class is represented by the common mineral pyrite and, on this account, it is known almost universally as the pyritohedral class. The symmetry symbol for this class, $2/m\bar{3}$, indicates that the a axes (cube edges of the lattice) have only twofold symmetry and are perpendicular to the principal mirror planes and that the cube-body diagonals are threefold rotoinversion axes. Therefore, it can be said that the face-diagonal directions [[110]] have no symmetry in this class. By trial, using the stereogram showing the symmetry elements of this class (Figure 2-60), you can readily see that the forms {100}, {110}, {111}, {211}, and {221} are identical to forms of the same indices in the holohedral class (Table 2-15), whereas the forms {hkl} and {$hk0$} have only half the number of equivalent faces as in the holohedral class. The

forms {210} (Figure 2-64a) and {120} (Figure 2-64b) are known as the positive pyritohedron and the negative pyritohedron; they are also known as pentagonal dodecahedra since they possess 12 faces, each with a pentagonal outline. The forms {321} (Figure 2-64c) and {312} are the positive diploid and the negative diploid, each with 24 faces. The diploid is the general form and lends its name to the traditionally accepted name for the class.

Again, we emphasize the fact that it is necessary to identify one of the two types of merohedral forms in order to identify the symmetry correctly. Unfortunately, the diploid occurs only rarely, and even where present, it is typically represented by small faces, (as would be expected from its rela-

Table 2-15.   *Isometric system**

Name	Diploidal (Pyritohedral)		Tetartoidal (Enantiomorphic)	
International symmetry symbol	$2/m\bar{3}$		23	
{100}	cube	6	cube	6
{110}	dodecahedron (rhombic)	12	dodecahedron (rhombic)	12
{111} {1$\bar{1}$1}	octahedron	8	+tetrahedron −tetrahedron	4 4
{$hk0$} {210} {$k\bar{h}0$} {1$\bar{2}$0}	+pyritohedron −pyritohedron	12 12	+pyritohedron −pyritohedron	12 12
{hhk} {221} {$h\bar{h}k$} {2$\bar{2}$1}	trisoctahedron	24	+deltohedron −deltohedron	12 12
{hkk} {211} {$h\bar{k}k$} {2$\bar{1}$1}	trapezohedron	24	+tristetrahedron −tristetrahedron	12 12
{hkl} {321} {$h\bar{k}l$} {3$\bar{2}$1}	+diploid	24	+right tetartoid −left tetartoid	12 12
{khl} {231} {$k\bar{h}l$} {2$\bar{3}$1}	−diploid	24	+left tetartoid −right tetartoid	12 12
Minerals	pyrite		ullmannite	

* Stereographic projections are shown in Figure 2-60.

94

tively complex indices). The pyritohedron with simpler indices and fewer faces is much more likely to be present. As an example, crystals of pyrite commonly occur as cubes, octahedra, or as combinations of these two forms; in outward form, these crystals do not display the definitive symmetry of pyrite. Pyrite also commonly occurs as pyritohedra or as combinations of the cube and pyritohedron or of the octahedron and pyritohedron (Figures 2-64d, e, and f); these are not definitive either. On the other hand, combinations of positive and negative pyritohedra, which are definitive, are virtually unknown. The cube and pyritohedron combinations have frequently developed together in oscillatory combination; that is, instead of having formed in such a way that each face is represented as a single face, the two faces occur as narrow alternating sections that give the appearance of a striated surface.

When the cube faces predominate in the striations, the result is a rough cube with striated faces; as can be seen in Figure 9-21a, each opposite pair of cube faces is striated in the same direction, which is perpendicular to the direction of striations on adjoining cube faces. This is a reflection of the fact that only the faces (210) and $(2\bar{1}0)$ will form striations on (100), only (102) and $(\bar{1}02)$ will form them on (001), and only (021) and $(02\bar{1})$ will form striations on (010). When the pyritohedron faces are predominant, the steplike faces (i.e., the cube faces) are typically almost cylindrical with the cylinder axes parallel to the striations. These relationships are distinctive.

The two classes of crystals in the isometric system with symmetry 432 and 23 are represented by only a few rather rare minerals; thus, we do not discuss them in this book.

Figure 2-63. Crystal forms of the hextetrahedral class $\bar{4}3m$. (a) Positive tetrahedron $o\{111\}$; (b) negative tetrahedron $-o\{1\bar{1}1\}$; (c) relation of positive tetrahedron to octahedron; (d) positive and negative tetrahedra; (e), (f) cube and positive tetrahedron; (g) positive deltohedron $\{221\}$; (h) positive tristetrahedron $\{211\}$; (i) positive hextetrahedron $\{321\}$.

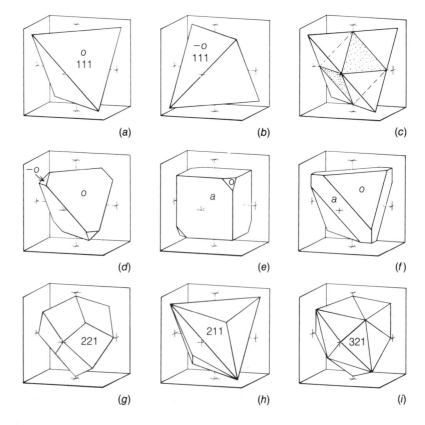

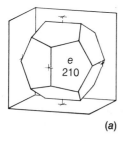

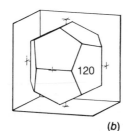

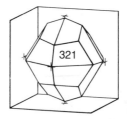

(a)

(b)

(c)

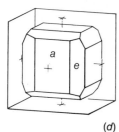

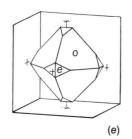

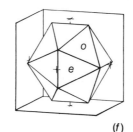

(d)

(e)

(f)

Figure 2-64. Crystal forms of the diploidal (pyritohedral) class $2/m\bar{3}$. (a) Positive pyritohedron e{210}; (b) negative pyritohedron {120}; (c) positive diploid {321}; (d) cube and pyritohedron; (e), (f) octahedron and pyritohedron. Form (f) approximates a regular 20-faced form; o and e faces are not identical in shape.

SPACE GROUP TERMINOLOGY

In the mineral descriptions in Part II, we give the space group designation as well as the crystal class for each mineral. The following résumé gives the fundamentals of these designations.

The 32 crystal classes can be considered, synonymously, as the 32 *point groups*, with each group being definable as a collection of symmetry operations that operate around a point that represents an intersection of the elements of symmetry of its class. It can be shown that if the 32 point groups are combined individually with the translation operations of each of the 14 Bravais space lattices, 73 different space groups can be defined. (Thus, a space group has operations that combine point group operations and translations.) If, in addition, glide planes and screw axes are added to the possible symmetry operations, the total number of space groups is increased to the maximum, 230. In other words, each of the 230 space groups represents a unique combination of symmetry elements with one of the 14 Bravais space lattices functioning as its lattice of translation.

The 230 space groups serve as the bases of structural descriptions. Two types of symbols—the Schoenflies symbols and the Hermann–Mau-

guin symbols—are used to designate the individual space groups. A complete list of these symbols is given in the *International Tables for X-Ray Crystallography* and has been reproduced in several other references. Hermann–Mauguin space group symbols, which are more widely used than Schoenflies symbols, are given in the mineral descriptions in Part II. They are based on the crystal class notations already outlined. Their content can be described as follows:

The first letter stands for the kind of space lattice:

P, primitive
A, face-centered on (100)
B, face-centered on (010)
C, face-centered on (001)
F, all face-centered
I, body-centered
R, rhombohedral

The remainder of the symbol gives the symmetry operations (which may be combinations of point group operations with translations):

1, 2, 3, 4, and 6: 1-, 2-, 3-, 4-, and 6-fold axes of symmetry

$\bar{1}, \bar{2}, \bar{3}, \bar{4}$, and $\bar{6}$: 1-, 2-, 3-, 4-, and 6-fold axes of rotoinversion

$2_1, 3_1, 3_2, 4_1, 4_2, 4_3, 6_1, 6_2, 6_3, 6_4$, and 6_5: screw axes; the symbol indicates the rotational repetition of the axis plus the fractional amount of translation that is coupled with the rotation. The latter is equal to the subscript divided by the number (e.g., 2_1 denotes a translation of $\frac{1}{2}$ of a lattice translation).

m: mirror plane of symmetry

a, b, c, n, and d: glide planes with respective translations of $a/2$, $b/2$, $c/2$, $\frac{1}{2}$ the diagonal, and $\frac{1}{4}$ the diagonal translation.

Two examples are: $P432$, which describes space group #207, a primitive lattice with axes of four-, three-, and two-fold symmetry; and $Cmc2_1$, which describes space group #36, a lattice that is face-centered on (001) and has a mirror plane of symmetry parallel to (100), a glide plane of symmetry with translation of $c/2$ parallel to (010), and a screw axis involving a 180° rotation about the c crystallographic axis plus a $\frac{1}{2}$ translation parallel to the c axis.

For a more complete treatment of space groups, your attention is directed to, for example, Bloss (1971).

IMPERFECTIONS OF CRYSTALS

One aspect of crystals and crystal growth that has been and continues to be of increasing interest involves a group of phenomena generally referred to as imperfections. Imperfections, which cause crystal structures to be less than homogeneous and perfect, are manifested by such features as macroscopic flaws and sporadic color differences (including zoning), by optical characteristics, by differences in calculated versus actual values for such properties as strength and electrical conductivity, and by aberrant X-ray, electron, and neutron diffraction patterns.

In some minerals, essentially the maximum allowable number of all the possible kinds of imperfections commonly exist. As a consequence,

some of the properties that are dependent on imperfections, rather than on inherent properties of the ideal material, have become those that are frequently used for identification purposes. Furthermore, some of the imperfection-dependent properties are the very ones that make certain crystals and crystalline materials useful and valuable. These properties, which are discussed in Chapters 3 and 4, include certain aspects of chemical reactivity, physical strength (plasticity, elasticity, etc.), and, as already mentioned, color.

Imperfections within crystals and crystalline grains typically involve a single atom, ion, or unit cell, or small groups of one or another of these moieties. Some defects, however, are rather complicated and extend throughout complete or large portions of individual crystals or grains. In addition, although these imperfections are usually characterized as physical, as distinct from chemical defects (e.g., lack of stoichiometry), they commonly serve to modify chemical bonds, to upset electronic neutrality, etc. in predictable manners. In fact, equations have been formulated for calculating such quantities as the equilibrium number of vacancies for certain compounds.

Several origins have been suggested for defects, especially for line defects. Among them are entrapment of impurities during rapid crystallization, side-by-side nucleation of layers at proper and improper sites, and control by previously formed defects. In any case, current data strongly suggest that perfect crystals (i.e., crystals without imperfections) will grow only under very few, if any, conditions at temperatures above absolute zero; this is true because the conditions that cause crystal growth also enhance the production of imperfections.

The diverse kinds of imperfections are usually classified as point, line, or plane defects. In the following paragraphs, we describe these kinds of defects briefly and note a few examples of the possible consequences of their presence.

Point Defects

There are three kinds of point defects: the *Frenkel defect*, the *Schottky defect*, and the *interstitial*

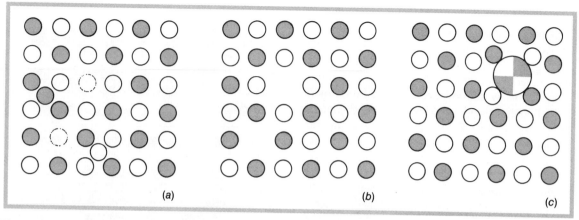

Figure 2-65. Point defect: (a) Frenkel (mislocation) defect; (b) Schottky (vacancy) defect; (c) interstitial (impurity) defect.

defect. As shown in Figure 2-65, the Frenkel defect involves the mislocation of an ion from its proper lattice site to a nearby site where it does not belong; although Frenkel defects have been found to involve anions as well as cations, the mislocated ion is typically a relatively small cation. The Schottky defect, which amounts to a vacancy in a lattice, is caused by the absence of an ion; in many cases, two nearby Schottky defects—one involving a cation, the other an anion—are paired to maintain electronic neutrality. The interstitial defect, sometimes referred to as an impurity defect, involves the presence of a substance that is foreign to the building blocks of the given structure; these defects involving either interstitial or substitutional foreign atoms are most common in substances with relatively low packing indexes.

Whereas neither the Frenkel nor the Schottky defect requires any deviation from the chemistry of the pure material, interstitial defects do, by definition, require chemical differences. In many cases, however, the compositional differences are so small that they cannot be detected by any generally employed method of chemical analysis; indeed, this is the case even for some of the minerals whose physical properties exhibit rather obvious influences of impurity defects.

When heated, point defects may become mobile and, so to speak, diffuse from one to another site. In some cases, this mobility has a significant effect on certain properties, such as plasticity, electrical conductivity, and overall susceptibility to diffusion. Perhaps the most frequently cited example of the effect of point defects on minerals is that involving certain minerals' colors. Each of the kinds of point defects can cause coloration, generally of a pastel hue, in minerals that are inherently colorless. It is widely recognized that certain colors often reflect trace amounts of certain impurities, some of which are most commonly present as mere traces causing interstitital defects.

Line Defects

Defects that involve rows of building blocks are called line defects or dislocations. The latter designation has been applied because, wherever present, these defects cause offsets of structures. In most cases, these defects distort crystal structures much more than point defects do. In fact, some of these defects can be seen in electron micrographs.

There are two limiting kinds of line defects—edge dislocations and screw dislocations. Most dislocations exhibit properties that are inter-

98   mediate between these two ideal types, and consequently they are called "mixed dislocations." As you can see in Figure 2-66: *Edge dislocations* involve partial planes of atoms or ions that extend through only portions of a crystal or crystalline grain; the individual dislocation may be defined as the line that coincides with the internal edge of the partial plane. *Screw dislocations* may be defined as rows around which some crystallographic plane appears to spiral (like the axis of a spiral ramp); this type of line defect has been shown to afford a means whereby crystal growth can proceed more or less continuously and therefore much more rapidly than it can as the result of interrupted, plane-by-plane growth.

It is quite apparent that the greatest distortion of the host material is centered around the line marking the edge or defining the axis of the screw. It is in those loci that the atomic arrangements are elastically distorted, and strain energies must be at their maxima. Line defects can strongly influence the physical behavior of crystalline materials. They commonly modify such properties as elasticity, ductility, and brittle fracture. As a consequence, they have gained the attention of and have been investigated by such groups as structural geologists (see, for example, Hobbs et al.), ceramists, and metallurgists (see, for example, Eisenstadt).

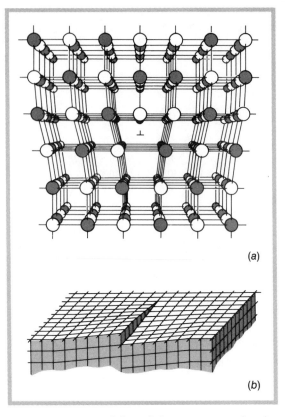

Figure 2-66.   Line defect (dislocations): (a) edge dislocation; (b) screw dislocation.

Plane Defects

There is little agreement as to what features should be classified as plane defects. Some crystallographers and mineralogists do not even use the term, and, among those who do, there are differences of opinion as to what phenomena should and should not be included.

Stacking faults and lineage structures—which are rather generally included—are briefly described in the next two paragraphs. Kinks, deformation band boundaries, and deformation lamellae (including Böhm lamellae) appear to reflect physical deformation in most descriptions, and thus are beyond the scope of our coverage.

Twinning is treated in the next section of this chapter.

A stacking fault can be defined as a planar discontinuity caused by one array's being displaced relative to its adjacent array. It is thought that this kind of imperfection usually forms during growth, but in some cases it apparently forms in response to deformation. Stacking faults are most common in layered structures, particularly those involving closest packing.

The concept of lineage structures was introduced by the American crystallographer M. J. Buerger. Under this hypothesis, crystals and crystalline grains are interpreted to be made up of essentially perfect inner portions surrounded by several more-or-less perfect zones with slightly different orientations (Figure 2-67). Despite their different orientations, each surrounding zone is considered to be individually traceable back to the "perfect" core and to be structurally continuous with it. Each "lineage" constitutes the boundary plane between two adjacent outer zones, however perfect in themselves, that are slightly twisted with respect to one another. Some crystallographers have described a lineage as comprising a series of edge defects that define a plane of discontinuity. Some lineage structures are indicated by differing extinction positions within single grains as they are viewed between crossed nicols.

AGGREGATES OF CRYSTALS AND CRYSTALLINE GRAINS

Most masses of minerals and rocks and their weathering products can be described as aggregates of crystalline grains. Many of these grains have no outward crystal form, but their crystalline character is evident from visible cleavage planes, from certain optical properties, or from X-ray diffraction studies. Internally, these grains have the

Figure 2-67. Plane defect—lineage structure. (After M. J. Buerger. 1934. "Lineage structure of crystals." *Zeitschrift für Kristallographie* . . . , **89**, 196.)

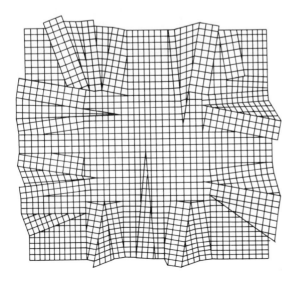

same regular crystalline character that well-formed crystals do. The individual grains range in size from those that are smaller than can be resolved using an optical microscope (with approximately 500X magnification) to those that are a few meters in greatest dimension. In many cases, the individual grains are oriented more or less at random in relation to one another; in other cases, they are related in regular ways, many of which are controlled by crystallographic relations. Under the final heading of this chapter, we describe one of the latter kinds of relationship, twinning.

TWINNING

Composite crystals of a single substance, in which the individual parts are related to one another in a definite crystallographic manner, are known as *twinned crystals*. The nature of the relationship between the parts of the twinned crystal is expressed in a *twin law*. Twin laws are often given specific names that are related to the characteristic shape of the twin, to a certain locality where such twin crystals were first found, to a mineral that commonly displays the particular twin law, etc. Many important rock-forming minerals, such as orthoclase, microcline, plagioclase, and calcite, commonly occur as twinned crystals. Indeed, for some minerals, twinned crystals are probably more common than crystals that are untwinned.

Most twinned crystals appear to consist of two or more crystals that are united with a symmetric interrelationship. In some twinned crystals, the orientation of one crystal individual is related to the other (or others) by a rotation of $360°/n$, where n is generally 2, about some rational lattice row that is present in both (or all) individuals. This lattice row, termed the *twin axis*, is typically a lattice row of simple indices. In other cases, the orientation of two individuals of a twinned crystal may be related by reflection across a lattice plane that is common to both individuals. This lattice plane, called a *twin plane*, is also typically of simple indices. If the two individuals of a twin meet along a plane, the

100

plane is referred to as the *composition* plane. Composition planes may or may not be twin planes.

In some twin crystals, particularly those of triclinic minerals, the perpendicular to the twin plane is not a lattice row and therefore cannot be defined in terms of rational indices. Such apparent twin axes are not twin axes and, therefore, they are not used to describe the twinning. Instead, these kinds of twin laws are described by defining the twin plane.

The twin plane is always parallel to a lattice plane and, therefore, a possible crystal face; it never, however, is a plane of symmetry in a single crystal of the mineral in question. Likewise, a twin axis is never a true symmetry axis in the equivalent single crystal.

Twinned crystals (Figure 2-68) may be described as: *simple twins* if composed of but two parts; *multiple twins* if more than two orientations are present; *contact twins* if a definite com-

position plane is present; and *penetration twins* if two or more parts of a crystal appear to interpenetrate each other with the surface between the parts being indefinable and irregular. Both contact and penetration twins may be multiple as well as simple. If three or more individuals are repeated alternately on the same twin plane (that is, if all twin planes are parallel in the twinned crystal), the result is *polysynthetic twinning*. If the individuals of polysynthetic twins are thin plates, the twinning is called *lamellar*. In plagioclase feldspar, lamellar twinning is quite commonly developed with individuals ranging from a few microns to a few millimeters thick. Some of the thinner types, with widths of individuals approaching that of unit cells, are only detectable with high resolution electron microscopes.

If the individuals of a multiple twin are related by twin planes that are different planes of the same crystal form, a cyclic twin results. In an analogous way, multiple penetration twins may

Figure 2-68.   Twinned crystals. (a) Simple contact twin—spinel octahedron twinned on ($\bar{1}\bar{1}1$), the stippled plane. (b) multiple (cyclic) twin—chrysoberyl cyclic twin on {130} as twin plane and composition plane; forms are c{001}, o{111}, and n{121}. (c) penetration twin—orthoclase Carlsbad twin; forms are b{010}, c{001}, m{110}, y{201}. (d) polysynthetic twinning —albite twinning in plagioclase; forms are b{010}, c{001}, M{1$\bar{1}$0}, and x{$\bar{1}$01}.

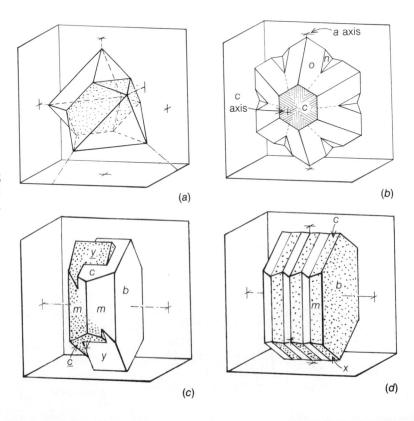

result if two equivalent lattice rows act simultaneously as twin axes; these twins are also cyclic in form.

Causes of Twinning

Some minerals are found as twinned crystals or grains in most occurrences, whereas other minerals rarely display twinning. It has been suggested by some authorities that the occurrence of twinning in a crystal depends on the dimensional properties of the lattice. This view is supported by the common occurrence of twinning in the orthorhombic carbonates (aragonite group) where the prism angle $(110)\wedge(1\bar{1}0)\approx63°$ is close to 60° and where multiple twinning results in the well-known pseudohexagonal twin crystals.

M. J. Buerger (1945) proposed a rational approach to resolving the causes of twinning that is based on considerations of the pertinent crystal structure. In essence, the proposal is a structural interpretation of Mallard's Law. Buerger summarizes as follows: "If the structure is of such a nature that, in detail, it permits a continuation of itself in alternative twin junction configuration without involving violation of the immediate coordination requirements of its atoms, the junction has low energy and the twin is energetically possible." As an example, the structure of aragonite, which is known to be readily susceptible to twinning, can be shown to be in agreement with Buerger's hypothesis (see Figure 2-69).

Effects of Twinning

Twinning, especially on a microscopic scale, may obscure the true symmetry of a crystal. In twinned crystals, the action of the twin plane is that of mirror reflection, and in the resulting twin crystal, one individual is the mirror reflection of the other. This fact has led to the statement that twinning apparently increases the symmetry of a crystal. This is not true, in the sense that we use the term symmetry with regard to crystals because the new "symmetry" plane is at the twin plane

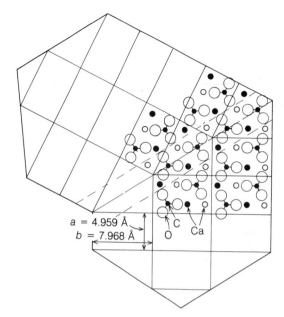

Figure 2-69. Twinning on $(1\bar{1}0)$ in aragonite. The lower right part of the diagram shows the lower half of a unit cell (along c) projected onto (001). The upper left part presents the upper half of the unit cell (also along c) after reflecting across $(1\bar{1}0)$ as twin plane. C atoms are small black circles; Ca atoms are medium-sized black and open circles; O atoms are large open circles. Note that the section of the crystal structure bounded by the broken lines is consistent with both of the individual structures—i.e., the atoms in this border zone have the same arrangement of atoms in the inner spheres of coordination as do corresponding atoms in a single crystal. Thus, the twinned structure in aragonite represents an energy state that is only slightly higher than that achieved by a single crystal. Consequently, it is to be expected that the perfect conditions of growth necessary to produce single crystals prevail only rarely and that twinning should be relatively common.

only, whereas a true crystallographic symmetry plane parallel to a certain set of lattice planes occurs at each and every lattice plane in the set. A true crystallographic symmetry plane, parallel to, for example, (010), coincides with each lattice plane of the set (010) in the structure and is not confined to only one or a few planes in a set (Figure 2-70).

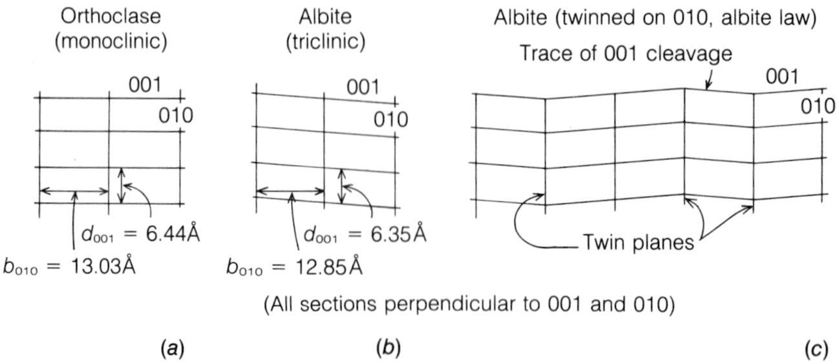

102

Orthoclase
(monoclinic)

Albite
(triclinic)

Albite (twinned on 010, albite law)

Trace of 001 cleavage

Figure 2-70. Sections through orthoclase and albite perpendicular to *a*. Part (c) shows the effect of the albite twin law (see Figure 15–17).

$d_{001} = 6.44$Å
$b_{010} = 13.03$Å

$d_{001} = 6.35$Å
$b_{010} = 12.85$Å

(All sections perpendicular to 001 and 010)

(a)

(b)

(c)

On many twin crystals, corresponding crystal faces on opposite parts of the twin meet in re-entrant angles, that is, with interior angles of other than 180° (see, for example, Figure 2-77). If the twinning is polysynthetic and lamellar, these re-entrant angles result in striations, such as the striations on cleavage faces of calcite (Figures 12-7a and b), dolomite (Figures 12-10a and b), and sphalerite, and also those on the {001} cleavage of plagioclase (Figure 2-68d), which are due to lamellar twinning with {010} as the twin plane. [It seems prudent to repeat here that, as previously noted for pyrite on page 94, re-entrant angles on a crystal can also be formed in response to, for example, oscillatory development of two crystal faces.]

Many substances tend to break readily along twin planes. Twinning is, in fact, one of the causes of parting, which closely resembles cleavage, in minerals. The mineral corundum, for example, shows no cleavage in untwinned crystals, but as crystals that are twinned polysynthetically, it shows excellent parting on {0001} and/or {10$\bar{1}$1} (Figure 10-7b).

In some minerals, the presence of twins has been correlated with certain structural geological phenomena.

Twinning and twinning tendencies may either promote or preclude the use of a mineral or other material in industry. For example: twins are desired in some metals because they enhance plastic deformation capabilities; but, twinning

according to certain laws precludes the use of quartz as either lenses or oscillators.

Genetic Kinds of Twinning

On the basis of their mode of origin, twin crystals may be classified into three types: *growth twins, transformation twins,* and *gliding twins.* Growth twins are primary, whereas transformation and gliding twins are frequently referred to as secondary.

Growth Twins

The aragonite twins alluded to in the subsection titled *Causes of Twinning* and described in the caption for Figure 2-69 are examples of growth twins. In essence, during growth—especially rapid growth in which early nucleation is connected with supersaturation—atoms (etc.) may assume positions appropriate to a twin orientation and thus initiate twin formation that ensues during subsequent crystal growth.

Transformation Twins

Transformation twins are formed during certain kinds of inversion of one polymorphic form of a substance into another polymorphic form of the same substance. Under certain environmental conditions, many elements and compounds may form different crystalline modifications or polymorphs (see Chapter 3). Polymorphs of a substance are stable under certain conditions, and some pairs of polymorphs of a substance are related by a transformation that becomes reversible when one of the conditions is changed. An interesting example of this is the high-low quartz transformation that takes place at 573°C. The transformation is rapid when the substance is heated or cooled through 573°C, the inversion temperature. The structures of these two forms of SiO_2 are closely related in symmetry. The high temperature form has a more symmetric arrangement of SiO_4 tetrahedra, with symmetry *622*, than the low temperature form, with symmetry *32*. In transforming from the high temperature form to the low temperature form, the atoms merely shift into a more compact arrangement with lower energy. The relation of the SiO_4 arrangements in the two forms is shown diagrammatically in Figure 2-71. In shifting from an arrangement with sixfold symmetry to one with threefold symmetry, two distinct orientations are possible. The two possible orientations of the low quartz structure are related by a rotation of 180° about the *c* axis (perpendicular to drawing).

As noted, this transformation takes place rapidly, as the temperature is changed, due to heat transmission through the crystal. Unless the crystal is extremely small, the inversion will commence at many centers within the crystal, and these centers may take on either of the orientations shown in Figure 2-71. If both orientations are present, the inverted crystal is twinned according to the Dauphiné Law (Figure 2-80). Under these conditions, the resulting twin may consist of interpenetrating portions of both orientations (Figure 2-72) rather than the simple contact twin shown in Figure 2-71. In the Dauphiné twin of low quartz, both individuals are of one enantiomorphic character—that is, either right- or left-handed. If such a twin is heated above 573°C, the twinning will disappear, resulting in a single crystal of high quartz. This crystal will be right- or left-handed, depending on the nature of its low quartz precursor. It has been found possible to remove Dauphiné twinning from quartz oscillator plates permanently by suitably controlled heat treatment. In the oscillator plate industry, Dauphiné twinning is often referred to as electrical twinning, in contrast to optical twinning (see the Brazil Law, p. 107).

Gliding Twins

These twins are also referred to as mechanical twins. The gliding that is involved should not be confused with the previously described and unrelated glide planes of symmetry.

Crystals and crystalline grains of many substances are known to yield plastically to directed stress by a process known as gliding. This gliding can be either of two types: *translation gliding* or *twin gliding*. Translation gliding involves a simple movement, by an integral number of lattice translations, along some plane within the crystal,

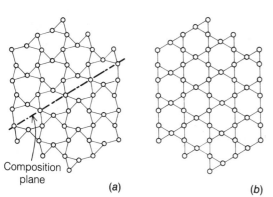

Composition plane

(a)

(b)

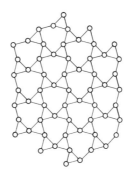

(c)

Figure 2-71. Arrangement of Si atoms in quartz projected on (0001). (a) Low-temperature quartz twinned on c as twin axis—Dauphiné Law. (b) High-temperature (beta) quartz, hexagonal. (c) Low-temperature (alpha) quartz, trigonal.

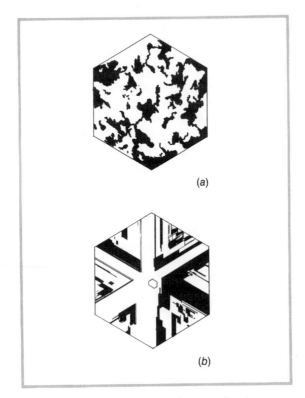

(a)

(b)

Figure 2-72. Basal sections of quartz showing complex interpenetration twinning of two individuals: (a) Dauphiné Law with irregular composition surfaces, (b) Brazil Law with plane composition surfaces.

scribing growth twins. Nor does it follow from the prevalence of growth twins in a certain mineral, according to a certain law, that the mineral will also show gliding twins according to that law. The prevalence of gliding twins is generally due to particular structural features that are not identical to those favoring growth twins. In order for gliding twins to occur, it must be possible for certain atoms to shift from their lowest energy positions to the twinned positions of moderately low energy, via troughs or valleys of moderate to high energy. The frequency of glide twinning should therefore depend on the energy levels at the high point in these troughs. Where atoms occur in energy depressions surrounded by high-energy barriers, crystals are unlikely to exhibit glide twinning but will instead break under stress.

It has been shown that with metal crystals, where twin gliding occurs with {111} as the twin plane in face-centered cubic metals and with {112} as the twin plane in body-centered cubic metals, the operation of twinning involves a minimum of distortion of the structure with no decrease in interatomic distances in the first instance and a decrease of 5.8 percent in the second. On the other hand, twinning on {112} in the first instance, or {111} in the second, would involve interatomic distances (during twinning) 42 percent and 33 percent shorter than normal, which would result in considerable distortion of the structure. This explanation appears to be adequate for the occurrence of twinning in these metals, as well as in complete accord with the general explanation just given.

Gliding twins are frequently encountered in calcite, with twin plane {01$\bar{1}$2}, especially in metamorphosed limestone; in dolomite, with twin plane {02$\bar{2}$1}, especially in metamorphosed dolostones; and in sphalerite, with twin plane {1$\bar{1}$1}. Gliding twins can be easily produced artificially in calcite by deforming a piece of coarsely recrystallized limestone or marble, or by applying suitable pressure with, for example, a knife blade on a cleavage rhombohedron of the mineral (Figure 2-73). The relation of the twinning to the crystal structure is shown in Figure 2-74.

whereby one part of the crystal is offset parallel to its other part; twin gliding occurs when the positions of the resulting twins can be related by a twin law involving either rotation or reflection, as opposed to translation.

Since the single crystal enjoys the state of lowest energy, twin gliding develops a higher-energy state; this higher energy is supplied at least in part by the work done in deforming the original crystal; that is, gliding twins acquire a higher state of energy as a result of their formation. [In contrast, it may be noted that growth twins have never achieved a state of lowest possible energy.]

The twin laws describing gliding twins in a mineral are not necessarily the same as those de-

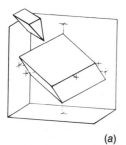
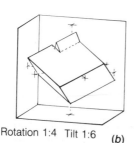

(a) Rotation 1:4 Tilt 1:6 (b)

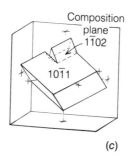

(c)

Figure 2-73. Twin gliding in calcite produced by pressure, composition plane ($1\bar{1}02$).

Examples of Twin Laws

The remainder of this section on twinning consists of a general outline of the kinds of twinning that are relatively common in well-known minerals of the six crystal systems.

Triclinic System

There are no symmetry planes or symmetry axes in triclinic crystals; therefore, any lattice row or lattice plane of simple indices may act as a twin axis or twin plane.

The plagioclase feldspars provide striking examples of twinning in this system, and two of the most common twin laws result from the lower symmetry of plagioclase and microcline as compared to monoclinic orthoclase. The *albite law*,

Figure 2-74. Relation between the atomic arrangement in calcite and twin gliding on ($01\bar{1}2$). [Redrawn from Pabst (1955).]

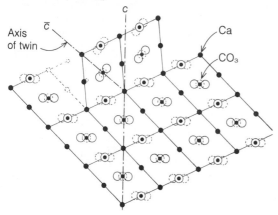

Axis of twin

$\bar{c}$ c

Ca

CO₃

with twin plane {010}, is represented as lamellar twins with composition surface (010), resulting in abundant striations on the {001} cleavage of most specimens of the plagioclase series (Figure 15-17). The composition surface is parallel to the other perfect cleavage {010} of feldspar. The *pericline law*, with twin axis b[010], also produces lamellar twinning; the composition surface is a rhombic section parallel to b. The orientation of the rhombic section varies substantially with the composition of the plagioclase; from 21° on one side of the a axis (between positive a and c), through a position parallel to a for andesine, to 18° on the other side of a in anorthite (between positive a and negative c). This lamellar twinning, rather frequently evident in hand specimens, results in a series of fine striations on {010} cleavage that will be inclined to the cleavage edge {001}:{010} by a small angle, again depending on the plagioclase composition. These are the most common types of twinning in plagioclase; they may occur together and also along with the simple twinning of the Carlsbad, Baveno, or Manebach laws.

In microcline, the albite and pericline laws combine to give two sets of polysynthetic twin lamellae that intersect at nearly right angles to form a grating or grid structure. The pericline lamellae are again parallel to b, but, in the case of microcline, they are also nearly parallel to c; thus the grid twinning is seen in sections that are nearly perpendicular to c. This feature is very striking when seen in the polarizing microscope (see Figure 15-16) and can sometimes be detected on the crystal faces {001} and {$\bar{1}$01} with a good hand lens.

106   Monoclinic System

Gypsum, pyroxene, and hornblende commonly form simple contact twins with twin plane {100}. In gypsum, these twin crystals are called swallow-tail twins (Figure 2-75b). Pyroxene very commonly shows lamellar twinning with {001} as twin plane, resulting in striations on the vertical faces {hk0} and parting on {001} (Figure 15-36d). This lamellar twinning can be produced by shearing stress.

Orthoclase crystals can twin according to several laws. The most clearly recognizable are: Carlsbad Law: twin axis c, composition face (010), or interpenetrating (Figure 15-15c); in these twin crystals, {001} of one individual and {Ī01} of the second individual very nearly coincide but are distinguishable by luster and cleavage (Figure 15-15c). Baveno Law: twin and composition plane {021}; simple contact twins in crystals elongated along a are nearly square prisms, since $(001) \wedge (021) = 44° 56\frac{1}{2}'$ (Figure 15-15d). Manebach Law: twin and composition plane {001}; these are typically simple contact twins (Figure 15-15e).

Orthorhombic System

All minerals of the aragonite group show twinning with {110} as twin plane, resulting in contact and, less commonly, penetration twins (Figures 2-69, 12-11b, and 12-12) that are cyclic, with three or more individuals, or polysynthetic. The twinned crystals of the cyclic type simulate hexagonal symmetry (Figure 12-12). [The prevalence of twinning in this group of minerals was discussed earlier.] Contact twins, simple or repeated in cyclic fashion, are also found in marcasite and arsenopyrite (Figure 9-28) with {110} as twin plane. Staurolite commonly forms cruciform twin crystals, which are penetration twins with twin plane {031} (Figure 15-56) giving a nearly right-angle cross or with twin plane {231} giving a sawhorse twin with an angle of about 60° (Figure 2-76).

In the rhombic pyramidal class, mm2 (hemimorphic), twinning may occur with twin plane {001}, as in hemimorphite.

Hexagonal System

Twinning is of rare occurrence in the common minerals of the holohedral class of the hexagonal system. On the other hand, as already noted, it is common in calcite, quartz, and other minerals of the merohedral symmetry classes.

Calcite forms twin crystals according to several laws:

1. The twin plane is {0001} (not a symmetry plane in calcite); this plane is also the composition plane (Figure 2-77a). Re-entrant angles are present about the equator of the crystal except when it is bounded by {10Ī0}. In that case, the twinning may be revealed by

Figure 2-75.   (a) Augite twin crystal, twin and composition plane (100); forms: a{100}, b{010}, m{110}, s{Ī11}. (b) Gypsum "swallow-tail" twin crystal, twin and composition plane (100), forms b{010}, f{120}, l{Ī11}.

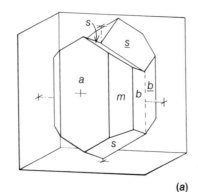

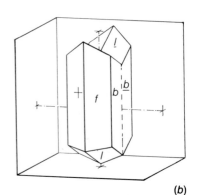

(a)

(b)

cleavage or by an apparent horizontal plane of symmetry.

2. More commonly, the twin plane and the composition face are $\{01\bar{1}2\}$ (Figure 2-74), generally repeated or lamellar (Figure 12-7b). This is the pressure-produced twinning previously described as originating by translation gliding.

Three kinds of twins are fairly common in quartz.

1. Brazil twins (optical twins), with $\{11\bar{2}0\}$ as twin plane, combine a right- and left-handed crystal in a complex penetration twin typically with plane composition surfaces (Figures 2-72b, 2-78b). Complex twinning of this type renders quartz crystals completely useless for optical or electrical purposes. This twinning can be detected in polarized light since the two parts of the twin rotate the plane of polarization in opposite directions.

2. Dauphiné twins (electrical twins), with c the twin axis, combine two right- or two left-handed individuals with very irregular composition surfaces (Figures 2-71a, 2-72a, 2-78a). In Dauphiné twins, the horizontal striations that commonly occur on the prism faces of quartz crystals are interrupted at an irregular line that marks the boundary between two individuals. The positive rhombohedron faces $r\{10\bar{1}1\}$ of one individual and the negative rhombohedron faces

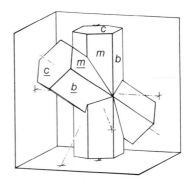

Figure 2-76.   Interpenetration twin of staurolite (sawhorse twin), twin plane $(\bar{2}31)$; forms: $c\{001\}$, $b\{010\}$, $m\{110\}$.

$z\{01\bar{1}1\}$ of the other will coincide on a Dauphiné twin; this results in an irregular line, which delimits the two parts of the twin, on most rhombohedron faces. When the trigonal trapezohedron is well developed, the right or left form will be repeated by twinning, giving 12 faces that simulate a hexagonal trapezohedron. This type of twinning also renders crystals useless for electrical work because it reverses the direction of the a axes in the two parts of the twin. It cannot be recognized in polarized light because it combines two crystal orientations with identical optical properties. It can be recognized in cut plates by etching in hydrofluoric acid or ammonium bifluoride and viewing the etched surface in a spotlight.

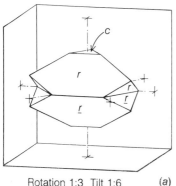

Rotation 1:3   Tilt 1:6   (a)

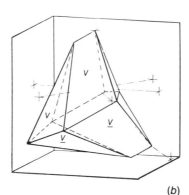

(b)

Figure 2-77.   (a) Calcite twin, twin plane and composition plane $\{0001\}$. (b) Calcite twin, twin plane and composition plane $(1\bar{1}02)$. Crystal forms $v\{21\bar{3}1\}$, $r\{10\bar{1}1\}$, $c\{0001\}$.

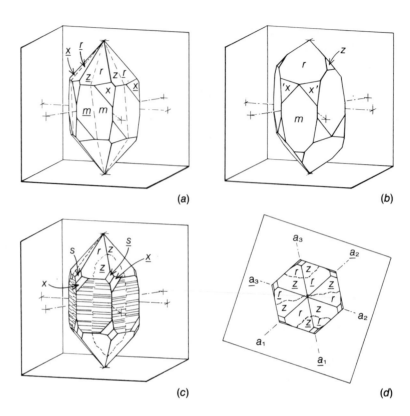

(a)

(b)

Figure 2-78. Low-temperature quartz. (a) Dauphiné twin, twin axis c, interpenetration of two right-hand crystals. (b) Brazil twin, twin plane {11$\bar{2}$0}, interpenetration of a right- and a left-hand crystal. (c) Dauphiné twin, twin axis c, interpenetration of two left-hand crystals; note interruption of striations on prism faces at irregular composition surface and parallelism of r and z. (d) Plan view of crystal in (c); forms: m{10$\bar{1}$0}, r{10$\bar{1}$1}, z{01$\bar{1}$1}, s{2$\bar{1}\bar{1}$1}, x'{51$\bar{6}$1}, 'x{6$\bar{1}5\bar{1}$1}; underlined letters indicate twin positions.

(c)

(d)

3. Japanese twins, with {11$\bar{2}$2} as both twin plane and composition plane, consist of two individuals combined in a contact twin; the c axes intersect at 84° 33', and one pair of prism faces is common to both parts of the twin (Figure 15-10). This twin law was named because of the prevalence of these twins at a locality in Kai province, Japan.

Tetragonal System

The most common examples of twinning among crystals in this system are seen in cassiterite and rutile; particularly in the latter, twinning is rather diverse and complex. In almost all cases, {011} is the twin plane. In cassiterite, simple contact (Figure 10-15b) or interpenetration twins are fairly common. In rutile, repeated twinning on {011} may produce geniculate (knee-shaped) forms (Figure 10-12) or polysynthetic twinning. Repeated twinning on the different equivalent faces

of {011} may result in complex sixlings or eightlings (Figure 2-79). Twinning is so common that the sharp knee-shaped offsets in columnar or thin prismatic crystals of rutile may be utilized as a diagnostic feature for identification.

Figure 2-79. Cyclic twin of rutile, twin plane {101}; forms a{100}, m{110}.

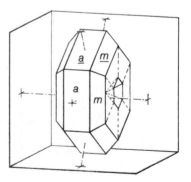

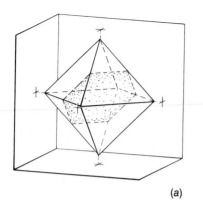
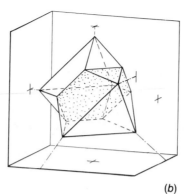

Figure 2-80. (a) Octahedron, showing the plane ($\bar{1}\bar{1}1$), which may act as a twin plane. (b) Octahedron twinned on ($\bar{1}\bar{1}1$) (stippled plane).

Isometric System

Twinning is common in crystals of the hexoctahedral class ($4/m\bar{3}2/m$) with {111} as twin plane or [[111]] as twin axis (*Spinel Law*). This law occurs in many crystals of this symmetry in addition to the minerals of the spinel group (Figure 2-80). Simple contact twins have (111) as composition plane. In magnetite, twinning is commonly lamellar, resulting in striations on {111} faces; such twinning is possibly the cause of the parting present in some magnetite. In galena, contact or penetration twins occur with {111} as twin plane; also, lamellar twinning on {114} results in striations that are visible on many cleavage surfaces. In sphalerite, twins are described (with reference to [111] as twin axis) as simple or multiple contact twins, as interpenetration twins, or as lamellar gliding twins on {111} due to directed pressure. Most fluorite twin crystals consist of interpenetrating cubes with [111] as twin axis (Figure 11-4*b*).

In the diploidal class, pyrite commonly twins with [110] as twin axis; two interpenetrating pyritohedra result in the so-called iron-cross twin (Figure 9-23). In this class, the [[110]] directions are not symmetry axes and therefore may act as twin axes.

SELECTED READINGS

Amorós, J. L., Buerger, M. J., and Canut de Amorós, M. 1975. *The Laue Method*. New York: Academic Press, 375 pp.

Bloss, F. D. 1971. *Crystallography and Crystal Chemistry, An Introduction*. New York: Holt, Rinehart & Winston, 545 pp.

Boisen, M. B., Jr., and Gibbs, G. V. 1976. "A derivation of the 32 crystallographic groups using elementary group theory." *American Mineralogist* **61**, 145–165.

Bradley, C. J., and Cracknell, A. P. 1972. *The Mathematical Theory of Symmetry in Solids. Representation Theory for Point Groups and Space Groups*. Oxford: Clarendon, 745 pp.

Bravais, Auguste. 1949. "On the systems formed by points regularly distributed on a plane or in space" (translated by A. J. Shaler). *Crystallographic Society of America*, Memoir 1, 113 pp.

Buerger, M. J. 1934. "The significance of 'Block Structure' in crystals." *American Mineralogist* **17**, 177–191.

Buerger, M.J. 1945. "The genesis of twin crystals." *American Mineralogist* **30**, 469–482.

Buerger, M. J. 1956. *Elementary Crystallography*. New York: Wiley, 528 pp.

110   Buerger, M. J. 1971. *Introduction to Crystal Geometry*. New York: McGraw-Hill, 204 pp.

Donnay, J. D. H. 1934. "The theory of determinants applied to crystallography." *American Mineralogist* **19**, 593–599.

Eisenstadt, M. M. 1971. *Introduction to Mechanical Properties of Materials*. New York: Macmillan, 444 pp.

Evans, R. C. 1964. *Introduction to Crystal Chemistry* (2nd ed.). Cambridge (England): Cambridge University Press, 410 pp.

Friedel, George. 1913. "Sur les symétries cristallines que peut révéler la diffraction des rayons Röntgen." *Comptes rendus de l'académie des sciences* [France], **157,** 1533–1536.

Frondel, Clifford. 1962. *System of Mineralogy of Dana* (7th ed.). New York. Wiley, v. 3, 334 pp.

Goldschmidt, Victor. 1913–1923. *Atlas der Kristallformen* (9 volumes). Heidelberg: Universitätsverlag Gmbtt.

Henry, N. F. M., and Lonsdale, Kathleen. 1952. *International Tables for X-Ray Crystallography*. International Union of Crystallography. Birmingham (England): Kynoch Press, 558 pp.

Hobbs, B. E., Means, W. D., and Williams, P. F. 1976. *An Outline of Structural Geology*. New York: Wiley, 571 pp.

Honess, A. P. 1927. *The Nature, Origin and Interpretation of the Etch Figures on Crystals*. New York: Wiley, 171 pp.

Janssen, T. 1973. *Crystallographic Groups*. New York: American Elsevier, 281 pp.

McLean, D. I. 1973. "Montreal Summer School on Grain Boundaries." *Metal Science Journal* **7,** 211–212.

Nassau, Kurt. 1980. *Gems Made by Man*. Radnor, Pa.: Chilton, 364 pp.

Newkirk, J. B., and Wenick, J. H. (editors). 1962. *Direct Observation of Imperfections in Crystals*. New York: Wiley-Interscience, 617 pp.

Phillips, F. C. 1972. *An Introduction to Crystallography* (4th ed.). New York: Wiley, 351 pp.

Schneer, C. J. (editor). 1977. *Crystal Form and Structure* (Benchmark Papers in Geology #34). Stroudsburg, Pa.: Dowden, Hutchinson and Ross, 368 pp.

Tunell, George, and Murdoch, Joseph. 1959. *Laboratory Manual of Crystallography for Students of Mineralogy and Geology* (2nd ed.). Dubuque, Ia.: Brown, 58 pp.

Wilson, A. J. C. 1962. *X-Ray Optics*. New York: Wiley, 147 pp.

Wolfe, C. W. 1953. *Manual for Geometrical Crystallography*. Ann Arbor, Mich.: Edwards Bros., 263 pp.

Yale, P. B. 1968. *Geometry and Symmetry*. San Francisco, Ca.: Holden-Day, 288 pp.

3 The Chemistry of Minerals

Mineral chemistry as a science was established in the early years of the nineteenth century. It closely followed and was largely dependent upon three things: (1) the proposal of the Law of Constant Composition by J. L. Proust (1754–1826) in 1799, (2) the enunciation of the atomic theory by John Dalton (1766–1844) in 1805, and (3) the development of accurate methods of quantitative chemical analysis.

Subsequently, mineral chemistry evolved, in partnership with petrology, to become the branch of the geological sciences widely referred to as geochemistry. As such, mineral chemistry now embraces thermodynamics, phase equilibrium investigations, and mineral synthesis, as well as continuing efforts involving the chemical analysis and description of minerals. Coordinated and correlated with crystallographic and physical investigations and considerations, it is also what many people currently characterize as crystal chemistry.

In this chapter, we consider chemical analyses and how they can be interpreted and used to determine the content of unit cells; the bonding of atoms and ions, with special emphasis on the ionic bond; Pauling's rules dealing with coordination; the structural classification of minerals; isostructuralism, isotypism, isomorphism, and solid solution; polymorphism and polytypism; pseudomorphism; and "noncrystalline minerals." We do not describe or discuss the makeup, shapes, and properties of atoms and ions; we assume prior knowledge of this fundamental information. Our related discussion of phase equilibria and thermodynamics is, for the most part, in Chapter 5. Our description of chemical properties used in determinative mineralogy is given in Chapter 6.

CHEMICAL ANALYSES

Considering the fact that the science of mineral chemistry is based on a knowledge of the composition of minerals, you need to understand the possibilities and limitations of chemical analyses of minerals.

A quantitative chemical analysis, no matter how it is made, aims to identify the elements present and to determine their relative amounts. It therefore becomes imperative that an analysis be complete; that is, all elements should be determined, and the amounts determined should correspond to the amounts actually present. Accuracy, of course, depends on the methods employed and the quality of the work of the analyst; even the best methods have a margin of error, although for some it may be insignificant. It is a truism that most analyses are only as good as the analysts who make them.

In the statement of an analysis, the amounts of elements present are expressed in percentages by weight. Therefore, the complete analysis of a mineral should total 100 percent. In practice, as a consequence of limitations on accuracy, a summation of exactly 100 is fortuitous; generally, a summation of between 99.5 and 100.5 is considered a good analysis. The mere fact that an analysis shows a total that lies between those figures, however, does not necessarily indicate that it is accurate; a good total may result from the balancing of plus and minus errors, even including such things as overlooking or misidentifying one or more elements. For example, the rather rare mineral bavenite, discovered in 1901, was described as a hydrated calcium aluminum silicate on the basis of an analysis with the good total of 99.72, whereas reexamination, 30 years later, showed that bavenite also contains beryllium, which was overlooked and apparently precipitated and weighed together with aluminum in the original analysis.

Interpretation of Analyses

In our definition of a mineral, we implied that a mineral has a characteristic chemical composition. A characteristic chemical composition can be expressed by a formula that indicates the elements that constitute the mineral and the proportions in which they are combined. Thus, the characteristic chemical composition of halite is expressed by the formula $NaCl$, which indicates that there are equal numbers of sodium ions and chlorine ions in halite. Similarly, the composition of brucite is expressed by $Mg(OH)_2$, which means that it is a compound consisting of one magnesium for each two hydroxyls; the composition for a certain tourmaline is

$$(Ca,Na)(Mg,Fe)_3Al_5Mg(BO_3)_3Si_6O_{18}(OH,F)_4,$$

which means . . . . Thus, formulas may be simple or complex, depending on the number of elements present and the way they are combined.

The basic data for assigning the correct chemical formula to a mineral are provided by its chemical analysis. However, this is not sufficient for all minerals. The chemical analysis shows what elements are present, and how much of each, but not how they are combined in the structure of the mineral. This last uncertainty is well exemplified by the role of water in an analysis. The water may have been adsorbed in the mineral powder and constitute an impurity; it may be present as water of crystallization; or it may have been formed from hydroxyl groups or hydrogen ions within the structure. As a general rule, small quantities of water lost at low temperatures (below 105°C) are assumed to be adsorbed, although this is not always valid. A correct decision as to how the water shown in the analysis is combined in the mineral may require extensive laboratory investigation. Fortunately, knowledge of the dimensions of the unit cell and the density of a mineral enables one to check the correctness of a suggested formula, as is described later in this chapter.

As we mentioned, the results of chemical analyses are expressed in weight percentages. To determine the formula of a mineral, these weight percentages must be converted to atomic proportions. This is done by dividing the weight percentage of each element in the analysis by the atomic weight of that element. Here, for a simple example, is the analysis of a specimen of marcasite from Jasper County, Missouri.

	Weight percent	Atomic weight	Atomic proportions
Fe	46.55	55.85	0.833 = 1
S	53.05	32.07	1.654 = 1.986
Total	99.60		

Within the experimental error of the analysis, the formula of marcasite becomes FeS_2.

The reverse procedure, that of calculating the percentage composition from the formula, is carried out as follows. The formula for marcasite is FeS_2; since the atomic weight of Fe is 55.85, and of sulfur 32.07, the gram-formula weight of marcasite is $55.85 + 2 \times 32.07 = 119.99$. Then

$$\%Fe = \frac{55.85}{119.99} \times 100 = 46.55$$

$$\%S = \frac{64.14}{119.99} \times 100 = 53.45$$

It is interesting to compare analyses of different specimens of marcasite with the theoretical composition calculated from the formula.

	1	2	3	4	5
Fe	46.55	46.55	46.53	47.22	46.56
S	53.45	53.05	53.30	52.61	53.40
	100.00	99.60	99.83	99.83	99.96

1: Calculated for FeS_2; 2: Jasper County, Missouri; 3: Joplin, Missouri; 4: Osnabrück, Germany; 5: Loughborough Township, Ontario.

All the analyses are good in terms of addition within the limits 99.5–100.5, but each deviates slightly from the composition calculated from the formula. It is rather generally thought, though not necessarily true, that these deviations can be attributed to imperfections of analytical procedures rather than to any deviation of marcasite from the formula FeS_2.

Within the experimental error, these analyses show that marcasite, whatever its source, has a fixed chemical composition. In this respect, marcasite is rather exceptional among minerals. Most minerals are not *fixed* in composition although each does have a *characteristic* composition that can be expressed by a formula. This is illustrated by the following analyses of sphalerite.

	1	2	3	4
Fe	0.15	7.99	11.05	18.25
Mn	—	—	—	2.66
Cd	—	1.23	0.30	0.28
Zn	66.98	57.38	55.89	44.67
S	32.78	32.99	32.63	33.57
	99.91	99.59	99.87	99.43

1: Sonora, Mexico; 2: Gadoni, Sardinia; 3: Bodenmais, Germany; 4: Isère, France.

These analyses show that sphalerite may be nearly pure zinc sulfide, or it may contain considerable quantities of iron and minor amounts of manganese and cadmium. The situation is clarified when the analyses are recalculated in atomic proportions:

	Atomic Weight	1	2	3	4
Fe	55.85	0.003	0.143	0.198	0.327
Mn	54.94	—	—	—	0.048
Cd	112.41	—	0.011	0.003	0.002
Zn	65.38	1.024	0.878	0.855	0.683
		1.027	1.032	1.056	1.060
S	32.07	1.022	1.029	1.017	1.047

For all the analyses, the atomic proportions of total metals to sulfur is ~ 1:1, corresponding to the formula ZnS, but with some Zn replaced by Fe, Mn, and Cd. Therefore, the formula ZnS is an oversimplification for the composition of sphalerite. Sphalerite can be practically pure ZnS (analysis 1), but it can also have one-third or more of the zinc replaced by iron. This phenomenon is common in minerals and, in this example, it can be indicated by writing the formula as (Zn,Fe)S, which shows that the total of Zn + Fe is 1 with respect to S = 1, but that the actual amounts of Zn and Fe are variable.

When you want to express the composition corresponding to a particular analysis as a formula, you first reduce the atomic proportions of the mutually replacing elements to decimal fractions of unity. Thus analysis 4 can be mathematically reduced and expressed as a formula (omitting the very small amount of cadmium) in the following way:

	Atomic proportions	Decimal fractions
Fe	0.327	$\dfrac{0.327}{1.060} = 0.31$
Mn	0.048	$\dfrac{0.048}{1.060} = 0.05$
Cd	0.002	
Zn	$\dfrac{0.683}{1.060}$	$\dfrac{0.684}{1.060} = 0.64$
S	1.047	$\dfrac{1.047}{1.047} = 1$

Formula: $(Zn_{0.64}Mn_{0.05}Fe_{0.31})S$

Therefore, the formula $(Zn_{0.64}Fe_{0.31}Mn_{0.05})S$ is a special case of the general formula (Zn,Fe)S, and it expresses the composition of a specific analyzed sample of sphalerite.

Another example of this kind of substitution is in the mineral olivine, for which the following analyses are recorded:

	1	2	3	4
FeO	8.58	31.48	47.91	58.64
MnO	0.20	0.22	0.41	0.85
MgO	50.00	30.50	18.07	8.49
SiO$_2$	40.99	38.11	33.72	31.85
	99.77	100.31	100.11	99.83

1: Sardinia; 2, 3, 4: Kangerdlugssuak, Greenland.

These analyses illustrate an especially significant limitation of the analytical technique when it is applied to oxygen-containing compounds. In effect, there is no analytical procedure for determining the total amount of oxygen in a compound. As a result, the analyses of such compounds are expressed in terms of the oxides of the individual elements, instead of in terms of the elements themselves; and instead of the weight percentages of the oxides being converted to atomic proportions, they are converted to molecular proportions by dividing the analytical figures by the formula weight of each oxide [e.g., formula weight of $SiO_2 = 28.09 + (2 \times 16.00) = 60.09$].

The preceding analyses of olivine differ so much one from another that you might well think they represent different minerals. Conversion into molecular proportions, however, shows that they are all variants of the same basic formula.

	Formula weight	1	2	3	4
FeO	71.85	0.119	0.438	0.667	0.816
MnO	70.94	0.003	0.003	0.006	0.012
MgO	40.31	1.240	0.757	0.448	0.211
		1.362	1.198	1.121	1.039
SiO$_2$	60.09	0.682	0.634	0.561	0.530

Despite the wide variation in the amounts of MgO and FeO, the ratio of molecular proportions of FeO + MnO + MgO to SiO_2 is always ~ 2:1, corresponding to the formula $(Mg,Fe)_2SiO_4$. If the minor quantity of manganese is neglected, you can express each analysis precisely in the form $(Mg_xFe_{1-x})_2SiO_4$ by merely normalizing the molecular proportions (e.g., in column 1 by dividing the 1.240 of MgO by 1.362 = 0.91) as follows: 1: $(Mg_{0.91}Fe_{0.09})_2SiO_4$; 2: $(Mg_{0.63}Fe_{0.37})_2SiO_4$; 3: $(Mg_{0.40}Fe_{0.60})_2SiO_4$; 4: $(Mg_{0.21}Fe_{0.79})_2SiO_4$.

In practice, however, many mineralogists and petrologists use an alternative set of calculations. As in the already outlined method, the chemical analysis values (column 1) are divided by the appropriate formula weights (column 2) to

	1 Chemical analysis	2 Formula weight	3 Molecular proportions	4 Number of oxygens	5 Cations on basis of 4 oxygens		
FeO	8.58	71.85	0.1194	0.1194	0.175		
MnO	0.20	70.94	0.0028	0.0028	0.004 } 1.999 ≅ 2 } 0.09		
MgO	50.00	40.31	1.2404	1.2404	1.820	0.91	
SiO$_2$	40.99	60.09	0.6821	1.3642	1.001 ≅ 1		
	99.77			2.7268			

so $(Mg_{0.91}Fe_{0.09})_2SiO_4$

obtain the molecular proportions (column *3*); those proportions are then multiplied by the number of oxygens per cation (e.g., by 2 for silicon and by 1 for magnesium) to get the oxygen equivalents (column *4*); the oxygen equivalents are normalized, in this case by multiplying by 4/2.7268 (and 2/2.7268 for the double oxide, SiO$_2$) to give the number of oxygens per formula unit (column *5*); and the cations other than Si are then normalized to their atomic ratio figure (e.g., for Mg 1.820 ÷ 1.999 = 0.91).

Another designation frequently used for indicating specific compositions in a mineral, having a range in its composition, states values in terms of percentages of its components. For olivine, for example, the components are Mg_2SiO_4 (forsterite, abbreviated Fo) and Fe_2SiO_4 (fayalite, abbreviated Fa). In this terminology, the composition corresponding to the preceding analysis can be expressed as Fo$_{91}$Fa$_9$. However, since the sum of the percentages must be 100, one component in these expressions can be omitted without creating any ambiguity; thus, olivine (Fo$_{91}$) is used to indicate olivine with a composition of 91 percent Mg_2SiO_4 and 9 percent Fe_2SiO_4. These percentages are molecular, so the composition by weight can be calculated as follows: the gram-formula weight of Mg_2SiO_4 contains 80.64 × [2(24 + 16)] grams of MgO and 60.09 × [28 + 2(16)] grams of SiO$_2$, and that of Fe_2SiO_4 contains 143.70 × [2(56 + 16)] grams of FeO and 60.09 grams of SiO$_2$. Hence the gram-formula weight of olivine (Fo$_{91}$) contains:

			Weight percent*
MgO	80.64 × 0.91 =	73.38	50.12
FeO	143.70 × 0.09 =	12.93	8.83
SiO$_2$	60.09 × 1 =	60.09	41.05
		146.40	100.00

*The difference between these figures and those given in column *1* of the original analysis is accounted for by the lack of consideration of Mn in this group of calculations.

Specific gravities for the analyzed olivines, given on page 475, are *1*—3.35; *2*—3.69; *3*—3.88; and *4*—4.16; thus, there is evidently a direct relationship between specific gravity and iron content. This relationship can be shown clearly by a graph of specific gravity versus composition (Figure 3-1). As shown, the plotting of Fo content on the abscissa and of specific gravity on the ordinate results in a straight-line graph that shows that the specific gravity of an olivine is directly related to its composition. Consequently, by using this graph, you can deduce the composition of a particular olivine specimen if you know its specific gravity or, conversely, you can deduce its specific gravity if you know its composition. For example, from the graph you can see that the densities of pure Mg_2SiO_4 and of pure Fe_2SiO_4 are 3.22 and 4.41, respectively—values

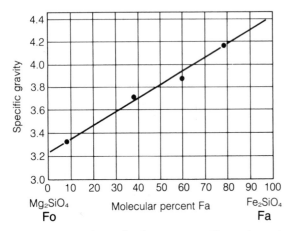

Figure 3-1.   Relationship between specific gravity and composition in the olivine series.

that do agree with those determined for the pure compounds.

Such graphs that correlate composition with physical and optical properties are available for many minerals of variable composition, and they are used widely for a rapid determination of composition. A practical example of their usefulness is the determination of niobium and tantalum content in minerals of the columbite-tantalite series from a density measurement; a chemical analysis for niobium and tantalum is difficult, time consuming, and requires an extremely well-equipped laboratory. On the other hand, a density measurement can be made rapidly with simple equipment. For this reason, density measurements, which are easily converted into specific gravity, are especially useful for a quick estimate of niobium and tantalum in samples of columbite and tantalite concentrates.

CHEMICAL COMPOSITION AND THE UNIT CELL CONTENT

The unit cell of any substance contains one or an integral multiple of chemical formula units. Hence, if the dimensions of the unit cell (as determined from X-ray measurements) and the specific gravity of the substance are known, the unit cell

content can be calculated. The dimensions of the unit cell give its volume, V, as follows:

Isometric	$V = a^3$
Tetragonal	$V = a^2c$
Hexagonal	$V = a^2c \sin 60°$
Orthorhombic	$V = abc$
Monoclinic	$V = abc \sin \beta$
Triclinic	$V = abc(1 - \cos^2\alpha - \cos^2\beta$
	$\quad - \cos^2\gamma$
	$\quad + 2\cos\alpha\cos\beta\cos\gamma)^{1/2}$

[It is noteworthy here, however, that some crystallographers designate cell dimensions in all the crystal systems on the basis of a, b, and c; under this scheme, the a^3 for the isometric system and the a^2c's for the tetragonal and hexagonal systems would each become abc.] In any case, the total weight of atoms in the unit cell (M) is given by $M = V \times G$, where G is the specific gravity of the mineral expressed as density in grams per cubic centimeter (g cm^{-3}). The dimensions of the unit cell are given in angstrom units (1 Å = 10^{-8} cm), so the volume V in cubic angstrom units must be multiplied by 10^{-24} to obtain M in grams; thus,

$$M = (V \times 10^{-24} \times G) \text{ grams}$$

Thence, if you know the cell weight M, you can calculate the number of atoms in the unit cell from the chemical analysis as follows: if a mineral contains P percent of a given element A of atomic weight N, the weight of A in the unit cell must be $PM/100$ and the actual weight in grams of an atom of A becomes $N \times 1.6602 \times 10^{-24}$ (the 1.6602×10^{-24} being the weight in grams of a hypothetical atom of atomic weight 1.0000). Therefore, the number of atoms of A in the unit cell is

$$\frac{PM}{100} \times \frac{1}{N} \times \frac{1}{1.6602 \times 10^{-24}} =$$

$$\frac{PVG \times 10^{-24}}{166.02 \times 10^{-24}N} = \frac{PVG}{166.02N}$$

This equation, of course, indicates that for a chemical analysis expressed in weight percentages of the elements, the number of atoms of each element in the unit cell can be obtained by dividing the percentage of the element P by the atomic weight N, (i.e., by converting the analysis to atomic proportions) and then multiplying each atomic proportion by the factor $VG/166.02$.

This procedure may be exemplified by the recalculation of analysis 4 of sphalerite on page 113 as follows:

	1	2	3	4	5
Fe	18.25	55.85	0.3268	1.22	
Mn	2.66	54.94	0.0484	0.18	
Cd	0.28	112.41	0.0025	0.09	$4.04 \cong 4$
Zn	44.67	65.38	0.6832	2.55	
S	33.57	32.07	1.0468	3.91	$3.91 \cong 4$
	99.43				

1: Chemical analysis; 2: atomic weight; 3: atomic proportions; 4: atoms in unit cell [i.e., atomic proportions multiplied by $(VG)/166.02$ with the unit cell edge being 5.41 Å and the density 3.92 $g\,cm^{-3}$—hence, the calculation becomes atomic proportions $\times$ $5.41^3 \times 3.92 \div 166.02$]; 5: atoms per unit cell.

Therefore, in the unit cell of sphalerite, there are four metallic atoms and four sulfur atoms—the small deviations from whole numbers being due to errors in the analysis, the cell dimensions, and the density. Thus, the unit cell contains four units of the usual formula (Zn,Fe)S. This number

of formula units in the unit cell is usually designated as Z.

The calculation of unit cell content from an analysis expressed in terms of oxides follows the same general procedure. The only variation is that the weight percentages of the oxides are first converted to molecular proportions by dividing by the molecular weights of the specific oxides; then, these are calculated into oxide contents per unit cell; and the oxide contents are converted into element and oxygen contents according to the proportions in each oxide. This is illustrated by the recalculation of analysis 1 of olivine on p. 114; see below.

Thus, in the unit cell of olivine, there are eight metal ions and four SiO_4^{-4} cation groups; so for the unit cell there are four units of the typical formula, $(Mg,Fe)_2SiO_4$, which means that $Z = 4$. From the calculations, it also follows that, conversely, if Z is known, the relationships can be used to check the correctness of the formula assigned to a specific mineral or to check the correctness of a measured specific gravity.

BONDING

The basic unit in crystal structures is the *atom* or *ion*, the ion being an atom carrying electric charge. An atom that has given up one or more electrons is said to be a *cation* and is indicated by a positive superscript, such as Zn^{+2}; an atom that has received one or more electrons is said to be an *anion* and is indicated by a negative super-

	1	2	3	4	5		
FeO	8.58	71.85	0.1194	0.70	Fe 0.70		
MnO	0.20	70.94	0.0028	0.02	Mn 0.02	8.01	
MgO	50.00	40.31	1.2404	7.29	Mg 7.29		
SiO₂	40.99	60.09	0.6821	4.01	Si 4.01		
	99.77				O 8.02 + 8.01 = 16.03		

1: Chemical analysis; 2: formula weight; 3: molecular proportions; 4: oxide contents per unit cell—molecular proportions $\times VG/166.02$ (when $V = 4.76 \times 10.21 \times 5.99$ and $G = 3.35$); 5: atoms per unit cell.

118   script, such as S^{-2}. An atom or an ion may act individually or be closely associated with other atoms or ions to form groups that behave as units; such groups—for example, $(CO_3)^{-2}$—are generally called radicals.

Although we recognize the fact that the hard sphere model of an atom is not realistic, that model remains instructive, and the descriptive terms and illustrations based on it greatly facilitate discussion. Thus, we shall use it.

An atom is made up of a very small, positively charged nucleus surrounded by one or more shells of electrons, the whole acting as a sphere whose effective radius is of the order of 1 Å. The radius depends not only on the nature of the element but also on its state of ionization and the manner in which it is linked to adjacent atoms or ions. The kinds of linkage are generally referred to as chemical bonding.

Four kinds of bonds provide a convenient basis for the consideration of crystal structures: the metallic bond, the covalent (or homopolar) bond, the ionic (or heteropolar or polar) bond, and the van der Waals bond. These bond types are not, however, mutually exclusive; that is, the bonding in many minerals and other inorganic crystalline substances may involve more than one kind of bonding. For example, the silicon-oxygen bonds in silica and the silicates are neither purely ionic nor purely covalent but intermediate (or hybrid) in nature. In essence, the structure assumed by any solid is such that the whole system of atomic nuclei and electrons tends to be arranged in a form with minimum internal energy. As far as the properties shown on Table 3-1 are concerned, a rule of thumb is that physical properties of crystalline substances are determined by their weakest bonds.

The Metallic Bond

This bond type is responsible for the cohesion of a metal. Metals consist of elements whose atoms readily lose their outer electrons. The crystal structure of a metal is determined by the packing of the positively charged atoms, the detached electrons being dispersed among the atoms and freely mobile. This electron mobility is responsible for the color, luster, and good electrical and thermal conductivity of metals. In minerals, metallic bonding is present in native metals and to some extent in a few sulfides and arsenides.

The atoms in metals are typically arranged according to one of two different arrays—one termed cubic closest packing [CCP(= FCC)], the other called hexagonal closest packing (HCP). As you can see in Figure 3-2, cubic closest packing is a periodic structure that can be described with a face-centered cubic unit cell, and hexagonal closest packing is based on a primitive hexagonal cell with two atoms per unit cell.

The Covalent Bond

The most stable configuration for an atom is one in which the outer shell of electrons is completely filled. This structure is the atomic structure of the inert gases and accounts for their almost complete lack of reactivity. One way for this configuration to be achieved is by two or more atoms sharing electrons in their outer shells. For example, the atoms in chlorine gas, Cl_2, are always linked in diatomic molecules; each chlorine atom has seven outer shell electrons, and the stable condition of eight electrons in the outer shell is reached by each atom sharing one electron with another atom:

$$:\!\ddot{C}l\cdot\ +\ \cdot\ddot{C}l\!: \ =\ :\!\ddot{C}l\!:\!\ddot{C}l\!:$$

Covalent bonds are involved in the formation of most radicals, are common in organic compounds, but are rare—at least in an unmodified form—in minerals. Perhaps the best mineralogical example is diamond, in which every carbon atom is surrounded by four other carbon atoms, each sharing one electron with the central atom. This pattern is repeated throughout the structure, and thus each crystal is in essence one giant molecule.

Table 3-1. Bond types and typical properties

Bond Type				
	metallic	covalent (=homopolar)	ionic (=heteropolar)	van der Waals (=residual =intermolecular)
Particles Involved				
	cations	atoms	ions	molecules (or noble gas atoms)
Nature of Bond				
	"cohesion" among free electrons and ions (resonating shared electrons)	sharing of electrons	electrostatic attraction (Coulombic forces)	weak forces
Bond Strength				
	variable	very strong	strong	weak
Cation Coordination				
	> 6	≤ 4	≥ 6	(0)
Packing				
	close	loose	intermediate	variable
Typical Properties				
Hardness:	hard/soft	hard	hard/moderate	soft
Strength:	variable ductile/maleable	fairly strong brittle	mechanically strong brittle	mechanically weak somewhat plastic
Melting Point:	moderate low to high	high	moderate to high	low
Thermal Expansion:	moderate	low	high	very high
Electrical Conduction:	good	poor	poor	poor
Others:	opaque metallic luster	transparent sub- or non- metallic luster	transparent nonmetallic luster	
	inherent color	colorless (or exotic color)	colorless (or exotic color)	

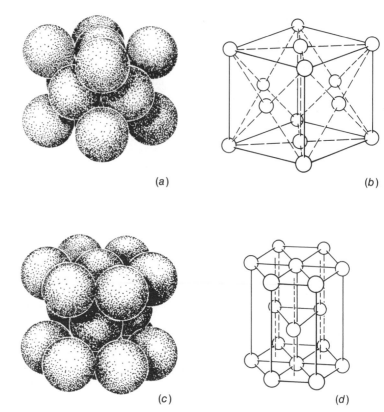

Figure 3-2.   Packing arrangements. (a) FCC lattice with atoms drawn to scale of cube edge; (b) FCC lattice, unit cube indicated by solid lines; (c) HCP lattice with atoms drawn to scale; (d) HCP lattice, unit cell indicated by heavy solid lines.

The Ionic Bond

Another way for an atom to achieve a completely filled outer shell of electrons is for it to gain or lose a sufficient number of electrons to reach the configuration of the nearest inert gas. Thus, a chlorine atom, by adding an additional electron, becomes a negatively charged ion with the electron configuration of argon. This type of adjustment clearly requires the presence of another atom that can provide the additional electrons; in other words, an atom that attains a stable configuration by losing electrons. Sodium is an element which, by losing one electron, becomes a positively charged ion with the electron configuration of neon. Sodium and chlorine therefore combine readily to give a structure of oppositely charged ions bonded by electrostatic attraction. There is, however, no pairing of individual positive and negative ions to give discreet molecules such as those we can distinguish in covalent compounds; instead, each ion is surrounded with ions of opposite charge, the number being determined largely by the relative sizes of the ions.

Ionic bonding is common in inorganic compounds and is therefore extremely important in the structure of minerals. In fact, practically all minerals, except the elements and sulfides, are predominantly ionic compounds. For this reason, we discuss the sizes of ions and the structure of minerals based on the crystal chemistry of ionic bonding in the next two major sections of this chapter.

The van der Waals Bond

This bonding, sometimes called residual bonding, is typically present in crystalline solids of frozen inert gases (helium, neon, argon, etc.). One consequence of these elements having completely filled outer electron shells is their inability to form bonds of the metallic, covalent, or ionic type. The force that does exist between such atoms is what is generally termed the van der Waals bond, force, or attraction. Although it is relatively easily accounted for by quantum mechanics, its explanation is otherwise rather complicated. Suffice it to say here that the van der Waals force is a result of electrical imbalances that exist because of an offcentering of positively charged atomic nuclei within their negatively charged cloud of surrounding electrons. Consequently, the attraction forces are quite weak, a condition that is reflected in the very low temperatures and high pressures required to condense the inert gases to liquids and solids. The bond is apparently of little significance in minerals, other than perhaps accounting for certain weaknesses within certain minerals.

As we have already mentioned, more than one bond type can occur in a single compound. In sulfur, for example, the atoms are covalently linked in rings of eight atoms—in effect, S_8 molecules—and these molecules are bound together in the crystal by van der Waals linkages. Physical properties, such as hardness and mechanical strength, are determined by the weakest bonds, which are the first to suffer disruption under increasing mechanical or thermal strain. In graphite, for example, the carbon atoms are covalently linked in sheets, and the sheets are linked by van der Waals bonds. The latter are weak and easily disrupted, hence the softness of graphite and its ready cleavability parallel to the sheets of carbon atoms.

According to the nature of the bonding, the structures of ionic compounds can be conveniently grouped into two types: the *isodesmic* and *anisodesmic*. In isodesmic compounds, the bonds are all of comparatively equal strength; these compounds comprise the oxides, hydroxides, and simple halides. In anisodesmic compounds, there is a pronounced difference in the strength of different bonds that results in the presence of discrete groups of atoms within the structure; such groups include the oxyacid anions, such as CO_3^{-2}, SO_4^{-2}, and PO_4^{-3}. In compounds of the latter kind, the groups act as units within their structures and bonding within these units is generally stronger than bonding between the units and external cations.

THE SIZES OF IONS

The fact that ionic structures are dominant among minerals is well demonstrated by the fact that about 90 percent of all minerals can be considered to be fundamentally ionic compounds. Therefore, the structures of a large percentage of all minerals can be said to depend on the numbers and sizes of the specific ions entering into their compositions. Ions can be considered to be approximately spherical with a radius characteristic of the element in question and the charge on the ion. Effective (not absolute!) ionic radii are given in Appendix B. It should always be kept in mind that published ionic radii values are not inviolable. In fact, the use of any published values for, for example, the prediction of interatomic distances is limited and thus must be used with great discretion. [Currently, the values calculated by Shannon and Prewitt (1969), as revised by Shannon (1977), and given in Appendix B are the ones used most widely by mineralogists.]

It merits special note that the hydrogen ion consists of a charged nucleus, the proton, which has no orbital electrons associated with it and therefore is exceedingly small; hence, it acts rather like a dimensionless center of positive charge. Consequently, the radius of the OH^{-1} ion is essentially the same as that of the O^{-2} ion; the hydrogen is, so-to-speak, embedded in the oxygen atom, and the OH group is effectively a sphere.

In any case, since the radius of an ion depends on atomic structure, the ionic radius of an element can be correlated with its position on the periodic table. In this regard, consider the following rules, which are generally valid, in conjunction with the radii tabulated in Appendix B.

1. For elements in the same group of the periodic table, the ionic radii increase as the atomic number of the elements increases.

Thus, for coordination number VI and valence of +2, the values for group II elements are: Be, 0.27; Mg, 0.72; Ca, 1.00; Sr, 1.18; and Ba, 1.35. These values, of course, are to be expected since, for elements in the same group of the periodic table, the number of electron orbits around the nucleus (and hence the effective radius) increases going down the column.

2. For positive ions of the same electronic structure, the effective radius for a given coordination number decreases with increasing charge.

As an example, we can consider the elements, for coordination IV, in the second horizontal tier in the periodic table (all have two electrons in the inner orbit and eight in the outer orbit): Na^{+1}, 0.99; Mg^{+2}, 0.57; Al^{+3}, 0.39; Si^{+4}, 0.26; P^{+5}, 0.17; and S^{+6}, 0.12. To state this generalization another way, in going across a horizontal tier (period) in the periodic table, the radii of the cations typically decrease. This decrease reflects the fact that as electrons are lost, the nucleus exerts a greater pull on those that remain, thus decreasing the effective radius of the ion.

3. For any element that can exist in several valence states (i.e., that can form ions of different charge), the ionic radius decreases with increasing positive charge.

An example is $^{IV}Mn^{+2}$, 0.66; $^{VIII}Mn^{+2}$, 0.96; $^{V}Mn^{+3}$, 0.58; $^{VI}Mn^{+4}$, 0.39; and $^{IV}Mn^{+6}$, 0.26. The cause responsible for rule 2 applies here also; that is, the loss of an electron causes the remaining electrons to be attracted more strongly by the nucleus, thus effectively contracting the outer electron orbits and decreasing the ionic radius.

4. Ionic radii, for an element that can assume different coordinations, increase with coordination number.

For example, for Al^{+3}, the values are IV, 0.39; V, 0.48; and VI, 0.53, and for Sr^{+2}, they are

VI, 1.18; VII, 1.21; VIII, 1.26; IX, 1.36; and X, 1.44.

The rare-earth elements provide an apparent contradiction to the first rule. The trivalent ions of these elements decrease in radius with increasing atomic number from $^{VI}La^{+3}$, 1.06 to $^{VI}Lu^{+3}$, 0.848. This remarkable feature, known as the lanthanide contraction, is a consequence of the building up of an inner electron shell instead of the addition of a new shell; as a result, the increasing nuclear charge produces an increased attraction on the outer electrons and an effective decrease in ionic radius. The lanthanide contraction also influences the geochemical activities of the elements following lutetium. Hafnium and tantalum have ionic radii almost identical to those of the elements directly above them in the periodic table—that is, zirconium and niobium, respectively—and, therefore, the two elements in each pair have almost identical crystallochemical properties.

THE STRUCTURE OF MINERALS

Considering the fact that oxygen is the most abundant element in the earth's crust, it is hardly surprising that practically all of the common minerals are compounds of oxygen. Therefore, the fact that oxygen has a large size (the O^{-2} ion has a radius of from 1.35 to 1.42 Å) is extremely significant. Not only is the earth's crust, which contains about 47 percent oxygen by weight, more than 90 percent oxygen by volume, but, in essentially all oxygen-containing minerals, the amount of oxygen far exceeds all the other elements on a volume basis. Therefore, it can be said that the structure in oxygen compounds is generally determined by the arrangement of the oxygen ions with the ions of the other elements, chiefly smaller cations, merely filling interstices.

Actually, the structure of any ionic compound is determined by both the size of its constituent ions and the charge on those ions, the latter being expressed by valence. Or, to state it otherwise, ionic structures are controlled by demands of both geometric and electrical stability. Geometric stability implies that the relative ionic sizes and the mode of packing must result in the ions' being held more or less rigidly within the structure. Electrical stability (i.e., neutrality) means that the sum of positive and negative charges on the ions must balance, both for a mineral grain as a whole and around individual ions. [These demands, by the way, give us a useful check on the correctness of a formula, especially for complex minerals such as the amphiboles.]

Pauling's Rules

Principles governing the geometric arrangement, electrical stability, and overall makeup of ionic crystals were codified by Linus Pauling in 1929 (see Selected Readings at the end of this chapter). Pauling's rules may be paraphrased and elaborated upon as follows:

1. A coordination polyhedron of anions is formed about each cation with the cation-anion distance equal to the sum of the radii of the two ions; hence, the coordination number, and consequently the configuration of the coordination polyhedron, is determined by the ratio of the radius of the cation to that of the anion.

Within ionic structures, each cation tends to surround itself with anions. The number of anions grouped around each cation depends on their relative sizes. For any given pair, the relative size is most clearly expressed by the *radius ratio*, which is the ratio of the radius of the cation to that of the anion. This ratio is often given as R_A/R_X with A denoting the smaller cation and X the larger anion. The number of anions that fit around each cation is known as the cation's *coordination number*. Thence, if it is assumed that ions act as spheres of given radii, the stable arrangements of cations and anions for particular radius ratios can be calculated on purely geometric grounds (Table 3-2). The perfect fit is one whereby adjacent anions are tangent to each other and to the cation they surround. The common types of coordination are illustrated in Figure 3-3. As we mentioned previously, oxygen is the most common anion; thus, when the term *coordination number* is used without qualification, it generally refers to the coordination by oxygen. The tetrahedra, octahedra, etc. are generally referred to as coordination polyhedra. Slight distortion in the geometry of actual coordination polyhedra is a common feature of mineral structures. On the other hand, irregular and unsymmetric polyhedra are somewhat atypical, although not uncommon for those surrounding alkali and alkaline earth cations.

Some cations occur exclusively in a particular coordination. Others—for example, aluminum, whose radius ratio lies near the theoretical boundary between two types of coordination (4 and 6)—may occur with either. In some cases, such coordination can be shown to be controlled by the temperature and pressure conditions at which crystallization of the containing mineral takes place. High temperatures and low pressures

Table 3-2. Relationship between radius ratio and coordination number for ions acting as rigid spheres

Radius Ratio	Arrangement of Anions Around Cation	Coordination Number of Cation
0.15–0.22	Corners of an equilateral triangle	3
0.22–0.41	Corners of a tetrahedron	4
0.41–0.73	Corners of an octahedron	6
0.73–1	Corners of a cube	8
1	Midpoints of cube edges	12

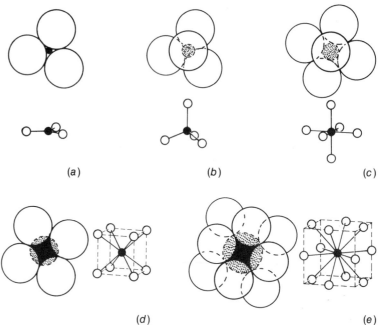

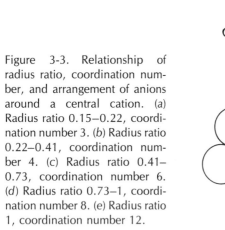

Figure 3-3. Relationship of radius ratio, coordination number, and arrangement of anions around a central cation. (a) Radius ratio 0.15–0.22, coordination number 3. (b) Radius ratio 0.22–0.41, coordination number 4. (c) Radius ratio 0.41–0.73, coordination number 6. (d) Radius ratio 0.73–1, coordination number 8. (e) Radius ratio 1, coordination number 12.

(a)   (b)   (c)

(d)   (e)

favor low coordination, whereas low temperatures and high pressures favor high coordination. Here again, aluminum is a good example; in typical high-temperature minerals, aluminum tends to assume fourfold coordination and substitute for silicon, whereas in minerals formed at lower temperatures, it occurs more commonly in sixfold coordination.

   2. *In stable ionic crystal structures, the total strengths of the electrostatic bonds that reach an anion within a coordination polyhedron from all neighboring cations equals the total charge on the anion.*

This rule, frequently referred to as the *electrostatic valence principle,* requires neutralization of charge to be closely localized around individual anions in ionic structures. Another way of viewing this rule is to note that a measure of the strength of this kind of bond can be determined by dividing the total valence charge on the anion by its coordination number (i.e., by the number of nearest neighbor cations). The resulting number is termed the electrostatic valency (e.v.).

An often-cited example is halite (NaCl), in which each Cl^- ion is surrounded by six Na^+ ions, and each Na^+ ion is surrounded by six Cl^- ions (see Figure 11-1). In this structure, electrostatic neutrality is maintained by assigning one-sixth of the negative charge of each anion to each of the surrounding cations and one-sixth of the positive charge of each cation to the surrounding anions. As previously noted, structures of this kind, in which all bonds are of essentially equal strength, may be termed *isodesmic.*

   3. *The sharing of edges, and especially of faces, by two coordination polyhedra decreases the stability of an ionic crystal structure.*

In essence, this rule reflects the fact that cations repulse each other; that is, they keep as far apart as possible, or shielded from one another, within a structure. Therefore, this effect is particularly marked for high-valence cations, for ions with low coordination numbers, and for polyhedra, the radius ratios of which are near the lower limit of stability.

As a consequence of this rule, the coordination polyhedra of stable ionic compounds tend not to share edges (and especially not faces) with other coordination polyhedra. And, in those that do contain polyhedra that are so linked, it can generally be shown that the shared edges are shortened.

4. In a crystal containing different cations, those of high valence and small coordination number tend not to share polyhedral elements with each other.

This rule, essentially an extension of rule 3, relates to repulsion, shielding, etc. involving different—rather than equivalent—cations. It emphasizes the fact that any sharing of coordination polyhedron anions tends to decrease the stability of a structure.

5. The number of essentially different kinds of constituents in a crystal structure tends to be small.

This rule, frequently referred to as the *principle of parsimony,* is true because there are only a few kinds of different sites, even in ionic crystals having relatively complex compositions and structures. Most mineral structures can be related to configurations involving closest packing of large anions with smaller cations fitting in the interstices. These holes, which provide three kinds of coordination sites (triangular, tetrahedral, and octahedral), will accept all of the common cations except the alkalis and alkaline earths. Hence, in many minerals, diverse cations occupy identical structural sites. [Although of somewhat different nature, this phenomenon may be exemplified by the amphiboles with the general formula $W_{0-1}X_2Y_5Z_8O_{22}(OH,F,Cl)_2$ in which: W may be Ca^{+2}, Na^+ and/or K^+ in the A site; X may be Ca^{+2}, Na^+, Mn^{+2}, Fe^{+2}, Mg^{+2}, and/or Li^+ in the $M4$ sites; Y may be Mn^{+2}, Fe^{+2}, Mg^{+2}, Fe^{+3}, Al^{+3}, and/or Ti^{+4} in the $M1$, $M2$, and $M3$ sites; and Z is Si^{+4}, Al^{+3}, and/or Ti^{+4} in the tetrahedral sites.]

The foregoing principles are, of course, fundamental to the crystal chemistry of minerals.

They express the conditions for low potential energy of the atoms and hence for high stability; that is, they indicate that only very stable compounds can occur as minerals, that less stable compounds either do not form in nature or soon decompose to minerals that are more stable under the existing conditions. They also provide a basis for classifying minerals.

Structural Classification of Minerals

As you will see in Chapter 7, we have adopted a chemical classification for the presentation of the mineral descriptions given in Part II. We also mention there that that classification has an obvious crystal structure basis even though its precursors were originally based solely on chemical composition. This correlation is shown in Table 3-3.

Most of the minerals with predominant metallic bonds are characterized by cubic closest packing [CCP(= FCC)] or hexagonal closest packing (HCP) arrangements as found in native metals (see Figure 3-2). In both arrays, each of the equally sized atoms is in contact with 12 other atoms.

The covalent structural arrangements are exemplified by the structures of diamond (Figure 8-12), sphalerite (Figure 9-6), and graphite (Figure 8-13). As shown, the atoms may be of one or more kind, and covalent bonds link each atom to its nearest neighbor. Note that the diamond and sphalerite structures involve tetrahedral coordination. That structure can, however, be viewed alternatively as consisting of two interpenetrating face-centered cubic lattices. The graphite structure is quite different; it consists of covalent bonded sheets held together by van der Waals attraction.

The ionic groupings of the oxides, hydroxides, and halides can be looked upon as essentially CCP and HCP arrangements of their anions with interstitial cations arranged in appropriate loci such that there is electrostatic balance and no "rattle"; that is, the cations are large enough to be

Table 3-3. The general classification of minerals

| | BASIS | |
| | Crystal Structure | |
Chemical Composition	Bond Type	Geometry
Native elements: metals	Metallic	Cubic or
semi-metals	Covalent	Hexagonal
nonmetals	Covalent	Close-packed
Sulfides (etc.)		
(and sulfosalts)	Metallic, covalent, and ionic	Diverse
Oxides		
(and hydroxides)	Ionic	Simple packing of spheres
Halides	Ionic	Simple packing of spheres
Carbonates		
(and nitrates, borates and iodates)	Ionic	Packing involving complex ions: $(XO_3)^-$
Sulfates		
(and chromates, molybdates, tungstates, etc.)	Ionic	$(XO_4)^-$
Phosphates		
(and arsenates, vanadates, etc.)	Ionic	$(XO_4)^-$
Silicates	Ionic	Complex groups (see Figures 15-2, 3, 4, & 5)

tangential or nearly tangential to their coordinating anions.

The ionic groupings involving complex anions (radicals) amount to regular arrangements in which the complex anions tend to occupy typical anion positions, while the cations again occur within appropriate open spaces. Most of the common radicals are tetrahedra [e.g., $(SO_4)^{-2}$, $(PO_4)^{-3}$, $(AsO_4)^{-3}$, $(WO_4)^{-2}$, and $(SiO_4)^{-4}$]; $(CO_3)^{-2}$ and $(NO_3)^{-1}$, however, are planar—that is, triangles. Although the ionic bonding in these compounds is combined with covalent bonding, the same electrostatic, space, and energy requirements as those in "simple" ionic substances prevail. For example, calcite has the halite (NaCl) structure with Ca^{+2} occupying positions analogous to those occupied by Na^{+1} and with $(CO_3)^{-2}$ in positions corresponding to those of Cl^{-1}.

One scheme rather frequently used to codify the diverse ionic arrangements (other than those of the silicates) designates the anions as X and the cations as A (or as A and B if two different cations/positions are present). Thence, the compounds are classified as, for example, AX, AX_2, or AB_2X_4. In addition, a number of schemes have been suggested for using meaningful superscripts. As already noted, one scheme that has gained rather wide acceptance gives the coordination number of the cation by Roman numerals; for example, $A^{VIII}B^{III}X_3$ indicates the A ion to have octahedral coordination and the B ion to have trigonal (planar) coordination. [This designation is, by the way, correct for minerals with the calcite structure.]

The silicates, most of which may be considered to consist predominantly of repetitively linked (polymerized) $(SiO_4)^{-4}$ tetrahedra, are classified according to the arrangements of their tetrahedra as shown on Table 15-1 and illustrated in Figure 15-1. This classification is described in the introduction to Chapter 15.

ISOSTRUCTURALISM, ISOTYPISM, AND ISOMORPHISM

These terms, generally used in their adjectival forms (isostructural, isotypic, and isomorphic) when applied to minerals, have been used in the literature in a number of different, and therefore frequently confusing, ways. We define and use them as follows:

Isostructural: Said of two or more minerals, or other crystalline substances, that have the same structure but different chemical compositions. That is to say, isostructural minerals have a one-for-one equivalence in the structural arrangement of their constituent atoms. Solid solution (see the following section) between isostructural minerals is neither required nor precluded. Examples of isostructural pairs of minerals are halite ($NaCl$) and galena (PbS), berlinite ($AlPO_4$) and quartz ($SiO_2 = SiSiO_4$), and tantalite ($FeTa_2O_6$) and brookite ($TiO_2 = TiTi_2O_6$).

Isotypic: Said of two or more minerals, or other crystalline substances, that have analogous structures. That is to say, isotypic minerals have closely related structures but do not exhibit a one-for-one equivalence of site occupancy. There are no implications with regard to chemistry. An example of an isotypic pair of minerals is halite ($NaCl$) and calcite ($CaCO_3$).

Isomorphic: Said of two or more minerals, or other crystalline substances, that have the same or similar crystal forms. This definition is the way Mitscherlich introduced the term in 1819. By implication, however, he indicated that such substances should have analogous formulas as well as crystallographic structure; that is, he suggested the term on the basis of crystals of KH_2PO_4, KH_2AsO_4, $(NH_4)H_2PO_4$, and $(NH_4)H_2AsO_4$, which he found to exhibit the same crystal forms, the same or similar interfacial angles between corresponding faces, and the same cleavage tendencies. Many isomorphic groups, such as the spinel group of minerals, are recognized.

The confusion to which we just alluded is based on the fact that some people have treated all, or some, of these terms (plus the term *solid solution*) as synonymous. We think that the terms should be clearly distinguished.

ATOMIC SUBSTITUTION AND SOLID SOLUTION

Many minerals are variable in their compositions. In fact, substitution of one element by another is the rule rather than the exception. When this phenomenon was first observed, it was described in terms of *solid solution* or *mixed crystals,* terms that implied the presence of molecules of two or more substances within a single homogeneous crystal. For example, common olivine was described as a solid solution of forsterite, Mg_2SiO_4 (Fo), and fayalite, Fe_2SiO_4 (Fa), so the precise composition of any specimen of olivine could be stated in terms of these components, as say Fo_{85}, with stands for $(Mg_{0.85}Fe_{0.15})SiO_4$.

This concept and terminology remain in general use, even though the light thrown on the structure of crystals by X-ray investigations has resulted in a revised interpretation: in an ionic structure, there are no molecules; rather, the structure is an infinitely extended three-dimensional network, and ions within the structure may, with restrictions described in a subsequent paragraph, be replaced by others without causing serious distortion of the structure. Consequently, because minerals usually crystallize from solutions that contain many ions in addition to those that are essential to any given mineral, some so-to-speak foreign ions are very likely to be incorporated into a mineral's structure. Thus, a solid solution (or mixed crystal) can be defined as follows:

Solid solution: A solid that has an homogeneous crystal structure in which some equivalent sites are occupied by different ions.

In solid solution by substitution, the size of the atoms or ions is a principal governing factor. It is not essential that the substituting ions have the same charge or valence as long as electrical neutrality is maintained by concomitant substitution elsewhere in the structure. Thus, in the albite ($NaAlSi_3O_8$)–anorthite ($CaAl_2Si_2O_8$) series, Ca^{+2} substitutes for Na^{+1}, and electrical neutrality is maintained by the coupled substitution of Al^{+3} for Si^{+4} and, similarly, in diopside ($CaMgSi_2O_6$), Mg^{+2}-Si^{+4} may be replaced in part by Al^{+3}-Al^{+3}. Such coupled substitutions are especially common in silicate minerals, and this fact made the interpretation of their compositions exceedingly difficult before the phenomenon was recognized and understood. Despite this kind of substitution, as a general rule, little or no atomic substitution takes place when the difference in charge on the ions is greater than 1, even if size is appropriate (e.g., Zr^{+4} does not substitute for Mn^{+2}, nor does U^{+4} replace Na^{+1}, and so on). The general lack of this kind of substitution is probably due, at least in part, to the difficulty in balancing the charge requirements by other substitutions. (Nonetheless, extensive solid solution not only can but commonly does occur between some substances that are not isostructural.)

The extent to which atomic substitution takes place is determined by the nature of the structure, the closeness of correspondence of the ionic radii, and the temperature of formation of the substance. The nature of the structure also appears to have considerable influence on the degree of atomic substitution; some structures, such as those of spinel and apatite, are well known for extensive atomic substitution, whereas others, such as that of quartz, show very little. To some extent, however, the latter is due to the lack of foreign ions of suitable size and charge. Ionic size has, of course, a fundamental influence on the extent of substitution, since the substituting ion must be able to occupy the lattice position without causing excessive distortion of the structure. From a study of many mixed crystals, it has been found that if the radii of substituting ions and the ions substituted for do not differ by more than 15 percent, a wide range of substitution can be expected at room temperature. Also, higher temperatures permit a somewhat greater tolerance (in this respect, solid solutions are analogous to aqueous solutions of salts, the solubility increasing with temperature). This property of increased atomic substitution at higher temperatures provides a means of estimating the temperature of mineral formation (geological thermometry). The point is that once the degree of atomic substitution has been determined for a given substance for different temperatures and pressures, the composition of its naturally occurring analog may be considered to indicate the temperature and pressure of formation. A number of examples of this kind of geothermometer are reported in the literature.

Since atomic substitution is generally greater at higher temperatures, it also follows that some solid solutions formed at high temperature may no longer be stable at lower temperatures. Therefore, a solid solution in which two different elements, A and B, are completely interexchangeable at high temperatures but not at lower temperatures may, upon cooling, tend to break down into two separate phases, one rich in A and the other rich in B. This kind of breakdown of a homogeneous solid solution is known as *exsolution*. An example is afforded by the alkali feldspars: at high temperatures, potassium and sodium are completely interchangeable—that is, for any composition in the system $KAlSi_3O_8$-$NaAlSi_3O_8$, there is a single-phase solid solution $(K,Na)AlSi_3O_8$; at ordinary temperatures, however, the degree of mutual replacement of Na and K is quite small. Thus, upon cooling, solid solutions of intermediate composition in this system generally break down into "intergrowths" of sodium-rich feldspar and potassium-rich feldspar; the mixtures are called perthite. [And the proportions of the two components provide still another geothermometer.]

A consequence of atomic substitution is that most minerals contain not only the elements characteristic of the particular species but also other elements that are able to fit into the crystal structure. For instance, dolomite is theoretically a simple carbonate of magnesium and calcium, but dolomites are found whose analyses show considerable iron and/or manganese contents. Traditionally, these dolomites were described as solid solutions of the carbonates of all these elements; today, it is more illuminating as well as more correct to consider them as products of the substitution of iron and manganese for magnesium. Nevertheless, we continue to use the traditional terms solid solution, mixed crystals, and solid solution series, since the terminology of atomic substitution has not yet provided expressions to take their place.

In interpreting chemical analyses of minerals and deriving their formulas, we must give due consideration to the effects of atomic substitution. In some minerals, such as olivine, substitution is comparatively simple and is readily elucidated by a study of analyses. The situation becomes progressively more complicated when substitution affects several elements and/or where coupled

substitution with elements of different valences takes place. Additional difficulty arises when one element can play dual structural roles and be present in two different coordinations, which is common in silicates where aluminum may be present in four-coordination replacing silicon and/or in six-coordination replacing cations such as Mg^2 and Fe^3. Interpretation of analyses of such minerals requires care and discrimination as well as a real understanding of the principles of crystal structure and the factors governing atomic substitution. The analysis of a pyroxene in Table 3-4 is a typical example. When this analysis is converted to molecular proportions in the usual way, there is no obvious relationship between the proportions of the different oxides. Conversion to atomic proportions and the arrangement of the elements in order of ionic size, however, permits grouping of the elements according to ionic size and coordination number. Aluminum evidently occurs in both fourfold and sixfold coordination so the amount in each is adjusted to give the best

Table 3-4. Interpretation of a pyroxene analysis

	Weight (percent)	Molecular Proportions	Atomic Proportions		Atomic Content of Unit Cell[b]	
SiO_2	48.40	0.806	Si^4	0.806	7.42	$8.00 = 4 \times 2$
Al_2O_3	3.95	0.039	Al^3	0.078	0.72 {0.58 / 0.14}	
TiO_2	0.27	0.004	Ti^4	0.004	0.04	$4.06 \sim 4 \times 1$
Fe_2O_3	3.90	0.024	Fe^3	0.048	0.44	
MgO	8.92	0.221	Mg^2	0.221	2.04	
FeO	10.52	0.146	Fe^2	0.146	1.34	
MnO	0.39	0.006	Mn^2	0.006	0.06	
Na_2O	0.46	0.007	Na^1	0.014	0.13	$3.94 \sim 4 \times 1$
CaO	23.20	0.414	Ca^2	0.414	3.81	
Total	100.01		O^{-2}	2.603[a]	23.96	$23.96 \sim 4 \times 6$

[a] $O = (2 \times 0.806) + (3 \times 0.039) + (2 \times 0.004) + (3 \times 0.024) + 0.221 + 0.146 + 0.006 + 0.007 + 0.414 = 2.603$

[b] Atomic proportions multiplied by $M/100 = VG/1.66 \times 100 = 9.2063$, where $V = 9.73 \times 8.91 \times 5.25 \times \sin 105°$ 50' and $G = 3.49$, giving $Z = 4$.

130 fit with the pyroxene formula, which for this specimen can be written

$$(Ca,Na)(Mg,Fe^{+2},Fe^{+3},Mn,Al,Ti)(Al,Si)_2O_6.$$

Thus, the specimen is intermediate in composition between diopside, $CaMgSi_2O_6$, and
hedenbergite, $CaFeSi_2O_6$, with minor substitutions of aluminum, ferric iron, manganese, titanium, and sodium. This characterization
should, of course, also correlate well with optical
and X-ray data for the mineral.

INTERSTITIAL AND DEFECT SOLID SOLUTION

Interstitial defects (see page 96) have sometimes
been termed interstitial solid solution, especially
those defects occurring in quantities that affect
the chemical analysis and, thus, the formula.
Several examples have been so described, for example, the hydrogen, carbon, boron, and nitrogen commonly occurring in metals and the sodium in cristobalite and tridymite. These and
other defects like them should not be referred to
in terms of ordinary substitutional solid solution.

The notable lack of stoichiometry in some
minerals was once described as reflecting solid
solution. For example, the presence of excess sulfur over FeS in analyses of pyrrhotite was once
attributed to the solid solution of sulfur in FeS.
Actually, the structure has a deficiency of Fe,
rather than an excess of S, and the deficiency of
Fe has been shown to be a manifestation of
ordered—that is, more or less regularly spaced—
vacancy defects. It is unfortunate that such phenomena as these have also been described by
some mineralogists under the unsuitable term defect solid solution.

POLYMORPHISM AND POLYTYPISM

An element or compound that can exist in more
than one crystallographic structure is said to exhibit polymorphism. Each polymorph has its own
physical properties and a distinct internal arrangement of its atoms and/or ions; that is, the
atoms and/or ions are arranged differently within
each of the different polymorphs of the same
chemical substance. Polymorphism is, therefore,
an expression of the fact that crystal structure is
not determined exclusively by chemical composition. Or, to state it otherwise, for some substances, the same atoms and/or ions in the same
proportions may assume more than one structure.

A polymorphic substance can be described
as dimorphic, trimorphic, etc., according to the
number of distinct crystalline forms it can have. A
simple illustration of dimorphism is the re-

Figure 3-4. Atomic arrangement in (a) diamond, (111) plane
horizontal; and (b) graphite,
(0001) plane horizontal. Note
the differences in the interatomic
distances in the two polymorphs;
the dashed lines in the graphite
structure indicate van der Waals
linkages, solid lines represent
covalent linkages. Small circles
indicate locations of centers of
carbon atoms, and bear no relation to their size.

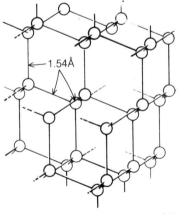

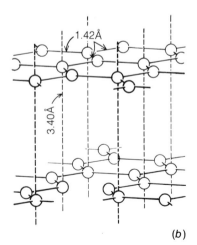

(a)

(b)

lationship between diamond and graphite (Figure 3-4). In diamond, each carbon atom is linked to four other carbon atoms by homopolar bonds, all the linkages being of equal strength, so that each crystal, as a whole, is a giant molecule; in graphite, each carbon atom is linked to three other carbon atoms by homopolar bonds, thus forming planar sheets, which, in turn, are joined by weak van der Waals forces.

Different polymorphs of the same substance are formed under different conditions of pressure, temperature, and/or chemical environment; hence the presence of a given polymorph in a rock will often tell us something about the conditions under which the rock was formed (e.g., see Figures 3-5 and 5-1). Notice, for instance, that the natural occurrence of diamond in igneous rocks implies an origin at considerable depths in the earth, where the combination of temperature and pressure is within the diamond stability field. Thus, it can be seen that diamond is actually unstable under the physical conditions in which it is found. The fact that it has not changed spontaneously into graphite is due to the infinitesimal rate of reaction that the energy relations nevertheless favor.

The change from one polymorph to another is generally termed an *inversion* or transformation. The rate of change can be essentially instantaneous or extremely slow, depending largely on the degree of reconstitution of the structure that is required. Two kinds of inversion, termed *displacive* and *reconstructive,* are recognized (Figure 3-6).

Displacive inversions are immediate and reversible at the transition temperature and pressure. Metastability is precluded; that is, a high-temperature form cannot be preserved. Some such changes do not involve the breaking of bonds between neighboring atoms or ions, but simply their bending, for example, low-quartz⇌high-quartz and low-leucite⇌high-leucite. Nonetheless, original crystallization as the high-temperature form can frequently be recognized from the nature of the crystals or from the twinning that so often results from inversions of

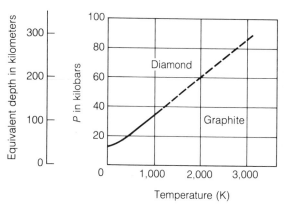

Figure 3-5. Diamond-graphite equilibrium curve, calculated to 1200 K, extrapolated beyond.

this kind. High-low polymorphs are characterized by the fact that the high-temperature form has higher symmetry than the corresponding low-temperature form.

Reconstructive inversions are sluggish. Consequently, some high-temperature polymorphs are quenchable as metastable phases because these transformations require the breaking of bonds in the structure and the rearrangement of the atomic or ionic linkages. Some of these transformations even require the presence of a solvent in order to obtain any appreciable rate of change. Reconstructed transformations are exemplified by the quartz⇌tridymite⇌cristobalite inversions.

In general, the high-temperature polymorph of a substance has a more open structure, and thus a lower density, than a low-temperature form. The open character of the structure is dynamically maintained at high temperatures by thermal agitations. It can also be statically maintained by the incorporation of foreign ions into the interstices of the structure. These foreign ions will often buttress the structure and prevent its transformation to a different polymorph when the temperature is lowered. Their removal may be necessary to permit inversion to the close-packed form that is stable at low temperatures. Some impure high-temperature polymorphs may be formed far below the normal stability range of

(a)   (b)   (c)

Figure 3-6.   Inversions between polymorphic forms. (a), (b), and (c) represent three different crystal structures that involve identical building blocks, which are indicated by the squares. Reconstructive inversions, such as those that might occur between (a) and (b) [or between (a) and (c)] involve rearrangements, breaking of original bonds, and formation of new bonds; hence, they are sluggish. Displacive inversions, such as the one that might occur between (b) and (c), generally involve only slight relative positional changes; hence, they are essentially instantaneous.

pure compounds and may survive indefinitely—a situation that is likely to arise in nature. (This phenomenon may be responsible for the formation and survival of cristobalite and tridymite under conditions in which the stable form of SiO_2 is quartz.) The occurrence of a high-temperature polymorph at ordinary temperatures does not, therefore, definitely indicate metastability.

Another interesting transformation that is sometimes considered to be a kind of polymorphism is termed *order–disorder*. Essentially, any degree of order–disorder may be quenched in. This phenomenon has been studied in alloys, in particular, because it has important effects on their physical properties; it is also common in minerals.

A simple example is an alloy of 50 percent Cu and 50 percent Zn. Two distinct phases of this alloy exist; in the disordered form, the copper atoms and the zinc atoms are randomly distributed over the lattice positions, whereas in the ordered form each element occupies a specific set of positions (Figure 3-7). The structures of the two forms are related, but the ordered one has lower symmetry than the disordered one. There is no definite transition point between the two forms. Theoretically, perfect order will be

achieved only at absolute zero, and, with increasing temperature, the degree of order gradually decreases to complete disorder above a certain temperature that is characteristic of the structure and composition of the substance. The relationship between microcline and orthoclase is one of order–disorder, the one aluminum atom and the three silicon atoms in $KAlSi_3O_8$ being disordered in orthoclase but ordered in microcline. The degree of disorder, which can be measured by X-ray diffraction, is temperature dependent, so temperatures of formation can be inferred from X-ray diffraction patterns of these minerals.

Laboratory investigations have shown that many substances show polymorphism if the conditions of temperature and pressure are sufficiently varied. Developments in equipment capable of withstanding both high temperatures and high pressures have led to the discovery of many previously unknown polymorphs. Along this line, it is interesting that the polymorph of SiO_2 called coesite was made in the laboratory before it was found in nature. Its high specific gravity (= 3.01) suggested that it requires extremely high pressures for formation. Subsequently, natural coesite has been found where such conditions existed—for example, in areas of severe impact where high velocity meteorites have collided with the earth. Thus, we see that such discoveries can be of geological significance. Today, many investigations at high pressures and temperatures, such as the ongoing research at the Geophysical Laboratory of the Carnegie Institution, are directed toward solving some of the problems relating to conditions and processes of the earth's interior.

Polytypism is another phenomenon that can be considered a variety of polymorphism. It involves stacking of identical layers in different sequences within a structure. As a result, polytypes have the same unit cell lengths in two dimensions but commonly have a different cell length in the third dimension, the one essentially perpendicular to the layers. Perhaps the best-known examples of polytypism among minerals are sphalerite–wurtzite, the micas, and some of the clay minerals. Each exhibits two or more well-known stacking sequences (see, for example, Figure 9-5). Polytypes are generally rather readily distinguished on the basis of their X-ray patterns.

PSEUDOMORPHISM

A mineral can be replaced by another mineral without any change in the external form. Such replacements are called *pseudomorphs,* and the

phenomenon is known as *pseudomorphism.* There are two main types: one in which no change of substance occurs; the other in which there is addition of some element or elements and removal of others.

The first type is that observed when one polymorph changes to another without change in external form. This specific type of pseudomorphism is known as *paramorphism,* and the replacing form is a *paramorph* of the replaced form (e.g., paramorphs of calcite after aragonite).

The second type—pseudomorphs in which the later mineral has been formed from the original mineral by a process of chemical change—may originate by: (1) the loss of a constituent (e.g., native copper after cuprite or azurite); (2) the gain of a constituent (e.g., gypsum after anhydrite and malachite after cuprite); (3) a partial exchange of constituents [e.g., goethite (limonite) after pyrite]; or (4) a complete exchange of constituents (e.g., quartz after fluorite).

The formation of a pseudomorph implies, of course, that the original mineral was no longer stable under changed physical and chemical conditions and thus was replaced by another mineral more suited to those conditions. The study of pseudomorphs can therefore provide valuable evidence toward deciphering the geological history of the rock containing them. They may, for example, indicate the nature and composition of circulating solutions that added or subtracted cer-

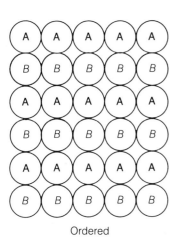

Ordered

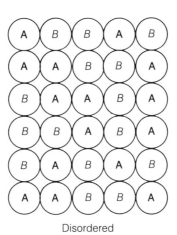

Disordered

Figure 3-7. Theoretical representation in one plane of order-disorder in a binary compound A*B*.

134   tain elements. Or, if the stability fields of the original mineral and of the pseudomorph are known, it may be possible to estimate the temperature and pressure conditions under which the change took place.

NONCRYSTALLINE MINERALS

Most definitions of the term *mineral* include a statement that implies that it is a crystalline solid. There are, however, a few naturally occurring solids that are not crystalline that are, nevertheless, of interest to mineralogists, examples are *metamict* minerals and some of the *amorphous* materials such as natural glasses and certain gel products.

Metamict minerals can be considered to be noncrystalline pseudomorphs after originally crystalline material: that is, these "minerals" were originally formed as crystalline compounds, but their crystalline structures have subsequently been destroyed. Metamict minerals are optically isotropic and do not diffract X-rays, which indicates an amorphous condition. They show no cleavage, are glassy or pitchy in appearance, and typically have a conchoidal fracture. On heating, they recrystallize, often with the evolution of so much heat that they become incandescent and glow brightly; the recrystallization is accompanied by an increase in density. Metamict minerals are radioactive from the presence of uranium and/or thorium, although the amount may be rather small (1 percent or less).

The development of the metamict condition in minerals involves the breakdown of crystal structure through bombardment by alpha particles ejected from disintegrating radioactive elements. It has been found, in fact, that strong alpha-particle and neutron bombardment in an atomic pile will render many crystalline substances metamict. Nonetheless, the presence of radioactive elements alone appears to be insufficient to induce the metamict state; for example, thorianite (ThO_2) is apparently never metamict and some minerals, such as allanite, may be either metamict or nonmetamict.

Natural glasses, which represent quenched rock or mineral melts, are described, along with macerals, in Appendix A.

Gels are formed when colloidal suspensions are flocculated. Colloidal suspensions are intermediate between true solutions and suspensions. Organic compounds with large molecules often form colloidal suspensions; inorganic compounds that are ordinarily insoluble in natural solutions may also form such suspensions. The diameter of the particles in a colloidal suspension ranges from approximately 10^{-3} to 10^{-6} millimeters. Some of the particles are crystalline, others are amorphous. The gels formed from colloidal suspensions may lose some or all of their water and thus form solids such as opal.

Naturally occurring substances that may form colloidal suspensions include the clay minerals (hydrated aluminum silicates) and hydrated iron, manganese, aluminum, and silicon oxides. In some cases, these substances retain their crystallinity even in the colloidal state. Therefore, colloidal suspensions may or may not yield gels upon flocculation. Even those that do yield gels generally transform into crystalline material within a comparatively short time. Some minerals that originally solidified as gels may be recognized by their habit; a relatively common manifestation consists of internal radiating fibers perpendicular to colloform external surfaces.

SELECTED READINGS

Bloss, F. D. 1971. *Crystallography and Crystal Chemistry*. New York: Holt, Rinehart and Winston, 545 pp.

Bragg, W. L., and Claringbull, G. F. 1965. *Crystal Structures of Minerals*. Ithaca, N.Y.: Cornell University Press, 409 pp.

Bunn, C. W. 1961. *Chemical Crystallography* (2nd ed.). Clarendon (England): Oxford, 509 pp.

Companion, A. L. 1979. *Chemical Bonding* (2nd ed.). New York: McGraw-Hill, 179 pp.

Cotton, F. A., and Wilkinson, Geoffrey. 1980. *Advanced Inorganic Chemistry* (4th ed.).

New York: Interscience (Wiley), 1396 pp.

Evans, R. C. 1964. *An Introduction to Crystal Chemistry* (2nd ed.). Cambridge (England): Cambridge University Press, 410 pp.

Fyfe, W. S. 1964. *Geochemistry of Solids.* New York: McGraw-Hill, 199 pp.

Huheey, J. E. 1978. *Inorganic Chemistry* (2nd ed.). New York: Harper & Row, 889 pp.

Mandarino, J. A. 1981. "Comments on the calculation of the density of minerals." *Canadian Mineralogist* **19,**531–534.

Pauling, Linus. 1960. *The Nature of the Chemical Bond* (3rd ed.). Ithaca, N.Y.: Cornell University Press, 644 pp.

Shannon, R. D. 1976. "Revised effective ionic radii and systematic studies of interatomic distances in halides and chalcogenides." *Acta Crystallographica,* **32A,**751–767.

Shannon, R. D., and Prewitt, C. T. 1969. "Effective ionic radii in oxides and fluorides." *Acta Crystallographica,* **25,**925–946.

4 The Physical Properties of Minerals

A close connection exists between the physical properties of a mineral, its crystal structure, and its chemical composition. Thus, the study of physical properties may enable you to make deductions about crystal structure and about chemical composition. In addition, physical properties may be of great technological significance because a mineral may have important industrial uses that depend on its physical properties. Two examples are the extreme hardness of diamond, which makes it a highly efficient abrasive, and the piezoelectric nature of quartz, which is the basis for its use in electronic equipment. Furthermore, physical properties are of great practical significance to, for example, the geologist or the prospector, in that they afford readily determined characteristics for mineral identification; several physical properties, which are more quickly and more easily determined than crystal structure or chemical composition, are uniquely diagnostic for certain minerals.

In this chapter, we describe many of the physical properties of minerals and how they correlate with a mineral's chemical composition and structure. A few of the uses that are directly dependent on physical properties are mentioned by way of illustration. Methods of determining physical properties are, for the most part, covered in Chapter 6, Determinative Mineralogy.

SPECIFIC GRAVITY

Specific gravity (*G* or *SG*) is a dimensionless number that can be defined as *the ratio of the density of a material to the density of water*; that is, it is the number of times that a given volume of a substance is heavier (or lighter) than an equal volume of water (strictly speaking, at 4°C). Unfortunately, the term *density*, which is specific and can be defined as the *mass per unit volume* (e.g., grams per cubic centimeter), is frequently used synonymously with specific gravity. One of the reasons for this confusion in terminology is the fact that water has a density of 1 gram per cubic centimeter (g cm^{-3}) and thus specific gravity is numerically the same as the value of density. We shall use the two terms as just defined.

The specific gravity of a substance is primarily dependent on a substance's chemical composition and crystal structure—that is, by the kinds of atoms/ions present and the way they are packed and bonded. To generalize: the heavier the atoms/ions, the higher the specific gravity; the closer the packing, the higher the specific gravity; and the stronger the bonding, the higher the specific gravity. Specific gravity also varies somewhat with varying temperature and pressure because changes in these conditions generally cause expansion or contraction. Therefore, the specific gravity of a pure substance with a fixed chemical composition and crystallizing with a specific structure should be constant at any stated temperature and pressure; and, indeed, careful measurements have shown this to be true. As an example, quartz, which is practically invariant in composition, has a constant specific gravity of 2.651 at ordinary temperature and pressure.

For a substance of variable composition crystallizing in a specific structure, the variation in specific gravity will depend almost completely on the mass of the individual atoms. For example, the specific gravity of olivine $(Mg,Fe)_2SiO_4$ increases with increasing replacement of the lighter magnesium atoms by the heavier iron atoms from 3.22 for Mg_2SiO_4 to 4.41 for pure Fe_2SiO_4 (see Figure 3-1). Similarly, in a group of isostructural compounds, the specific gravity will show a direct relationship to the mass of the atoms present, as you can see in the following tabulation for the aragonite group:

Mineral	Composition	Specific Gravity	Atomic Weight of Cation
Aragonite	$CaCO_3$	2.93	40.08
Strontianite	$SrCO_3$	3.78	87.63
Witherite	$BaCO_3$	4.31	137.36
Cerussite	$PbCO_3$	6.58	207.21

In other words, the specific gravity (and density) of a substance can be said to correlate its chemistry and its crystallography and, thus, to reflect the nature of the atoms in the structure and the manner in which they are packed. Therefore, if the dimensions of a unit cell have been measured and the number and kinds of atoms in the unit cell are known, it is possible to calculate the density and thus the specific gravity. The procedure was described in Chapter 3 where it was used to determine the number of formula units (Z) within a unit cell from the substance's cell dimensions, formula, and measured density.

A further application of X-ray crystallography to density (and to other dependent physical properties) is in calculating the factor known as the packing index.

Packing index: The percentage of the total volume of a substance that is occupied by atoms/ions.

It is derived from crystal structure data in the following way:

$$\text{packing index} = \frac{\text{volume of ions}}{\text{volume of unit cell}} \times 10$$

The concept of a packing index assumes that ions behave like spheres that support each other in a crystal structure and that, even with the same ions in the same numbers, different modes of packing, which have different volume requirements, are possible. For ionic compounds, packing indices range between about 3 and 7; that is, in actual crystal structures, between 30 percent and 70 percent of the volume is occupied by the atoms. The relationship between packing index and density is readily seen in the differences between polymorphs of individual substances (Table 4-1).

The temperature and pressure conditions of formation of a mineral control its polymorphic form and thus the manner in which its constituent atoms/ions are packed. In general, conditions of high pressure favor formation of polymorphic forms with high densities whereas conditions of high temperature favor looser packing and for-

Table 4-1. Relationship between packing index and specific gravity in polymorphs

		Specific Gravity	Packing Index
TiO_2	rutile	4.25	6.6
	brookite	4.14	6.4
	anatase	3.90	6.3
Al_2SiO_5	kyanite	3.63	7.0
	sillimanite	3.24	6.2
	andalusite	3.15	6.0

mation of polymorphic forms with relatively low densities (see Figure 5-1).

Methods commonly used for determining the specific gravity of a mineral are outlined in Chapter 6. Among those methods, the matching of the specific gravity of a mineral with that of a heavy liquid is generally the most accurate.

Another application of heavy liquids in mineralogy is their use for the separation of individual minerals or groups of minerals from mixtures. This technique is important in sedimentary petrology, since the so-called heavy minerals in a sediment—that is, those with density greater than the common minerals quartz, the feldspars, calcite, and dolomite—may provide valuable evidence as to the source of the sediments and the conditions under which they were deposited.

Separation of different minerals by differences in density is also an important ore dressing technique for preparing concentrates of certain valuable minerals. Sometimes heavy liquids are employed—not the heavy liquids generally used in research, which are far too expensive for large-scale use, but heavy media that are suspensions of extremely finely ground heavy minerals (such as magnetite or galena) in water. More often, however, mechanical devices, such as vibrating tables, are used to separate mixtures of minerals of different specific gravities. Also, as is evident, recovery by gold panning and "jigs" is based on differences in specific gravity.

OPTICAL PROPERTIES

The optical properties of minerals comprise a wide variety of phenomena—reflection and refraction, luster, diaphaneity, color and streak, luminescence—each of which is dealt with in a summary fashion in this chapter. Additional information is given, especially with regard to optical mineralogical methods, in Chapter 6.

Reflection and Refraction

When a ray of light in air impinges obliquely on the surface of a nonopaque solid, part of the light is reflected back into the air (the reflected ray) and part enters the solid (the refracted ray) (Figure 4-1). The direction of the reflected ray is governed by the Law of Reflection, which states that the angle of reflection r^1 is equal to the angle of incidence i and that the reflected and incident rays lie in the same plane. The light that passes into the solid is known as the refracted ray since its path is bent or refracted from the path of the incident ray. The relationship between the paths of the incident and refracted rays is known as the Law of Refraction, or Snell's Law, having been discovered in about 1621 by Willebrod Snellius

Figure 4-1. Reflection and refraction of a light ray at an interface; i = angle of incidence, r = angle of refraction, r' = angle of reflection.

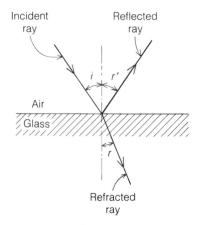

(1591–1626), professor of mathematics at Leyden in Holland. It states that for any given substance, the index of refraction n can be defined as follows:

Index of refraction n: the ratio of the sine of the angle of incidence i to the sine of the angle of refraction r:

$$n = \frac{\sin i}{\sin r}.$$

It was later proved that the index of refraction is also the ratio of the velocity of light in air to the velocity of light in the solid, so that if V is the velocity of light in air, and v the velocity in the solid, then

$$n = \frac{V}{v}.$$

The velocity of light in air is 300,000 km sec^{-1}; if the velocity of light in a substance is 200,000 km sec^{-1}, the refractive index is 300,000/200,000, or 1.5. Most solids have refractive indices of between 1.4 and 2.0.

The refractive index of a substance is related to its chemical composition and crystal structure, just as its density is; in fact, specific gravity (G) and refractive index (n) may be correlated by the following approximate equation

$$\frac{n - 1}{G} = K$$

where K is a constant related to the composition of the substance.

There is also a close connection between optical properties and the crystal structure of a solid. In isometric and noncrystalline substances, the velocity of light is the same in all directions, and hence the refractive index is the same for all directions; such substances are said to be optically *isotropic*. In all other substances, the velocity of light varies according to its direction of vibration; such substances are said to be optically *anisotropic*. A ray of light entering an anisotropic substance is split into two rays vibrating at right angles to each other, generally traveling with different velocities and thus having different refractive indices. The difference in refractive indices, known as the birefringence, is typically quite small (e.g., for quartz it is 0.009) and essentially unobservable except by instruments. In some minerals, however, the difference is relatively large (e.g., for calcite it is 0.172), and thus a single spot observed through a cleavage fragment of any of these minerals appears doubled.

The relationship between refractive indices and crystallography can be visualized best by extending lines in all directions from the center of a crystal or a crystalline grain with the length of each line proportional to the refractive index for that vibration direction. The resulting figure is known as the indicatrix (Figure 4-2). For non-crystalline and isometric substances, the form of the indicatrix is a sphere since the refractive index is the same in all directions. For substances crystallizing in the tetragonal and hexagonal systems, the indicatrix has the form of a rotation ellipsoid in which all sections perpendicular to one axis are circular, this axis coinciding with the c axis of the crystal. This form results from the fact that all rays traveling in the direction of the c axis have the same velocity since their vibrations are in the plane of the horizontal axes, which are equivalent in these systems. For this reason, substances crystallizing in the tetragonal and hexagonal systems are said to be *uniaxial*. For substances crystallizing in the orthorhombic, monoclinic, and triclinic systems, the indicatrix has lower symmetry, in agreement with the lower crystallographic symmetry; it is a triaxial ellipsoid. A property of such an ellipsoid is that it has only two circular sections, all others being ellipses; rays traveling at right angles to the circular sections have the same velocity no matter what the direction of vibration in these sections. The two directions at right angles to the circular sections are known as the optic axes. For this reason, orthorhombic, monoclinic, and triclinic substances are said to be optically *biaxial*.

The orientation of the indicatrix in a crystal is related to the crystallographic symmetry. In tri-

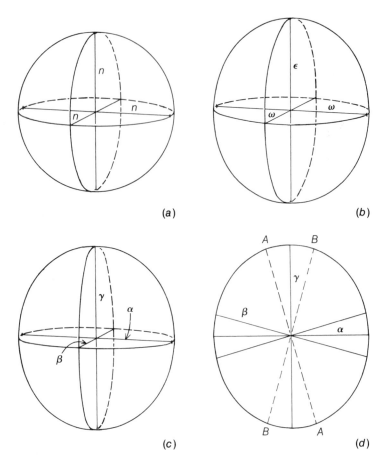

Figure 4-2. The optical indicatrix for (a) isotropic, (b) uniaxial, (c) biaxial substances. In (a), the indicatrix is a sphere whose radius is proportional to n, the refractive index of the substance. In (b) the indicatrix is a rotation ellipsoid, the horizontal equatorial section being a circle with radius proportional to ω, one of the principle refractive indices, and with the vertical axis proportional to ϵ, the other principal refractive index; ϵ may be greater or less than ω, and its vibration direction is always parallel to the c crystallographic axis. In (c), the indicatrix is a triaxial ellipsoid, the lengths of the principal axes being proportional to α, the smallest index of refraction in the substance, γ, the greatest index, and β, the intermediate index. In (d) is represented the $\alpha\gamma$ section of the ellipsoid; AA and BB are the optic axes, at right angles to the two circular sections of radius β.

clinic substances, the positions of the three principal axes of the indicatrix are independent of the directions of the crystallographic axes. In monoclinic substances, one axis of the indicatrix coincides with b, the axis of symmetry, whereas the other two axes of the indicatrix are in the ac plane but independent of a and c. In orthorhombic substances, the three axes of the indicatrix coincide in position with the three crystallographic axes. In hexagonal and tetragonal substances, the unique axis of the indicatrix coincides with crystallographic c, and the circular sections perpendicular to the unique axis contain the crystallographic a axes.

Because the optical properties of a substance are so closely related to its crystallographic symmetry, optical studies have been used widely to

determine the crystal system of minerals that do not occur in well-formed crystals. Optical properties, especially refractive indices, are among the most valuable determinative properties for mineral identification. For this reason, several of the simpler optical mineralogical techniques are outlined in Chapter 6, and certain optical properties of minerals are given in Part II and tabulated in Part III.

Luster

Luster, an optical property closely related to reflection and refraction, can be defined as *the appearance of a material in reflected light*. Two classes of luster are prevalent and recognized in minerals—those termed *metallic* and those termed *nonmetallic*. Strictly speaking, however, no sharp division can be made between the two classes, and minerals that appear to be neither, or either, are usually said to be *submetallic*.

The impression of luster is produced by the amount and nature of light reflected from the surface of a mineral. It depends largely on the character of the surface and the quantity of reflected light. It is quite apparent that a smooth cleavage surface will reflect more light than an uneven fracture will, even if both are on the same mineral; in fact, every detail of surface configuration is important. It is also quite obvious that different crystallographic directions, especially in anisotropic minerals, are likely to absorb different percentages of light and consequently reflect different amounts of incident light; thus, even different surfaces of individual specimens of a single mineral may have different lusters.

In general, the greater a mineral's refractive index, the higher the luster. This relationship has been described in different ways (compare the following descriptions with the values given in Table 4-2).

Metallic luster: Minerals that absorb visible radiation strongly, being opaque or nearly opaque even in very thin fragments (although they may be transparent to infrared radiation), generally have metallic luster. Their refractive indices are 3 or greater. The native metals and most of the sulfides are in this group.

Submetallic luster: Minerals with refractive indices between 2.6 and 3, most of them being nearly opaque or opaque, generally have submetallic luster. Examples are cuprite ($n = 2.85$), cinnabar ($n = 2.9$), and hematite ($n = 3.0$).

Nonmetallic luster (several varieties of which are recognized):

Adamantine luster: The brilliant luster typical of a diamond. It is characteristic of minerals with refractive indices between 1.9 and 2.6. Examples are zircon ($n = 1.92–1.96$), cassiterite ($n = 1.99–2.09$), sulfur ($n = 2.4$), sphalerite ($n = 2.4$), diamond ($n = 2.45$), and rutile ($n = 2.6$). The combination of a yellow or brown color with refractive indices in this range produces a *resinous* luster, a luster like that of resin.

Vitreous luster: The luster of glass. It is characteristic of minerals with refractive indices between 1.3 and 1.9. This range includes about 70 percent of all minerals, comprising nearly all the silicates, most other oxysalts (carbonates, phosphates, sulfates, etc.), the halides, and oxides and hydroxides of the lighter elements such as Al and Mg.

Greasy, waxy, silky, pearly, and dull lusters: Variants of nonmetallic luster, caused by the character of the reflecting surface. Diamonds often have a somewhat greasy luster, evidently as the result of a microscopically rough surface that scatters the reflected light. Cleavage surfaces of halite have a vitreous luster when fresh, but they take on a greasy or waxy appearance after exposure to damp air, which produces a slightly roughened surface. The greasy luster common in nepheline is due to a beginning alteration. Cryptocrystalline and amorphous minerals, such as chalcedony and opal, commonly have a waxy luster. Minerals occurring in

Table 4-2. *Luster and index of refraction (according to Povarennykh, 1964*)*

Luster	Index of Refraction	Example (mean index)
subvitreous	1.3–1.5	fluorite (1.434)
vitreous	1.5–1.8	topaz (1.620)
subadamantine	1.8–2.2	zircon (1.964)
adamantine	2.2–2.7	diamond (2.417)
adamantine-splendent	2.7–3.4	cinnabar (3.093)

*For opaque to subtranslucent minerals, Povarennykh also uses the following terms and examples: submetallic, ilmenite; metallic, molybdenite; and metallic-splendent, gold.

142

parallel-fibrous aggregates, such as asbestos and some varieties of gypsum, are said to have silky luster. Transparent minerals with layer-lattice structures and accompanying perfect lamellar cleavage have characteristically pearly luster produced by reflection from successive cleavage surfaces; examples are talc, the micas, and coarsely crystallized gypsum. Porous aggregates of a mineral, such as the clays, scatter incident light so completely that they seem to be without luster and are described as dull or earthy.

It can be shown that a mineral's refractive index and light absorption, and thus its reflectivity (luster), can be roughly correlated with its predominant bonding as follows: the indices of refraction are high, moderately high, and low for materials with predominantly metallic, covalent, and ionic bonding, respectively; the absorption of light is high, moderately low, and low for materials with these bonds, also respectively; consequently, the luster tends to be high, moderate,

Figure 4-3. The electromagnetic spectrum (to right), with colors of the visible light portion shown on the expanded section (to left).

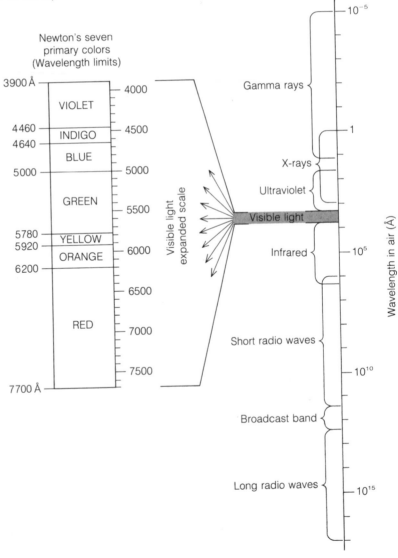

and low for materials with these kinds of bonds, also respectively.

Luster can also be correlated roughly with color and light transmission in that most minerals that have dark colored streaks and/or are opaque have metallic or submetallic lusters; most minerals that have white or light colored streaks and/or are transparent have nonmetallic lusters. Within the major classes, however, luster appears to be independent of color.

The luster of minerals has an economic aspect, as is evidenced by the use of minerals as gemstones. The qualities of beauty attributed to several gemstones include luster as well as color and transparency. Luster, in fact, is largely responsible for the brilliance of a gemstone so that, other things being equal, the higher the luster (and refractive index) of a gemstone, the greater its brilliance and its beauty.

Light Transmission (Diaphaneity)

The degree to which a material transmits light is generally termed *diaphaneity*. A few people incorrectly call this property opacity.

As you can see in small grains, or along the thin edges of large grains, nearly all minerals are either transparent or opaque. Many minerals, however, cannot be seen to be transparent in any but extremely small grains, although they do transmit light; they are generally termed *translucent*. The widely subscribed-to definitions of these three adjectives are as follows:

Transparent: Said of materials capable of transmitting light and through which an object can be seen clearly.

Translucent: Said of materials capable of transmitting light but through which an object cannot be discerned, except possibly in outline and generally distorted. (Most translucency depends on either absorption of light because of a material's dark color or scattering of light within the material.)

Opaque: Said of materials that are incapable of transmitting light.

Several other terms are also used to describe diaphaneity; for example, semitransparent (etc.), subtranslucent (etc.), and transopaque. In addition, it warrants mention here that the term *opaque* is sometimes applied to energy other than that of the visible spectrum.

Diaphaneity is correlative with luster and consequently with, for example, predominant bond type. In general, materials with metallic lusters are opaque, whereas those with nonmetallic lusters are not.

Color and Streak

In most minerals, the impression of color is produced by the selective absorption of some of the wavelengths that constitute white light, the resultant color being, in effect, white light minus the absorbed wavelengths. When white light, which consists of all of the colors in the visible part of the electromagnetic spectrum (Figure 4-3), strikes the surface of a crystalline material, some of the wavelengths are transmitted and reflected while others are absorbed. When nearly all wavelengths are transmitted, the substance appears colorless or white; when nearly all wavelengths are absorbed, it appears dark or nearly black; when a relatively narrow band or mixture of relatively narrow bands of wavelengths are transmitted (the others being absorbed), some color is perceived by the normal eye—that is, the color perceived is complementary to the color absorbed. Two examples are the green of an emerald, which represents transmission in the 5000–5500 Å range, and the purplish-red of a fine quality ruby, which represents transmission of a mixture of wavelengths in the red-orange (~6000–7000 Å) and blue (~4400–4800 Å) ranges.

Strictly speaking, however, as Nassau (1978, p. 219) has stated: "Specific causes for color in the majority of minerals are in fact not known, and detailed investigations, often requiring pre-

144 cision spectroscopy, impurity analysis down to ppm levels, magnetic resonance, and sometimes radiation and even laboratory synthetic procedures may be necessary for unambiguous conclusions." We consider several of the complexities related to color of minerals to be beyond the scope of this book. The brief treatment, which follows, is based largely on Nassau's just mentioned article.

The causes of color in minerals, diverse as they are, all involve electronic processes, the energies of which correspond to the wavelengths of light. Most of the processes require external stimulation—for example, impinging light—and involve changes in the distribution of energy, such changes being associated with electron shifts within constituent atoms, ions, or crystal imperfections.

Minerals that have a constant and characteristic color are generally termed *idiochromatic,* or their color is termed *inherent.* In these minerals, the color-absorbing qualities—typically ions or groups of ions that absorb certain wavelengths—are essential to the makeup of the mineral. Minerals whose color may differ from specimen to specimen are called *allochromatic,* or their color is referred to as *exotic.* In these minerals, color absorbing mechanisms do or do not occur in any given specimen. [In addition, some minerals exhibit apparent color or colors that are not true color but, instead, a "play of color(s)" produced by certain physical effects; these should be characterized as *pseudochromatic.* Their appearance depends on such phenomena as diffraction or scattering of wavelengths of light by, for example, specially oriented inclusions.]

Each type of color, but especially inherent color, can be used as one of the more useful determinative physical properties of a mineral. You should keep in mind, however, the fact that such use of color often requires both experience and sharp discrimination.

In the same article, Nassau lists four "formalisms" and twelve causes to account for color in minerals (Table 4-3). These can be described briefly as follows, with the letters and numbers in parentheses in the text corresponding to those in Table 4-3.

A. Under the crystal field theory (see, for example, Cotton, 1964), colors are accounted for by shifts of unpaired electrons. These unpaired electrons are associated with either the ions of the transition metal elements (Ti, V, Cr, Mn, Fe, Co, Ni, and Cu) or with crystal imperfections (defects). The involvement of transition elements depends on their having outer labile electrons,

Table 4-3. *Causes of color in minerals*

Formalism	Cause of Color	Mineral Examples
A. Crystal field theory	1. Transition metal compound	azurite, goethite, spessartine
	2. Transition metal impurity	citrine quartz, ruby corundum
	3. Color centers	purple fluorite, amethyst, smoky quartz
B. Molecular orbital theory	4. Charge transfer	cordierite, kyanite, vanadinite
	5. Organic material	amber, coral, lignite
C. Band theory	6. Conductor	copper, gold, silver
	7. Semiconductor	cinnabar, galena, pyrite
	8. Doped semiconductor	blue and yellow diamond
D. Physical optics theories	9. Dispersion	"fire" in gems (e.g., diamond)
	10. Scattering	asterism, aventurescence, chatoyancy
	11. Interference	iridescence (e.g., on bornite)
	12. Diffraction	opalescence, labradorescence

After Nassau, 1978; several additional mineral examples are given by Nassau.

which are especially susceptible to shifts because they are held so lightly by their nuclei. In response to light, these electrons shift from their ground state to an excitation level, thereby absorbing certain visible wave lengths while permitting others to be transmitted and thus perceptible. The ions or groups of ions that produce characteristic colors are known as *chromophores*. (Some chromophores may be responsible for two or more completely different colors, apparently because of differences in the ways they are affected by different structural environments; furthermore, some chromophores appear to be ineffective in the presence of certain other chromophores.)

1. Transition metal elements that are present as essential constituents give inherent color to their containing minerals. (There is, however, one qualification: the valence state of the transition metal must be appropriate; for example, Cu^+ and Cr^{+6} have no labile electrons so their compounds may be colorless.) Examples of minerals whose color depends on essential transition element constituents and which are described in Part II are: crocoite and uvarovite—chromophore chromium; azurite, chrysocolla, cuprite, and malachite—copper; almandine, goethite, and olivine—iron; rhodochrosite, rhodonite, and spessartine—manganese; and autunite and carnotite—uranium.

2. Transition metal elements that are present as impurities in minerals that are colorless, or nearly so, when pure, commonly give exotic color to their containing minerals. The classic example is corundum, which is colorless when pure Al_2O_3 but also occurs as red ruby and pink, yellow, green, and blue sapphire. Other examples are: emerald, green grossular, pink topaz, and green tourmaline—chromophore chromium; chrysoberyl, yellow idocrase, and jadeite—iron; andalusite, spodumene (var. morganite), and pink tourmaline—manganese; chrysoprase—nickel; and apophyllite—vanadium.

3. Lattice imperfections (defects) within minerals that are colorless, or nearly so, may also give exotic color to their containing minerals. Defects that are responsible for color are widely referred to as *color centers*. Those dependent on the presence of an extra electron and those involving the absence of a normally present electron have been termed *electron color centers* and *hole color centers*, respectively. This color production also involves the excitation and shift of unpaired electrons and the consequent energy shifts, which, in turn, cause selective absorption and transmission of certain wave lengths of the visible part of the electromagnetic spectrum. Examples are purple fluorite, in which unpaired electrons may be entrapped in positions normally occupied by fluorine ions, and smoky quartz, in which hole color centers are created by the presence of aluminum that replaces silica and is electronically neutralized by the presence of an alkali element or a hydrogen ion.

B. Under the molecular orbital theory (see, for example, Ballhausen and Gray, 1964), certain colors of minerals result from charge transfers between electrons that are orbiting around more than one ion or atom. Such charge transfers are most likely to occur in minerals with predominantly ionic and/or covalent bonds.

4. Charge transfers that cause light absorption attendant on energy produced by electron "hopping" are responsible for the colors of some minerals with predominantly ionic bonds. Examples include magnetite with charge transfers between Fe^{+2} and Fe^{+3} and vanadinite with charge transfers between V^{+5} and O^{-2}. (Note that the former is a metal-to-metal transfer, whereas the latter is a metal-to-nonmetal transfer.)

5. This "organic-compound" cause as described by Nassau (op. cit.) is apparently of little, if any, importance so far as minerals are concerned. Thus, it is not treated here.

146

C. Under the band theory, color production is related to electrons that float—that is, electrons that are detached and freely mobile—within their containing structures and which, as a consequence, can comprise bands of energy. The shapes of the energy bands and the gaps between them are responsible for the colors formed under this hypothesis.

6. Energy band differences in conductors are responsible for colors (and also for lusters and opacities) of native element metals and natural alloys. The colors are manifestations of slight absorption differences that take place because of differences in the shapes of the included energy bands. Minerals whose colors are so formed include native copper, gold, and silver, and natural alloys such as meteoritic nickel-iron and irodosmine.

7. Band gaps—or more correctly, their associated energy gaps—also account for the colors of some semiconductor minerals. These minerals have structures with predominantly covalent bonds. The color perceived is based on the energy gap, which is, in turn, correlative with the band gap. The band gap color sequence—from narrow to wide (i.e., from $\sim 1\frac{1}{2}$ to $> 3\frac{1}{2}$ electronvolts)—is black, red, orange, yellow, and colorless. (The colorlessness is due to the fact that there is no absorption of visible electromagnetic energy.) Examples of minerals of this category are: cobaltite, galena, marcasite, and pyrite—narrow band gap; cinnabar, greenockite, orpiment, proustite, pyrargyrite, realgar, and sulfur—medium band gap; and diamond, sphalerite, and zincite—wide band gap.

8. Band gap coloration of "doped semiconductors" is accounted for by the presence of trace impurities within the just mentioned broad band gaps. Only a few minerals appear to owe their color to such a relationship. Two examples are the yellow color of some diamonds, which is caused by the presence of nonclustered nitrogen atoms $(N:C \cong$

$1:100,000)$, and the blue color of, for example, the Hope Diamond, which is caused by the presence of boron (with a few boron atoms per few million carbon atoms). Broadly simplified, the explanations are as follows: in the yellow diamonds, nitrogen (with five available electrons) substitutes for carbon (with four available electrons), thus adding an additional electron that will cause a donor level energy charge and consequently a color absorption within a band gap; in the blue diamonds, boron (with three available electrons) substitutes for carbon (with four available electrons), thus creating an electron hole that will cause an acceptor level energy change and consequently a different band gap.

D. Colors that depend on physical optics are not, as previously mentioned, colors in the strict sense. Nonetheless, they are often perceived as colors and hence merit brief discussion here.

9. Dispersion of light does not add color, it merely tends to enhance color in some materials—for example, in cut gemstones. It amounts to the degree that white light is separated into the spectral colors when it passes through, for example, a prism. Dispersion, which is responsible for the characteristic often termed *fire*, generally varies directly with refractive index.

10. Scattering of light, though it generally affects luster more than it does color, may be responsible for appearances such as those generally referred to as adularescence, aventurescence, and chatoyancy. Much such scattering of light is caused by tiny inclusions or structural discontinuities, typically those with preferred orientations. Some of the inclusions and structural discontinuities can be seen and identified with a handlens or a microscope; others are submicroscopic. In some cases, the mineral appears to have been homogeneous when originally crystallized and the so-called impurity to have been produced by exsolution; for instance,

the brownish red color of some feldspars is very likely due to the presence of sub-microscopic hematite, probably produced by exsolution of ferric iron that replaced aluminum in the feldspar structure when it first crystallized.

11. Interference of light that is macro-scopically observable is essentially restricted to minerals containing thin films of air in cracks (e.g., "iris quartz") and minerals exhibiting surface tarnish (e.g., chalcopyrite and bornite). The colors seen, which are often compared to the iridescent films produced by oil on water, depend on the thickness of the film, the refractive index of the film material, and the character of the incident light.

12. Diffraction of light causes the colors perceived for some opal and labradorite. The principle involved is that which accounts for the combined scattering and interference of electromagnetic waves by diffraction gratings with appropriate spacings. Thus, the spacing of the domains, whatever their character and origin, that constitute the effective grating and the angle of observation are responsible for the color(s) perceived.

Several minerals not belonging to the isometric crystal system selectively absorb visible wavelengths differently in different directions. The result is that these minerals exhibit different colors when light is transmitted through them along different directions. This phenomenon, which is called *pleochroism*, is generally seen during microscopic examination using polarized light. Nonetheless, pleochroism can be observed in some mineral specimens by using only sunlight or common tungsten light bulbs. For example, cordierite (also called dichroite because of this feature) can be seen to transmit purple light when viewed in one direction and light yellow when viewed at right angles to that direction. It appears, in fact, that the Vikings recognized this property and used grains of cordierite and its reaction to sunlight, which is partially plane-polarized, as a navigation aid.

Some colors of some minerals are changed when the minerals are submitted to heat treatment or atomic bombardment (e.g., exposure to X-rays, gamma rays, or neutron beams). Examples of such changes include the heating of brown zircons to render them blue, the heating of yellow topaz to change it to pink, and the atomic bombardment of off-color diamonds to make them light green or bluish. (The change in diamonds is generally restricted to an outer zone that is only a few tens of microns thick.) Heat treatments presumably cause the addition of sufficient energy to exceed that necessary for holding electrons in vacancy defects. In some cases, the original colors, either natural or artificially induced, can be restored by another of these processes. A general rule of thumb is that heat treatment will reverse results of atomic bombardment and vice versa. For example, smoky quartz can be made colorless by heating, and topaz that has been made pink by heating may be "reyellowed" by gamma-ray bombardment.

As with other properties of minerals, color and investigations relating to color have many applications. Several of these deal with gemstones and were determined on the basis of empirical observations and experimentation long before the pertinent theories were even hypotheses; the fundamental explanations of some of these applications are not well understood even today. Applications include several diverse things other than the heat treatments and atomic bombardment we just mentioned: Some minerals, because of their pleochroism, have to be cut according to a certain crystallographic orientation to exhibit their desired colors; for example, rubies with their table facets perpendicular to c exhibit their preferred purplish-red color, whereas rubies cut parallel to c are orange-red. Porous varieties of some minerals can be stained; for example, most brightly colored agates and turquois pieces have been soaked in solutions that have deposited pigments within their pores. There are, of course, several other diverse uses of mineral pig-

148   ments; one is in paint. Also, some of the knowledge gained from investigations of mineral color have had a great influence on glass technology.

Streak is the color of the fine powder of a mineral. It is often used as a diagnostic property for mineral identification (see Chapter 6). It may be used to distinguish between inherent and exotic color in most minerals.

Luminescence

Luminescence is the emission of light as the result of any process other than incandescence. Such light is dependent on energy conversions. The kinds of energy thus far shown to cause luminescence include those usually referred to as electromagnetic (optical and nuclear), electric, mechanical, chemical, and biochemical energy. Actually, at least 15 different kinds of luminescence, each named on the basis of the source of excitation, have been recognized and described (see, for example, Marfunin, 1979; note especially the extensive bibliography, which includes articles published as recently as 1976).

The luminescence most frequently observed in minerals is generally termed *photoluminescence*. It is usually stimulated by irradiation with ultraviolet light and is best observed in total darkness. *Fluorescence* is the emission of light at the same time as the irradiation; *phosphorescence* is continued emission of light after the irradiation is terminated. Both natural and man-made materials that exhibit luminescence are widely referred to as *phosphors*.

The luminescence of materials has long been of scientific interest and has been found to have many practical applications. Luminescence of minerals can be a valuable aid in determinative mineralogy, in prospecting, and in mineral dressing. It also has several geological applications, such as that of a tracer in hydrological studies. In addition, luminescence is the basis of a number of modern illumination techniques that use the fluorescence of inorganic compounds, many of which are mineral analogs (e.g., $CaWO_4$, $CaCO_3$, and $ZnSiO_4$). Luminescence serves an important role in certain electron microprobe investigations. The list could go on and on. . . .

The fundamental law of luminescence, Stokes' Law, states that the wavelength of the luminescence that is emitted is longer than the wavelength of the exciting radiation or, to state it otherwise, the energy of the luminescence that is emitted is less than the energy of the exciting radiation. Two examples can be cited: the absorption of invisible radiation in the ultraviolet, X-ray, or gamma-ray part of the spectrum with the simultaneous emission of the energy as visible light, which is called fluorescence; and the absorption of visible light that results in emission in the infrared part of the spectrum, which is known as *infrared luminescence*. This latter emission is, of course, invisible and hence must be detected by special equipment that is sensitive to infrared radiation.

Most substances that fluoresce at ordinary temperatures no longer do so above 500–600°C. On the other hand, nearly all organic compounds and many inorganic compounds do fluoresce at low temperatures, such as at the temperature of liquid nitrogen (77 K). It also is noteworthy that a mineral that fluoresces under the electron probe may not fluoresce in a luminoscope, that one that fluoresces in luminoscope may not fluoresce in short wave ultraviolet, one that fluoresces in short wave ultraviolet may not fluoresce in long wave ultraviolet, etc.

The phenomenon of luminescence is linked with lattice disturbances that are the result of either imperfections or the presence of foreign ions that function as activators. In many instances, it has been shown that the foreign ions are substituting for major elements in the given structure; for example, in willemite, the activating ion is Mn^{+2} substituting for Zn^{+2}, and in scheelite, it is Pb^{+2} substituting for Ca^{+2} and Mo^{+6} for W^{+6}. The former explains why Mn-bearing willemite from Franklin, New Jersey, fluoresces, whereas Mn-free willemite from many other localities does not fluoresce.

In essence, the production of luminescence involves the following dependent processes (see Figure 4-4): a quantum of energy is absorbed in response to excitation by, for example, ultraviolet light or X-ray irradiation; electrons in the activators are raised to higher, excited levels; subsequently, when they fall back to their ground state, the energy that is released is in the form of the appropriate wavelength of electromagnetic radiation (light). The best-known activators are the transition metals.

Sometimes the absorbed energy is "frozen in" and is released only upon heating of the material; this process is known as *thermoluminescence.* The thermoluminescence of most minerals is thought to be due to long term exposure to radiation from traces of radioactive impurities. Such radiation apparently produces sufficient energy to free electrons inside their mineral hosts. Thence, although most of these freed electrons quickly return to their ground state, a few of them are trapped in sites from which they escape only when assisted by thermal vibrations and thus are able to return to their loci of origin. Crystal defects at which electrons can be trapped include vacant negative ion sites, interstitial cations, impurities, dislocations, and structural discontinuities. The rate of escape of electrons from their traps, and consequently the intensity of the thermoluminescence, is temperature dependent. With traps of a single depth, the thermoluminescence will increase in brightness as the temperature is raised until a maximum brightness is attained, and then it will diminish to zero when all of the previously entrapped electrons have been released. A plot of thermoluminescent intensity against temperature—the so-called *glow curve*—will show a glow peak, from which the depth of the trap can be calculated. If traps of different depths are present, the glow curve will exhibit a corresponding number of glow peaks. The usual procedure is to observe the luminescence while the mineral (or rock) is slowly heated, under carefully controlled conditions, up to about 475°C. Because glow peaks have been removed naturally as well as in the laboratory, relatively common rock-forming minerals with more than one glow peak (e.g., calcite) have been used in geothermometry. In addition, thermoluminescence of minerals has been used as a

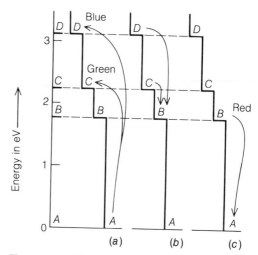

Figure 4-4.   Energy-level diagram for unpaired electrons of chromium ions in ruby: (a) during absorption, the ion may go from A, the ground state, to C or D but not to B; (b) from C or D, it may go to B but not directly back to A; (c) from B it may return to A. (By correlating the energy values attained with those of the electromagnetic spectrum given in Figure 4-3, prove to yourself that as a consequence of a, the bluish red color of ruby will be produced and, from c, there will be a deep red fluorescence.) (By permission, from Nassau, Kurt, 1964, *Growing Synthetic Crystals:* © Bell Telephone Laboratories, Murray Hill, N.J.)

stratigraphic tool, in depth-of-burial studies, in pressure-effect investigations of diverse rocks, and for age dating of materials, such as lava flows, that have once been heated. Also, thermoluminescent dosimeters that are capable of recording nuclear explosions have been mounted in man-made earth satellites for surveillance.

Another interesting kind of luminescence that is met with occasionally in minerals is termed *triboluminescence.* This luminescence is induced by pressure, crushing, scratching, or rubbing. Typically only a flash, some triboluminescence has been shown to represent ionization of nitrogen atoms of the air around the grains being disintegrated, whereas other triboluminescence is due to the presence of certain activators within the grains. It has been observed in many minerals, including aragonite, barite, calcite, dolomite,

150 fluorite, gypsum, halite, lepidolite, quartz, rutile, and sphalerite. [An interesting geological application is the distinguishing of dolomite from magnesite on the basis of their triboluminescent colors by striking them sharply with a hammer in the dark; this application has been used to great advantage in a mine at Kilmar, Quebec.] A possibly related phenomenon, termed *crystalloluminescence*, has several of the triboluminescent minerals emitting light as they crystallize from melts or solutions.

CLEAVAGE, PARTING, AND FRACTURE

If a crystalline grain is struck or crushed—that is, subjected to sufficient stress to exceed its elastic and plastic limits—the grain will break. If the grain breaks along plane surfaces related to the crystal structure, it is said to show cleavage; if it breaks along planes not controlled directly by the crystallographic structure of the mineral, it is said to exhibit parting; if it breaks irregularly, it is said to show fracture.

Cleavage

Cleavage, as a reflection of internal structure, can be treated quantitatively using bond strength values and data relating to the distribution of atoms and/or ions. This treatment, of course, derives from the fact that a mineral possesses cleavage because the strength of the bonding within its structure is selectively weak along certain planes. This relation is particularly apparent in minerals that have layered structures in which the bonding within layers is very strong, whereas that between layers is relatively weak. In such structures, we find perfect unidirectional cleavage parallel to the layers. Examples are graphite and the phyllosilicates (the subclass of the silicates in which the SiO_4 groups form sheet structures), as typified by talc and the micas (see Figure 4-5). In graphite, the carbon atoms are covalently bonded in sheets, and the sheets are linked together by weak van der Waals' forces that are easily disrupted. A similar situation is present in talc; strongly bonded sheets are linked by weak residual forces. In the micas, however, the sheets, which are similar to those in talc, are linked by potassium ions; nonetheless, these linkages are also much weaker than the bonding within the sheets so the single perfect cleavage in the micas follows the planes occupied by the potassium ions. As is apparent, these examples indicate not only that cleavage is a function of internal structure but also that cleavage is related only indirectly to chemical composition. This latter fact is further corroborated by the observation that within an isostructural group, the cleavage tends to be the same for all species although it may differ in quality from one species to another. This fact is well shown by the isometric substances with the halite structure, all of

Figure 4-5. Cleavage. Perfect unidirectional cleavage parallel to {001} of mica, locality unknown (15 cm across).

which have perfect cubic cleavage; that is, they have three equal cleavages at right angles to one another. These isometric substances include not only halite (NaCl), sylvite (KCl), and many other univalent halides, but also periclase (MgO) and other bivalent oxides, and galena (PbS), a sulfide. Furthermore, the perfect rhombohedral cleavage of the calcite group (see Figure 4-6) can be seen to reflect a similar structure; in the calcite structure, the Ca plays the part analogous to that of Na, whereas the CO_3 group acts like the Cl (and, of course, there is hexagonal—trigonal—rather than isometric symmetry).

Cleavage is always parallel to a face or a possible crystal face, generally one with simple indices. It can occur virtually an infinite number of times parallel to a given direction, down to the thicknesses of atomic layers. In descriptions, the positions of cleavage planes are noted by the appropriate crystal form name (e.g., octahedral) or its Miller index (e.g., $\{111\}$).

A mineral may show cleavage in more than one crystallographic direction. In most cases, the quality is different in the different directions, thus enabling us to distinguish one cleavage from another. The quality of cleavage is described by such terms as *perfect, good, distinct,* and *indistinct. Perfect cleavage* is present when a mineral breaks giving a smooth lustrous surface; most minerals with a perfect cleavage break with difficulty in other directions (e.g., calcite and muscovite). A mineral is said to have *good cleavage* when it breaks easily along the cleavage but can also be broken transverse to the cleavage (e.g., feldspar). Minerals with *distinct cleavage* break most readily along the cleavage, but they also fracture rather easily in other directions. Consequently, what appear to be individual cleavage surfaces are really a series of steps (e.g., scapolite). Minerals with *indistinct cleavage* generally fracture as readily as they cleave, so careful inspection may be necessary even to recognize the cleavage (e.g., beryl). Most cleavage is easily recognized in thinsections cut for optical studies of minerals and rocks because cutting and grinding of the sections cause the minerals to break

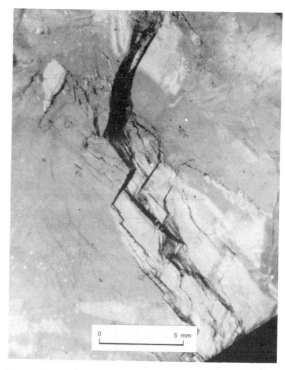

Figure 4-6. Cleavage. Perfect rhombohedral cleavage $\{10\overline{1}1\}$ in calcite. Photograph shows a broken surface magnified as it would appear through a handlens. (Courtesy of B. J. Skinner.)

along their cleavages, which then appear as cracks that are more or less straight, with the degree of development dependent on the quality of the cleavage. Cleavage is a very useful diagnostic property because it often affords a guide to the symmetry of a substance (see Chapter 6).

A few additional generalizations relating cleavage to mineral structures are noteworthy: most minerals with appreciable metallic character to their bonds do not cleave well, if at all; minerals with chiefly covalent bonds may or may not cleave—in those that do, the cleavage planes coincide with planes involving fewest bonds per unit area; most minerals with appreciable ionic bonding tend to have one or more good cleavages. Among the silicates; those with isolated tetrahedra and those made up of three-dimensional networks tend not to cleave; those with isolated groups typically exhibit only indistinct to distinct cleavage; and those with chain and sheet structures tend to cleave between the chains and sheets but not through them. [And, of course, amorphous materials do not cleave.] Interestingly, when ionic minerals cleave, the cleavage fragments are electrically neutral. Also, when carefully measured, the energy required to cleave minerals is considered by some physicists to afford the structurally important value generally termed surface energy.

From the standpoint of technical application, cleavage is an exceedingly important property, and industrial use of many minerals depends on it. The ability of muscovite to be cleaved into very thin sheets, together with its dielectric properties, is the basis of its use in electrical equipment. A pencil is an efficient writing implement because the perfect cleavability of graphite causes small cleavage flakes of the mineral to be rubbed off so they adhere to paper. The lubricating qualities of graphite and talc depend on their softness and the ease with which the minerals break along cleavage surfaces.

Parting

Parting, which resembles cleavage and is sometimes referred to as *false cleavage*, is separation along planes of weakness caused by twinning, deformation, or the presence of oriented inclusions or of approximately parallel exsolution lamellae; each is, in some cases, accentuated by incipient alteration. Like cleavage, parting is typically parallel to possible crystal faces; unlike cleavage, it consists of a finite number of discrete, relatively widely spaced planes. Parting does not have to conform to symmetry requirements of the material involved. It may or may not be present in any particular specimen of a given mineral, although several minerals commonly exhibit parting whereas others rarely, if ever, do. In some specimens, parting is present in more than one direction; for example, much corundum has both

$r\{10\bar{1}1\}$ and $c\{0001\}$ parting. Especially noteworthy examples of parting are those controlled by the composition planes of polysynthetic twinning parallel to r in corundum and those controlled by included exsolution lamellae parallel to the basal pinacoid $\{001\}$ of pyroxenes such as augite, diopside, pigeonite, and hypersthene.

Fracture

Strictly speaking, *fracture* includes all breakage in minerals; in general, however, the term is limited in application to rupture patterns that result in the formation of irregular surfaces—that is, to ruptures that bear no consistent, well-defined angular relationship to crystallographic directions (see Figure 4-7). Fracture generally starts along a scratch or a crack and continues along concentrations of dislocations, flaws, inclusions (including fluid inclusions), or the like. It generally takes place in substances that have about the same strength in all directions (i.e., in those in which the density and strength of bonds are similar or equal in all directions) or in noncleavage directions in minerals that have cleavage. Terms applied to diverse kinds of fracture are given in Chapter 6.

Figure 4-7. Fracture. Conchoidal fracture of a quartz pebble. (Courtesy of B. J. Skinner.)

TENACITY

Tenacity, also termed *tensile strength*, is the resistance that a material offers to mechanical deformation or disintegration; that is, it is the resistance that the atoms and ions of a material offer to being separated when submitted to processes that tend to cause bending, breaking, crushing, or cutting. Along with the already discussed cleavage, parting, and fracture, additional terms frequently used to describe diverse expressions of tenacity in minerals are elastic, plastic, brittle, malleable, sectile, flexible, and tough. These qualities, as generally applied in determinative mineralogy, are also described in Chapter 6.

Although many of the just-listed properties (elasticity, etc.) are important to geophysical investigations, in most cases, geophysicists deal with rocks rather than with minerals. Therefore, a large percentage of the published values for geological materials are for rocks rather than for minerals (see, for example, Clark et al., 1966).

The general form of the curve that relates elasticity (recoverable deformation), plasticity (permanent deformation not involving macroscopic rupture), and brittleness (deformation involving fracture) of substances is well known. Fortunately, a few noteworthy generalizations can also be given because of investigations made by materials scientists—especially by ceramists, metallurgists, and solid state physicists. Before listing these generalizations, however, it seems only prudent to mention that some of of the pertinent nomenclature is not altogether consistent from discipline to discipline, from author to author, or even from one time to another within the same discipline. Consequently, one must be extremely careful to determine exactly what is meant in any given presentation of data and in discussions or interpretations of those data. (For example, some scientists have attached certain values to ductility and brittleness, whereas others use the terms in a strictly qualitative sense, *and some workers' elastic moduli and constants are other workers' compliances and stiffnesses.*) Furthermore, for a number of values (so-called constants), different values have been determined by different methods, so one must be extremely careful not to fall into the proverbial trap of comparing "artichokes with prunes." It appears that much of this source of confusion should be superseded in the near future; this is true because it is now known that many of these constants (e.g., stiffnesses) can be determined by sending sound waves into the material being tested and then measuring the transmitted wavelengths that correspond to "established states" of the material (see, for example, Wooster, 1973).

The generalizations include the following:

1. As should be expected, essentially all of the mathematical expressions related to tensile strength phenomena (e.g., Hooke's Law, the Lamé constant, Poisson's ratio, and Young's modulus) have been shown to be related to bonding, structure, and/or surface energy.

2. Elasticity, as measured by the transmission of elastic waves, is found to be generally highest for substances with ionic bonds, relatively high for those with covalent bonds, low for those with metallic bonds, and lowest for materials with appreciable van der Waals bonding.

3. Most elastic properties are anisotropic, even in some materials that belong to the isometric crystal system.

4. Plasticity is, for the most part, the result of movements in response to shear parallel to specific lattice planes. In some cases, however, the movements are controlled by concentrations of dislocation defects.

5. For substances with the same structures and/or bond types, shear moduli have been found to vary indirectly with interatomic distances—that is, the greater the interatomic distance, the lower the shear modulus. Also, for the materials tested, at any given interatomic distance, the shear moduli are (with a few exceptions) highest for substances with covalent bonds, lowest for those with ionic

154

bonds, and in between for those with metallic bonds (see, for example, Gilman in Austin, 1963). This correlates rather well with deductions made with regard to the relative amounts of stress required to exceed the elastic limits for materials with different kinds of bonds.

6. Malleability is limited to metallic substances with metallic bonding.

7. Brittleness has been found to be dependent on the presence of small cracks (e.g., discontinuities caused by concentrations of defects). Under stress, these cracks tend to propogate through the mineral, thus causing it to fail.

8. Toughness, which can be defined as the amount of energy that can be absorbed before a material fractures, is generally much higher for polycrystalline aggregates (rock) than for individual crystal grains; for example, the typical carbonado, which is an intergrowth of many small diamond grains, is much more difficult to break than individual diamond crystals are.

9. Tenacity is not related directly to hardness; note, for example, that diamond (H-10) has a much lower tensile strength, because of its ready cleavage, than jade (H-6), which typically comprises extremely tough polycrystalline masses.

HARDNESS

Hardness of a mineral is generally defined as its resistance to scratching (i.e., abrasion). It is, of course, another property that manifests the internal cohesiveness of a material. For some materials, such as copper, hardness measures plastic deformation; for other materials, such as most brittle minerals, it measures stress required to initiate rupture. In any case, relative hardness has been employed as a useful diagnostic property

ever since the beginning of systematic mineralogy, and in 1824 it was given qualitative precision by Friedrich Mohs, who proposed a scale for measuring degree of hardness.

It is a tribute to the perspicacity of Mohs that the scale he established is still used today in unaltered form. In his original description, Mohs mentions that he endeavored to make the intervals on the scale as nearly equivalent as possible and, at the same time, to select common minerals for the individual units. He was aware that the interval between corundum and diamond is greater than that between other units on the scale, but minerals of intermediate hardness were (and still are) unknown. He was also aware that hardness varies somewhat with crystallographic direction, many minerals being softer on cleavage surfaces than in other directions; he therefore recommended finely crystalline specimens, rather than coarse cleavages, of talc and gypsum as standards.

Many procedures for quantitative determinations of hardness have been devised, and their results show that the intervals on Mohs' scale are not equal, as has been frequently stated; rather, they increase progressively from hardness (H) of 1 through 2, 3, 4, . . . up to hardness of 9. They also indicate that, in line with quantitative measurements made on metals, this scale closely approximates the way most present-day scientists and engineers would probably establish a hardness scale. [The quantitative methods sometimes used actually measure plastic and elastic deformation as reflected by indentation. The best-known methods are the Brinell, Knoop, Rockwell, and Vickers tests; the main differences among them relate to the shapes of the indenting heads used, the mode of measuring the indentations produced, and the calculations employed to convert the measurements to the desired quantities (see, for example, Eisenstadt, 1971). With the exception of the rather extensive use of Vickers hardness tests on opaque minerals, none of these methods has been used often or on many minerals.] Hence, although Mohs' scale is qualitative, it is considered well suited for comparing relative hardnesses in minerals and is unlikely to be superseded in the near future.

Although the establishment of exact relations between hardness and any inherent characteristics or properties of minerals has been somewhat handicapped by lack of an absolute quantitative scale of reference, a number of correlations have been found—most with exceptions yet to be resolved. Correlations have been made with both chemical composition and structure, particularly bonding.

Hardness and chemical composition have been related according to the following generalizations:

1. Minerals of the heavy metals, such as gold, silver, copper, mercury, and lead, are soft, few with hardnesses exceeding 3 (exceptions are platinum, $H = 4$–$4\frac{1}{2}$ and iron $H = 4\frac{1}{2}$).

2. Most sulfides and sulfosalts are relatively soft ($H < 5$); the sulfides of iron, nickel, and cobalt are exceptions.

3. Most hydrous minerals are relatively soft ($H < 5$).

4. Most anhydrous oxides and silicates are hard ($H > 5\frac{1}{2}$); those of the heavy metals are exceptions.

5. The halide, carbonate, sulfate, and phosphate minerals are relatively soft ($H < 5\frac{1}{2}$).

Correlations between hardness and mineral structure, which for the most part pertain to minerals within isomorphic groups, have been expressed by the following generalizations:

1. Hardness is greater the smaller the atoms or ions.

2. Hardness is greater the greater the valence or charge of the constituent cations.

3. Hardness is greater the greater the packing density.

All these generalizations are interrelated; for example, the one dealing with sizes of atoms and ions (1) and the one dealing with packing density (3) are both related to interatomic distances and consequently to bond density. Therefore, it also seems worthwhile to note the two following generalizations:

1. Hardness varies inversely with the interatomic distance—that is, the cation–anion separation.

2. Hardness varies directly with bond density.

The effect of ionic size can be seen most clearly in an isomorphous group where the structure is the same in all species. Thus, the calcite group comprises carbonates of divalent metals ranging in ionic size from Ca (1.00 Å) to Mg (0.57 Å); hardness increases with decreasing ionic size, from 3 for calcite to $4\frac{1}{2}$ for magnesite. Another good example is furnished by the marked contrast between hematite, Fe_2O_3 ($H = 6$), and corundum, Al_2O_3 ($H = 9$).

The effect of valence or charge can be seen most clearly by comparing compounds with the same structure and similar ionic sizes. Thus, nitratite ($NaNO_3$) and calcite have the same structure, and the ionic sizes of Ca^{+2} and Na^+ are very similar, but the hardness of nitratite is 2, whereas that of calcite is 3. A greater difference exists between the isomorphous minerals niter, KNO_3 ($H = 2$), and aragonite ($H = 4$), because in these minerals there is not only the difference in charge between potassium and calcium but also a considerable difference in ionic radii ($K^+ = 1.37$ Å, $Ca^{+2} = 1.00$ Å).

The effect of density of packing is well shown by the relationship between density and hardness of different polymorphic pairs. Examples are calcite ($G = 2.71$, $H = 3$) and aragonite ($G = 2.93$, $H = 4$) and quartz ($G = 2.65$, $H = 7$) and tridymite ($G = 2.26$, $H = 6\frac{1}{2}$). The same correlation between hardness and density exists between hardness and packing index; the greater the packing index, the greater the hardness. Thus, for example, a packing index greater than 6 generally,

156

though not in all cases, means a hardness of 6 or more.

Since the bonding of a crystal structure is typically different in different directions, the hardness of a mineral may be expected to differ somewhat with crystallographic direction. Such variation is generally quite small, but in a few minerals, it is considerable. Thus, on the {100} cleavage surface of kyanite, the hardness is $4\frac{1}{2}$ in the direction of the c axis (i.e., parallel to the chains of silica tetrahedra) and $6\frac{1}{2}$ in the direction of the b axis (i.e., perpendicular to the chains). Also, on c{0001}, calcite has a hardness of $2\frac{1}{2}$, whereas on r{10$\bar{1}$1}, its hardness is 3. Such differential hardness in the case of diamond is of great practical importance; the {111} surface in a diamond crystal is the hardest surface known to man. This fact, coupled with diamond's {111} cleavage, enables diamond dust to be used to abrade all but octahedral planes of other diamonds.

There are, of course, several second-order generalizations; for example, the lesser the hardness, the lower the melting temperature. Also, with polymorphic pairs, but with exceptions, the ones formed at higher pressures tend to be denser and thus harder than the ones formed at lower pressures. And, hardness may increase with decreasing temperature; for example, the hardness of ice becomes approximately 6 at $-40°C$. In addition, a number of apparent anomalies are known; for example, hardness values for intermediate members of some solid-solution series are slightly greater than the hardness values for either end member.

Hardness, considered in conjunction with other properties related to strength, is important in gemology and also in the choice of minerals and other materials for use as abrasives, jewel bearings, and pressure equipment. Hardness is also important to several ore dressing considerations.

MAGNETIC PROPERTIES

Only a few minerals are attracted strongly by a simple bar or horseshoe magnet. Of these, the most common are magnetite, Fe_3O_4, pyrrhotite, $Fe_{1-x}S$, and maghemite, generally considered to be a polymorph of Fe_2O_3. Some specimens of magnetite and maghemite are themselves natural magnets and thus will attract, for example, iron filings. When suspended, these specimens will orient themselves with their appropriate axes trending north-south; such specimens, generally called *lodestones*, were used in the earliest forms of compasses.

Actually, nearly all minerals are affected by a nonuniform magnetic field; that is, they react mechanically, to some greater or lesser extent, whenever they are placed in a magnetic field. Although it has been shown that there are five kinds of reactions—termed *diamagnetism, paramagnetism, ferromagnetism, ferrimagnetism,* and *antiferromagnetism*—only three different kinds are easily distinguished:

Diamagnetism: The slight repulsion whereby a material moves, or tends to move, toward regions of weaker magnetic field strength; theoretically, this property is possessed by certain domains of all minerals.

Paramagnetism: The slight attraction whereby material moves, or tends to move, toward regions of stronger magnetic field strength; this tendency varies directly with the strength of the applied magnetic field.

Ferromagnetism: The strong attraction toward regions of strong magnetic field strength; these materials are also magnetic even outside a magnetic field, although most become more magnetic within such a field.

Macroscopically, ferrimagnetic materials react in the same way that ferromagnetic materials do, and antiferromagnetic materials exhibit no magnetism.

The understanding of magnetism in minerals and other crystalline materials is less than complete. Some of the especially pertinent aspects of magnetism follow:

1. Individual atoms are magnetic—both protons within nuclei and electrons in their surrounding orbital clouds are magnetic. The

overall magnetism is dependent on the particles' spins; the magnetism of the electrons also involves orbital spin.

2. The magnetism within nuclei is essentially negligible in that it is overshadowed by the magnetism associated with the surrounding electrons.

3. Whereas atoms with even numbers of electrons generally have their electrons with opposite spins (etc.) paired and their magnetism thus nullified, atoms with an odd number of electrons are magnets, albeit infinitesimal ones.

4. Those atoms that are most effective in promoting para- or ferro- (and ferri-) magnetism are Fe^{+2}, Fe^{+3}, Mn^{+2}, Mn^{+3}, Mn^{+4}, Cr^{+3}, Co^{+2}, Ni^{+2}, Cu^{+2}, V^{+3}, U^{+4}, Dy^{+3}, Er^{+3}, Gd^{+3}, Nd^{+3}, and Ce^{+2}, especially the first ten; in other words, atoms with unpaired electrons within inner clouds are the most magnetic.

5. Diamagnetism reflects the nearly exclusive presence of paired electrons.

6. Paramagnetism depends on a reorientation of some magnetic atoms or larger magnetic domains so that they become parallel, or nearly so, and thence cause even larger domains to possess a noticeable magnetic moment; some anisotropic minerals are susceptible to notably different degrees of such reorientation along different directions.

7. Ferromagnetism manifests a locked-in parallelism of magnetic domains; the locked-in parallelism may be disoriented when a ferromagnetic mineral has its temperature raised above its Curie point, for example, above 585°C for magnetite and above 770°C for native Fe. [The Curie point is "the temperature above which thermal agitation prevents spontaneous magnetic ordering . . . (i.e.) the temperature at which the phenomenon of ferromagnetism disappears and the substance becomes simply paramagnetic" (A.G.I. *Glossary of Geology,* 2nd ed.).]

8. Antiferromagnetism is caused by the coupling of ferromagnetic domains, the coupling nullifying their inherent magnetic moments.

9. Ferrimagnetism is the combination of ferro- and antiferromagnetism with a resulting balance giving a ferromagnetic effect. For example, magnetite is ferrimagnetic in that its Fe^{+3}'s are coupled and thereby nullify each other's magnetic moments, whereas its Fe^{+2}'s are not and thus render the mineral strongly magnetic.

10. Most ionic and covalent compounds tend to be dia- or paramagnetic: native metals in which there are clouds of delocalized electrons are ferromagnetic (e.g., iron) or paramagnetic (e.g., copper). Most iron-bearing minerals are either para- or ferromagnetic. Several iron-free minerals (e.g., beryl) are also paramagnetic.

Magnetism in minerals has several geological and economic applications. Magnetic separations, in which electromagnets are used to produce high-intensity magnetic fields, are used both in the research laboratory and in ore dressing plants for separating pure concentrates from diverse mixtures of minerals. Sensitive instruments are capable not only of separating paramagnetic minerals from diamagnetic minerals but also of separating two paramagnetic minerals from each other (e.g., biotite from hornblende). The magnetic properties of minerals are also used in geophysical prospecting by employing the magnetometer, an instrument that measures variations in the earth's magnetic field. Such magnetic surveys are valuable for locating ore bodies, in detecting changes in rock type, and in tracing formations with specific magnetic properties. One advantage of surveys using magnetometers is that they can be carried out rapidly and easily from an aircraft. Similar instruments have also been installed in man-made satellites and used to measure, for example, variations in the earth's magnetic field. Another application is in the field of geomagnetism, one of the major, relatively new subfields of geophysics. Investigations in this field

158

are based on the following premise: the relative positions of land masses through geologic time can be reconstructed on the basis of the assumption that the direction of the magnetic field of a mineral at the time of its formation is (and was in the geologic past) controlled by the earth's magnetic field. Uses of geomagnetism in polar wandering studies and in geochronology are well described by Strangway (1970) and McElhinny (1973).

ELECTRICAL PROPERTIES

On the basis of their electrical properties, minerals can be divided into two groups—the *conductors* and the *nonconductors*. Some scientists, however, subdivide conductors into conductors, *per se*, and *semiconductors*. The electrical nature of many minerals can be correlated directly with bonding; for example, conductors are minerals with metallic bonds, and nonconductors (sometimes referred to as dielectrics or insulators) are minerals with covalent or ionic bonds. Electrical conductivity in semiconductors can be described conveniently under the band theory, previously referred to in the section "Color and Streak." The native metallic minerals—such as copper, gold, iron, platinum, and silver—are conductors; some of the sulfides and oxides are semiconductors; nearly all of the other minerals are nonconductors.

Electrical conductivity is a measure of the ease of transfer of an electric charge from one point to another within a material (e.g., a mineral) that is placed in an electric field. The transfer requires the presence of charge carriers, which are valence electrons in most minerals. Valence electrons are free to act as carriers where they occur delocalized around the outer shells of metal atoms within minerals having metallic bonds; valence electrons cannot participate in this way where they are, so to speak, locked in place because of sharing (e.g., as in minerals having covalent bonds) or where there are specific

transfers (e.g., in minerals having ionic bonds). In a few cases, however, electrical conductivity in minerals is dependent on the migration of ions and/or of holes along defects within a mineral's structure; this kind of conductivity may take place in minerals that are inherently nonconductors, as well as in those that are intrinsic conductors. Within true conductors (including semiconductors), electrical conductivity tends to be measurably anisotropic in anisotropic minerals. Also, the electrical conductivity of minerals can vary with temperature and pressure.

In some nonconducting minerals, it is possible to induce electrical charges by changes in temperature (*pyroelectricity*) or directed pressure (*piezoelectricity*). Opposite charges (positive versus negative) are developed at opposite ends of one or more polar axes: for pyroelectricity, at the ends of a major polar axis; for piezoelectricity, at the ends of any polar axis. For pyroelectricity, the end of the polar axis that becomes positively charged is known as the analogous pole, whereas the end that becomes negatively charged is known as the antilogous pole. The charges, as manifested on the mineral's surface, are measurable in terms of charge per unit area.

Pyroelectricity was first observed in gemmy tourmaline crystals being transported from Ceylon by Dutch traders, who noticed that a crystal dropped in warm ashes attracted ash particles at one end but not at the other end. Pyroelectricity can develop in minerals of only the ten hemimorphic crystal classes (*1, 2, m, mm2, 4, 4mm, 3, 3m, 6,* and *6mm*), each of which is noncentrosymmetric and has one major polar axis—that is, a crystal axis with different forms and properties at its opposite ends. (It is noteworthy that *pyromagnetism* can be developed in minerals of the same crystal classes.)

False pyroelectricity, which may develop in piezoelectric minerals that do not belong to the ten crystal classes we just mentioned, closely resembles true pyroelectricity both from the standpoint of cause and macroscopic effect. The false version, however, is an expression of the thermal stresses and deformation that result from unequal thermal expansion rather than an expression of electronic transfer of electricity. In other words, false pyroelectricity is due to nonuniform heating, whereas true pyroelectricity is due to uniform heating.

Piezoelectricity was discovered in 1880 by the brothers Pierre and Jacques Curie when they observed that quartz crystals subjected to properly directed pressure develop positive and negative charges at opposite ends of a axes. In the following year, Gabriel Lippmann suggested that such crystals would become mechanically deformed if they were subjected to an electrical field; his suggestion was confirmed by the Curies. The effect, however, remained a laboratory curiosity until World War I, when experiments were made in transmitting and detecting submarine sound waves by using plates of piezoelectric quartz. Subsequent applications, especially as oscillators, have been numerous and diverse, particularly for synthetic quartz (see Figure 4-8).

Theoretically, any substance that lacks a center of symmetry is piezoelectric, although some may be only slightly so; minerals of 20 crystal classes ($1, 2, m, mm2, 222, 4, \bar{4}, 4mm, \bar{4}2m, 422, 3, 3m, 32, 6, \bar{6}, 6mm, \bar{6}m2, 622, 23$, and $\bar{4}3m$) are included. The causative mechanical forces can be either compressional or tensional; compression, of course, produces conduction in the reverse direction from that produced by tension. Furthermore, as already noted, when electricity is applied to minerals that develop piezoelectricity, the minerals will contract or expand; a phenomenon which is termed the *converse piezoelectric effect*.

Both pyroelectricity and piezoelectricity in minerals can be detected and demonstrated as is noted in Chapter 6. In some cases, the detection has served as a valuable adjunct in determining the appropriate crystal class for a mineral (see page 58).

Static electricity, sometimes referred to as frictional electricity, can be produced by rubbing some nonconductors with silk or fur; this characteristic of amber has been observed for many centuries. [In fact, the word electricity is derived from

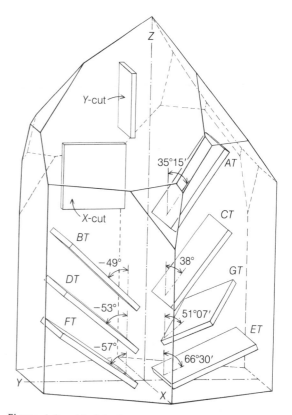

Figure 4-8. Model of a quartz crystal, showing the different directions of cutting to produce oscillator plates of different characteristics.

the Greek name for amber—ηλεκτρον.] Many minerals exhibit this effect. It is rather curious that some gemstones show this property after being cut and polished even though they do not exhibit the property in their rough state.

THERMAL PROPERTIES

The presentation of a comprehensive theoretical treatment of the thermal properties of minerals is beyond the scope of this book because it involves rather lengthy and, for some, fairly formidable mathematical derivations as well as a thorough understanding of several aspects of quantum mechanics [for additional information see, for example, Ziman, 1967].

Thermally dependent properties—such as pyroelectricity, pyromagnetism, polymorphic inversion, thermoluminescence, and color changes—are treated in their appropriate sections. Several other properties—such as specific gravity, certain optical properties, thermal coefficient of expansion, surface tension, and the moduli of elasticity—are also known to vary with temperature. In addition, the following facts seem worthy of note:

1. Although thermal expansion of and heat transfer through crystalline materials tend to correlate with the materials' crystal systems, either property may exhibit anisotropism even within minerals of the isometric system.

2. Thermal expansion can be simultaneously positive along one axis and negative along another axis within one and the same mineral specimen (for thermal expansion data for several minerals, see Skinner in Clark, 1966).

3. Melting temperatures can be calculated in terms of different elastic forces (e.g., the velocity of sound) and volume.

When heated at a uniform rate, many minerals undergo one or more changes below the temperatures at which they form chemically equivalent melts. Most of the changes involve absorption or evolution of heat; that is, the changes are endothermic or exothermic reactions, respectively. The kinds of changes that can be identified and measured are: (a) incongruent melting (e.g., orthoclase to leucite plus melt); (b) loss of part or all of a mineral's water of constitution (e.g., gypsum goes to hemihydrite, which, in turn, goes to anhydrite, see figure 13-8); (c) loss of gaseous components of anion radicals e.g., CO_2 evolves from carbonates; and (d) polymorphic changes of structure (e.g., inversion of low-temperature quartz to high-temperature quartz). The methods usually employed in the identification and characterization of these reactions are differential thermal analysis and thermogravimetric analysis (see Table 6-4).

Mineralogic uses of thermal properties include the determination of melting temperatures (in the past, often referred to as a fusibility scale) in the area of macroscopic determinative mineralogy generally referred to as blowpipe analysis and the employment of the thermal conductivity "probe" for distinguishing diamonds from diamond simulants (thermal conductivities, in watts per square centimeter per degree kelvin, are 10–30 for diamonds and 0.1 or less for diamond simulants).

SURFACE PROPERTIES

Surfaces of crystals and crystalline grains can represent growth, cleavage, parting, or fracture. In any case, the parts of a structure at or within a few atomic layers of the surface of a crystalline grain are not in an equilibrium state in the same sense that parts well within a crystalline grain are. This is so because surface and near-surface atoms are in environments that are nonuniform from the standpoint of their interatomic and/or interionic distances, the number of bonds per atom/ion, probable structure distortions, and (in at least some cases) cation shielding effects. As a result, surfaces are characterized by an excess of free energy, generally termed surface energy, and the constituent atoms and/or ions there tend to behave differently from those elsewhere within crystalline units. Surface energy can be calculated or it can be measured by determining such diverse values as heat of fusion, heat of sublimation, or the energy required for cleavage. Thermodynamically equivalent to Gibbs free energy of the surface, the quantities are usually expressed in newtons per meter (= dynes per centimeter). These surface tension values are used in several thermodynamic calculations. Like many mineral properties, surface tension may be anisotropic; it is dependent on the crystallographic structure adjacent to the surface. Several minerals have been shown to have different surface tension values for

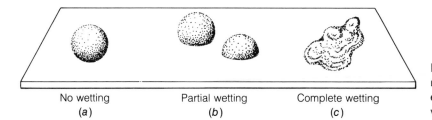

No wetting
(a)

Partial wetting
(b)

Complete wetting
(c)

Figure 4-9. Wettability. Schematic diagram showing differences in appearance of different wettability characteristics.

surfaces with different crystallographic orientations; in general, closely packed planes have lower surface tension values than more loosely packed planes have.

A few surface properties are also dependent on surface configuration; for example, different cleavage planes and different fracture surfaces can have quite different shapes and absolute surface areas. Actually, however, essentially all surfaces are irregular on the atomic level; in fact, even highly polished surfaces can be shown to comprise almost innumerable hills and valleys with reliefs of a hundred to a few thousand angstrom units. The relief of a given surface can be qualitatively described or, if necessary, measured by a needle profile meter or by one of the gas absorption methods.

Surface properties are important to many technologies, especially those in which metals or ceramics are used. Surface properties influence a number of physical and chemical properties and processes—for example, adhesion, cohesion, friction, lubrication; chemical reactivity, solution, nucleation; brittle fracture, diffusion, and wettability. The next to the last, diffusion, bears on petrogenetic considerations that relate to inter- versus intragranular movements of ions and fluids within and through rocks. As an example, we know that as long as other conditions are kept constant, diffusion will take place much more rapidly along grain boundaries (i.e., along junctures of surfaces) than through individual grains. The last property, wettability, finds widespread commercial application in separating certain economic minerals from their gangue mineral associates. Wettability is the relative ease with which a surface can be coated with water (Figure

4-9). According to this property, minerals can be divided into two groups: *lyophile* minerals are those that are easily wetted, and *lyophobe* minerals are those that are not easily wetted. Naturally, there are all degrees of wettability between extremely lyophile and extremely lyophobe minerals. Minerals with ionic bonding are generally lyophile, whereas minerals with metallic or covalent bonding are lyophobe.

This difference in surface properties has been applied for many years in the separation of diamonds from accompanying heavy minerals, such as garnet. Diamond-bearing rock is crushed, and a concentrate of the diamonds and other heavy minerals is separated by mechanical means. This concentrate is then washed over tables coated with thick grease. The lyophile minerals such as garnet are readily wetted and washed away, while the lyophobe diamonds are not wetted and thus stick in the grease from which they are easily recovered.

The principal application of differences in surface properties is, however, in the ore dressing technique known as flotation. Flotation is used primarily for the separation of sulfide minerals from gangue minerals, and separation of individual sulfide minerals from mixtures. In general, the sulfide minerals are lyophobe, whereas the typical gangue minerals (quartz, calcite, etc.) are lyophile. The finely crushed ore is mixed with water, to which are added small amounts of oil (which is attracted to and covers the sulfide grains) and a foam-producing compound. Air is then blown through the mixture, and the foam that is produced carries the sulfide minerals with it, while the gangue sinks. Variations in the type and amount of oil, in the foaming agent, and in other

162

(a)

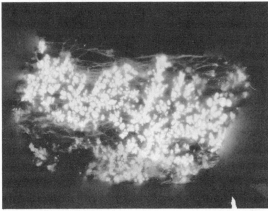

(b)

Figure 4-10.   Photograph (a) and autoradiograph (b) of uraninite in feldspar, from pegmatite at Grafton, New Hampshire (~10 cm across). In the photograph, the uraninite is black; in the autoradiograph it is white. (Courtesy of the American Museum of Natural History.)

conditions enable selective flotation whereby even a complex ore containing several different sulfide minerals can be completely separated into diverse mineral concentrates. In fact, even different lyophile minerals can be separated from one another by flotation techniques.

RADIOACTIVITY

Essentially all relatively easily detected radioactivity in minerals is linked with the presence of uranium or thorium. The other radioactive isotopes (e.g., ^{40}K and ^{87}Rb) typically defy detection except by highly sensitive instruments—for example, instruments used in specialized laboratories and instruments used for certain types of drill hole logging.

Each radioactive isotope (radionuclide) decays at a constant rate—generally referred to as its half-life—that is unaffected by temperature, pressure, or the nature of the containing compound. The decay is accompanied by the emission of energetic particles and/or radiation: alpha particles, which are identical to helium nuclei (^4He); beta particles, which are the same as electrons (e^-); gamma rays, which are a type of electromagnetic radiation with wavelengths equal to those of short X-rays; plus or minus other energetic particles [e.g., positrons (e^+)] and atomic de-excitation phenomena. Therefore, radioactivity can be detected readily by the emissions, either because of their effect on photographic film (Figure 4-10) or by means of an instrument such as a Geiger-Müller counter or a scintillometer.

Any mineral that contains a fairly large amount of certain radionuclides is subject to structural changes, including those as radical as the destruction of its coordination polyhedra or of even larger units of its structure. Such modifications are thought to take place because of size and/or charge differences between the original atom and one or more of its daughter nuclides. That is to say, even if the original element (e.g., ^{238}U) and the stable end product (e.g., ^{206}Pb) have sizes and charges such that they can occupy identical structural positions, the original containing structure still tends to become disrupted during the existence, especially during any relatively long-term existence, of any intermediate daughter nuclide (e.g., ^{226}Ra) that has a notably different size and/or charge. As previously noted, metamict minerals (see page 4) result from such structural disruptions.

Impelled alpha particles can also cause mechanical breakdown or, because of their oxidizing tendency, chemical change. Pleochroic haloes (e.g., those surrounding radioactive minerals, such as thorium-bearing zircon in such minerals as biotite, cordierite, and hornblende) bear witness to alpha particle bombardment.

As a consequence of the fact that the rate of decay of radioactive isotopes is known, the age of a radioactive mineral can be calculated once the amounts of, for example, uranium, thorium, and lead isotopes are determined (as long as the mineral contains no primary lead and has not been altered, leached, or had any loss or gain of the original nuclides or daughter nuclides other than those due to radioactive decay). Fresh specimens of radioactive minerals can, therefore, be of great scientific value because of the information they can provide on geological age.

The development of atomic energy has resulted in a worldwide search for radioactive minerals, especially uranium-bearing minerals because uranium has found the greatest use. The ease with which radioactive minerals can be detected by scintillometers and Geiger-Müller counters has greatly simplified the search. Two important aspects of the post-World War II flurry of activity as far as mineralogy is concerned were the discovery of many new uranium minerals and the thorough investigation of many previously rather poorly known species.

SELECTED READINGS

Austin, A. V. (editor). 1963. *Mechanical Behavior of Crystalline Solids*. National Bureau of Standards, Monograph 59, 113 pp.

Ballhausen, C. J., and Gray, H. B. 1964. *Molecular Orbital Theory*. New York: W. A. Benjamin, 273 pp.

Birch, Francis, et al. 1942. *Handbook of Physical Constants*. Geological Society of America, Special Paper 36, 325 pp.

Burns, R. G. 1970. *Mineralogical Application of Crystal Field Theory*. London: Cambridge University Press, 224 pp.

Chemical Rubber Co. 1980–1981. *Handbook of Chemistry and Physics* (61st ed.). Cleveland: CRC Press.

Clark, S. P., et al. 1966. *Handbook of Physical Constants*. Geological Society of America, Memoir 97, 587 pp.

Cotton, F. A. 1964. "Ligand field theory." *Journal of Chemical Education*, **41**, 466–476.

Eisenstadt, M. M. 1971. *Introduction to Mechanical Properties of Materials*. New York: Macmillan, 444 pp.

Klemens, P. G. 1958. "Thermal conductivity and lattice vibration modes." *Solid State Physics*, **7**, 1–98.

Parrott, J. E., and Stuckes, A. D. 1978. *Thermal Conductivity of Solids*. New York: Academic Press, 157 pp.

Marfunin, A. S. 1979. *Spectroscopy, Luminescence and Radiation Centers in Minerals* (translated by V. V. Schiffer). New York: Springer-Verlag, 352 pp.

McElhinny, M. W. 1973. *Paleomagnetism and Plate Tectonics*. Cambridge (England): Cambridge University Press, 357 pp.

McKie, Duncan, and McKie, C. H. 1974. *Crystalline Solids*. New York: Wiley, 628 pp.

Nassau, Kurt. 1978. "The origins of color in minerals." American Mineralogist, **63**, 219–229.

Povarennykh, A. S. 1964. "On the scale of lustre of minerals and the chemical bond." In Battey, M. H., and Tomkeieff, S. I., *Aspects of Theoretical Mineralogy in the U.S.S.R.* New York: Macmillan, 488–495.

Strangway, D. W. 1970. *The History of Earth's Magnetic Field*. New York: McGraw-Hill, 168 pp.

Williams, D. E. G. 1966. *The Magnetic Properties of Matter*. New York: Elsevier, 232 pp.

Wooster, W. A. 1973. *Tensors in Group Theory for the Physical Properties of Crystals*. Clarendon (England): Oxford, 344 pp.

Wulff, John. 1964–1966. *The Structure and Properties of Materials* (4 volumes). New York: Wiley.

Ziman, J. M. 1960. *Electrons and Phonons*. Oxford (England): Oxford University Press, 554 pp.

Ziman, John. 1967. "The thermal properties of materials." *Scientific American*, **217**, no. 3, 181–188.

5 The Genesis of Minerals

In some instances, the identification of a mineral is an end in itself. More often, it is merely an initial step in deciphering the process and conditions under which the mineral was formed.

The controls of mineral formation are chemical (including biochemical) and physical; in the main, they are the identity and proportions of the elements present and the extant conditions of both temperature and pressure. This chapter gives the fundamental information required to understand the basic concepts of mineral genesis; in turn, that understanding gives an appreciation of the occurrences and associations of minerals, as given in, for example, the mineral descriptions in Part II. Included are a brief consideration of the controls of mineral genesis in the light of the phase rule; a description of some exemplary phase equilibrium diagrams; a review of the chemical makeup of the earth, particularly its crust; and a résumé of the relatively common rocks and mineral deposits that constitute the earth's crust. Our treatments are brief because of space limitations and the fact that these aspects of mineralogy fall in the realm of geochemistry and petrology, both of which are covered in detail in several books and journal articles (see Selected Readings at the end of this chapter).

MINERAL FORMATION AND THE PHASE RULE

The three factors—temperature (T), pressure (P), and chemical composition (X)—that control the formation of a given mineral may be expressed quantitatively by a basic principle of physical chemistry that is known as the *phase rule*:

In any system at equilibrium, the number of phases (P) plus the number of degrees of freedom (F) is equal to the number of components (C) plus 2 (i.e., $P + F = C + 2$).

Pertinent definitions follow:

System: An arbitrarily chosen, finite volume of material. (It may be the contents of a tiny capsule, a hand specimen of a rock, or a complete geological unit.)

Equilibrium: The condition of minimum energy for a system. (Under equilibrium conditions, the constituent phases will exist together in a stable relationship indefinitely as long as the external conditions (*P, T,* and *X*) remain the same.)

Phase: A physically and chemically homogeneous, mechanically separable part of a physical–chemical system.

Degree of freedom (also known as *variance*):   A variable which, if changed, does not result in a change in the equilibrium assemblage or the state of the system. [The degrees of freedom generally considered for geologically important systems are temperature (*T*), pressure (*P*), and quantity of a component (*X*).]

Component: One of the minimum number of chemical species necessary to define the composition of each of the phases of a system; that is, it is an independent chemical entity that defines a system. (As you might suspect, components are often difficult to establish in practice. Among other things, systems are commonly open in nature; that is, components can enter and/or leave the system while it is adjusting itself toward equilibrium. In general, the components considered are the elements or compounds used to express the composition of a system—for example, in the systems shown in Figures 5-1 and 5-2, the components are SiO_2, and $NaAlSiO_4$ and SiO_2, respectively, and in Figure 5-5, the two components are $NaAlSi_3O_8$ and $CaAl_2Si_2O_8$.)

If we rewrite the equation $P + F = C + 2$ in the form $P = C + 2 - F$, it emphasizes the fact that the maximum number of phases (e.g., minerals) increases with the number of components. In other words, assuming equilibrium, the greater the chemical complexity of a rock, the larger the number of minerals that may occur within it. As

you can also see, the maximum number of phases will occur when $F = 0$, so that $P = C + 2$. In a large number of geological environments, however, the condition of an absolutely constant temperature and pressure is highly unlikely; for example, both *T* and *P* tend to vary rather considerably during the crystallization of a magma or the metamorphism of a rock. As a consequence, in mineralogy, we are often dealing with products of divariant systems in which $F \geq 2$. Hence, if $F \geq 2$ is substituted into the phase rule, $P \leq C$; this equation—known as the *mineralogical phase rule*—can be expressed by words as follows:

> In a system of n components under specified conditions of pressure and temperature, the maximum number of mutually stable minerals does not exceed n. (And, as indicated by the equation, it may be less than *n*!)

Two examples follow:

The system with n = 1.   A sample consisting of any individual element is a one-component system. Sulfur, for example, can exist in at least two distinct solid phases, one orthorhombic and one monoclinic; each phase is individually stable over a range of temperature and pressure conditions. Hence, *P* and *T* can be varied independently within a set range without changing the number of phases; thus $F = 2$ and, in the general case, $P = C + 2 - 2 = 1$. Below their melting points, however, at any fixed pressure, the two phases can coexist in equilibrium only at some specific temperature (e.g., at atmospheric pressure, the temperature is 95.5°C). Hence, $F = 1$, so, in this case, $P = C + 2 - 1 = 2$.

As another example, below the temperature of dissociation to Si and O, SiO_2 is a one-component system for which a number of phases (*i.e.*, polymorphs) are known (Figure 5-1). Each polymorph is the stable form over a considerable range of temperature and pressure ($P = C + 2 - 2 = 1$); two polymorphs can coexist in equilibrium under some conditions of temperature and pressure ($P = C + 2 - 1 = 2$)—for example, tridymite and quartz can coexist at 1 atmosphere pressure and 867°C; and a maximum

166

of three polymorphs can coexist in equilibrium under certain set conditions ($P = C + 2 - 0 = 3$)—for example, tridymite, cristobalite and β quartz are in equilibrium at ~4 kilobars pressure and 1470°C. As a consequence, only one form of SiO_2 is found in most rocks, and the rocks containing more than one form are generally considered to represent metastability.

The system with n = 2. The mineralogical phase rule predicts that in a two-component system, the maximum number of minerals will be two ($P = C = 2$). An example is the system $NaAlSiO_4$-SiO_2 (Figures 5-2a and b). [You should, however, keep in mind the fact that in such TX (temperature-composition) diagrams, P (pressure) is constant, generally at 1 bar, and thus not a degree of freedom. Consequently, the lines (areas in three dimensions) and points (lines in three

Figure 5-1.   Phase diagram for unary system SiO_2. (After Boyd, F. R. and England, J. L. 1960. "The quartz-coesite transition." *Journal of Geophysical Research,* **65,** 749–756.)

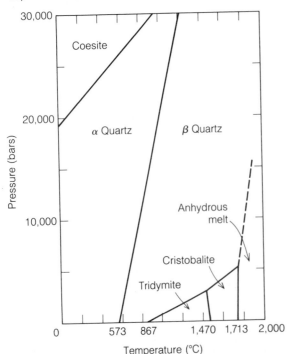

dimensions) of such a diagram represent restrictions on degrees of freedom that seldom obtain.] The following minerals are possible in this system: nepheline ($NaAlSiO_4$), quartz or other silica polymorphs (SiO_2), albite ($NaAlSi_3O_8$), and jadeite ($NaAlSi_2O_6$). In igneous rocks, there are two associations:

> nepheline + albite
>
> albite + quartz (or tridymite).

In metamorphic rocks, there are four associations:

> albite + quartz
>
> jadeite + quartz
>
> jadeite + albite
>
> jadeite + nepheline

Generally speaking, then, in any rock that has reached equilibrium, we find only two of the four possible minerals of this system. [The last pair, jadeite + nepheline, is rare; the best-documented occurrence probably represents metastability.]

Two points, in particular, warrant elaboration:

1. The phase rule does not preclude the existence of a rock containing, for example, nepheline + albite + tridymite (Figure 5-2a); it merely indicates that such a rock either does not represent equilibrium or is unlikely in terms of degrees of freedom.

2. The presence of additional minerals—for example, leucite and zircon in a predominantly nepheline + albite rock—does not invalidate the phase rule; instead, it indicates that more components should have been chosen—in this case, $KAlSiO_4$ and ZrO_2.

The second of these points emphasizes the significance of the mineralogical phase rule to the petrology of essentially all rocks except detrital sedimentary rocks, most of which represent nonequilibrium chemical conditions. For instance, consider a typical granodiorite; its major components are SiO_2, Al_2O_3, Fe_2O_3,* FeO,* MgO,

*Note that different valence states constitute different components.

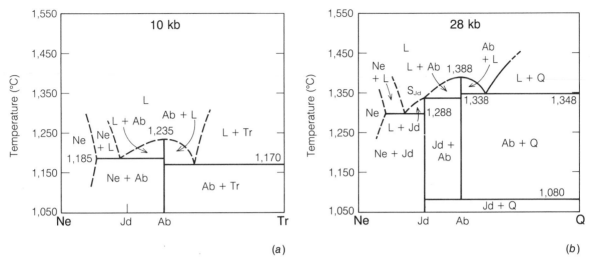

Figure 5-2. The system Ne-Tr (or Q)—i.e., NaAlSiO$_4$-SiO$_2$: (a) at 10 kb; (b) at 28 kb. Ne, nepheline; Jd, jadeite; Ab, albite; Q, quartz; Tr, tridymite; L, liquid (melt). (After Bell, P. M. and Roseboom, E. H. 1969, "Melting relationships of jadeite and albite to 45 kilobars with comments on melting diagrams of binary systems at high pressures." Mineralogical Society of America, Special Paper No. 2, 151–161.)

CaO, Na$_2$O, K$_2$O, and H$_2$O ($C = 9$), and its major phases are quartz, oligoclase-andesine, microcline, and hornblende ($P = 4$). As the phase rule implies, P does not exceed C. Furthermore, even if the common accessory minerals (sphene, apatite, and magnetite) *and* the required additional components (TiO$_2$ and P$_2$O$_5$) are added, there is still conformity with the rule ($P = 7 < C = 11$). As this example suggests, it is often true that the actual number of minerals (phases) is less than the maximum number allowable under the mineralogical phase rule. In fact, the definitions of both components and degrees of freedom are far from clear in many rock systems. For instance, even the minimum number of components can frequently be constituted in more than one way.

PHASE EQUILIBRIUM DIAGRAMS

As shown in the figures of this chapter, phase equilibrium diagrams identify the phases that exist for a given system under the conditions indicated; that is, the diagrams are graphic presentations of chemical systems, and they show how stability fields for diverse phases correlate with the given variables. Although some phase diagrams that pertain to minerals have been prepared on the basis of thermodynamic calculations, most represent results of laboratory investigations of systems equilibrated under carefully controlled temperatures, pressures, and compositions. If used with discretion, the information given in these diagrams can greatly aid mineralogists and petrologists in their interpretations of the geneses of minerals and rocks. (And, these diagrams, as well as phase diagrams for nongeological systems, also find important applications in ceramics and metallurgy.) A number of books outline the methods used in constructing and interpreting phase diagrams (e.g., Levin et al., 1964; Ehlers, 1972; and Ernst, 1976). The following sections will introduce you to some of the more frequently encountered relations shown on phase diagrams.

Unary Diagrams

Phase diagrams for one-component systems are exemplified by the diagram for SiO$_2$ (Figure 5-1).

168

Figure 5-3.   Binary eutectic; the diopside (Di)-Anorthite (An) system. See text for discussion. (After Osborn, E. F. 1942 "The system CaSiO$_3$-diopside-anorthite." *American Journal of Science*, **240**, 751–799.) [The *Lever Rule*, which must be understood for correct interpretation of certain aspects of many phase diagrams, can be explained as follows: The relative amount of the coexisting crystalline and melt phases can be found for any given composition at any given temperature in the two-phase field between the liquidus and the solidus. The amounts are proportional to the lengths of the two segments of the appropriate isothermal tieline that connects the phases involved; the segment lengths are fixed by the intersection of the tieline and the line defining the bulk composition of the material in question (the term

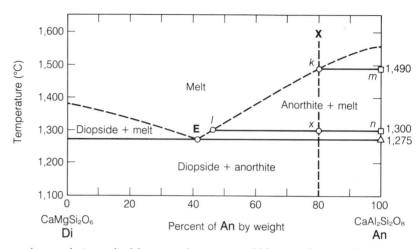

lever rule is applied because the proportions are relative to the lengths of the segments—arms— with the just-noted intersection as the fulcrum); the phase nearer the intersection constitutes the greater relative quantity. Two examples for the bulk composition **X** are as follows: at 1490°C (*k-m*), there would be an infinitesimal amount of crystalline anorthite within a liquid just slightly different (i.e., slightly depleted in An) from the original melt; and, at 1300°C (*l-m*), the proportions would be *l-x* : *n-x* = crystalline An : melt— that is, approximately 62 percent anorthite and 38 percent melt.]

As you can see on that diagram, at atmospheric pressure, α quartz is the stable polymorph up to 573°C, β quartz is stable between 573°C and 867°C, . . . and an anhydrous melt forms above 1713°C. Also, it is evident that the system is divariant when one phase is stable ($F = C - P + 2 = 1 - 1 + 2 = 2$), univariant along the boundary curves where two phases can coexist stably ($F = 1 - 2 + 2 = 1$), and invariant at the junctions, known as triple points, where three phases can exist in equilibrium ($F = 1 - 3 + 2 = 0$). Respective examples are: the α-quartz field, within which both the temperature and pressure are degrees of freedom; the boundary between the α-quartz and β-quartz fields, where pressure is a function of temperature (i.e., there is only one independent variable); and the triple point for the tridymite, cristobalite, and β-quartz fields, where any change of temperature or pres-

sure will result in a reduction in the number of phases that coexist in equilibrium. It also is noteworthy, however, that quenching from high *PT* conditions can give rise to nonequilibrated phases.

Binary Diagrams

Four general features that are shown on two-component (and other multicomponent) equilibrium diagrams are referred to as eutectics, peritectics, solid solution, and solvi. Figures 5-2 through 5-6 include one or more of these relations in systems of significance to mineralogy and petrology. As is generally the case, in each of these diagrams, temperature (*T*) is indicated on the ordinate, composition (*X*) is shown on the abscissa, and pressure (*P*), which has constant value, is specified if it is other than 1 atmosphere (see Figure 5-2).

Binary Eutectic

Figure 5-3 illustrates a binary eutectic. The diopside ($CaMgSi_2O_6$)-anorthite ($CaAl_2Si_2O_8$) system is part of a ternary system, which was originally termed the haplobasaltic system. [This designation was given to direct attention to the fact that this system may be considered to represent basalt in terms of its simplest, Fe-free composition.] In the diagram, the heavy broken line is the liquidus, the heavy solid line is the solidus, and point E is the eutectic.

Liquidus: The line (or surface, if P is a variable) above which the system is completely liquid and below which a solid with or without a liquid is in equilibrium.

Solidus: The line (or surface) representing the TX conditions below which the system is completely solid and above which a liquid with or without a solid is in equilibrium.

Eutectic point: The lowest melting temperature for any mixture. (For this binary system, it is 1275°C, the temperature at which the melt and the two solid phases are at equilibrium.)

As you can see in Figures 5-8, 9, and 10, in multicomponent systems, the liquidus and solidus are surfaces rather than lines, and more than two solid phases can be in equilibrium with a melt at a eutectic point.

If a melt with composition X (on Figure 5-3, $An_{80}Di_{20}$) is cooled so that equilibrium crystallization occurs, the following changes will take place: at ~1490°C, anorthite crystals will start to crystallize; between 1490°C and 1275°C, anorthite crystallization will continue, thus impoverishing the melt phase of its An component; at 1275°C, the remaining melt will have attained the composition corresponding to that of the eutectic, and crystals of anorthite and diopside, in the proportion of the eutectic mixture (An : Di = 42 : 58), will crystallize simultaneously. In all likelihood, the resulting crystalline mass would

consist of relatively large grains of anorthite (formed above 1275°C) surrounded by finer grains of anorthite and diopside (formed at 1275°C); in any case, the overall composition would be 80 weight percent anorthite and 20 weight percent diopside.

If a crystalline aggregate of the same composition ($An_{80}Di_{20}$) is heated, the reverse sequence of changes would take place under equilibrium conditions—that is, melting of the eutectic mixture at 1275°C until all of the diopside (as well as some of the anorthite) has disappeared and then melting of the remaining anorthite, which is present in excess of the eutectic mixture, until total melting has taken place at 1345°C.

Compositions on the diopside side of the eutectic behave in an analogous manner. If, however, a melt of the eutectic composition (the special case) is cooled, only liquid will exist down to 1275°C, at which temperature diopside and anorthite will crystallize simultaneously; if a crystalline aggregate having the eutectic proportions is heated, both minerals will be entirely melted, with total melting being achieved at 1275°C.

The reason that such mixtures of minerals have lower melting points than do the pure phases is beyond the scope of our coverage. Discussions of this phenomenon are given in, for example, Ernst, 1976.

Binary Peritectic

Figure 5-4 illustrates a binary peritectic. The leucite (Lc, $KAlSi_2O_6$)-silica(Tr, SiO_2) system is one of the classic systems with a peritectic. (Note that the eutectic portion of this system, which has not been determined, is shown by a dashed line.)

Peritectic point: A TX point on the liquidus below which an already formed crystalline phase can no longer exist in equilibrium with a melt.

That is, the peritectic point represents the conditions under which the already formed crystalline phase reacts, either partly or completely, with the remaining liquid to form a completely

170   different crystalline phase with composition intermediate to that of the preexisting solid and the remaining liquid. The peritectic is designated by P in Figure 5-4.

In this system, any bulk composition from $Lc_{\sim 59}$ to Lc_{100} will, so to speak, pass through the peritectic. The end products that would result from the cooling of different melts within this compositional range are as follows: Lc_{100} yields leucite; $Lc_{\sim 78}$ to Lc_{100} yields leucite and K-feldspar; $Lc_{\sim 78}$ yields K-feldspar; and $Lc_{\sim 59}$ to $Lc_{\sim 78}$ yields K-feldspar and tridymite.

The sequences of changes that would take place when melts with compositions to the left of P are cooled are as follows: At the appropriate temperature of the liquidus for each melt, leucite crystals will start to crystallize; at the liquidus (= solidus) temperature for Lc_{100} material (i.e., at $1686° ± 5°C$), that melt will be crystallized completely as leucite; from the appropriate liquidus temperatures down to $1150° ± 20°C$, melts of the other compositions will undergo continuous changes involving crystallization of leucite, decrease in the proportion of liquid : leucite crystals, and SiO_2-enrichment (Lc-impoverishment) of the remaining melt; at the peritectic temperature ($1150° ± 20°C$), the leucite plus melt mixtures will undergo reactions whereby some (for Lc_{78-100}) or all (for Lc_{59-78}) of the leucite is converted to K-feldspar and, as a consequence, the Lc_{78-100} product will consist of appropriate proportions of leucite and K-feldspar, the Lc_{78} material will be wholly K-feldspar, and the Lc_{59-78} material will consist of K-feldspar plus melt; between $1150° ± 20°C$ and $990° ± 20°C$, K-feldspar will continue to crystallize from the original Lc_{59-78} material, further enriching the melt in silica; and, at $990° ± 20°C$, the remaining melt will have had its composition changed to that of the eutectic mixture, and K-feldspar and tridymite will then crystallize together, thus using up all of the remaining melt at that temperature.

Heating of crystalline aggregates (other than Lc_{100}) with compositions to the left of P will result

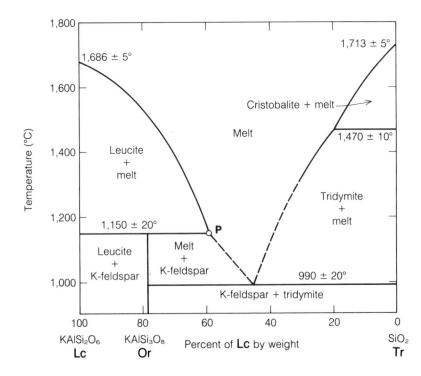

Figure 5-4.   Binary peritectic; the leucite (Lc)-silica (Tr) system. See text for discussion. (Dashed line is indicative of the fact that the eutectic has not been determined.) (After Schairer, J. F., and Bowen, N. L. 1947. "Melting relations in the systems Na_2O-Al_2O_3-SiO_2 and K_2O-Al_2O_3-SiO_2." American Journal of Science, 245, 193–204.)

in a phenomenon termed *incongruent melting*.

Incongruent melting: Melting whereby a substance does not go directly to a melt of its own composition; instead, it goes through an intermediate stage.

[In this system, K-feldspar goes to leucite plus melt, rather than to a $KAlSi_3O_8$ melt, at the temperature of the peritectic.]

Binary Solid Solution

Figure 5-5 illustrates a binary solid solution. The albite($NaAlSi_3O_8$)-anorthite($CaAl_2Si_2O_8$) solid-solution diagram is the best-known system, and perhaps the only truly binary system, in silicate petrology. [Even in this system, however, the solidus is still in the process of confirmation.]

For this system, different rates of cooling of a melt may lead to the formation of different end products: quenching will yield a glass; theoretically, slow cooling will result in the formation of homogeneous crystals; and intermediate rates of cooling should lead to the formation of zoned crystals. The bulk chemistry of each of these products will, of course, have the same composition as the original melt; the zoned crystals attain the bulk composition by having their more calcic cores compensated for by more sodic outer zones.

When crystals are formed, the first-formed crystals are more calcic than the melt. For example, a melt with the composition An_{50}, X on Figure 5-5, would first yield crystals of $An_{\sim85}$, Y. Theoretically, during slow cooling, the first formed and subsequently formed crystals would react continuously with the remaining liquid, thus maintaining equilibrium, until all of the melt had been used up, and the system would then consist wholly of homogeneous crystals with the composition An_{50}. (This situation is seldom achieved either in nature or in the laboratory.) As we have already stated, during more rapid (i.e., "intermediate rate") crystallization, equilibrium between already formed crystals and the remaining liquid is not maintained; consequently, the crystals are commonly zoned with, for example,

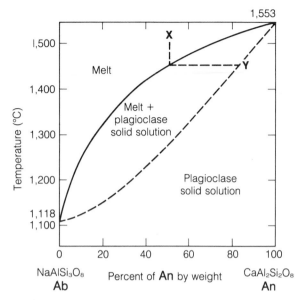

171

Figure 5-5. Binary solid solution: the albite (Ab)-anorthite (An) system. See text for discussion. (After Bowen, N. L. 1913. "The melting phenomena of the plagioclase feldspars." *American Journal of Science*, 35 (4th ser.), 577–599.)

their outer zones more sodium-rich than the original melt. Such zoning reflects the fact that preservation of early formed, relatively calcium-rich plagioclase leads to both sodium-enrichment of the melt and ensuing crystallization of relatively sodium-rich plagioclase, so that the bulk composition of the crystals is the same as the composition of the original melt. The zoning may be gradational, well defined, or any combination of both.

The Solvus

Figure 5-6 illustrates the solvus. In the subsolidus part of the albite($NaAlSi_3O_8$)-K-feldspar ($KAlSi_3O_8$) system, the relationship termed *solvus* is exhibited.

Solvus: In binary systems a line in *TX* space (or a surface in *PTX* space) that separates a higher temperature field, in which there is single homogeneous solid solution, from a lower

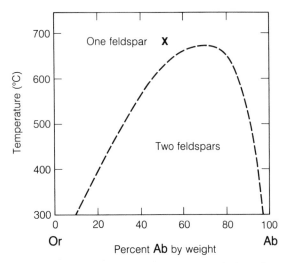

Figure 5-6. Solvus; subsolidus part of the albite (Ab)-orthoclase (Or) system. See text for discussion. (After Tuttle, O. F. and Bowen, N. L. 1958. "Origin of granite in the light of experimental studies in the system $NaAlSi_3O_8$-$KAlSi_3O_8$-SiO-H_2O." Geological Society of America, Memoir 74, 153 pp.

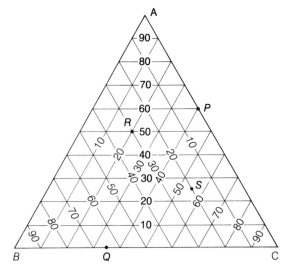

Figure 5-7.   The representation of compositions in a three-component system. *A* represents 100 percent of component *A*, *B* 100 percent of component *B*, *C* 100 percent of component *C*. *P* represents 60 percent of *A*, 40 percent of *C*; *Q* represents 65 percent of *B*, 35 percent of *C*; *R* represents 50 percent of *A*, 30 percent of *B*, 20 percent of *C*; *S* represents 25 percent of *A*, 20 percent of *B*, 55 percent of *C*.

temperature field, in which two phases may form by exsolution from the solid solution.

The diagram indicates, therefore, that a homogeneous alkali-feldspar that is stable at high temperatures will undergo exsolution to two feldspars, typically to form a perthite (see page 403), if held at or slightly below the appropriate temperature of the solvus for a sufficiently long time for exsolution to take place. For example, an alkali-feldspar with composition $Ab_{50}Or_{50}$ (*X* in the diagram) would, upon slow cooling, exsolve to two feldspars at temperatures $\leqq 630°C$.

Ternary Diagrams

Phase equilibrium relations have been determined for several three-component systems. Figure 5-7 and its caption explain how compositions are plotted on these diagrams. Figures 5-8 and 5-9 are ternary diagrams; their captions explain how they can be interpreted. Note that in the

triangular diagram, pressure is constant and temperature, which is variable, is shown by contours. The triangular prism, block diagram is constructed merely to show an alternative presentation of the same information. In any case, as you can see, interpretation involves application of the principles already outlined for binary diagrams.

Other Kinds of Phase Diagrams

Many systems have been investigated at diverse pressures. The resulting diagrams can be interpreted individually in the manner already outlined. In many cases, however, two or more diagrams for a single system—each under a different isobaric condition—are considered together (e.g., Figures 5-2a and b); from these, you can interpret phase relations that may occur at, for example, different depths below the earth's surface.

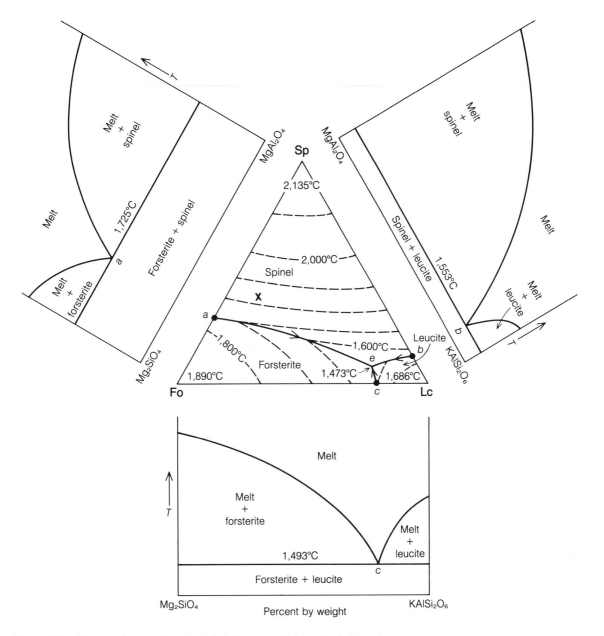

Figure 5-8. Ternary diagram for spinel (Sp)-forsterite (Fo)-leucite (Lc), and associated binary systems. Cooling of melt with composition **X** would first give spinel; next, spinel and forsterite would crystallize in eutectic proportions; finally, at 1473°C (the ternary eutectic for the system), the spinel and forsterite would be accompanied by crystallization of leucite. Complete crystallization would occur at this temperature. (After Schairer, J.F. 1955. "The ternary systems leucite-corundum-spinel and leucite-forsterite-spinel." *Journal of American Ceramic Society,* **38,** 153–158.)

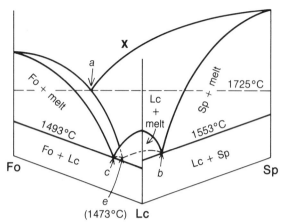

Figure 5-9. Temperature-composition prism model (pressure held constant) of system shown in Figure 5-8.

scriptive in nature. Nonetheless, genetic inferences can often be made on the basis of their relations. The most frequently used diagrams of this type are named—on the basis of their apical components—*ACF, A'KF,* and *AFM* diagrams. [In *ACF* diagrams $A = Al_2O_3 + Fe_2O_3 - (Na_2O + K_2O)$, $C = CaO - 3.3P_2O_5$, and $F = FeO + MgO + MnO$. In *A'KF* diagrams $A' = Al_2O_3 + Fe_2O_3 - (Na_2O + K_2O + CaO)$, $K = K_2O$, and $F = FeO + MgO + MnO$. In *AFM* diagrams $A = Al_2O_3$, $F = FeO$, and $M = MgO$.] The calculations and construction of these diagrams are outlined in, for example, Turner (1981).

Still another kind of diagram, generally called an Eh-pH diagram, is sometimes used in sedimentary petrology and in studies dealing with chemical weathering and with groundwater solution and deposition. These diagrams correlate mineral stability with Eh and pH of pore solutions under given temperature-pressure conditions.

Phase relationships for systems with more than three components have been investigated. Quaternary systems under given *PT* conditions can be plotted in three-dimensional space by using tetrahedra, each face of which is a ternary system (Figure 5-10); systems with more than four components cannot be expressed solely by graphic means.

Another kind of diagram that is related to phase diagrams was introduced by the Finnish petrologist Pentti Eskola; they are generally termed *three-component triangular diagrams.* The originally introduced diagram and also some more recently introduced versions are used widely, especially by metamorphic petrologists. In essence, the overall compositions of different metamorphic facies (minus almost universally present quartz and plagioclase and accessory mineral contents) are keyed to three chemical components that define the apices of the triangles; actual mineral phases are then plotted; tie lines are drawn between the phases known to coexist in diverse rocks of the depicted facies (e.g., see Figure 5-11). In nearly all of these diagrams, there are, of course, approximations that render them less than quantitatively accurate; that is, the diagrams are, strictly speaking, de-

Figure 5-10. Quaternary system—a hypothetical system *A-B-C-D*—phase regions are volumes; two solids can coexist with a liquid, three solids can coexist with a liquid, or four solids can coexist. (After E. G. Ehlers. *The Interpretation of Geological Phase Diagrams.* © 1972 by W. H. Freeman and Company, San Francisco.)

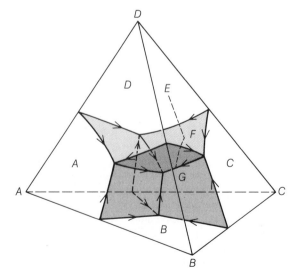

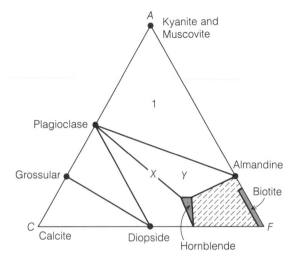

Figure 5-11. *ACF* diagram for the amphibolite facies. Modified after Turner (personal communication, 1981). Phases that coexist in equilibrium are at the ends of individual lines having no indicated intermediate phases or at the apices of triangles not circumscribing any smaller triangle.

The reduction-oxidation potential, Eh, is expressed as the potential of a half-cell as compared to the standard hydrogen half-cell; the acidity (or basicity) of a solution, pH, is expressed as the negative $\log_{10}$ of hydrogen activity. Several of these diagrams are given and explained by Garrels and Christ (1965). An example is shown in Figure 5-12.

CHEMICAL COMPOSITION OF THE EARTH'S CRUST

As is evident from the preceding discussion of the phase rule and phase equilibrium diagrams, the controls of mineral formation are bulk chemistry, pressure, and temperature. Of these, bulk chemistry is the ultimate control. Therefore, the identity of most of the minerals that we see depends on the overall composition and the general distribution of elements in the earth's crust.

Determination of the relative abundances of the different elements in the earth's crust is fraught with difficulties because the crust is not homogeneous, and we can sample it only at the surface or at relatively shallow depths in mines or boreholes. Fortunately, however, such geological processes as mountain building and denudation have resulted in the surface exposure of rocks and mineral deposits formed at depths of up to several kilometers, and there also are geophysical data from which the properties and hence the compositions of subsurface rocks can be extrapolated. Thus, when we take all of the pertinent considerations into account, we can make fairly reliable estimates of crustal abundances of the elements.

A first step is to calculate the distribution of rock types within the crust. This calculation has been done in various ways. One of the recent results is given in Table 5-1. On the basis of these and most other calculations, the average composition of the crust becomes, in effect, the appropriately weighted composition of igneous and metamorphic rocks. Thus, the next step is the conversion of rock distribution data into chemical data.

From the many thousands of chemical analyses that have been made of rocks, it has been well established that only a few elements (and for nearly all rocks, the same elements) are present as the essential components. Consequently, the analysis of nearly any rock can be considered reasonably complete if the following constituents are determined: SiO_2, TiO_2, Al_2O_3, Fe_2O_3, FeO, MnO, MgO, CaO, Na_2O, K_2O, P_2O_5, H_2O, and CO_2. In practice, the rarer elements are often not determined unless there is some mineralogically based reason to expect them (e.g., if a rock is zircon-rich, ZrO_2 might also be determined) or unless the analysis is being made in conjunction with research dealing with minor or trace element behavior. Nonetheless, during the last half century, many, many rocks have been analysed for minor and trace elements as well as for major elements; and, from these analyses, the average chemical makeup of the crust has been determined (Table 5-2).

As is evident from Table 5-2, only eight elements—O, Si, Al, Fe, Ca, Na, K, and Mg—are

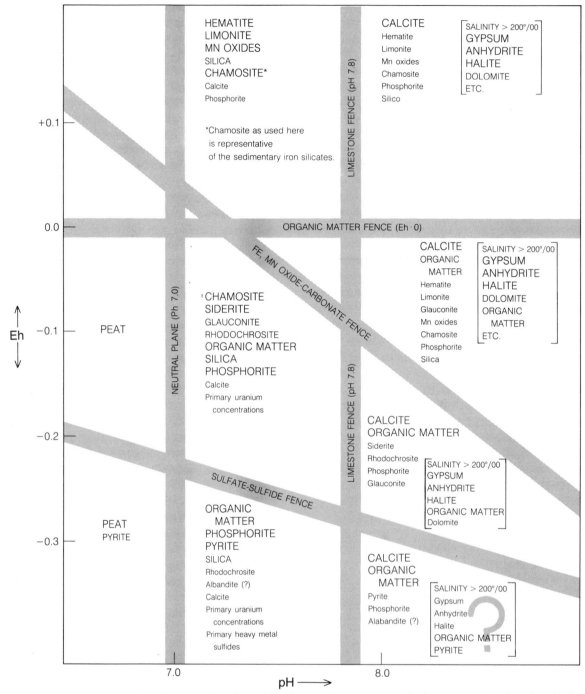

Figure 5-12   Qualitative classification, based on Eh and pH, of several minerals that occur as (or in) chemically precipitated sediments. (After Krumbein, W. C. and Garrels, R. M. 1952. "Origin and classification of chemical sediments in terms of pH and oxidation-reduction potentials." *Journal of Geology*, **60**, 1–33.)

present in amounts greater than 1 percent, and these make up nearly 99 percent of the earth's crust; four more—Ti, H, P, and Mn—are present in amounts between 0.1 percent and 1 percent; the remaining elements together make up the remaining less than 0.5 percent of the crust. It is also evident that of the major elements, oxygen is absolutely predominant; it makes up about one-half of the crust by weight. And, this predominance is even more marked when the weight percentages are recalculated to atom percentages and to volume percentages (Table 5-3). As shown in Table 5-3, in terms of number of atoms, oxygen exceeds 60 percent and, in terms of the volume, oxygen makes up more than 90 percent of the total volume that is occupied by the elements of the crust. Hence, for all practical purposes, the crust as a whole may be considered to be a packing of oxygen anions that are bonded by silicon and ions of the common metals. It is also obvious from this table that the dominant minerals of the earth's crust must be quartz and silicates and alumino-silicates of iron, magnesium, calcium, sodium, and potassium.

As you can also see from Table 5-2, some of the elements that have long been known and used by man are less abundant than several of the relatively little known elements. Thus, for example, copper is less abundant than zirconium, lead is comparable in abundance to gallium, mercury is rarer than most of the rare earths, etc., etc. As you might expect, the occurrence of each particular element depends on its crystal chemistry. Thus, if an element forms one or more minerals that occur in quantity in a specific geological environment, you would expect the mineral(s) to have attracted attention and eventually to have been used; and, indeed, the longest known and most familiar elements are those that are major constituents of easily recognized minerals that occur in local concentrations and are readily converted into useful materials. For example, lead, although of comparatively low abundance, occurs as the sulfide galena in local concentrations and has been known and used since prehistoric times. And, on the contrary, gallium forms no minerals

Table 5-1. *Distribution of rock types, expressed in volume percent, within the earth's crust*

Igneous Rocks	Volume Percent	
basalts, etc.	42.5	
granites	10.4	
granodiorites and diorites	11.2	
syenites	0.4	
peridotites	0.2	
		64.7
Sedimentary Rocks		
sandstones	1.7	
clays and shales	4.2	
carbonates and evaporites	2.0	
		7.9
Metamorphic Rocks		
gneisses	21.4	
schists	5.1	
marbles	0.9	
		27.4
Totals	100.0	100.0

Data from Ronov, A. B., and Yaroshevsky, A. A., 1969, in Hart, P. J. (ed.) *The Earth's Crust and Upper Mantle.* Geophysics Monograph 13, American Geophysical Union.

in which it is present in significant amounts, is little known, and has found no common use.

The lack of gallium minerals depends on the fact that in ionic size (see Appendix B) and general chemical properties gallium is so similar to aluminum that most of it substitutes for aluminum in alumino-silicate minerals. Elements with this mode of occurrence are generally referred to as *dispersed elements*, which means that these elements, rather than being concentrated as major constituents in any particular mineral(s), are systematically dispersed in solid solution, substituting for some major constituent, in one or more

Table 5-2.   Average amounts of the elements, except for the rare gases and short-lived radioactive elements, in rocks of the earth's crust, expressed in grams per ton (or parts per million).

Atomic Number	Element	Crustal Average	Atomic Number	Element	Crustal Average
1	H	1,400	45	Rh	0.005
3	Li	20	46	Pd	0.01
4	Be	2.8	47	Ag	0.07
5	B	10	48	Cd	0.2
6	C	200	49	In	0.1
7	N	20	50	Sn	2
8	O	466,000	51	Sb	0.2
9	F	625	52	Te	0.01
11	Na	28,300	53	I	0.5
12	Mg	20,900	55	Cs	3
13	Al	81,300	56	Ba	425
14	Si	277,200	57	La	30
15	P	1,050	58	Ce	60
16	S	260	59	Pr	8.2
17	Cl	130	60	Nd	28
19	K	25,900	62	Sm	6.0
20	Ca	36,300	63	Eu	1.2
21	Sc	22	64	Gd	5.4
22	Ti	4,400	65	Tb	0.9
23	V	135	66	Dy	3.0
24	Cr	100	67	Ho	1.2
25	Mn	950	68	Er	2.8
26	Fe	50,000	69	Tm	0.5
27	Co	25	70	Yb	3.4
28	Ni	75	71	Lu	0.5
29	Cu	55	72	Hf	3
30	Zn	70	73	Ta	2
31	Ga	15	74	W	1.5
32	Ge	1.5	75	Re	0.001
33	As	1.8	76	Os	0.005
34	Se	0.05	77	Ir	0.001
35	Br	2.5	78	Pt	0.01
37	Rb	90	79	Au	0.004
38	Sr	375	80	Hg	0.08
39	Y	33	81	Tl	0.5
40	Zr	165	82	Pb	13
41	Nb	20	83	Bi	0.2
42	Mo	1.5	90	Th	7.2
44	Ru	0.01	92	U	1.8

Table 5-3. *The eight most common chemical elements in the earth's crust*

	Weight Percent	Atom Percent	Volume Percent
O	46.60	62.55	91.7
Si	27.72	21.22	0.2
Al	8.13	6.47	0.5
Fe	5.00	1.92	0.5
Ca	3.63	1.94	1.5
Na	2.83	2.64	2.2
K	2.59	1.42	3.1
Mg	2.09	1.84	0.4

relatively common minerals. Two other dispersed elements are rubidium, which is always dispersed in potassium minerals, and hafnium, which is typically dispersed in zirconium minerals. [It is for this reason that, for example, Rb (or Rb_2O) and K (or K_2O) might be dealt with as a single component in phase equilibrium considerations.]

Geochemical Classification of the Elements

An element is often quite specific as far as the type of minerals it forms. Some elements, such as gold and the platinum metals, nearly always occur in the native state; others, such as copper, zinc, and lead, typically occur as sulfides; still others (e.g., the alkali and alkaline earth metals) generally occur as oxygen compounds, especially as silicates or alumino-silicates, and are never found in the native state or as sulfides; and, the inert gases, as the name implies, form no minerals at all. A useful, though admittedly little-used, geochemical and mineralogical classification of the elements categorizes them into four groups (Table 5-4):

1. *Lithophile* elements occur mainly in oxygen compounds.

2. *Chalcophile* elements occur mainly as sulfides.

3. *Siderophile* elements occur mainly as native elements.

4. *Atmophile* elements are gaseous elements that do not readily form compounds and therefore occur mainly in the atmosphere.

Some elements may be considered to belong to more than one group since the type of compounds that an element forms is dependent not only on the nature of the element but also on the conditions of formation—that is, the temperature, the pressure, and the identity of other elements that are present. For instance, most of the iron in the crust is present as oxides or silicates, but in chemical environments with low oxygen and high sulfur, iron sulfides are formed, and in highly reducing environments, when little sulfur is present, native iron may be produced. Such variations are indicated in Table 5-4 by showing the element in parentheses under the group of secondary affinity.

The geochemical character of an element is governed largely by the electronic configuration of its atoms, which, in turn, controls the type of compounds the element can form. Hence, geochemical character is closely related to position in the periodic table, as you can see in Table 5-5. Lithophile elements are those that ionize readily or form stable oxyanions, such as CO_3^{-2} and PO_4^{-3}; the bonding in their compounds is largely ionic in character. Chalcophile elements ionize less readily and thus tend to form covalent compounds with sulfur (and selenium and tellurium, where present). Siderophile elements do not readily form compounds with oxygen or sulfur; metallic bonding is the normal condition. As is evident, a knowledge of the geochemical character of an element will enable you to predict the type of minerals the element is likely to form and, carried one step further, the associations in which it is likely to occur.

Mineralogical Composition of the Earth's Crust

As we have just implied, the geochemical character and the abundance of an element are largely

Table 5-4.   Geochemical classification of the elements

Lithophile	Chalcophile	Siderophile	Atmophile
Li, Na, K, Rb, Cs	Cu, Ag	Pt, Ir, Os	Inert gases
Be, Mg, Ca, Sr, Ba	Zn, Cd, Hg	Ru, Rh, Pd	N, (O)
B, Al, Sc, Y	In, Tl, Pb	Au, (Fe)	
Rare earths	As, Sb, Bi		
C, Si, Ti, Zr, Hf	S, Se, Te		
Th, P, V, Nb, Ta	Ni, Co, (Fe)		
O, Cr, W, U	Mo, Re, (Mn)		
H, F, Cl, Br, I	(Ga), (Ge), (Sn)		
Fe, Mn, Ga, Ge, Sn			
(Mo), (Cu), (Zn), (Pb)			
(Tl), (As), (Sb), (Bi)			
(S), (Se), (Te), (Ni), (Co)			

responsible for both the number of minerals the element forms and the geological environment in which the minerals may form and occur. On first consideration, it seems rather remarkable that the earth's crust, which is made up of more than 80 elements (excluding the short-lived radioactive ones), contains only some 3000 different minerals, most of which are rare. There are, of course, a far greater number of inorganic compounds, but most of them do not occur as minerals; only the most stable ones can occur as minerals because less stable compounds either do not form in nature, or after forming, they soon decompose. Another limitation on the number of minerals is the already mentioned fact that geological processes fail to separate dispersed elements from the more abundant mineral-forming elements with similar radii and charges. Also, the 15 rare-earth elements form very few minerals—for one thing, individual rare-earth elements are never differentiated; for another, they typically occur as dispersed elements with calcium. Indeed, altogether, the rare-earth elements form fewer minerals than does, for example, antimony, an element that is much less abundant than most of the individual rare-earth elements. Consequently, the mineralogy of the crust is much simpler than you might expect from its elemental composition, and

the mineralogical variation is even more limited for most specific geological environments.

On a broad scale, three major kinds of environment can be recognized: magmatic, sedimentary, and metamorphic. Each can be divided into subsidiary environments according to the variety of physical and chemical conditions. The mineralogy of most rocks depends on the temperatures and pressures of crystallization (or recrystallization) and the variation in chemical composition of the magmas, solutions, and/or preexisting rocks from which they were derived.

Magmatic environments are characterized by high-to-moderate temperatures, a wide range of pressure, and a fairly limited range in chemical composition. *Sedimentary environments* are characterized by a moderate range of temperatures (generally between 0°C and 40°C), essentially constant pressure (approximately atmospheric), and diverse materials that can be physically or chemically deposited materials derived from preexisting rocks and/or their weathering products. *Metamorphic environments* cover a wide range of temperature and pressure and include the materials of preexisting rocks of any kind.

There are also several rocks, and mineral deposits, and mineral occurrences that do not fit

Table 5-5. *Geochemical classification of the elements in relation to the periodic table*

H																	He
Li	Be	B	C										N	O	F		Ne
Na	Mg	Al	Si										P	S	Cl		A
K	Ca	Sc	Ti	V	Cr	Mn	Fe	Co	Ni	Cu	Zn	Ga	Ge	As	Se	Br	Kr
Rb	Sr	Y	Zr	Nb	Mo		Ru	Rh	Pd	Ag	Cd	In	Sn	Sb	Te	I	Xe
Cs	Ba	La-Lu	Hf	Ta	W	Re	Os	Ir	Pt	Au	Hg	Tl	Pb	Bi			
			Th		U												

Lithophile: **Na (boldface)**
Chalcophile: <u>Cu</u> (underlined)
Siderophile: Au (Roman)
Atmophile: *N* (italic)

well into these three categories. Four are especially noteworthy as far as mineral genesis and occurrence are concerned. They are (1) pegmatites, (2) veins and products of wallrock alteration, (3) diagenetic rocks, and (4) minerals produced by weathering and groundwater activities. Most pegmatite masses and many veins have been deposited by volatile-rich fluids derived from or formed as a result of magmatic activity; deposition from magmatically generated or heated fluids and solutions has also occurred around fumaroles and hot springs. Furthermore, some of these fluids have caused alteration of nearby country rock, and the result is widely referred to as *wallrock alteration. Diagenesis* is the term applied to low-temperature/low-pressure changes that frequently take place in sediments prior to, and often contributing to, their lithification (i.e., their conversion into rock); the main diagenetic processes that involve mineral formation are replacement, recrystallization, and cementation. Weathing can cause chemical changes such as solution, oxidation, reduction, and hydration of minerals: original minerals may be decomposed; new minerals may be formed; mineral matter in solution may be added to a rock; minerals may be dissolved and subtracted from a rock. Nearly all chemical weathering involves groundwater solutions. These same solutions can also be responsible for formation of minerals such as those that line vugs, those that fill fissures, those that constitute speleothems, and those that are deposited as duricrusts and hardpans.

As you may deduce from these brief accounts about rock and mineral genesis, many mineral associations can be explained on the basis of the phase rule and be correlated with the results of experimental investigations of quasi-geological systems. Indeed, the already available and to-be-determined data of this type also provide information about equilibrium versus nonequilibrium relations, temperatures and pressures of formation, etc. Actually, of the listed occurrences, the minerals of all except the detrital sedimentary rocks can be considered to have had at least a tendency to have come into chemical equilibrium with the environment in which they were formed. And, even the minerals of the detrital suites are predictable in that most of them are minerals that are physically resistant and chemically stable under conditions of weathering, transportation, and deposition.

When you realize that nearly all minerals fall within these relatively few genetic categories, each of which can be characterized by a few

182   mineral associations, it becomes amply clear that insight into and familiarity with mineral associations are extremely important in both the identification of minerals and the understanding of their origins. Such insight is best gained by working with minerals, especially those still attached to their surroundings, and by first becoming familiar with and thence keeping abreast of the literature of mineralogy, petrology, economic geology, and geochemistry.

SELECTED READINGS

Bateman, A. M. 1950. *Economic Mineral Deposits* (2nd ed.). New York: Wiley, 916 pp.

Carmichael, I. S. E., Turner, F. J., and Verhoogen, John. 1974. *Igneous Petrology*. New York: McGraw-Hill, 739 pp.

Dietrich, R. V., and Skinner, B. J. 1979. *Rocks and Rock Minerals*. New York: Wiley, 319 pp.

Ehlers, E. G. 1972. *The Interpretation of Geological Phase Diagrams*. San Francisco, Ca.: W. H. Freeman and Company, 280 pp.

Ernst, W. G. 1976. *Petrologic Phase Equilibrium*. San Francisco, Ca.: W. H. Freeman and Company, 333 pp.

Garrels, R. M., and Christ, C. L. 1965. *Solutions, Minerals and Equilibria*. New York: Harper & Row, 450 pp.

Levin, E. M., Robbins, C. R., and McMurdie, H. F. (compilers). 1964. *Phase Diagrams for Ceramists*. Columbus, Oh.: American Ceramic Society, 601 pp.

Lindgren, Waldemar. 1933. *Mineral Deposits*. New York: McGraw-Hill, 930 pp.

Mason, Brian, and Moore, Carleton. 1982. *Principles of Geochemistry* (4th ed.). New York: Wiley, 368 pp.

Pettijohn, F. J. 1975. *Sedimentary Rocks* (3rd ed.). New York: Harper & Row, 628 pp.

Ramdohr, Paul. 1979. *The Ore Minerals and Their Intergrowths* (2nd ed. in English). 2 volumes. New York: Pergamon, 1205 pp.

Stanton, R. L. 1972. *Ore Petrology*. New York: McGraw-Hill, 713 pp.

Turner, F. J. 1981. *Metamorphic Petrology* (2nd ed.). New York: McGraw-Hill, 524 pp.

Williams, Howel, Turner, F. J., and Gilbert, C. M. 1982. *Petrography* (2nd ed.). San Francisco, Ca.: W. H. Freeman and Company, 640 pp.

6 Determinative and Descriptive Mineralogy

Determinative mineralogy and descriptive mineralogy constitute an interdependent pair: determinative efforts are based on descriptions; descriptions are based on determinative efforts. Nonetheless, some of the information and procedures given in this chapter are used most frequently in determinative mineralogy, whereas others are currently used almost exclusively in descriptive or applied mineralogy. Our coverage includes the description of properties that can be seen or determined easily without using laboratory equipment; an outline of the most widely used laboratory methods—optical mineralogy, X-ray diffraction, and electron probe microanalysis; and a tabulation, for reference, of several special techniques, most of which require sophisticated equipment and/or specialized training to use or to interpret.

MACROSCOPIC IDENTIFICATION

The identification of a mineral in hand specimen is for many persons as much an art as it is a scientific procedure. The recognition of an unknown mineral may be instantaneous, or it may require careful, in some cases, time-consuming tests. The difference, of course, depends on the identity of the mineral, the quality of the specimen, and the knowledge, experience, and skill of the observer. Several schemes have been devised for mineral identification; one such scheme is outlined in Chapter 16.

Actually, identification of minerals is relatively simple. Compared to, for example, animal and plant species, the total number of mineral species is small and, furthermore, of the approximately 3000 known mineral species, most are so rare that they are seldom encountered except in collections. Nonetheless, to some beginning students, even the limited number of minerals to be identified in an elementary course poses an apparently formidable problem. The fact that this mental hurdle is more apparent than real is demonstrated by the facility with which experienced mineralogists and many mineral collectors can often name an apparently nondescript specimen

183

184 on sight or after performing only a simple test or two. The reason they can do this, however, warrants emphasis; it is because of their familiarity with many, many specimens. Because of that familiarity, they are able instinctively to sum up the characteristics of an unknown specimen and subconsciously to compare those characteristics with their mental picture of innumerable specimens they have previously identified. A good analogy is the way people recognize others whom they have seen several times—by the sum total of the person's physical features, dress, voice, gait, and other characteristics.

Nonetheless, in the beginning, logical schemes of identification often serve as useful guides (see Chapter 16). Next, careful examination of several different specimens of a species soon reveals the properties that are suitably diagnostic. Thence, the best training is the repeated (and repeated) study of minerals in collections and in the field. With the experience thereby gained, many short cuts become apparent. Subsequently, even when initial examination does not serve to identify a specimen, it frequently limits the possibilities to a comparatively few minerals so that the next step requires the selection of only one or a few suitable diagnostic tests to complete the identification.

The properties described in the following sections are arranged in line with the identification scheme suggested in Chapter 16.

PHYSICAL PROPERTIES

Luster

Luster, which is discussed in detail in Chapter 4, is a useful diagnostic property. The identifying terms generally used are largely self-explanatory.

Minerals can be divided into two main categories on the basis of their luster—those that are *metallic* and those that are *nonmetallic*. An imperfect metallic luster is sometimes referred to as *submetallic*; on some determinative tables, minerals with submetallic lusters are included with both metallic and nonmetallic minerals. Minerals with metallic luster are generally opaque; chemically, they are most of the native metals, the sulfides, and the metallic oxides. Minerals with nonmetallic luster are generally transparent or translucent; chemically, they include minerals of the other classes.

Different varieties of nonmetallic luster can be recognized. The more frequently noted ones are *vitreous*, the luster of broken glass; *adamantine*, the brilliant luster of diamond; *resinous*, a luster like that of resin; *silky*, the appearance often termed *sheen*, which is typical of fibrous minerals; *pearly*, the semiglossy luster of the inside of an oyster shell, a luster that is common on cleavage surfaces of some minerals such as the feldspars; and *dull* or *earthy*, the low reflectance of compact masses of some minerals.

Several terms are used to describe special appearances in reflected light. Some of the more frequently used terms are (alphabetically): *adularescence*, *asterism*, *aventurescence*, *chatoyancy*, *iridescence*, *labradorescence*, *opalescence*, and *schiller effect*. Most refer to some play-of-color effect whereby there is an apparent variation of color when the material is viewed from different angles. As noted in Chapter 4, these effects can be shown to be dependent on a mineral's high dispersion or on inhomogeneities due to the presence of such things as specially spaced and/or sized lamellar structures, fibrous or platy inclusions, voids, or structural discontinuities. Some surficial films—tarnishes—give similar appearences. A few of these phenomena are noted in the appropriate mineral descriptions in Part II, and they are included in the Remarks column of the determinative tables.

Color and Streak

As noted in Chapter 4, the use of color as a diagnostic feature often requires experience as well as discrimination. *Inherent color*, which depends on the essential constituents of a mineral, is constant and characteristic; *exotic color*, which may or may not be present in a given specimen of a mineral, is variable. Therefore, inherent color is a useful diagnostic feature, whereas exotic color may be misleading. In the tables in Chapter 16, we note frequently exhibited exotic colors as well as inherent colors.

In many cases, streak—the color of the fine powder of a mineral—is useful in distinguishing between exotic and inherent colors. Most exotic colors, which are much more common in non-metallic than in metallic minerals, are not perceptible in mineral powders; for example, dark-colored plagioclase feldspars give white streaks.

Fine powders of minerals used for checking their streaks can be obtained by grinding, scratching, or otherwise bruising the surface of a mineral, or by drawing the mineral across a piece of white unglazed porcelain—usually called a *streak plate*. (A streak plate cannot be used for minerals that are harder than the plate—$H \sim 6\frac{1}{2}$.) Streak is more constant and hence more reliable than color as perceived for larger specimens because it eliminates spurious effects such as those attributable to grain size differences. Streak is most useful for characterizing minerals with metallic lusters.

Hardness

Mineral hardness is usually expressed in terms of Mohs' scale:

Hardness	Mineral	
1	talc	
2	gypsum	
		fingernail
3	calcite	
4	fluorite	
5	apatite	
		knife blade
6	feldspar (orthoclase or microcline)	and glass
7	quartz	
8	topaz	
9	corundum	
10	diamond	

As noted on the scale, most fingernails have a hardness of about 2.5, and most knife blades and pieces of common window glass have hardnesses of about 5.5. In addition, copper coins have a hardness of 3, and most high quality metal files have a hardness of about 6.5.

Hardness testing sets can be purchased from many mineral dealers and hobby shops. They consist of pencil-like rods with sharp chips of minerals of known hardnesses mounted on their ends. For many identifications, determining that the mineral is soft ($H < 2.5$), hard ($H > 5.5$), or intermediate (H between 2.5 and 5.5) is all that is necessary.

As an example of how the scale can be used, a mineral with $H = 6.5$ would scratch feldspar, would be scratched by quartz, and would neither scratch nor be scratched by the typical high quality metal file. You must, however, keep two precautions in mind:

1. The hardness given for a mineral is that for a smooth clean surface, such as a crystal face or a cleavage plane; therefore, superficial coatings of weathered or altered material may give deceptively low hardness values.

2. Fine-grained masses of a mineral may be disaggregated when checked for hardness, thus giving the false impression of having been scratched. In some cases, this false impression can be avoided by using the unknown as the scratcher rather than the material being scratched.

In addition, a difference in hardness with direction is significant for a few anisotropic minerals, such as for kyanite.

Specific Gravity

The specific gravity (G) of a crystalline substance is a fundamental property and thus characteristic of the substance. As such, it is a valuable diagnostic property. Additionally, its determination has the advantage of being completely non-destructive—except, of course, for minerals that are readily soluble.

186

The accurate determination of the specific gravity of a mineral requires considerable care because there are numerous possibilities for error that must be guarded against. The more serious of these are: errors inherent in the method used, errors arising from inhomogeneity of the sample, and errors introduced by the observer. With regard to the first, it is important to select the technique that is best suited to give accurate results for the material available. With regard to the second, it is often difficult to obtain large pieces of homogeneous material—even of a mineral that is available in considerable amounts—because of inclusions of foreign material; therefore, best results are usually obtained by working with small amounts that are chosen as carefully as possible (preferably small grains whose purity has been checked by microscopic examination). Great difficulty also arises in working with fine-grained porous materials (e.g., the clay minerals); air trapped in the pores give a fallaciously low value. Special procedures must be used to remove the trapped air from these minerals. Usually, the air is removed by boiling the material in either water or the displacement liquid used. The third kind of error, operator error, can be obviated only by exercising extreme care.

Of the methods available for determining the densities of solids, the following are particularly suitable for minerals:

1. The weight is measured directly, the volume by the Principle of Archimedes.

2. The weight is measured directly, the volume from the weight of liquid displaced in a pycnometer.

3. The specific gravity is measured by direct comparison with heavy liquids having known (or subsequently determined) specific gravities—the suspension method.

In the first method, the volume is determined by measuring the apparent loss of weight when a weighed fragment of the mineral is immersed in a suitable liquid. The fragment displaces an amount of liquid equal to its own volume, and its weight is apparently dimished by the weight of the liquid displaced. If W_1 is the weight of the fragment in air, and W_2 the weight of the fragment in liquid of specific gravity L, the specific gravity of the unknown solid, G, is:

$$G = \frac{W_1}{W_1 - W_2} L \;.$$

Water is often used as the displacement liquid because it is readily available, and, since its specific gravity is 1 or close to 1, the factor L can be eliminated in routine determinations. Water, however, is not the most suitable liquid for accurate determinations because it has a high surface tension and does not wet most solids readily; consequently, it can have a damping effect on the fine wire, that is often used to hold the specimen, and, in many instances, bubbles are tenaciously held by the solid and thereby give rise to low specific gravity values. [Some people add a pinch or so of detergent to the water to reduce the surface tension; for most purposes, this addition does not change the density of water enough to make it necessary to deal with any correction factor.] Alternatively, organic liquids of known density and high purity, such as toluene or carbon tetrachloride, whose surface tensions are one-third or one-fourth of that of water, can be used. In any case, this method is one of the simplest for determining specific gravity, and, if homogeneous pieces of sufficient size are obtainable, it is one of the most accurate. Practically, any moderately precise balance can be used to determine the specific gravity of fair-sized fragments. Such fragments are often not obtainable, however, so special balances have been devised to be effective for small pieces. In addition, most of the devices have been developed so that they give direct and rapid determination of specific gravity. An example is the Jolly balance, in which the weight of the fragment in air and the weight in the liquid are measured by the extension of a spring, thereby removing the necessity of using a set of weights. And, for very small fragments, the Berman balance can be used; it is a torsion microbalance that weighs up to 25 milligrams with a precision of 0.01 milligram and has an attachment enabling rapid weighing both in air and in a suitable liquid. Small grains weighing only a few milligrams can be controlled for purity by microscopic examination and can be handled easily with tweezers. Many mineralogists consider this to be an especially good way to identify heavy opaque minerals.

The pycnometer used in the second method is simply a small, stoppered glass bottle that holds a definite volume of liquid (determined by measuring the weight of liquid of known specific gravity required to fill it). The volume of a known weight of solid is determined from the weight of the displaced liquid. If

G = specific gravity of the solid

L = specific gravity of the liquid used

W_1 = weight of the pycnometer empty

W_2 = weight of the pycnometer with solid

W_3 = weight of the pycnometer filled with solid and liquid

W_4 = weight of the pycnometer filled with liquid

then

$$G = \frac{L(W_2 - W_1)}{(W_4 - W_1) - (W_3 - W_2)}.$$

The pycnometer method requires good technique for accurate results. Care must be taken that air is not trapped in and between the grains of the solid. A serious source of error is often the variation in volume of liquid within the pycnometer; this variation can occur simply as a result of seating the stopper at different depths before and after the mineral is placed in the bottle; slight differences in the forces with which the stopper is inserted are the cause. Also, for accurate results, a fair quantity of homogeneous material is required. On the whole, it is best to avoid the pycnometer if another technique that will yield the accuracy required is available. With certain materials, such as friable fine-grained clays, however, it may be the only feasible macroscopic method.

The principle of the third method—using heavy liquids—is very simple. The grains of the mineral are immersed in a suitable liquid to determine whether they sink or float. If they float, the liquid is gradually diluted with a miscible liquid of lower density until, when mixing is uniform, the grains neither sink nor float. If the grains sink initially, the liquid is made more dense by adding a miscible liquid of higher density until equality is reached. The density of the liquid at equality between liquid and solid is then determined by one of the standard procedures, generally by using a Westphal balance or by using a set of (glass) density standards. A particular advantage of this third method is that small grains can be used, and several of them can be compared at the same time; if some are impure, this fact will be immediately revealed by a variation in density from grain to grain.

Suitable liquids for determining the densities of many minerals are:

Bromoform, $CHBr_3$; G = 2.89.
Acetylene tetrabromide (tetrabromethane), $C_2H_2Br_4$; G = 2.96.
Methylene iodide, CH_2I_2; G = 3.32.
Clerici solution, a saturated aqueous solution of equal amounts of thallous malonate and thallous formate; G = 4.2 (at room temperature).

[Clerici solution must be used with caution. It is both toxic and corrosive.]

To vary their densities, you can dilute the organic liquids with acetone (G = 0.79) or xylene (G = 0.86), and the Clerici solution with water. Other fluids that can be used, and their specific gravities, are listed in the periodically published *Handbook of Chemistry and Physics*.

Habit

Some mineralogists refer to the geometric shape of a crystal as its crystal form. We do not; we

188   restrict our use of the term *crystal form* to the following definition:

Crystal form: Those crystal faces that are equivalent in a particular symmetry class.

The term *habit* has also been used with different meanings. We define it as follows:

Habit: The shapes of commonly occurring crystals and/or of aggregates of crystalline grains.

Consequently, habits can be characterized by such terms as *octahedral, prismatic, singly terminated,* and *twinned,* and/or by such terms as *bladed, fibrous,* and *radiated.*

When a mineral occurs as a crystal, its shape may suffice to identify the mineral without any additional tests. The hexagonal prisms typically terminated by unequally developed rhombohedra that are so characteristic of quartz can be cited as an example. Sometimes, in fact, it is not even necessary to see a free crystal; for example, the rounded triangular cross sections of tourmaline crystals immediately identify them.

It is important, however, to remember that crystals can be, so to speak, distorted by the unequal development of different faces, which tends to obscure their true symmetry. For example, it is extremely unusual to find a cubic crystal that is a perfect, symmetric cube. Also, some minerals sometimes crystallize in forms that give an appearance of false symmetry; for example, the simple crystals of the adularia variety of orthoclase can simulate rhombohedra. In identifying distorted crystals, you should keep in mind the fact that although the appearance may be unusual, the interfacial angles remain the same, and like faces are physically alike in type of luster, in striations, and so on. Quartz, for example, commonly occurs as platy crystals resulting from the strong development of one pair of prism faces; however, the angle between each pair of prism faces is 60°, and each prism face generally shows the characteristic horizontal striations.

Twinning is often a critical diagnostic feature. Some common minerals have highly characteristic twin crystals—for example, the penetration twins of fluorite, the right-angle and diagonal twins of staurolite, and the latticelike twinned aggregates of cerussite. It is important to remember that repeated twinning often results in an apparently higher symmetry than that possessed by the mineral. Aragonite, for example, often occurs in crystals that appear to be hexagonal prisms, as a result of repeated twinning on {110}; fortunately, there is no reason to misidentify these crystals as calcite, since calcite, although hexagonal, only infrequently crystallizes in simple hexagonal prisms.

The recognition of polysynthetic twinning by the striations on cleavage planes is particularly important in identifying plagioclase. It is almost the only simple way to distinguish it from orthoclase. The rare mineral amblygonite shows similar polysynthetic twinning lamellae on cleavage surfaces and occurs in some pegmatites together with plagioclase; however, it is readily differentiated by its density and by simple chemical tests. Calcite and corundum may also show similarly striated surfaces, but each is easy to distinguish from plagioclase on the basis of its hardness ($H = 3$, 9, and ~ 6, respectively).

Many mineral specimens are aggregates of imperfect crystals or crystalline grains. In many cases, crystallographic controls and conditions of formation have led to the production of rather distinctive aggregates. Several of these aggregates are so common that they have been given names. Many of the designations require no explanation—for example, bladed, blocky, capillary, columnar (or prismatic), compact, concentric, equant (or equidimensional), fibrous, globular, granular, hairlike, lamellar (or tabular), mammillary, massive, microcrystalline, mossy, needlelike, nodular, radiated, splintery, stalactitic, and wirelike. Others are as follows:

Acicular: Needlelike or thornlike.

Botryoidal: *See* colloform.

Colloform: Said of radiated, and in some cases also concentric, groups that are expressed by external spheroidal forms that resemble bunches of grapes, kidneys, etc.; it includes botryoidal, mammillary, reniform, and tuberose of most mineralogists.

Coralloid: See helicitic.

Cryptocrystalline: Dense, in some cases almost glassy, because it is made up of submicroscopic grains.

Dendritic: Branching like deciduous trees.

Drusy: Coated by numerous, closely spaced, small (typically < 1mm) crystals.

Filiform: Threadlike or hairlike.

Foliated: See micaceous.

Helicitic: Twisted and curved, and in some cases interlaced, like branching coral; it includes corralloid of some mineralogists.

Micaceous: Consisting of thin platelike lamellae.

Oolitic: Consisting of small ellipsoids or spheroids that resemble fish roe.

Pisolitic: Similar to oolitic but with spheroids that are larger—typically pea size to small plum size.

Plumose: Consisting of featherlike scales.

Reniform: See colloform.

Reticulated: Consisting of a network of crisscrossed, bladelike masses.

Stellate: Arranged like a star.

Tuberose: See colloform.

Cleavage, Parting, and Fracture

The properties of cleavage, parting, and fracture were discussed in Chapter 4, so only those aspects frequently considered in mineral identification are mentioned here.

Cleavage is the tendency of a mineral to break in definite directions, along planes that are parallel to possible crystal faces. The quality of a cleavage can be described as *perfect, good, distinct,* and *indistinct,* as noted on page 151. The direction of cleavage is described by the face to which it is parallel, either by the indices or by the name of the face. In the isometric system, {111} cleavage is also known as octahedral cleavage; in the tetragonal and orthorhombic systems, it would be pyramidal; in the monoclinic system, prismatic; and in the triclinic system, pinacoidal or pedial. Similarly, {110} cleavage is dodecahedral cleavage in isometric minerals and prismatic cleavage in the other systems, except the triclinic for which it is pinacoidal or pedial. The number of directions of cleavage for any crystallographic form depends on the number of faces in that form; thus, {110} cleavage in isometric minerals is manifested by six cleavage directions, whereas in tetragonal, orthorhombic, or monoclinic minerals, it means two directions, and in triclinic crystals, it means one direction. A careful study of cleavage can thus help in determining the crystal system of a mineral with no well-developed crystal faces. Some examples follow: A mineral with a single direction of cleavage cannot belong to the isometric system, since every form in that system has more than two faces. A mineral with three directions of cleavage—all of unequal quality—probably belongs to the orthorhombic, monoclinic, or triclinic systems; if the three directions are at right angles to each other, the system must be orthorhombic. Given three equal cleavages corresponding to a cube, a hexagonal prism, or a rhombohedron, the choice can be resolved by their angular relationship (all three at 90° to one another, cubic; all three at 60°, hexagonal prismatic; otherwise, rhombohedral). Four equal cleavages indicate an octahedral, tetrahedral, or rarely tetragonal or orthorhombic dipyramidal, cleavage. Six equal cleavages are characteristic of dodecahedral cleavage.

Some minerals rather commonly exhibit parting, which at first glance is not easily distinguished from cleavage. As we have already noted, parting, like cleavage, is typically parallel to possible crystal faces. Unlike cleavage, however, parting consists of a finite number of

190 discrete, relatively widely spaced planes. Furthermore, parting does not have to conform to symmetry requirements of the mineral involved, although in most cases it does. Also, parting may or may not be present in any particular specimen of a given mineral although some minerals rather commonly exhibit the property whereas others very rarely, or never, do. Many twinned crystals (and grains) part along their composition planes.

Some minerals without cleavage and some composite masses of fine-grained cleavable minerals show characteristic fractures. Although many terms have been used to describe diverse kinds of fracture, those most frequently used are still only qualitative:

Conchoidal: Smooth and curving.

Even: More-or-less flat.

Hackly: Rough, jagged, and sharp-edged.

Splintery (or fibrous):   Resembling split wood.

Uneven (or irregular):   Rough.

Depending on one's viewpoint, fracture and indistinct cleavage may or may not be considered to merge in aspect as well as in appearance.

Tenacity

Tenacity, as discussed in Chapter 4, is the resistance that a material offers to mechanical deformation or disintegration when the material is submitted to bending, breaking, crushing, or cutting. Macroscopically, minerals are generally referred to as either *brittle* (readily crushed to a fine powder) or *malleable* (capable of being rolled or hammered into plates). In addition, a mineral may be referred to as *sectile* if it can be cut into shavings with a knife (e.g., stibnite), *flexible* if it can be bent easily (e.g., thin plates of gypsum), and *elastic* if, after bending, it springs back into its original form (e.g., thin plates of micas). Also, some minerals are termed *tough*, indicating that they are difficult to break (e.g., jadeite in the form of massive—polycrystalline—jade).

Special Physical Tests

There are certain special tests that, on a macroscopic basis, are applicable to only a few minerals. Those that are fairly frequently used follow.

Luminescence

Some minerals emit light when they are energized by some energy other than incandescence. Macroscopically, luminescence is generally checked by placing a mineral under a long wave (3200–4000 Å) or short wave (centered around ~ 2535 Å) ultraviolet light. Minerals may react the same way or quite differently to the short versus long wave sources: some minerals fluoresce under one but not the other; some emit different colors under the different wave lengths; etc.

Although many minerals fluoresce (see Marfunin, 1979), the following are especially noteworthy:

1. The mineral fluorite (CaF_2) always comes to mind in connection with fluorescence; the term *fluorescence* was derived from the mineral name. Most fluorite fluoresces in blue; the fluorescence is generally due to the substitution of rare-earth elements for a small amount of calcium.

2. Calcite is another mineral that is commonly fluorescent, generally in red, pink, or yellow; its fluorescence is generally caused by the presence of Mn but in some cases, it is caused by the presence of organic impurities (porphyrins).

3. Most scheelite fluoresces white or bluish white, but its fluorescence becomes increasingly yellowish as Mo^{+6} replaces W^{+6}; consequently, its fluorescent color is employed as a semiquantitative test for the amount of molybdenum replacement.

4. Some secondary uranium minerals and uranium-bearing opal have a strong green or yellow-green fluorescence; also, the bead

obtained by fusing uranium-bearing material with sodium fluoride is fluorescent and provides a rapid and sensitive test for uranium.

Fluorescence, however, must be recognized as an unpredictable phenomenon. Be sure to keep the following points in mind:

1. Many localities are characterized by certain typical, fluorescent minerals—for example, the green-fluorescing willemite and the red-fluorescing calcite of Franklin, New Jersey.

2. Within the same locality, the typically fluorescent minerals may also occur as nonfluorescent varieties.

3. The characteristic fluorescence of certain minerals from one locality does not necessarily apply to the same minerals from other localities. (This, of course, is a variation of 2).

Thus, except for those who have access to highly specialized instrumentation, the effects of fluorescence may be best considered as sometimes useful in characterizing some minerals, but it is seldom definitive.

Magnetism

From the standpoint of macroscopic tests, a mineral is either noticeably attracted to or not attracted to a hand magnet. (Many mineralogists and geologists magnetize their pocket knives or nail files and thence use them to check this property.) Native iron, magnetite, most pyrrhotite, and maghemite are the only at all common minerals that are attracted to such magnets. Most iron-bearing minerals, however, if heated strongly in air, become magnetic as a result of the formation of magnetite; hence, this test is a useful one for the presence of iron.

Electrical Properties

Electrical properties, also described in Chapter 4, are seldom used in macroscopic determinative

mineralogy. In rare cases, however, the identification of, for example, tourmaline is verified by checking its pyro- and/or piezo-electricity.

Thermal Properties

The ease with which a mineral melts is often termed *fusibility*. As such, it was one of the properties usually checked in the formerly popular blowpipe analysis procedures.

Radioactivity

Minerals that contain uranium and thorium give off radiations that are readily detected by a Geiger–Müller counter or a scintillometer. With a little practice, it is possible to use one of these detectors to make an estimate of the amount of uranium and thorium in a specimen and thus identify the mineral. For example, a specimen of allanite, in which uranium and thorium contents are rather low, will give a much weaker response on the counter than a specimen of similarly appearing uraninite of approximately the same size and held at the same distance from the counter.

Radioactivity can also be detected by autoradiographic methods. Although several different detecting media are currently used, photographic film is still the one most widely used, probably because of its general accessibility. Autoradiography is particularly adaptable to the study of aggregates that contain both radioactive and nonradioactive components. In fact, the method has become so specialized by the introduction of different emulsions that one can now differentiate, for example, between alpha and beta radiation (see Bowie *in* Zussman, 1977).

Mineral Associations

Before we proceed to the chemical (generally destructive) tests, it seems only prudent to direct your attention back to Chapter 5 where mineral occurrences and associations are treated, albeit briefly. These aspects frequently aid mineralogists and geologists in their mineral identification efforts. In fact, association frequently serves as the single most useful clue to a mineral's identity. Experience enables both scientists and collectors to know what minerals to expect (and also those not to expect) in nearly any given geological environment. For example, a pale-colored brownish mica in a magnesium-rich rock is probably phlogopite, not one of the other micas. It is not possible to enumerate all of these clues in this book; many, however, are referred to in the descriptions in Part II. Others will be learned by examining rocks and minerals in collections, and better still, in the field.

CHEMICAL TESTS

The observations and tests outlined to this point may be termed nondestructive because a mineral is not decomposed by them. Contrariwise, chemical tests are destructive.

If, after all the physically based observations and tests have been made, the identity of a specimen is still in doubt, some relatively simple chemical test or brace of tests can frequently be performed to resolve the doubt. In most cases, the chemical tests used for identification purposes are both few and simple; this is true because, in most cases, careful observation of the physical properties will have already narrowed the range of possibilities to a small number of species, so that careful consideration of the chemical properties of those species will indicate clearly the test or tests required to provide unambiguous identification.

Most field geologists carry neither the equipment nor the supplies required to make more than simple chemical tests. In most cases, they select those that they do carry on the basis of knowing or anticipating what minerals they are likely to encounter. Thence, they check pertinent solubility/reaction-based properties according to the nature of the reaction and, in some cases, by observing the taste, odor, and/or coloration of either the solvent or the specimen. Examples of each of these kinds of tests are described in succeeding paragraphs. A few field geologists and others still employ additional, relatively simple, chemical tests carried out by using the blowpipe. Blowpipe analysis procedures are given in Penfield (1904) and are outlined in the first edition of this book.

Solubility

Although different minerals are soluble or insoluble in different solvents, only two solvents are widely used in field identification procedures: HCl of diverse concentrations and water. Some of the water-soluble minerals have distinctive tastes, as noted in the next section. A generally applicable test involving HCl is as follows: a small amount of the powdered mineral is placed in a test tube, and about 5 cc of dilute (1:1) hydrochloric acid is added. If no reaction takes place, the test tube is heated. If still no reaction takes place, the test is repeated using concentrated HCl. The following reactions are characteristic (see Table 6-1):

1. Many minerals are completely soluble without effervescence; they include many oxides and hydroxides, some sulfates, and some phosphates and arsenates. Abundant iron gives a yellow solution; copper minerals give a blue or greenish blue solution; cobalt minerals give a pink solution.

2. Solubility with effervescence occurs when the mineral contains a potentially gaseous component—for example, the carbonates. All carbonates dissolve in HCl with effervescence; some, such as calcite and aragonite, do so in cold acid, but many do so only when the HCl is heated. Some sulfides dissolve in HCl and give off H_2S, which is readily recognized by its rotten egg odor. Oxides of manganese give off chlorine, a greenish poisonous gas, from hot concentrated hydrochloric acid.

3. Decomposition with the separation of an insoluble material is characteristic of many silicates. The insoluble material is generally silica, and it may separate as a fine powder or as a jellylike mass; in the latter case, the mineral is said to gelatinize.

In some cases, all that is desired is a simple test to distinguish one mineral from another. An example is the frequently employed test that is used to distinguish calcite from dolomite: dilute HCl (in this case, $HCl : H_2O = \sim 10 : 90$) dissolves calcite with brisk effervescence, whereas it does not dissolve dolomite unless the mineral has been freshly powdered or the acid heated, and even that reaction is a relatively quiet smoldering effervescence.

Taste and Odor

Taste and odor are sometimes referred to as sensate properties. Minerals that are water-soluble may have the following tastes: astringent or puckering (e.g., alums such as melanterite), bitter (e.g., sylvite), cooling (e.g., nitratite), and salty (e.g., halite). Also, a few minerals adhere to one's tongue (e.g., kaolinite).

A few minerals give off odors when, for example, they are heated, ground, or acted upon by some solvent. Examples are: acrid or sulfurous (e.g., chalcocite upon heating), fetid (e.g., pyrrhotite with HCl), and garliclike (e.g., arsenopyrite upon grinding). Perhaps the most commonly used odor tests relate to the fact that, after

Table 6-1. *Solubility of minerals in HCl*

	Metallic Luster		
Soluble in HCl	Soluble in HCl with difficulty	Soluble in HCl evolution of chlorine	Soluble in HCl with evolution of H_2S
Goethite (limonite)	Hematite Ilmenite Magnetite	Pyrolusite Manganite Hausmannite Braunite	Stibnite Galena Pyrrhotite Sphalerite
	Nonmetallic Luster		
Soluble in HCl	Soluble in HCl with formation of silica gel	Decomposed by HCl leaving silica residue	Soluble in HCl with evolution of CO_2
Cryolite Zincite Brucite Colemanite Gypsum Jarosite Apatite Turquoise Carnotite Tyuyamunite Crocoite	Anorthite Nepheline Sodalite Cancrinite Olivine Willemite Hemimorphite Datolite Analcime Natrolite Laumontite	Leucite Rhodonite Wollastonite Pectolite Scapolite Cordierite Biotite Serpentine Garnierite Chrysocolla Stilbite Chabazite Heulandite	All carbonates

being breathed on, clay minerals emit a musty odor and, upon being scratched, sphalerite gives off H_2S, which smells like rotten eggs.

Coloration of Solvents and Specimens

When some minerals are dissolved by or react with acids or other solutions, the liquid becomes colored and/or the specimen is coated with a colored residue. For example, as we have already noted, copper minerals that are soluble in HCl impart a blue or bluish green color to the resulting solution, and HCl-etched calcite is stained purple when subsequently immersed or painted by Harris' hematoxylin. Several staining techniques have been developed, primarily for petrographic study of mixtures of minerals (rocks). Two examples are the procedure used to distinguish calcite from dolomite (see Friedman, 1959) and that used to tell potassium feldspar from plagioclase feldspar (see Bailey and Stevens, 1960).

WIDELY USED LABORATORY METHODS

Three widely used laboratory techniques for mineral identification are described briefly in the remainder of this chapter: They are optical mineralogy, X-ray diffraction, and electron probe microanalysis. Each of these techniques, or the results obtained through its use, is utilized by students as well as by professional mineralogists and petrologists.

Extensive discussion of the theories and instrumentation fundamental to these methods is beyond the scope of this book. Such discussions are available in many books and articles; see, for example, the pertinent articles and extensive bibliographies given in Zussman, 1977, and also the Selected Readings whose titles indicate treatment of these topics. We shall, for the most part, merely describe the general techniques and the kinds of information they provide.

OPTICAL MINERALOGICAL DETERMINATIONS

The general relationships between crystal structure and optical properties given in Chapter 4 are summarized in Table 6-2. As indicated by the table and also by the appropriate determinative tables in Chapter 16, the most useful optical properties for identifying most minerals are optical class and refractive index.

For the first group of procedures outlined in this section, we assume (1) that the observer is using a polarizing (commonly called petrographic) microscope (Figure 6-1); (2) that the grains being observed are are about 0.1–0.2 millimeter in greatest dimension (i.e., 70–150 mesh); and (3) that the grains are immersed in a liquid that is on a glass slide and is covered by a thin coverslip. Other setups can be used; for ex-

Table 6-2.   *Correlation of crystal systems and optical classes*

Crystal System	Optical Class		Refraction
Isometric	Isotropic		Single refractive index
Tetragonal Hexagonal	Anisotropic	Uniaxial (+ or −)	Two refractive indices
Orthorhombic Monoclinic Triclinic		Biaxial (+ or −)	Three refractive indices

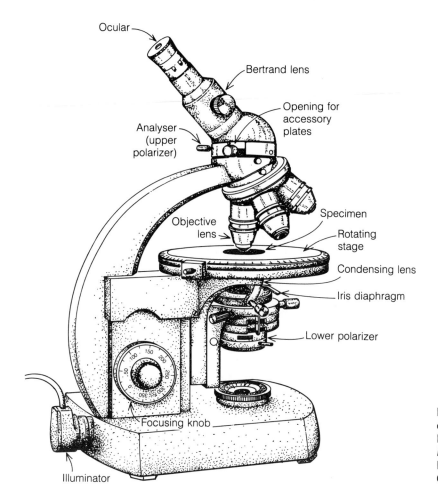

Ocular

Bertrand lens

Opening for accessory plates

Analyser (upper polarizer)

Objective lens

Specimen

Rotating stage

Condensing lens

Iris diaphragm

Lower polarizer

Focusing knob

Illuminator

Figure 6-1. The polarizing microscope. (Modified after W. R. Phillips, 1971, *Mineral Optics: Principles and Techniques*. San Francisco: W. H. Freeman and Company.)

ample, the mineral in question can be surrounded by other minerals in a thinsection (the name given to specially ground rock slices that are ~ 0.03 millimeter thick). In these other cases, the suggested procedures require only slight modifications in the microscopist's thinking.

Determination of Optical Class

Anisotropic substances are easily distinguished from isotropic substances when grain mounts are observed in the polarizing microscope with the analyzer inserted. Liquids and isotropic substances are dark and remain so upon rotation of the microscope stage, whereas most anisotropic grains show interference colors and become dark (extinguish) four times at intervals of 90° in a complete rotation of the stage.

To determine whether an anisotropic mineral (or other material) is uniaxial or biaxial, the usual procedure is to find a grain that will give a definitive interference figure. Differently oriented grains exhibit different interference figures, most of which can be used once a person becomes a competent optical mineralogist. One general type, however, which is usually easy to find, suffices to determine whether a substance is uniaxial or biaxial. We strongly urge students to familiarize themselves with this kind of figure and its correct interpretation before they go on to other figures. The type we recommend consti-

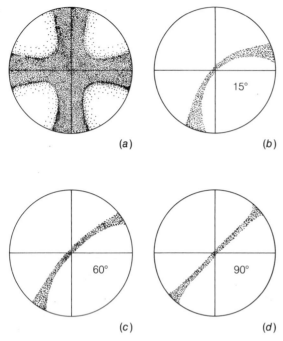

Figure 6-2. Optic axis figures: (a) uniaxial figure; (b) biaxial figure exhibiting a small 2V; (c) biaxial figure, moderate 2V; (d) biaxial figure, large 2V.

tutes a group generally termed *optic axis figures*. These figures are given by grains that can be recognized in a mount because they are the grains that show very weak interference color (i.e., they appear dark gray) even at 45° from their extinction positions. To obtain an interference figure, we must observe the grain between crossed polarizers, in strongly convergent light, with a high-power objective lens. This means that the analyzer is inserted, the substage convergent lens is moved into place, and a 40X to 50X objective lens is used. The interference figure is then seen when the ocular lens is removed or the accessory Bertrand lens is inserted. For a uniaxial substance, the figure is a black cross (Figure 6-2a). For a biaxial substance, the figure may have any of several diverse appearances, such as those illustrated in Figures 6-2b, c, and d. The darkened areas—the uniaxial cross and the biaxial bars—are both referred to as isogyres. The curvature of the biaxial isogyres are expressions

of the angle between the optic axes, generally designated 2V (Figure 6-3).

The optical character—that is, whether the mineral is positive or negative—can be determined from these interference figures by using an accessory plate. For example, if a gypsum plate is inserted, the isogyre (black cross) of a uniaxial figure becomes purplish red, and opposite pairs of quadrants are colored blue and yellow directly adjacent to the area of the isogyre. The relative positions of the blue and yellow spots indicate the optic sign (Figures 6-4a and b). Although most microscopes are set up so the colors are arranged as shown in this figure, some microscopes are set up in an opposite sense. Therefore, it is prudent to establish the relationship between the blue and yellow spots for a particular microscope and its accessory plate by observing a known interference figure, such as that given by a cleavage flake of biotite, which is biaxial (actually pseudo-uniaxial—that is, the two biaxial isogyres are so close to each other that they resemble a uniaxial cross) and has a negative sign. As the diagrams in Figures 6-4c and d indicate, interpretations of the effects of a gypsum plate on biaxial interference figures, no matter what their axial angle, also follow a logical pattern.

If the optic sign of an anisotropic mineral is not found by using the method just suggested, it can be determined by other means. One possi-

Figure 6-3. Visual estimation of 2V for biaxial optic axis figures can be made by comparing them with this diagram.

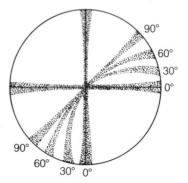

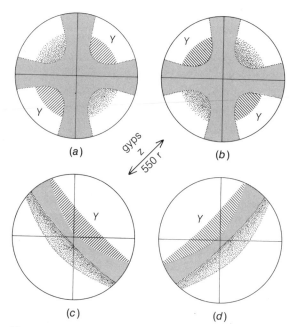

Figure 6-4. Optic sign determination. Accessory plate (gypsum, ~ 550 retardation plate) has its slow ray vibration direction NE-SW. Lined pattern indicates yellow; light stippling indicates blue; dark stippling indicates position of isogyres. (a) uniaxial negative; (b) uniaxial positive; (c) biaxial negative; (d) biaxial positive.

bility, which involves the determination of a mineral's indices of refraction, is described at the end of the following section.

Determination of Refractive Index

Each mineral has a characteristic index (or indices) of refraction. Consequently, the determination of refractive index values constitutes a frequently used procedure in mineral identification. Of the several techniques available for such determinations, the immersion method has found the most general application and the widest use. This method is a comparative procedure whereby the mineral (or other nonopaque crystalline solid) is immersed in liquids of known refractive index

until a match is found. It is an extremely powerful method because it can be applied to minute grains, and it is available to anyone who has access to a microscope and a set of liquids of known refractive indices. A polarizing microscope, though preferable, is not required. Sets of specially prepared index oils can be procured from many scientific supply houses, or they can be made up by using various, readily available liquids with known indices of refraction. Indices for several liquids are listed in the *Handbook of Chemistry and Physics*.

The technique of measuring the refractive index of a nonopaque, isotropic mineral is extremely simple. All you need to do is to find a liquid of known refractive index that has the same or nearly the same refractive index as the mineral. When mineral grains are immersed in a liquid with the same refractive index, the grains become practically invisible; that is, light, in passing from the liquid into the solid and from the solid into the liquid, is not refracted, so the edges of the solid grains can not be seen. This is true because the whole liquid-solid mixture acts as a homogeneous medium in its effect on light.

The procedure is to immerse the mineral particles in a drop or so of a liquid of known refractive index on a glass slide, and to observe the grains under low or moderate magnification (e.g., 7.5X ocular and 10X objective lenses) with transmitted light. If the grains are invisible, or nearly so, the mineral has the same, or nearly the same, index of refraction as the liquid; if the grains show up plainly, their refractive index differs considerably from that of the liquid. Thence, in the latter case, other liquids of different refractive indices are tried until a liquid is found in which the grains are invisible or very nearly so, and hence the index of refraction of the grains becomes known.

The search for the matching liquid is not as laborious and time-consuming as one might suppose because it is possible, by observing the *Becke line effect* (Figure 6-5), to tell whether the refractive index of the liquid is higher or lower than that of the mineral. With experience, one can even estimate approximately how much higher or lower the refractive index is (Figure

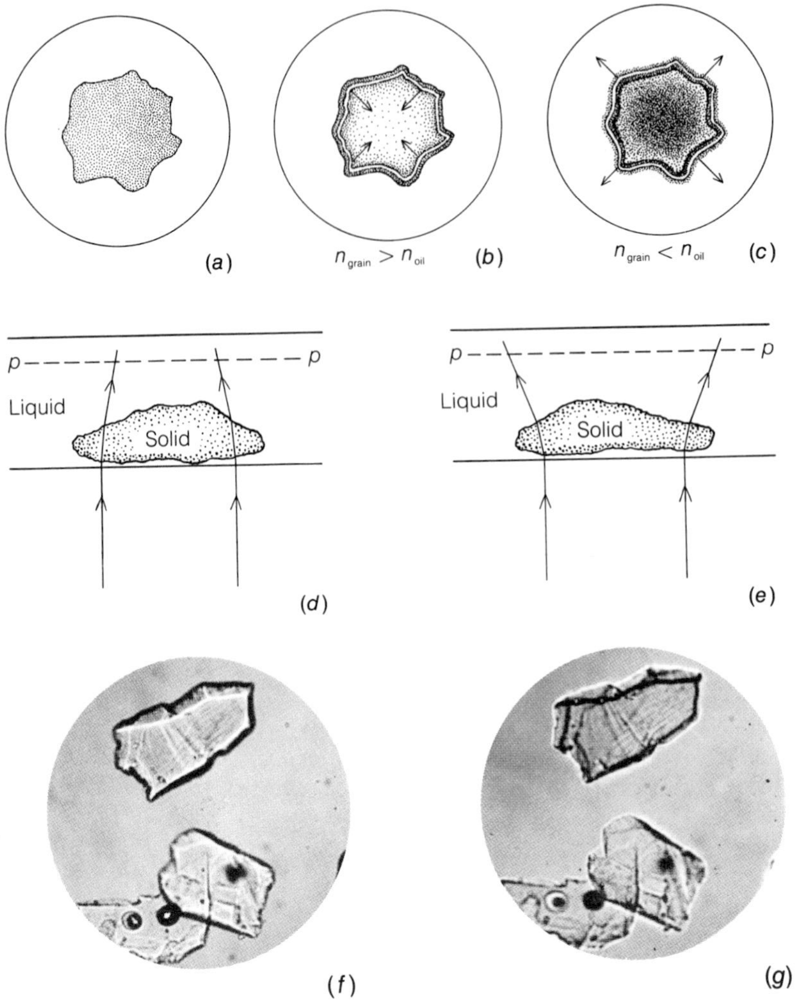

Figure 6-5.   The Becke Line. After careful focusing, upon raising the tube of the microscope (or lowering the stage), the Becke line moves into the medium with the higher index (a), (b), (c). The explanation for this is shown in (d) and (e). Examples of what is actually observed are given in (f) (index of refraction of fragment is greater than that of the surrounding medium) and (g) (index of refraction of fragment is less than that of the surrounding medium). (Photographs in (f) and (g) are used by permission, from W. R. Phillips, 1971, *Mineral Optics: Principles and Techniques*, W. H. Freeman and Company, San Francisco.)

6-6). The use of the Becke Line is as follows: if you focus a grain sharply, a line of light (the Becke line) can be seen at the edges of the grain (the line is clearest when the substage diaphragm is partly closed); if then you slightly raise the lens system (or lower the stage), the line will appear either to contract within the grain or to expand beyond the grain. The rule is that, using this procedure, the Becke line moves into the material with the higher index of refraction. Therefore, if it moves into the grain, the grain has a higher index of refraction than the liquid, whereas if it moves into the liquid, the liquid has a higher index of refraction than the grain. The shape of the grain does not matter because the Becke line coincides with, or is parallel to, the outline of the grain. Therefore, determination of the refractive index is just as easy for irregular fragments as for well-formed crystals.

The identification of anisotropic substances by determination of the refractive indices is somewhat more time-consuming than for isotropic substances because, in anisotropic substances, the velocity of light, and hence the refractive index, differs with the direction of vibration of light within the substance (see Figure 4-2). Consequently, grains differing in crystallographic orientation give different refractive indices, and thus one and the same grain will give different values for its refractive index according to its orientation relative to the direction of vibration of the light passing through it. With a polarizing microscope, the direction of vibration of the light passing through the object is controlled by the substage polarizer, and it is either N-S or E-W. Thus, by rotating the stage of the microscope, you can test the refractive index of a grain for different vibration directions.

In the identification of anisotropic substances, it is usually easiest to determine the ω index for uniaxial substances and the β index for biaxial substances. Grains that appear dark throughout a 360° rotation and exhibit an optic axis interference figure, as just described, give these vibration directions.

If necessary, the other refractive indices, ϵ for a uniaxial mineral and α and γ for a biaxial mineral, can be determined by examining grains that show the highest interference colors (i.e., maximum birefringence) when the analyzer is inserted. These grains are the ones that have come to rest in the plane of the ϵ- and ω-vibration directions of uniaxial minerals and of the α- and γ-vibration directions of biaxial minerals. A grain of uniaxial substance showing maximum birefringence will give the index of refraction for ω in one extinction position and for ϵ in the ex-

tinction position at 90° to the first. Thence, if you know whether the lower refractive index is for ω or for ϵ, you also know the optic sign of the mineral. ω is the lower index and ϵ is the higher index for optically positive minerals, whereas the reverse is true for optically negative minerals. The identity of ϵ versus ω can be established by various means. One commonly used method is based on the fact that many (but not all!) tetragonal and hexagonal minerals have cleavages parallel to the c crystallographic axes and therefore tend to break into tabular grains that are elongated parallet to c. For these grains, a characteristic called the sign of elongation is the same is the optic sign of the mineral. The *sign of elongation* is determined by the following steps (using the polarizing microscope with the analyzer inserted): rotate the stage of the microscope to one of the extinction positions for the mineral grain; from the extinction position, rotate the stage 45° in the direction that will make the longer dimension of the grain parallel to the slow ray of the inserted accessory plate (this is marked on the plate); insert the accessory plate and observe the color change of the grain or, if the grain shows several birefringent colors, of some specific part of the grain. If the gypsum plate is used and the part of the grain with a gray-to-white birefringent color is observed, blue indicates positive elongation and yellow indicates negative elongation. (With other plates, an apparent increase in birefringence—that is, the appearance of a higher interference color—indicates positive elongation, whereas an apparent decrease in birefringence indicates negative elongation.) For a biaxial substance, α is, by

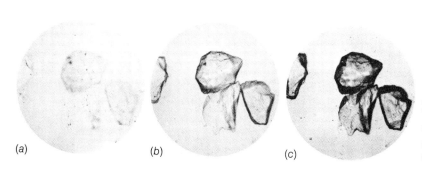

(a)   (b)   (c)

Figure 6-6.   Relief is the degree of visibility of a mineral grain within its surrounding medium. (a) Low relief; (b) moderate relief; (c) high relief. (Photographs are used by permission from W. R. Phillips, 1971, *Mineral Optics: Principles and Techniques*, W. H. Freeman and Company, San Francisco.)

definition, the lowest index and γ is the highest. If the β refractive index has also been determined, the sign becomes known; in biaxial positive substances, β is closer to α than to γ whereas in biaxial negative substances, the reverse is true. Thus, even if it is not possible to observe an interference figure, by careful study of the refractive indices it is still possible to determine whether a substance is uniaxial or biaxial and also to determine its optic sign.

Other Optical Properties

Other optical observations can contribute materially to the identification of a mineral. Some examples follow.

1. Cleavage versus lack of cleavage is usually revealed by crushed fragments; that is, minerals with good cleavage tend to be broken into fragments with straight edges (e.g., the amphiboles, the pyroxenes, the feldspars, and the trigonal carbonates). In some cases, cleavage directions and/or angles between cleavages can be measured.

2. Color may be significant (be sure, however, to observe color when the analyzer is *not* inserted!). Note in particular whether the color changes upon rotation of the stage. If it does, the mineral is said to exhibit pleochroism, and the different colors should be correlated with the corresponding refractive indices and vibration directions.

3. Extinction angle, a characteristic of monoclinic and triclinic minerals, is often a useful diagnostic property. If the mineral shows an identifiable crystallographic direction, measure the angle between that direction and the nearest extinction position; this angle is the *extinction angle*. Properly used, this parameter can be used for such things as distinguishing between monoclinic amphiboles and most of the monoclinic pyroxenes; the common amphiboles have maximum extinction angles of less than 25°, whereas most pyroxenes have maximum extinction angles near 40°.

4. Birefringence, the difference between the least and greatest refractive indices of an anisotropic substance, has already been mentioned. It is, as noted, expressed by interference colors; the higher the color or order of color, the higher the birefringence. Although this manifestation of birefringence is an often used characteristic in thinsection investigations, it can only be used with extreme care—and then, only in a general way—in the study of grain mounts. This is so because the interference color is controlled by the thickness of the grain through which the light is transmitted as well as by the orientation of a grain. These controls account for the observations that, in many grain mounts, different grains show different interference colors and that some individual grains, if lenticular in shape, show a series of different interference colors (with the higher ones on the thicker parts of the grains).

5. Dispersion is the expression of a mineral's capability to separate white light into colors of the visible spectrum. It depends on differences in indices of refraction for different wavelengths of the spectrum. Dispersion is so great for some anisotropic minerals that they do not exhibit well-defined extinction positions between crossed polarizers; instead, they appear to go through a sequence of darkened spectral colors over a few degrees of rotation of the stage of the microscope. Isogyres of these minerals also tend to be less than well defined; typically, they are fringed by red and violet on opposite sides of their curvature. Further description of the different kinds of dispersion—generally expressed as $r > v$ or vice versa in complete descriptions of optical data—is beyond the scope of this book. Nonetheless, we do have one additional note: dispersion is an especially important property of transparent gemstones and, for that purpose, it is often expressed by its appropriate numerical value;

that value is the difference between the mineral's index of refraction measured in red light with a wavelength of 6867 Å and that measured in violet light with $\lambda = 4308$ Å. Thus, for example, the dispersion of diamond is 0.044, and that of quartz is 0.013.

Other Optical Techniques

There are, of course, other techniques, such as those using special light sources, that can be used along with the polarizing microscope. These techniques are covered in optical mineralogy courses and books. For those students who may never be involved in such courses, we describe briefly a few other pieces of equipment that have been devised to measure certain optical properties and also a few additional accessories for the polarizing microscope that facilitate certain operations and/or add to its capabilities.

The Reflecting Microscope

The study of opaque minerals by use of vertically incident polarized light is generally termed *ore microscopy*. The microscopic setup used is often referred to as a *metallographic microscope* because it also finds wide use in investigations of metals and alloys. Many polarizing microscopes can be readily adapted for reflecting microscopy by adding appropriate accessories.

High-quality polished surfaces are required. Consequently, the method is seldom used for identification purposes per se. Instead, it finds its greatest use in observing the interrelations of different opaque minerals within mineral aggregates.

Mineral differentiation is made on the basis of properties such as color, reflectivity, isotropic or anisotropic character, and hardness. Colors are, for the most part, shades of gray, so color contrasts between coexisting minerals are generally more useful than the colors themselves. In some essentially opaque minerals, the incident light may penetrate the surface and be reflected from cleavages and fractures, thus giving a character-

istic internal reflection, the color of which rather commonly resembles the individual mineral's streak. When the stage is rotated, the color and reflectivity of isotropic minerals are unchanged, whereas either or both tend to change for anisotropic minerals. Also, when the analyzer is inserted, isotropic minerals remain dark or nearly dark on rotation, whereas anistropic minerals display bireflectance (which is analogous to birefringence in transparent minerals) with four positions of extinction during a 360° rotation. Hardness is revealed by slight differences in relief produced during polishing, by employing a microhardness indentation apparatus, or by observing a phenomenon that closely resembles the Becke line in transparent minerals. Simple michrochemical tests are often performed and observed in ore microscopy. These and other procedures of reflecting microscopy are summarized by, for example, Bowie and Simpson in Zussman (1977), and by Craig and Vaughan (1981).

Tri-Axial Rotation Equipment

These devices are ones that can be attached to the stage of a polarizing microscope to enable the operator to rotate a mineral grain at angles to the plane of the stage of the microscope. Essentially all of the optical properties of a single grain can be determined by using such devices. The two most frequently used rotation stages are the universal stage (U-stage) and the spindle stage. The U-stage can be used to study minerals as free grains or as grains in thinsections; the free grains can be permanently mounted or within, for example, an index of refraction oil. The spindle stage is used only for free grains. Step-by-step procedures for the use of these stages are given in a number of articles and books (e.g., Emmons, 1943, and Bloss, 1981).

The Refractometer

A few pieces of equipment, currently used chiefly by gemologists and jewelers, can be used for determining certain diagnostic optical properties of many mineral specimens, especially those that

202

are transparent or subtransparent crystals. The refractometer, which translates critical angle data into refractive indices with values up to about 1.81, is an example.

Critical angle: The least angle of incidence for which there is total reflection when an electromagnetic wave—for example, visible light—tends to pass from a denser into a less dense medium.

That is, when a wave passes from a denser into a less dense medium, it is refracted away from the normal to the interface; when the incident angle reaches a certain value, the wave no longer crosses the interface but is totally reflected within the denser medium; the angle of incidence at which refraction is 90° to the normal is the critical angle. It is fixed by the refractive indices of the media involved.

In air, the sine of the critical angle equals the reciprocal of the pertinent index of refraction of the material being measured. As you might suspect, this value is an important consideration in establishing angles between facets on gem stones.

With a little experience, you can also use the refractometer to determine whether a mineral is isotropic or anisotropic and, if the mineral is known to be uniaxial, its optic sign.

The Polariscope

This simple instrument consists of a light source that is polarized and which, with a second polarizer, can be analyzed; that is, it is essentially a lensless polarizing microscope. Any transparent or translucent mineral can be classified as isotropic or anisotropic by noting its behavior when it is rotated between the polarizers with their directions of light transmission at right angles to each other. Just as with the polarizing microscope, if the specimen is dark (no matter what its orientation) throughout a rotation of 89° or more, it is isotropic, whereas, if it flashes light and dark when rotated, it is anisotropic.

The Dichroscope

This device is still another simple one that is used primarily for the inspection of gemstones. It simplifies observation of pleochroism.

X-RAY DIFFRACTION

The effects of X-ray diffraction by mineral structures constitute the basis of a generally applicable method for identifying minerals and other crystalline substances. Whereas qualitative chemical tests do not distinguish between different compounds of the same elements, quantitative chemical analyses do not distinguish between polymorphs, and refractive index measurements (etc.) are not applicable to opaque minerals, X-ray methods can be used to identify all crystalline materials. Furthermore, only minute amounts of material are required for X-ray identification; a microscopic sliver of a crystal or grain or a milligram or less of powder generally suffices.

The two generic kinds of X-ray diffraction techniques used for determinative and descriptive mineralogy are usually referred to as powder methods and single-crystal methods. For the most part, powder methods are used for identification, whereas single-crystal methods are used for structural analyses. [Powder data, however, can also provide some insights into structural features.] It is also noteworthy that various pieces of auxiliary equipment permit determination of additional information—for example, the way a mineral behaves under diverse temperature and pressure conditions.

In addition, the fluorescent radiation emitted by elements under X-ray bombardment is now used widely in both qualitatively and quantitatively detecting the constituent elements in minerals and rocks. This is true because this radiation is the basis for both X-ray fluorescence spectroscopy and elemental analysis using the electron microprobe.

The Production of X-Rays

X-rays are produced when high-speed electrons strike the atoms of any substance. A modern X-ray tube consists essentially of a heated filament, which provides a convenient source of electrons, and a metal target enclosed in an evacuated chamber. A directed high potential moves electrons at high speed from the filament (cathode) to the target, and X-rays result from two types of interaction of the electrons with the atoms of the target material.

The X-ray spectrum includes both continuous, or "white," radiation and line spectra of characteristic wavelength. White radiation is produced as a result of the sudden reduction of energy of the bombarding electrons as they pass through the strong electric fields near the nuclei of atoms of the target. Since a high potential V [of the order of 50 kilovolts (kV)] is maintained between the cathode and the target and electrons of charge e are attracted to the target, an electron will have an energy eV when it reaches the target.

If the energy eV is converted into one quantum of energy, the wavelength λ of the resultant radiation is given by

$$eV = \frac{hc}{\lambda}$$

where h is Planck's constant and c is the velocity of light. Thus $\lambda = hc/eV = 12.4/V$, where V is in kilovolts and λ is in Ångstrom units. This relation gives the minimum wavelength produced. For example, at 50 kV, the minimum wavelength is 0.25 Å. The continuous spectrum starts at this minimum wavelength, regardless of the target material, reaches maximum intensity at a slightly higher wavelength, then drops off in intensity with increasing wavelength (Figure 6-7).

In the generally accepted model of the atom, electrons surround the nucleus in definite shells. X-radiation in the form of a line spectrum is produced when an emitted electron strikes an atom of the target and dislodges an electron from one or more of the inner shells; consequently, the atom is in an excited state. When so excited, outer shell electrons—which have higher characteristic energies than inner shell electrons— shift to the vacancies in the inner shells with the resulting loss of energy ($\Delta E = E_{outer} - E_{inner}$) taking the form of emitted X-radiation. The X-radiation is, of course, of a characteristic energy (and, therefore, wavelength) for any shift within a given element. If a K-shell vacancy is filled by an L electron, characteristic $K\alpha$ radiation is emitted; if a K-shell vacancy is filled by an M electron, characteristic $K\beta$ radiation is emitted; and, if L-shell vacancies are filled by outer shell electrons, characteristic L-series radiation is emitted. Thus, atoms of a given element emit X-radiation of several wavelengths, each of which is characteristic of that element. And, the emitted wavelengths of X-radiation can be used to identify any element in question; this, in fact, is the basis of microprobe analysis (see page 214).

The most commonly used X-ray tube consists

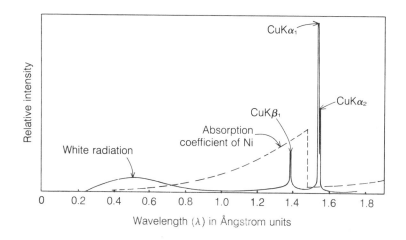

Figure 6-7. (Solid line) Distribution of intensity of X-ray emission by a copper target. (Broken line) Absorption coefficient for nickel.

204 of an evacuated glass chamber fitted with a filament, a water-cooled metal target, and two or four "windows" of beryllium or mica. The target is grounded and a rectified, high-voltage current is supplied to the filament. The X-rays are emitted from the face of the target, the maximum intensity being in the form of a flat cone having an angle of about 6° to the target surface.

White X-radiation in the range 0.25–0.8 Å is readily produced from a tungsten target using a potential of 50 kV. The K-radiation from molyb-denum, copper, nickel, cobalt, iron, or chromium targets, using potentials of 20–60 kV, is the most useful for X-ray diffraction studies. The wavelengths of the K lines of these metals are given in Table 6-3.

Note that each metal yields three wavelengths—$K\alpha_1$, $K\alpha_2$, and $K\beta_1$—of which the first two are close together. The intensity of α_1 is about double the intensity of α_2 and several times that of β_1. The spectrum for copper (Figure 6-7) is emitted when the applied voltage exceeds a certain

Table 6-3.   X-ray target materials and filters

Target						Filter	
Element	Atomic Number	Line	Wavelength (Å)	Excitation Potential (kV)	Suitable Working Potential (kV)	Element	Thickness (mm)
Mo	42	$K\alpha_1$	0.70926	20.0	90	Zr	0.108
		$K\alpha_2$	0.71354				
		$K\beta_1$	0.63225				
		A.E.[a]	0.6197				
Cu	29	$K\alpha_1$	1.54050	9.0	50	Ni	0.021
		$K\alpha_2$	1.54434				
		$K\beta_1$	1.39217				
		A.E.	1.3802				
Ni	28	$K\alpha_1$	1.65783	8.3	50	Co	0.018
		$K\alpha_2$	1.66168				
		$K\beta_1$	1.50008				
		A.E.	1.4869				
Co	27	$K\alpha_1$	1.78890	7.7	45	Fe	0.018
		$K\alpha_2$	1.79279				
		$K\beta_1$	1.62073				
		A.E.	1.6072				
Fe	26	$K\alpha_1$	1.93597	7.1	40	Mn	0.016
		$K\alpha_2$	1.93991				
		$K\beta_1$	1.75654				
		A.E.	1.7429				
Cr	24	$K\alpha_1$	2.28962	6.0	35	V	0.016
		$K\alpha_2$	2.29352				
		$K\beta_1$	2.08479				
		A.E.	2.0701				

[a] Absorption edge.
Data from N. F. M. Henry, H. Lipson, and W. A. Wooster, 1951. *The Interpretation of X-Ray Diffraction Photographs.* Macmillan, London.

critical value (9.0 kV for copper). The intensity ratio of the lines is independent of applied voltage, thus the $K\beta$ radiation cannot be weakened by alteration of voltage. As the applied voltage is raised above the critical excitation voltage, the output of K-radiation and also of white radiation increases. In practice, there is an optimum voltage for best contrast between line spectrum and background; for copper it is 30–50 kV.

When a beam of X-rays passes through a substance, the beam is partly transmitted, partly scattered, and partly transformed into other forms of energy. For monochromatic X-radiation, the intensity I of the original wavelength transmitted by an absorbing substance of thickness t is related to the intensity I_0 of the incident beam by the relation $I = I_0 e^{-\mu t}$, where μ is a constant known as the linear absorption coefficient . Absorption can also be expressed as a mass absorption coefficient, μ_m, which is given by the preceding equation when t is expressed in grams per square centimeter. The mass absorption coefficent of an element increases rapidly as the wavelength of the radiation increases; μ_m is roughly proportional to λ^3. Therefore, it is important that X-ray tube windows be made of elements of low atomic number and be as thin as possible, especially for transmission of the longer X-ray wavelengths.

When μ_m is plotted against wavelength of radiation, a smooth curve cannot be drawn through all the points. In the curve for nickel (Figure 6-7), the absorption increases to a maximum at a certain value of λ; the absorption edge falls off sharply and then increases again slowly. The occurrence of absorption edges can be explained easily in terms of the theory of X-ray emission. K-series spectra are produced when a K electron is dislodged from its shell; this can be achieved by electron bombardment or by absorption of X-rays. The quantum of energy of the exciting radiation must be greater than the energy required to dislodge the K electron. If λ is larger than a certain value, the quantum hc/λ may be too small; as λ decreases, however, the quantum will increase to a value that is sufficient. The result is as follows: (1) The absorption will increase suddenly at the absorption-edge wavelength be-

cause almost all of the energy of the incident rays is taken up in dislodging the K electrons. (2) Characteristic K radiation will be emitted by the irradiated sample. (This is called fluorescent radiation, and its wavelength is always longer than that of the incident radiation.)

The occurrence of absorption edges is valuable in X-ray diffraction work in reducing the intensity of certain wavelengths. The peak absorption for nickel (Figure 6-7) is at a wavelength that lies between the α and β wavelengths for copper; thus, nickel foil can be used as a filter, reducing the intensity of $K\beta_1$ to about 1/600 of that of $K\alpha_1$, since nickel has a very low absorption for wavelengths in the $CuK\alpha$ region. The intensity of the white radiation is also reduced. The wavelength of the absorption edge for each target metal and the appropriate filter material are given in Table 6-3.

In some minerals, the absorption characteristics of the constituent metals require selection of a special target material. The absorption edges of cobalt and iron are slightly higher than the $K\alpha$ wavelength of copper; therefore, Cu-radiation is unsuitable for use with minerals high in cobalt or iron, whereas Co-radiation or Fe-radiation are quite satisfactory. Owing to the high absorption of Cu-radiation by cobalt and iron, strong fluorescent radiation is produced, giving serious background darkening on diffraction films. Therefore, X-radiation should not be used for minerals containing large amounts of elements of atomic number 2 or 3 less than that of the target metal producing the radiation.

Diffraction by a Row of Atoms

Some of the X-rays incident on a fragment of a crystalline material are scattered by the constituent atoms. For a diffracted beam to result, the scatterings from many regularly spaced atoms must reinforce one another. Reinforcement of two or more waves results when the waves are in phase or when they differ in phase by a whole number of wavelengths. The scattering of X-rays by a regularly spaced row of atoms and their rein-

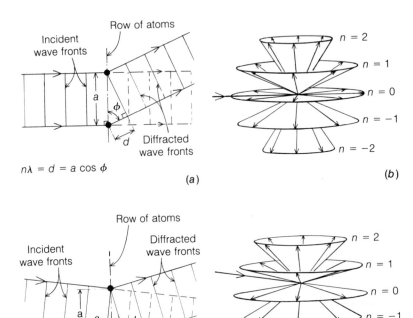

Row of atoms

Incident
wave fronts

a

ϕ

d

Diffracted
wave fronts

$n\lambda = d = a \cos \phi$

(a)

$n = 2$
$n = 1$
$n = 0$
$n = -1$
$n = -2$

(b)

Row of atoms

Diffracted
wave fronts

Incident
wave fronts

a θ ϕ

e d

$n\lambda = e + d = a \cos \theta + a \cos \phi$

(c)

$n = 2$
$n = 1$
$n = 0$
$n = -1$
$n = -2$

(d)

Figure 6-8.   Scattering of X-rays by a row of identical, equally spaced atoms. $\lambda =$ the wavelength of X-rays, $a =$ the unit period of the row, $n = 0$ or a whole number.

forcement into a diffracted beam are shown in Figure 6-8. In Figure 6-8a, an incident beam of X-rays is shown striking a row of identical atoms with unit period a at right angles. The diffracted rays are given off at various particular angles ϕ such that $n\lambda = a \cos \phi$, where n must be 1, 2, 3, etc. The diffracted rays with $n = 0, 1, 2, 3$, etc. form cones of diffraction with interior angle 2ϕ (Figure 6-8c). Figures 6-8c and d show the result when the incident beam is not perpendicular to the row. Note that part of $n = 0$ diffractions "reflect" from the row, with the angle of reflection being equal to the angle of incidence. The angle ϕ of the zero-order diffracted cone depends only on the incident angle θ and is independent of the period a.

Diffraction by Planes of Atoms. The Bragg Law

In crystals, the rows of regularly spaced atoms are arranged parallel to one another and are equally spaced, the whole array forming a plane lattice (Chapter 2). Figure 6-8d shows one ray of the diffracted cone with $n = 0$ reflected from a point-row with an angle ϕ equal to the glancing angle θ of the incident beam. When an X-ray beam strikes a crystal plane, it makes a particular angle of incidence to each point-row in the plane, and each row produces a diffracted cone with $n = 0$. These diffractions from different point-rows reinforce one another only where the different conical surfaces intersect. In general, any two such cones intersect along a straight line (see Figure 6-9). The diffracted beam resulting from cooperation of the zero-order cones from all point-rows in the plane is a straight line, a "reflection" of the incident beam in which the angle of reflection is equal to the angle of incidence. Note that the periods of the point-rows and their arrangement in the plane do not affect the angle of reflection.

In crystals, the planes of atoms are arranged parallel to one another at a regular repeat spacing, thus forming a crystal lattice. Figure 6-10

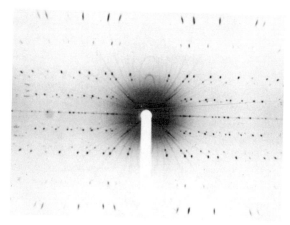

Figure 6-9. X-ray diffraction pattern (rotation method) from quartz crystal rotated about c axis (vertical). Cu*K* radiation, three layer lines due to each of Cu*Kα* and Cu*Kβ* radiation, radius of film cylinder 28.55 mm, original length of film about 17 cm.

when the path difference between reflections from adjacent identical planes is equal to a whole number of wavelengths of the X-rays in use. For example, on Figure 6-10, the path difference between *AXD* and *BYE* is *GYH*. $GY + YH = 2d \sin \theta$, where *d* is the interplanar spacing between regularly spaced identical planes and θ is the glancing angle. Thus, a diffracted beam will follow the direction *XD* or *YE* if

$$n\lambda = 2d \sin \theta,$$

where *n* is a whole number and λ is the wavelength of the X-rays used. This relation is called the Bragg Law.

The X-Ray Powder Method of Identification

As we have already noted, one application of X-ray diffraction in mineralogy is the identification of an unknown by its powder diffraction pattern. A useful and diagnostic pattern can be obtained from any substance that possesses an ordered internal structure. The powder pattern can be used for identification by comparing it with tabulated data. The details of the method depend somewhat on the kind and amount of material available and the kind of information desired. Generally useful equipment consists of an X-ray tube with a copper target and nickel-foil filters (although an interchangeable X-ray tube

depicts a vertical section through a crystal lattice. Each point-row *PP'*, *QQ'*, *RR'* is equally spaced, so the points form a lattice plane. In three dimensions, the rows *PP'*, *QQ'*, and *RR'* are point-rows of identical lattice planes perpendicular to the page. It was shown earlier that a single plane *PP'* can produce a diffracted beam from an X-ray beam incident at any angle. For a regular stacking of such planes to produce a diffracted beam, the rays diffracted from each plane must reinforce one another. Thus, a diffracted beam results

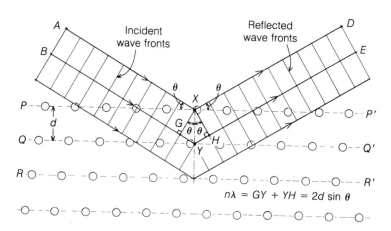

$$n\lambda = GY + YH = 2d \sin \theta$$

Figure 6-10. Diffraction of X-rays by equally spaced, identical planes of atoms governed by the Bragg Law, $n\lambda = 2d \sin \theta$.

Figure 6-11. Cylindrical X-ray powder cameras. 57.3-mm diameter (left), 114.6-mm diameter (right). In each, the pinhole is to the left, the exit port trap to the right. On top, the taller knurled knob enables centering of the specimen while it is being observed through the pinhole with a lens (black cap); the other knurled knob clamps the film.

with an iron target is preferable for minerals containing much cobalt, iron, or manganese), and one or more cylindrical powder cameras of 57.3-mm or 114.6-mm diameter (Figure 6-11), which can be mounted on stable tracks in line with the windows of the X-ray tube. The cameras should be fitted with an entrance pinhole, which defines the X-ray beam; an exit port, which catches the undiffracted beam; a device for holding the film tight against the inside of the metal cylinder; and a rotating spindle with a sample mount that permits centering the sample in line with the axis of the spindle and the pinhole. The camera must have a light-tight cover and a cap to cover the exit port. This cap should include a fluorescent screen to aid in aligning the camera with the X-ray tube and a thick piece of lead-glass to absorb the beam but permit observation of the fluorescent screen. The track and camera-mounting brackets should permit adjustment of the camera to obtain the most intense beam through the pinhole.

Camera Geometry

A good X-ray powder photograph requires a large number of orientations of very small grains of the mineral. To ensure all possible orientations, the powdered sample should be rotated at a slow speed (1 rpm). During rotation, grains will come into position to give the appropriate angle for many planes of spacing d and diffracted beams will result. [Alternatively, sometimes it is advantageous, if not imperative (because of amount of available sample), to use a gandolfi sample mount, which enables the rotation of a single grain about two axes simultaneously and nearly randomly, thus producing a powder diagram.] A diffracted beam will be bent from the direct beam by an angle 2θ, as determined by the Bragg Law. Diffracted beams from the same plane will form a cone (with internal angle 4θ), which will intersect a flat film set perpendicular to the direct beam in a circle with radius $s(100)$ or $s(110)$ (Figure 6-12). As shown in Figure 6-13, they will intersect a

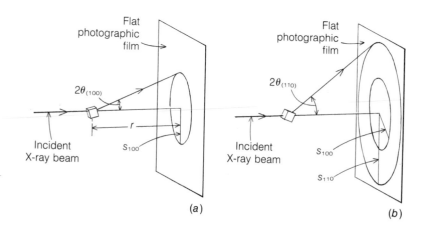

Figure 6-12. Diffraction by one grain in a powder sample giving a diffraction from (100) (a) and (110) (b). The radius of the small circle is s for (100); the radius of the larger circle is s for (110); r is the specimen-to-film distance.

short cylinder, presented by the typical powder film, in two arcs equidistant on either side of the direct beam trap (A). Thus, the semidiameter s of a powder diffraction ring on the film is proportional to 2θ, and $\tan 2\theta = s/r$ for a flat film, where r is the specimen-to-film distance. From measurement of s and r, we can determine θ and, knowing the wavelength of the X-rays used, we can obtain d by using the Bragg Law.

The cylindrical powder camera gives a most useful record, for several reasons: (1) a cylindrical film provides a much fuller record of the diffractions from a powder specimen than does a flat film; (2) the specimen-to-film distance can be deduced from the circumference length of the film,

and thus any film shrinkage can be taken into account; (3) the diffracted beams always strike the film perpendicular to its surface; and (4) it enables the use of fast film normally made with sensitive emulsion on both sides. The most useful method of film mounting is the Straumanis mounting. Two holes are punched in the film to permit the collimating pinhole and exit ports to project into the camera as close as possible to the sample so as to minimize air scattering of the direct beam. Two clips in the camera press the film tight against the interior cylindrical surface. The holes in the film are for clearance only; they have no fixed positions with respect to the diffraction pattern formed on the film.

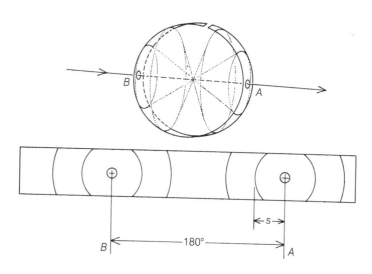

Figure 6-13. Diffractions in relation to a cylindrical powder camera film. At right, forward reflections (diffraction) forming nearly circular arcs centered at A. At left, back reflections (diffraction) forming nearly circular arcs centered at B (the entrance pinhole).

Most cylindrical cameras record diffractions from $\theta = 5°$ to $\theta = 80°$; special collimating pinhole construction can extend this range. Those diffractions with $\theta < 45°$ are usually referred to as forward reflections (or diffractions); those diffractions with $\theta > 45°$ are referred to as back reflections (or diffractions). The former appear as arcs of circles (Figure 6-13) centered about the exit port (A); the latter form arcs about the entrance pinhole (B). A diffraction with $\theta = 45°$ will be a straight line parallel to the axis of the camera. If points A and B are the centers of the forward- and back-diffraction rings, respectively, the distance from A to B is proportional to 180° (of 2θ), and the distance s from A to each diffraction ring is proportional to 2θ for that ring. Cylindrical cameras of any reasonable diameter can be used, but those of 57.3-mm $(180/\pi)$ and 114.6-mm $(360/\pi)$ diameter are especially convenient. In the former, the semi-circumference is 90 mm and the length s mm = $\theta°$; in the latter, the semi-circumference is 180mm and the length s mm = $2\theta°$. If the distance AB on the film differs slightly from 90 mm or 180 mm, a small proportional correction can be applied to the length s mm, which will correct for the effective film radius (or the film-to-specimen distance). The θ value for each diffraction ring can, along with the wavelength of X-rays, be used to obtain the interplanar spacing of the set of planes that caused the diffraction according to the Bragg Law. The resulting d values are in angstrom units, which are the units in which X-ray wavelengths are expressed.

It should be noted that the diffraction rings or lines on a powder film have different densities or degrees of darkness. A detailed explanation of this phenomenon is beyond the scope of this book. It suffices to say that the density of the lines is primarily a function of the arrangement of the atoms and of their atomic number. Thus, the dimensions of an array of atoms in a crystal determine the distribution of diffraction lines, whereas the kinds and arrangement of atoms determine the relative intensities of the lines. Therefore, the description of a powder pattern of a mineral should include both the interplanar spacings (or θ

values) and the relative intensities of the lines. Since the intensities cannot be measured in absolute units, it is usual to express them on a relative scale, using 100 for the darkest line and grading down to 5 for a just-visible line. Ten degrees of darkness are readily distinguishable by eye. [Certain (hkl) values may not be represented among the observed reflections, owing to the very low intensity of the diffracted beam, to a nonprimitive lattice, or to the presence of screw axes or glide planes of symmetry.]

Preparation of Sample

To obtain useful powder patterns, a sample must be ground to minus 200 mesh. This requires, especially with hard minerals, vigorous grinding in an agate or alundum mortar. The powder is then mixed with a drop of collodion solution (colorless fingernail polish is convenient) and rolled into a spindle 0.1–0.3 mm in diameter (or a small sphere, when a very small amount of material or only platelet-shaped particles are available). The hardened spindle is then mounted in the powder camera, coaxial with the camera axis and in line with the collimating pinhole (a sphere must be mounted on the tip of a glass fiber). Useful powder photographs can also be obtained by scraping a little powder off the surface of the mineral grain and picking it up on the tip of a greased glass fiber. In any case, it is important to obtain as pure a sample as possible.

Method of Measurement

Several procedures can be used to identify a mineral from its powder photograph. If previous examination has suggested a possible identification, and if a file of identified powder photographs is available, direct comparison will confirm or deny the identification. Frequently, however, it is necessary to determine the d values and relative intensities for the individual reflections and to compare them with published data, usually the J.C.P.D.S. (see Selected References at end of this chapter) data file (see Figure 6-14). Transparent

d 4-0784	2.355	2.039	1.230	2.355	Au					
I/I₁ 4-0784	100	52	36	100	Gold					

Rad. Cu λ 1.5405 Filter N₁		d Å	I/I₁	hkl	d Å	I/I₁	hkl
Dia. Cut off Coll.		2.355	100	111			

Rad. Cu   λ 1.5405   Filter N₁
Dia.   Cut off   Coll.
I/I₁   d corr. abs.?
Ref. Swanson and Tatge, JC Fel. Reports, NBS (1950)

Sys. Cubic (f.c.)   S.G. O_H^5 - Fm3m
a₀ 4.0786 b₀   c₀   A   C
α   β   γ   Z 4
Ref. Ibid.

d Å	I/I₁	hkl	d Å	I/I₁	hkl
2.355	100	111			
2.039	52	200			
1.442	32	220			
1.230	36	311			
1.1774	12	222			
1.0196	6	400			
0.9358	23	331			
.9120	22	420			
.8325	23	422			

εa
2V   n ωβ   ε γ   Sign
Ref.   Dₓ 19.302 mp   Color

Sample purified at NBS laboratory and is about 99.997% Au.
At 26°C.
To replace 1-1172, 1-1174, 2-1095

Figure 6-14. Card 4-0784 for gold. (From *X-Ray Powder Data File*, published by the American Society for Testing Materials, by permission).

plastic scales for specific camera diameters and X-ray wavelength, from which *d* values can be read directly, are used widely. Otherwise, the positions of the lines can be measured by mounting the film on an illuminated screen with a scale and cursor (Figure 6-15). The spacing for each line can be converted to θ on the basis of camera radius, and θ can be converted to *d* by using the equation developed on page 207. More often, however, tables that give *d* values that correspond to 2θ values are used.

The Powder Diffractometer

Today, a counter that is sensitive to X-radiation is frequently used in lieu of film to record the positions of diffracted beams. Scintillation counters are generally used, but proportional Geiger or solid-state detectors are also used. A flat sample is mounted in line with a collimating slit, and the scintillation counter (or other detector) is moved radially about the sample center through the angle 2θ from the direct-beam direction. The plane surface of the sample rotates through the angle θ. By connecting the counter through suitable electronic circuits to recording equipment, the intensity of radiation received by the counter as it moves through the angle 2θ can be recorded by pen on a moving-paper chart. The diffractions

from the sample show up as peaks whose heights above the background level are proportional to the intensity of the diffracted beam (see Figure 6-16). Diffractometer systems are now available where several samples can be processed auto-

Figure 6-15. Measuring screen suitable for measuring and comparison of X-ray diffraction powder films from 57.3-mm diameter cameras. Top film, gold, giving measurements as in Table A-2, a = 4.078 Å, F-lattice. Lower film, sperrylite, PtAs₂, isometric, a = 5.967 Å, P-lattice. The double lines around the hole at the left are due to α_1-, and α_2-radiation (CuKα_1 = 1.54050, CuKα_2 = 1.54434).

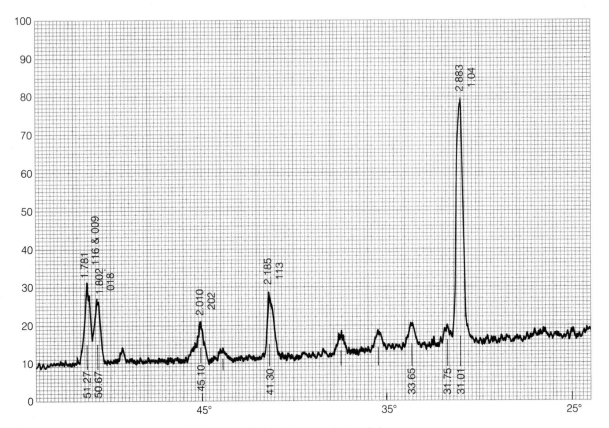

Figure 6-16.   An X-ray pattern of the mineral dolomite, as obtained from a powder diffractometer unit. 2θ values are indicated below the record (e.g., 31.01 for the highest peak); *d* values (e.g., 2.883) and Miller indices (e.g., 104) for the crystal planes that yielded the peaks are given above the record.

matically: their diffraction patterns are recorded and stored in computer memory; the patterns are interpreted by computer programs; and identifications are provided. This equipment is convenient, but it is much more expensive than film cameras and usually requires a larger sample. [A typical sample consists of about 0.1 cm³ of powder mounted on a microscopic slide. Frequently, the powder is put on the slide as a slurry that, when dried, adheres to the slide.]

Identification of Materials

As we have already mentioned, the spacings and intensities from an unknown substance can be compared with the J.C.P.D.S. data file; the file consists of cards (Figure 6-14) and search manuals and is also available in other forms, such as data banks for computers. Each card lists interplanar (*d*) spacings and their intensities, plus other parameters and properties of the mineral (or other substance) to which they refer. The cards are also available in book form, currently in six volumes. The search volumes, which list the major lines, are especially helpful in the identification of an unknown specimen; instructions for their use are given in the volumes. Once an entry in one of the search indexes is found to correspond to the lines for the unknown, the file card indicated in the index can be compared to the whole pattern. In a few cases, more than one card may be indicated and warrant comparison. In addition, you should keep in mind the possibility that you may be led astray because of technical difficulties; for example, different methods of sample preparation, camera design, and the X-radiation used can result in variations in the observed intensities as compared with those listed in the file, and oversized or off-centered powder mounts can result in high standard errors in the measured spacings, especially for the strong lines with low θ angles. Furthermore, in spite of all precautions to obtain a pure sample, a powder sample may include more than one mineral, especially if the sample has been obtained from a fine-grained aggregate; and, if the sample is a mixture, the resulting X-ray diffraction pattern will have two or more patterns superimposed.

[Such patterns can sometimes be sorted out. How would you go about doing this?]

Nonetheless, a match can nearly always be found for any mineral that one is likely to find.

Interpretation of X-Ray Powder Patterns

The powder pattern yields the relative intensities of the diffractions from a crystalline material. Measurement of the pattern plus appropriate calculations give the absolute values for the interplanar spacings of the crystal structure. The powder pattern also provides indirect information about the symmetry of the structure; the dimensions of a unit cell of the structure can be deduced from the spacings, and the indicated geometry of the unit cell will then suggest its probable symmetry. The description of all the procedures used for these determinations is beyond the scope of this book; the following solution for isometric substances is given to serve as an example of these solutions.

For isometric substances, a graphic solution of $d = a/\sqrt{h^2 + k^2 + l^2}$ is a simple matter leading to (hkl) values for each set of diffracting planes and thence to a value for the cube edge a. A graph

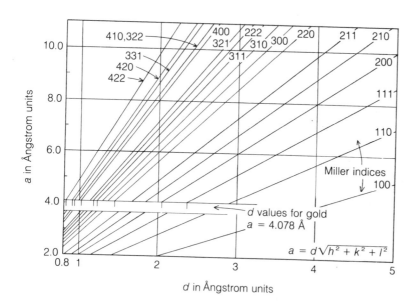

Figure 6-17. Cube-edge dimension a related to spacing values d for values of (hkl). The horizontal strip gives observed d values for gold, adjusted to give indices (hkl) and $a = 4.078$ Å.

214

of the type shown in Figure 6-17 can be used. The observed d values are plotted on a strip of stiff paper according to the d scale used in making the graph. Then the strip of paper is slid vertically on the graph, being kept parallel to the d scale, until each mark representing an observed d value coincides with one of the diagonal lines representing one possible combination of (hkl). The value of a at the level of the strip is the edge length of the cubic unit cell. The smallest value of a that gives a complete match should be taken. The (hkl) indices of each observed diffraction can be read from the graph, and the value of a can be computed readily from the formula.

Single-Crystal Techniques

Single-crystal diffraction photographs yield more extensive data than is available from powder patterns. In fact, the data so obtained can be used to determine essentially all structural information about a mineral (or other crystalline substance). As the name suggests, these X-ray patterns are of single crystals or homogeneous crystalline grains, which are generally only fractions of millimeters across. The best known methods follow:

The *Laue method* is primarily of historical interest. Reflections from a specially oriented, stationary crystal are recorded on a flat plate that is perpendicular to the incident X-ray.

The *rotation method* is used to determine the magnitude of lattice translations. Layer lines are recorded for a crystal that is rotated around a translation of the lattice (e.g., any intersecting edge between two nonparallel crystal faces or cleavage faces) such that the axis of rotation is perpendicular to the incident X-ray beam and coincident with the axis of the cylindrical film upon which the reflections are recorded. Thus, the pattern (e.g., Figure 6-9) represents the heretofore alluded-to cones of diffraction for the lattice row that coincides with the common axis of the cones (which, in turn, coincides with the axis of rotation of the crystal). When the film is developed and laid flat, the diffraction spots lie along

straight lines, each of which is a layer line corresponding to an n cone, as shown on Figure 6-8. [Consequently, cell dimensions, which can be used for such things as indexing powder patterns, can be determined. This, of course derives from the fact that d values are functions of a, b, and c; α, β, and γ; and h, k, and l.]

The *Weissenberg method* is also used to determine cell dimensions. Diffraction spots for a single layer line ($hk0$ or $hk1$, etc.) are recorded for a crystal (or grain) with the crystal and film oriented the same as for the rotation method but, during exposure to the X-ray beam, the crystal is turned on its axis and the recording film is simultaneously translated parallel to that axis. The resulting spots, on the basis of their positions, can be indexed readily by graphic methods.

The *precession method* most commonly used involves a coupled motion of a crystal (or grain) and a recording film that is flat during its exposure to the X-rays refracted by the crystal. The resulting pattern (e.g., Figure 6-18) is an undistorted representation of a particularly oriented section of the reciprocal lattice (see page 44). As you can see, the spots can be indexed, essentially by inspection, because they are in regular sequences along straight lines.

ELECTRON PROBE MICROANALYSIS

Three microbeam techniques—electron probe microanalysis, scanning electron microscopy, and ion microprobe analysis—are used in mineralogical investigations. Electron probe microanalysis, in particular, has found increasing use and application since it became readily available in the late 1950s; today, its use dominates studies dealing with mineral chemistry. In addition, many—probably most—of the newly discovered minerals found during the last several years have been discovered with the probe, and several other "new" minerals have been described largely on the basis of probe analyses. To the present, the other two techniques have found

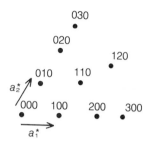

030
020
120
010   110
a_2^*
000   100   200   300
a_1^*

(a)

(b)

Figure 6-18. (a) Precession photograph of fluorapatite. The crystal was oriented with the six-fold axis perpendicular to the film. The discrete spots, X-ray reflections, form a hexanet (In this case, a "reciprocal" lattice). The streaks are caused by "white" radiation and should be disregarded. (b) Labeled reproduction of a portion of the precession photograph showing (1) the unit translations of the lattice and (2) the Miller indices of some reflections. (Photograph courtesy of D. R. Peacor.)

many fewer applications to mineralogy; they are used for special studies in only a few laboratories.

Electron probe microanalysis is, as previously noted, based on the fundamental principle that X-rays characteristic of a material's (e.g., a mineral's) elemental composition are emitted when the material is bombarded by high-energy electrons. [The specimen area that is generally analyzed is in the range of 10,000 Å (= 1 micron = 0.001 mm) across.] Hence, when the electron beam is focused on the area to be analyzed, the emitted X-rays provide wavelength (and energy) data that are used to identify the emitting elements, and their intensities are used to calculate the elements' concentrations.

In its simplest form, the required instrument can be considered to consist of a scanning electron microscope (SEM) with its electron detector replaced by an X-ray spectrometer. In practice, most systems consist of an electron-optical system capable of focusing an electron beam, a movable stage in which the specimen ($\pm$ standards) is mounted, an optical microscope through which the specimen can be viewed during analysis, and appropriate detectors and recording equipment for measuring the characteristic radiations and their intensities (Figure 6-19).

The electrons must be given energy sufficient to cause excitation and emission of X-rays from the specimens to be analyzed. The electron beam must be able to be focused on extremely small areas. The optical system must be precisely centered and of high enough magnification to permit viewing and positioning of the specimen for analysis. The detectors must be accurate and their measurements need to be recorded in easily translatable modes.

The two most widely used kinds of instruments use the wavelength dispersive spectrometer (WDS) or the energy dispersive spectrometer (EDS). With WDS, the emitted X-rays are dispersed by a curved crystal that is arranged to satisfy the Bragg equation ($n\lambda = 2d \sin \theta$). The setup must, of course, be such that the angle θ can be varied and also such that crystals with different d values can be substituted for one another. Scans are made and mapped for individual elements. [Elements with atomic numbers of 11 (Na) and greater can be analyzed quantitatively if they are present in amounts exceeding about 50–100 ppm.; elements with atomic numbers of 4 (Be) to 11 can be analyzed at the same levels of concentration but generally with less accuracy.] With EDS, the emitted radiations are measured on the basis of their energies. Multiple scans are unnecessary because radiations from all elements are detected simultaneously. Thence, they are sorted electronically according to their energy

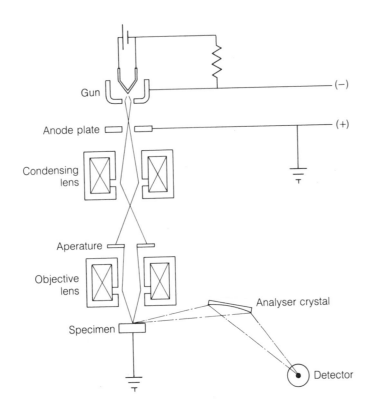

Figure 6-19. Schematic diagram of an electron probe microanalyzer.

ranges, which, of course, means that in a given time period, more specimens can be analyzed by EDS than by WDS. It also means that each complete analysis relates unequivocally to a given area of the specimen being analyzed.

Specimens to be analyzed must be given high-quality polished surfaces. In many cases, the surfaces must also be cleaned and coated. Polished thinsections are frequently used; this use has the added advantage that one and the same sample can also be studied and characterized optically.

The special advantages of the microprobe in mineralogical research are: (1) it enables the analyst to make a quantitative chemical analysis of grains as small as a few microns in diameter; (2) grains can be viewed during analysis, enabling the analyst to make a correlation of the chemical composition—including the stoichiometry—with features such as zoning and exsolution phenomena; (3) individual grains in a rock or ore can be analyzed in thinsections or polished sections, without the necessity of time-consuming, often essentially impossible, mineral separations; (4) several analyses can be made in a comparatively short time; and (5) the method is essentially non-destructive since, during analysis, the specimen is not significantly altered or used up. [There are, however, exceptions to this last advantage; temperature-sensitive minerals, such as hydrated compounds, may be decomposed by the heat of the electron beam.]

Several techniques are available for visualizing and recording microprobe data. One of the more illuminating techniques is the electron beam scanning picture. These "pictures" are made by photographing oscilloscope images that result when the electron beam scans the area of the sample selected for examination and the X-ray signals given off by specific elements are detected and displayed. Figure 6-20 illustrates the results of this procedure for a microscopic inter-

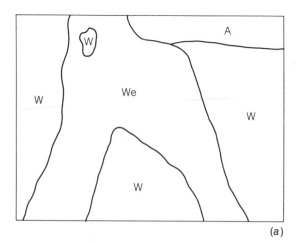

(a)

Figure 6-20. Electron beam scanning picture for Zn, Ca, Mn, and Fe of an intergrowth of willemite (We), wollastonite (W), and andradite (A) from Franklin, New Jersey (width of field is 0.5 millimeters). The concentrations of the elements are indicated by the density of spots. Zn is an essential component in willemite, and it is absent from the other minerals; Ca is an essential component in wollastonite and andradite, whereas it is absent in willemite; Fe is present only in andradite; Mn is present in minor amounts in willemite (about 1 percent MnO) and in greater amounts in andradite and wollastonite (about 4 percent MnO); and the small brighter areas in wollastonite are exsolved bustamite with about 12 percent MnO.

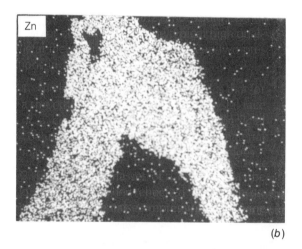

(b)

(c)

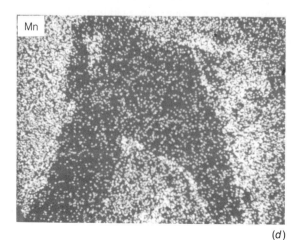

(d)

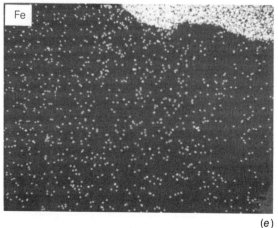

(e)

218   growth of willemite, wollastonite, and andradite from Franklin, New Jersey.

In nearly all cases, however, the output data are processed to make them truly meaningful. Although the calculations can be made directly or by using hand calculators, almost universal use is made of computers. In some cases, small computers are dedicated for such use; more commonly, large multipurpose computers are used with terminals located in the probe laboratory.

In the third paragraph of this section, we noted that "in its simplest form, the . . . [electron microprobe] can be considered to consist of a scanning electron microscope (SEM) . . . ." SEM was once used almost exclusively for morphological (i.e., topographic) studies of small areas. Today, it is being used increasingly in mineralogy because (1) SEM combined with an X-ray detection device is useful for rapid qualitative analysis, and (2) SEM combined with EDS can be used for quantitative analysis. One of the chief advantages is that specimens that cannot or should not be polished can be submitted to these analyses. [Granted, to the present, quantitative

Table 6-4.   *Other methods of analysis used in mineralogy*

Name (Common Abbreviation)	Remarks
Differential thermal analysis (DTA)[a,b]	Records exothermic and endothermic reactions caused by dehydration, oxidation, decomposition, and polymorphic transformation, and/or melting that occur on heating below 1000°C; may be done simultaneously with TG; may be done at high pressure; destructive.
Infrared spectroscopy (IR)[a,b]	Records spectra influenced by crystal structure and bonding. Identifies minerals, order-disorder, polymineralic substances, glasses, and mineraloids; distinguishes structural OH from water of crystallization; reflection or emission methods may be used; painstaking sample preparation; destructive (sample is powdered).
Spectrographic analysis[b] [Also called optical emission spectrometry]	Determines chemical content—typically qualitative to semiquantitative; usually used for trace elements; destructive.
"Classical chemical analysis"[b]	Classical analytic technique; generally takes 2–5 days; destructive.
X-ray fluorescence spectroscopy (XRF)[a,b,c]	Determines chemical content; for elements of atomic number ≥ 9, measures concentrations from a few parts per million to 100 percent; imprecise for major components; field units are available; generally nondestructive. (Under favorable circumstances, it may be used to distinguish valence states and coordination numbers.)
Atomic absorption spectrophotometry[a,b]	Determines chemical content (in 1.0 ppm–0.1 percent range) good for some 30 plus elements, especially metals; accuracy generally fair; best range is between those of NAA and XRF; most instruments require liquid—that is, a dissolved mineral sample.
Auger electron spectroscopy (AES)[b,c]	Determines structural and analytical data; determines atom positions *in outer few atomic layers*; information may be important to, for example, flotation process considerations; essentially non destructive.
Differential scanning calorimetry (DSC)	Similar to DTA but measures energy of thermal reactions; only good up to 700°C; destructive.

[a] Method is described in Zussman, 1977.   [b] Method is described in Nicol, 1975.   [c] Method is described in Marfunin, 1979.

analyses on nonpolished specimens are generally thought to be less than accurate.]

In addition, it seems noteworthy here that the electron microprobe is often used for the creation and observation of cathodoluminescence (i.e., luminescence caused by electron bombardment). Cathodoluminescence often provides information that can be translated into special definitions of such things as composition, symmetry, and coordination within minerals; and, as previously mentioned, it can be used to detect extremely small concentrations of certain elements and

compounds. The probe is especially well adapted for such use because its optical system permits viewing of a specimen while it is being excited by electrons.

SPECIAL TECHNIQUES

A number of special techniques used in determinative and/or descriptive mineralogy are given in Table 6-4. We think that students should be aware of the existence of these methods even

Table 6-4. *Other methods of analysis used in mineralogy (continued)*

Name (Common Abbreviation)	Remarks
Electron microscopy: Scanning (SEM)[b] Transmitted (TEM)[a,b]	(See p. 215) Resolves both chemical and structural features at high resolution, approaching 1Å in rare cases. Used for features below optical resolution—for example, may show defects; STEM gives analytical data with 100 Å resolution; sample preparation may be tedious; nondestructive.
Electron paramagnetic resonance (EPR)[c]	Measures electronic structure, bonding, etc.; identifies and describes radiation electron hole centers and free radicals; gives data for paramagnetic ions and concentrations of unpaired electrons; Cr^3, Mn^3, Ni^2, Cu^2, (etc.) give information at room temperature—others only at low temperatures. (May distinguish ions constituting impurity defects from those in inclusions.)
Electron spectroscopy (ESCA)[b,c]	Determines chemical content—an alternative method to qualitative XRF; H is not detectable; essentially nondestructive.
Evolved gas analysis (EGA)[a,b]	Determines amount and composition of volatiles evolved during heating; may be coupled with DTA, DSC, mass spectrometry, or TG; destructive.
Fission-track radiography[a]	Measures radioactivity; uses muscovite or manufactured emulsions to record tracks from thermal-neutron induced fission of ^{235}U; sensitivity to ~1 ppb U; may be used for some other radioactive isotopes; nondestructive.
Flame emission spectrometry and photometry[b]	Determines chemical content; used primarily for alkali and alkaline earth elements; requires liquid (dissolved mineral) specimen; destructive.
Ion probe microanalysis[b]	Individual atom analysis; can analyze low atomic number elements, even H; essentially nondestructive. Involves mass spectrographic analysis of atoms "blasted off."
Low-energy electron diffraction (LEED)[b] Mass spectrometry [spark source][b]	Determines structural and analytical data; remarks for AES apply. Determines isotope or element content; H, He, and Li not detectable; especially good for trace amounts; may analyze all but major constituents in single run; destructive.

(continued)

[a] Method is described in Zussman, 1977.   [b] Method is described in Nicol, 1975.   [c] Method is described in Marfunin, 1979.

though many of the required pieces of equipment are not widely available. Such awareness should lead to increased use of these methods in mineral investigations.

Although a few of these methods may suffice individually to identify a mineral, none results in a complete characterization. Therefore, in nearly all cases, a combination of two or more of these methods, or of one or more of these methods with optical determinations and X-ray diffraction, is generally employed to gain the information desired.

Table 6-4.   Other methods of analysis used in mineralogy (continued)

Name (Common Abbreviation)	Remarks
Mössbauer spectroscopy (NGR)[c] [Also called nuclear gamma resonance]	Gives complete characterization of iron—for example, its valence, coordination, bonding; also has limited applications to other selected elements, such as Sn.
Neutron activation analysis:[a,b] Instrumental (INAA) Radiochemical (RNAA)	Determines chemical content (in ppb–1 percent range); requires nuclear reactor, accelerator, or isotopic source for the activating neutrons; especially useful in determining O (though accuracy may be suspect), Na, and rare-earth elements, but is also useful for most other elements in trace amounts; generally destructive.
Nuclear gamma resonance (NGR)[c]	See Mössbauer spectroscopy.
Nuclear magnetic resonance [broad line] (NMR)[c]	May aid study of, for example, behavior of water in minerals (e.g., in zeolites and clays), coordination (e.g., of B in borosilicates), and crystal field gradients; applicable to relatively few nuclei.
Nuclear quadruple resonance (NQR)[c]	To date applied largely to As, Sb, and Bi minerals; determination of their bonding, coordination, etc.; cannot be used on isometric minerals; nuclei studied must make up > 3–10 percent.
Optical absorption spectrometry[b] [Also called calorimetric analysis; technically is visible/UV spectrophotometry]	Fair to good for quantitative analyses in low concentration levels; destructive.
Raman spectroscopy[c]	Determines crystal structure and bonding; complementary to infrared spectroscopy; amounts to absorption and reemission of intense, monochromatic radiation (e.g., from a laser); destructive.
Thermogravimetry (TG)[a,b]	Measures weight changes (e.g., loss of H_2O) during heating; generally coupled with DTA; destructive.
Thermoluminescence[a]	Gives glow curve for thermoluminescent materials; see page 149.
Thermomechanical analysis[a]	Records effect of heating on deformation of materials under stress; seldom used for minerals; destructive or nondestructive, depending on level of temperature attained.
Others	There are several modifications and combinations of the listed techniques, only a few of which are noted. There also are several infrequently employed techniques that have been designed to measure some additional parameters; for example, measurements have been made of changes in optical, electrical, and magnetic properties (and even of emitted sound effects) during controlled heating and cooling of minerals.

[a] Method is described in Zussman, 1977.   [b] Method is described in Nicol, 1975.   [c] Method is described in Marfunin, 1979.

Each of the tabulated techniques is described in the footnoted articles. Attention is also directed to the bibliographies given with these articles.

SELECTED READINGS

Azároff, L. V., and Buerger, M. J. 1958. *The Powder Method in X-Ray Crystallography.* New York: McGraw-Hill, 342 pp.

Bailey, E. H., and Stevens, R. E. 1960. "Selective Staining of K-feldspar and plagioclase on rock slabs and thin sections." *American Mineralogist,* **45**, 1020–1025.

Bloss, F. D. 1980. *The Spindle Stage: Principles and Practice.* New York: Cambridge University Press, 340 pp.

Buerger, M. J. 1964. *The Precession Method in X-Ray Crystallography.* New York: Wiley, 276 pp.

Chemical Rubber Company. 1980–1981. *Handbook of Chemistry and Physics* (61st ed.). Cleveland: CRC Press,

Craig, J. R., and Vaughan, D. J. 1981. *Ore Microscopy and Ore Petrography.* New York: Wiley-Interscience, 406 pp.

Cullity, B. D. 1978. *Elements of X-Ray Diffraction* (2nd ed.). Reading, Mass.: Addison-Wesley, 555 pp.

Emmons, R. C. 1943. *The Universal Stage.* Geological Society of America, Memoir 8, 205 pp.

Friedman, G. M. 1959. "Identification of carbonate minerals by staining methods." *Journal of Sedimentary Petrology,* **29**, 87–97.

Goldstein, J. I., and Yakowitz, Harvey (editors). 1975. *Practical Scanning Electron Microscopy (Electron and Ionic Microprobe Analysis).* New York: Plenum, 582 pp.

Henry, N. F. M., Lipson, H., and Wooster, W. A. 1960. *The Interpretation of X-Ray Diffraction Photographs* (2nd ed.). New York: Macmillan, 282 pp.

Hutchinson, C. S. 1974. *Laboratory Handbook of Petrographic Techniques.* New York: Wiley, 526 pp.

Joint Committee on Powder Diffraction Standards. 1974–1980. *Powder Diffraction File.* (This file consists of several items, such as the Hanawalt and Fink search manuals, data cards, data books, microfiches, and magnetic tapes, and including separate mineral data books and search manuals.) Swarthmore, Pa.: Joint Committee on Powder Diffraction Standards.

Marfunin, A. S. 1979. *Spectroscopy, Luminescence and Radiation Centers in Minerals* (translated by V. V. Schiffer). New York: Springer-Verlag, 352 pp.

Nicol, A. W. (editor). 1975. *Physicochemical Methods of Mineral Analysis.* New York: Plenum Press, 508 pp.

Penfield, S. L. 1904. *Determinative Mineralogy with an Introduction on Blowpipe Analysis* [by G. J. Brush] (16th ed.). New York: Wiley, 312 pp.

Phillips, W. R., and Griffen, D. T. 1981. *Optical Mineralogy, The Nonopaque Minerals.* San Francisco, Ca.: W. H. Freeman and Company, 677 pp.

Smith, D. G. W. (editor). 1976. *Short Course in Microbeam Techniques.* Mineralogical Association of Canada, Short Course Handbook, v. 1, 186 pp.

Wahlstrom, E. E. 1979. *Optical Crystallography* (5th ed.). New York: Wiley, 488 pp.

Zussman, Jack (editor). 1977. *Physical Methods in Determinative Mineralogy.* New York: Academic Press, 720 pp.

7 The Systematics of Mineralogy

Approximately 3000 mineral species are now known. Use of the qualifying adverb *approximately* is necessary because (1) the exact number of species depends on the definition of the term *mineral species* and how it is interpreted and (2) the number increases continually because additional species are discovered, currently at the rate of about 50 minerals per year.

In any case, nearly all minerals fit into the classification scheme that is outlined here, and this classification scheme is used as the basis for the descriptions in Part II of this book.

CLASSIFICATION OF MINERAL SPECIES

To deal systematically with minerals, it is necessary to have a classification scheme. The purpose of classification is to bring like things together and to separate them from unlike things—to give order to arrays of data, thus increasing their utility.

In the early years of the development of mineralogy as a science, from about 1750 to 1850, many systems of classification were proposed, some based on chemical criteria, others on physical criteria. Of the former, a system originally devised by the Swedish chemist Berzelius gradually became generally accepted. This system groups the mineral species into major divisions, or classes, according to the nature of the anionic group present. As adapted for this book, these classes are as follows:

    I. Native elements

   II. Sulfides (including sulfosalts)

  III. Oxides and hydroxides

  IV. Halides

  V. Carbonates, nitrates, borates, iodates

  VI. Sulfates, chromates, molybdates, tungstates

VII. Phosphates, arsenates, vanadates

VIII. Silicates

It can be seen that this classification, although originally based purely on chemical principles, has a definite significance in terms of crystal structure. The native elements include the metals (with metallic type of bonding) and the semimetals and nonmetals (with covalent bonding); the sulfides include some compounds with metallic bonding, most of those with covalent bonding, and a few with ionic bonding; practically all of the species in the remaining classes have ionic bonding to a greater or lesser degree. In the oxides and hydroxides and in the halides, the structures are simple packing of positive and negative ions; in the remaining classes, complex anions are present: XO_3 groupings in the carbonates, nitrates, and borates; SO_4 groupings in the sulfates; PO_4 groupings in the phosphates; and SiO_4 and more complex groupings in the silicates.

Thence, the individual classes can be subdivided on chemical or structural grounds; thus Class I, the native elements, is divided into two subclasses—the metals and the semimetals and nonmetals; the silicates are divided into six subclasses according to the structural linkage (polymerization) of their SiO_4 tetrahedra; etc.

Next, there is division into groups. Most groups include species that are closely related both chemically and structurally (e.g., the feldspar group and the amphibole group). In some cases, however, groups consist of species with some paragenetic similarity (e.g., the feldspathoid group and the zeolite group). Attention is directed to the report covering the amphibole group (Leake, 1978), approved by the International Mineralogical Association's (IMA's) Commission on New Minerals and Mineral Names (see Selected Readings at the end of this chapter); this report is representative of one of the kinds of activity of the commission.

Groups may be divided into species or series on the basis of variability in chemical composition. For example, the zeolite group comprises a number of individual species of limited variability, whereas the pyroxene group includes the orthorhombic and monoclinic series, each of which is further divisible into individual species.

As we indicated in the first paragraph of this chapter, the concept of the mineral species is subject to diverse interpretations. Consequently, there are several rather obvious inconsistencies in mineral nomenclature. The following question points up an example. Should all members of a solid-solution series be given the same name or several different names? In practice, the difficulties that have arisen have been overcome by rather flexible use of the term *mineral species*. Though perhaps unscientific, this resolution is probably best. In fact, it can be argued that the use of any rigid definition would almost surely lead to confusing, if not illogical, ends. In the future, few, if any, new problems should develop; this is true because nearly all major journals now require screening and prior approval of names by the previously mentioned IMA commission.

Some species include specimen types that have been further subdivided into either subspecies or varieties. Most subspecies are based on arbitrary compositional divisions within the range of composition established for the species (e.g., labradorite can be considered to be a subspecies of plagioclase). Most varieties have distinctive physical properties (e.g., amazonite is a green variety of microcline) and/or distinctive chemical compositions (e.g., fuchsite is a chromium-bearing variety of muscovite). During the last few decades, however, the use of separate names for chemical varieties has been more or less abandoned; such varieties are, instead, distinguished by descriptive adjectives. Thus, fuchsite is called chromian muscovite, a usage which simplifies nomenclature and provides information as to structure and composition. This usage was proposed by W. T. Schaller in 1930 and is adopted in the seventh edition of *Dana's System of Mineralogy*. The adjective modifiers are formed by adding the suffix *an* (or *ian* or *oan* to represent different valences) to the name of the substituting element. Thus, magnesite with some Mg replaced by Fe is referred to as ferroan magnesite.

224 The classification outlined here is not as detailed as that used in the seventh edition of *Dana's System of Mineralogy*. Our classification does not, for example, provide specifically for some minerals of rather rare occurrence—selenates, selenites, tellurates, tellurites, antimonates, antimonites, arsenites; these minerals can be included with the sulfates. There also are some naturally occurring salts of organic acids (mainly oxalates) and a few solid hydrocarbons that are considered to be minerals; these substances, which require two additional classes, are not included in this book. [On the other hand, we do provide brief descriptions of some of the more common and widespread nonminerals that are sometimes misidentified as minerals; these descriptions are given in Appendix A, Natural Glasses and Macerals.]

THE NAMING OF MINERALS

The names used for some minerals are very old, so old that their origins are lost in antiquity. In the first century AD, Pliny listed a number of native or easily reduced elements, common ore minerals, and gem minerals. With the development of mineralogy in the latter part of the eighteenth century, rival systems of nomenclature grew up. Carl von Linne (1707–1778), better known by his latinized name Carolus Linnaeus, applied the same binomial nomenclature that he developed for plants and animals to minerals; this system was adopted and extended by J. D. Dana in the first (1837) and second (1844) editions of *The System of Mineralogy*. Thus, the genus *Baralus* included *B. ponderosus* (barite), *B. prismaticus* (celestite), *B. fusilis* (witherite), and *B. rubefaciens* (strontianite). In the third edition of *The System of Mineralogy* (1850), however, Dana completely abandoned that nomenclature and adopted the procedure, which was already current in Europe, whereby a single name was used for each mineral. This use is now universal.

So, how are minerals named? The most straightforward answer is: however the original describer wishes, with no requirements or system other than that the name usually, but not always, ends in *ite*. Many names have been derived from Greek or Latin words that give some information about the mineral—for example, its color (albite, from the Latin *albus*, meaning white), crystal form or habit [sphene (= titanite), from the Greek *sphen*, meaning wedge], or density (barite, from the Greek *barys*, meaning heavy). Others have come from words that relate to the mineral's chemical composition—for example, calcite and zincite and even babefphite [for BaBe (PO_4)-(F,O)]. Descriptive suffixes have been used—for example, *clase* (from the Greek *klasis*, meaning fracture) in such names as oligoclase, orthoclase, and plagioclase; and *phyllite* (from the Greek *phyllon*, meaning a leaf) in anthophyllite, chalcophyllite, and pyrophyllite; etc. (see, for example, Mitchell, 1979).

These methods were admirable because they gave some indication of the nature of the mineral; they soon became limited in application, however, because there are far too few distinctive properties to go around. Hence, the practice of naming minerals after persons,—often mineralogists, but also mineral collectors, mine owners and officials, public figures, et al.—has grown. These names have been based on the person's given name (e.g., cliffordite for Clifford Frondel), family name (e.g., fleischerite for Michael Fleischer), or even both names (e.g., tombarthite for Tom. F. W. Barth). In addition, some names refer to the locality at which a mineral was first found, e.g., anglesite for the island of Anglesey off the coast of Wales; aragonite for Aragon, Spain; bytownite for Bytown (now Ottawa), Ontario, Canada; and zunyite for the Zuni mine, San Juan County, Colorado. [But, alas, to the present there is no gesundheit(e).] For recommendations relating to the naming (and describing) of newly discovered minerals, see the articles marked with an asterisk (*) in the Selected Readings.

In a few cases, two or more names have been applied to the same species. In such cases, the standard procedure is to adopt the name first applied to the species (the rule of priority). Other difficulties have arisen in various ways, especially when different names have become current in different countries. In English-speaking countries, the usage in *Dana's System of Mineralogy* is generally accepted as standard. That standard, with a few modifications based on recommendations of the IMA commission, is followed in this book.

SELECTED READINGS

Dana, E. S. 1892. *System of Mineralogy of Dana* (6th ed.). New York: Wiley, 1134 pp.

*Donnay, Gabrielle, and Fleischer, Michael. 1970. "Suggested outline for new mineral descriptions." *American Mineralogist,* **55**, 1017–1019.

*Fleischer, Michael. 1970. "Procedure of the International Mineralogical Association Commission on New Minerals and Mineral Names." *American Mineralogist,* **55**, 1016–1017.

Fleischer, Michael. 1980. *1980 Glossary of Mineral Species.* Tucson, Ariz.: Mineralogical Record, 192 pp.

Leake, B. E. 1978. "Nomenclature of amphiboles." *American Mineralogist,* **63**, 1023–1052.

Mitchell, R. S. 1979. *Mineral Names: What Do They Mean?* New York: Van Nostrand Reinhold, 229 pp.

Palache, Charles, Berman, Harry, and Frondel, Clifford. 1944 and 1951. *The System of Mineralogy of James Dwight Dana and Edward Salisbury Dana, Yale University, 1837–1892* (7th ed.), Vols. 1 and 2. New York: Wiley, 834 pp., 1124 pp.

Schaller, W. T. 1930. "Adjectival ending of chemical elements used as modifiers to mineral names." *American Mineralogist,* **15**, 566–574.

Part II
DESCRIPTIONS

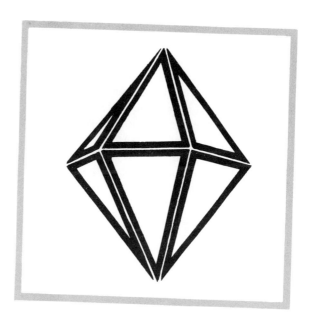

MOST OF THE MINERALS described in the following chapters are those rather frequently encountered in rocks and mineral deposits. The descriptions, as noted in Chapter 7, are arranged according to the major chemical classes—the native elements, sulfides and sulfosalts, oxides and hydroxides, the halides, carbonates (etc.), sulfates (etc.), phosphates (etc.), and the silicates. Both group descriptions and properties of individual mineral species are given. The group descriptions outline group characteristics and provide comparisons of certain properties of group members; the properties given for individual species refer the reader to group descriptions (where given) and present additional data, including those helpful in the identification of individual species by macroscopic and microscopic means. In addition, diagnostic macroscopic features that can be used to distinguish each mineral from other minerals that it may closely resemble are summarized, and brief statements about the mineral's occurrence, associations, and common uses (if any) are given. Most occurrence and mineral association data, which are extremely helpful in distinguishing some minerals from others they resemble, are not repeated under the diagnostic features heading; therefore, this information should be read in conjunction with that entry. Unlike many descriptive mineralogy compilations, we do *not* list a whole series of individual localities; information of this type, which is ever changing and thus primarily historical in nature, can be found in references such as the Dana/Ford . . . *Textbook of Mineralogy*, the first edition of this book, and in original accounts that have been published in national and state survey reports and in periodical publications, etc.

To facilitate use of the descriptions, we have arranged the data for the individual minerals consistently. Some descriptions, however, do not include all categories of data: for example, tenacity is given only for those minerals that do not have the typical normal brittleness of most mineral substances; common forms and angles are given only for those minerals for which they are likely to be found useful in macroscopic examinations;

and statements about the chemistry and/or alteration of several minerals are not given because they seem superfluous. However, redundant data are given under some headings; for example, the crystal system, class, and space group are all given for most minerals. The arrangement of data is as follows:

NAME, Chemical formula. (With a few exceptions, the names and formulas given by Fleischer, 1980, are given.)
Reference to group description (if one is given).
Crystal system; Class; Space group (Hermann–Mauguin symbol).
Cell dimensions (in Ångstrom units); Interaxial angles for triclinic (designated α, β, and γ) and monoclinic (β) minerals; *Cell content* in formula units (Z). (Cell dimensions given for minerals that are members of solid-solution series are not referenced to any particular composition; hence, they should be considered only as examples.)
Habit.
Twinning.
Common forms and angles.
Cleavage, parting, and fracture.
Hardness.
Tenacity.
Specific gravity.
Color and streak.
Luster and light transmission.
Optical properties (for nonopaque minerals).
    isotropic or anisotropic [uniaxial or biaxial and optic sign, and optic axial angle (2V) for biaxial minerals].
    refractive index(es): *n* for isotropic minerals; ω and ϵ for uniaxial minerals; α, β, and γ for biaxial minerals.
    other diagnostic properties (e.g., pleochroism).
Chemistry (including alteration).
Diagnostic features. (These features include miscellaneous properties—such as taste, feel, and fluorescence—and frequently determined macroscopically discernible properties. Blowpipe and chemical tests, other

than simple and frequently used ones, are not included. These are given in, for example, the first edition of this book.)
Occurrence, associations, and use(s).

For determining the identity of an unknown mineral, you can use the determinative tables (and the suggestions for using them) given in Chapter 16. Tables to aid both macroscopic and microscopic identification are included.

Most of the general information in Part II has been taken from original sources, from the first edition of this book, and from the general references listed in the following Selected References. All but the recently revised space group notations are those given and referenced by Strunz (1970) or Pierrot (1979). Most of the optical data are taken from the compilation of Phillips and Griffen (1981). The remarks about the present-day occurrence of crystals, which are given under "Habit," pertain to field occurrence (not availability from dealers); these entries represent the admittedly subjective opinions of Carl A. Francis, Anthony R. Kampf, John S. White, and R. V. Dietrich. The terms applied—common, uncommon, rare, and nonexistent (or nearly so)—refer to the availability of crystals as compared to the abundance of the mineral, not to any absolute availability; for example, diamond crystals are indicated as common, even though diamond is not of common occurrence, because most diamonds occur as crystals.

SELECTED READINGS

Dana, E. S. 1892. *System of Mineralogy of Dana* (6th ed.). New York: Wiley, 1134 pp.

Deer, W. A., Howie, R. A., and Zussman, J. 1962–1978. *Rock-Forming Minerals.* New York: Wiley, 5 vols.

Dietrich, R. V. 1969. *Mineral Tables. Hand-Specimen Properties of 1500 Minerals.* New York: McGraw-Hill, 327 pp.

Dietrich, R. V., and Skinner, B. J. 1979. *Rocks and Rock Minerals.* New York: Wiley, 319 pp.

Fleischer, Michael. 1980. *1980 Glossary of Mineral Species.* Tucson, Ariz.: Mineralogical Record, 192 pp.

Ford, W. E. 1932. *A Textbook of Mineralogy* (4th ed.). New York: Wiley, 851 pp.

Jensen, M. L., and Bateman, A. M. 1979. *Economic Mineral Deposits* (3rd ed.). New York: Wiley, 593 pp.

Palache, Charles, Berman, Harry, and Frondel, Clifford. 1944 and 1951. *The System of Mineralogy of James Dwight Dana and Edward Salisbury Dana, Yale University, 1837–1892* (7th ed.), Vols 1 and 2. New York: Wiley, 834 pp., 1124 pp.

Pierrot, R. M. 1979. *Chemical and Determinative Tables of Mineralogy* (without the silicates). New York: Masson, 591 pp.

Phillips, W. R., and Griffen, D. T. 1981. *Optical Mineralogy: The Nonopaque Minerals.* San Francisco: W. H. Freeman and Company, 677 pp.

Ramdohr, Paul. 1980. *The Ore Minerals and Their Intergrowths* (2nd ed.; English translation). New York: Pergamon, 1207 pp.

Roberts, W. L., Rapp, G. R., Jr., and Weber, Julius. 1974. *Encyclopedia of Minerals.* New York: Van Nostrand Reinhold, 693 pp.

Stanton, R. L. 1972. *Ore Petrology.* New York: McGraw-Hill, 713 pp.

Strunz, Hugo. 1970. *Mineralogische Tabellen.* Leipzig: Akademische Verlagsgesellschaft Geest & Portig K. -G., 621 pp.

8 Class I: Native Elements

Although several chemical elements occur as minerals in the earth's crust, none of them constitutes large masses of rocks. Nonetheless, many elements, such as gold, silver, copper, carbon (diamond and graphite), and sulfur, have been found in sufficient quantities to have attracted the attention of man from earliest times up to the present, and important amounts of each of these elements are still obtained from deposits containing them as native elements. In the case of silver, copper, and to a lesser extent, sulfur, however, the native elements are becoming less and less important as economic sources. In addition to these five elements, platinum and its related metals, which were not recognized until the eighteenth century or later, are also recovered in substantial quantities from native element minerals or natural alloys. Other elements have been found in the native state but are, in essence, mineralogical curiosities.

The important native elements can be classified as follows:

Metals
      Gold Group
        Gold, Au

Silver, Ag
Copper, Cu
Platinum Group
    Platinum, Pt
    Palladium, Pd
    Platiniridium, (Pt, Ir)
Iron Group
    Iron, Fe
    Nickel-iron, (Ni,Fe)

Semi-Metals and Nonmetals
    Arsenic Group
        Arsenic, As
        Antimony, Sb
        Bismuth, Bi
    Sulfur Group
        Sulfur, S
    Carbon Group
        Diamond, C
        Graphite, C

(a)

(b)

Figure 8-1. (a) Face-centered cubic structure as found in metals of the gold group; the atoms are actually larger than shown and touch along the face diagonals as shown in Figure 8-4. Each atom has 12 neighboring atoms touching it. (b) Body-centered cubic structure, illustrating the arrangement in iron.

GOLD GROUP

The gold group comprises the well-known metals gold, silver, copper, and lead; the first three are common in small quantities in the earth's crust, whereas lead is a mineralogical curiosity. These native metals are alike in structure with their atoms lying on the points of a face-centered cubic lattice; that is, they are based on a face-centered cubic closest packing arrangement* (Figures 8-1a, 8-2a, 8-3a, 8-4, and 8-5). Each atom is closely coordinated with twelve neighboring atoms, and the unit

* Two hexagonal polytypes have also been described for silver. The three polytypes are referred to as silver-*3C*, *-2H*, and *-4H*.

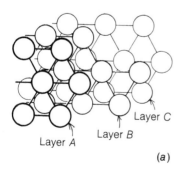

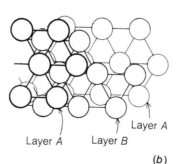

(a)

(b)

Figure 8-2. (a) Face-centered cubic packing showing the atoms as spheres about one-half actual size relative to the space and the layer sequence *ABC*, *ABC*, . . . along a threefold axis. (b) Hexagonal close packing, showing the atoms as spheres about one-half actual size relative to the space and the layer sequence *AB*, *AB*, . . . along the sixfold axis.

232

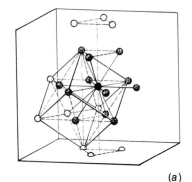

(a)

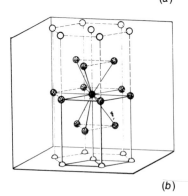

(b)

Figure 8-3. (a) Face-centered cubic arrangement showing the twelvefold coordination and the relation of the layer sequence ABC, ABC, . . . to a face-centered cubic lattice. (b) Hexagonal cell in the hexagonal close packed arrangement showing twelvefold coordination (as in a) and AB, AB, . . . sequence, which is distinct from the face-centered cubic arrangement in diagram a.

cube of the structure contains four equivalent atoms. The length of the edge of the cubic unit, the a length, which depends on the radius of each atom, is given in the following tabulation, along with other physical constants for the pure metals.

Crystals of these metals are rare; each forms dendritic and arborescent growths that commonly have a crystal form at many, if not all, of their extremities. These metals also have several similar physical properties; for example, each is soft, malleable, ductile, a good conductor of heat and electricity, opaque to light, and has a high specific gravity.

Crystal system, class, and space group: Isometric [For silver, this is the 3C polytype; hexagonal polytypes 2H (a = 2.93, c = 4.79) and 4H (a = 2.93, c = 10.11) have also been described.]; $4/m\bar{3}2/m$; Fm3m.

	Au	Ag	Cu	Pb
Cell dimensions:	a = 4.0786	a = 4.0862	a = 3.6150	a = 4.9505
Cell content:	Z = 4	Z = 4	Z = 4	Z = 4
Hardness:	$2\frac{1}{2}$–3	$2\frac{1}{2}$–3	$2\frac{1}{2}$–3	$1\frac{1}{2}$
Specific gravity:	19.3	10.5	8.94	11.37
Color:	Yellow	White	Light rose	Gray-white

As found in nature, these metals are rarely pure; typically, they contain various other metals and semi-metals in solid solution. As you might expect from the similarity in their cell sizes, a nearly continuous series of solid solutions exists between gold and silver. Gold and copper are also mutually soluble. Silver and copper, however, are almost insoluble in each other.

Mercury can also be listed with the gold group, but only because it is commonly alloyed with gold or silver. In addition, native mercury occurs sparingly in nature as isolated drops in the upper oxidized zones of some cinnabar deposits (e.g., in California) and also with cinnabar in a few volcanic regions.

Gold, Au

See also the preceding tabulated data.

Habit: Crystals uncommon; typically roughly octahedral, dodecahedral, or cubic; in parallel groups and twinned aggregates. Commonly leafy, dendritic, filiform, or spongy; also massive, in rough, rounded, or flattened grains ("nuggets"—see Figure 8-6) or scales.

Twinning: Common on {111}, generally repeated.

Cleavage and fracture: None. Hackly.

Hardness: $2\frac{1}{2}$–3.

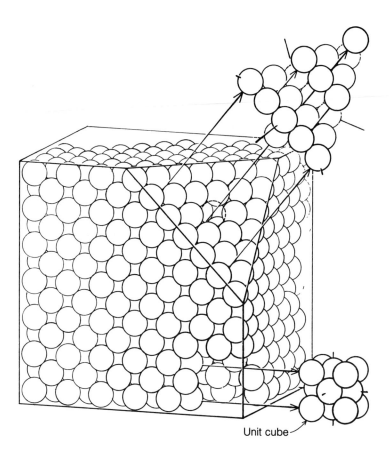

Figure 8-4. Face-centered cubic packing showing the atoms as spheres in their true size relative to the space. The face-centered cubic unit contains four atoms, and the simple unit containing only one atom is a rhombohedron.

Unit cube

Tenacity: Very malleable, ductile, and sectile.

Specific gravity: 19.3, less with contents of silver, copper, and other alloying elements.

Color and streak: Gold-yellow when pure, whiter due to alloyed silver, orange-red due to copper.

Luster and light transmission: Metallic. Opaque except in very thin foil, which transmits blue and green light.

Chemistry: Silver substitutes for gold in the structure, and a complete series extends from pure gold to pure silver. Native gold commonly contains silver up to 10–15 percent. The name *electrum* is applied to natural gold with 20 percent or more of silver. Other metals—palladium, rhodium, copper, bismuth, and mercury—can be dissolved in gold to a minor extent.

Diagnostic features: Color, malleability, and sectility. In small grains, it may often be confused with pyrite or chalcopyrite, hence the common designation *fool's gold* for these minerals. Pyrite can be distinguished by its superior hardness, and chalcopyrite by its brittle nature. Small flakes of biotite partially weathered to a golden color may also be mistaken for gold, especially when wet; these partially

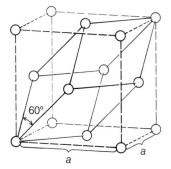

Figure 8-5. Face-centered cubic packing showing the simple unit cell, which contains only one equivalent atom, to be a rhombohedron with edge = $a\sqrt{2}$ and $\alpha = 60°$.

234

Figure 8-6. Gold nugget from Spotsylvania County, Virginia. Longest dimension is ~ 13 cm. (Courtesy of the Smithsonian Institution, National Museum of Natural History.)

altered flakes of mica can be distinguished by their relatively low specific gravity and their brittleness. Gold is insoluble in acids except aqua regia. It forms an amalgam with mercury.

Occurrence, associations, and uses: Gold is present in the earth's crust and in sea water to the extent of about four parts in a hundred million. Most of the gold that occurs in the crust is in the native state. In addition, compounds with tellurium occur at a number of localities, and compounds with bismuth, antimony, and selenium occur rarely. Gold occurs in notable amounts in two types of deposits: (1) in hydrothermal veins and (2) in placers, both those that are unconsolidated and those that are consolidated. Other kinds of gold deposits include, for example, the colloidal-size gold that is disseminated in clay aggregates at Carlin, Nevada—gold that is thought to represent deposition from hot springs.

Most gold of hydrothermal origin is in quartz veins, commonly with pyrite, other sulfides, and gold-silver tellurides—rarely with lead and mercury tellurides, scheelite, tourmaline, ferroan dolomite (commonly called ankerite), sericite (very fine-grained muscovite), or the green chromian sericite; minor amounts of gold occur in massive hydrothermal deposits along with pyrite, chalcopyrite, pyrrhotite, and sphalerite. The latter gold is often recovered from these deposits as a valuable by-product. The gold of placer deposits has been concentrated from hydrothermal deposits as a consequence of its resistance to both chemical and physical weathering and its high specific gravity. Placer gold occurs as rounded or flattened grains and nuggets, in many cases in association with other heavy and resistant minerals. Whereas gold has been obtained from earliest times, largely from unconsolidated placers, the mining of hydrothermal veins and consolidated placers has become increasingly important since the middle of the nineteenth century. The most important single source of gold has been the consolidated and hydrothermally altered quartz conglomerates of the Witwatersrand, Transvaal, South Africa.

Gold is perhaps best known for its use as bullion, which is kept in reserve for monies issued. It also is well known because of its use in jewelry, gold leaf coatings and inlays, and dentistry. Industrial and scientific uses include gold plating of electronic components (especially those employed in high-speed computers) and gold capsule sample containers used in high-temperature/high-pressure research of, for example, mineral systems.

Silver, Ag

See also the data given in the introduction to this group of minerals.

Habit: Crystals uncommon; roughly cubic, octahedral, or dodecahedral; commonly arborescent and wirelike; also massive, as scales, leafy forms, or thin plates in vein fractures.

Twinning: Common on {111}.

Cleavage and fracture: None. Hackly.

Hardness: $2\frac{1}{2}$–3.

Tenacity: Sectile, ductile, and malleable.

Specific gravity: 10.1–11.1, differing with dissolved gold, copper, or other metals.

Color: Silver-white, commonly gray to black due to tarnish.

Luster and light transmission: Metallic. Opaque.

Chemistry: Native silver may contain gold in substitution for the silver atoms in the structure (see Gold). However, there appears to be a break in the high silver part of the series. Mercury, arsenic, and antimony also may occur dissolved in native silver. The antimonian silver, *allargentum* is hexagonal.

Diagnostic features: Color, tarnish, sectility, and hackly fracture. Silver is soluble in nitric acid; HCl added to solution gives a white curdlike precipitate. It is readily tarnished, with the formation of silver sulfide, by either H_2S fumes or solutions of sodium sulfide.

Occurrence, associations, and uses: Native silver occurs in the crust in two principal ways: (1) as small amounts in the oxidized zone of ore deposits and (2) as deposits from hydrothermal solutions. It occurs with sulfides, zeolites, calcite, barite, fluorite, and quartz at, for example, Kongsberg, Norway; it occurs with arsenides and sulfides of nickel and cobalt, silver minerals, calcite, and barite at, for example, Cobalt, Ontario; it occurs with uraninite and nickel-cobalt minerals at, for example, Great Bear Lake, Canada; and it is associated with native copper on the Keweenaw Peninsula of upper Michigan.

In recent years, much of the total silver production has been obtained as a by-product of gold refining and of lead-zinc mining. Silver, frequently noted as earth's finest electrical conductor, has many diverse uses. Among these are its use in photography, electronic components, refrigeration, jewelry, silverware, and coinage. Because of its generally perceived value and uses, it is also highly susceptible to hoarding.

Copper, Cu

See also the data given in the introduction to this group of minerals.

Habit: Crystals rare; cubic or dodecahedral; typically massive filiform or arborescent (Figure 8-7); also as pseudomorphs after, for example, cuprite and azurite.

Twinning: Common on {111}.

Cleavage and fracture: None. Hackly.

Hardness: $2\frac{1}{2}$–3.

Tenacity: Ductile and malleable.

Color: Light rose on fresh surface, quickly changing to copper-red, then brown.

Luster and light transmission: Metallic. Opaque.

Chemistry: Native copper generally contains small-to-trace amounts of silver, arsenic, iron, bismuth, or antimony.

Diagnostic features: Color, malleability, sectility, and hackly fracture. Copper dissolves readily in nitric acid, the resulting solution being greenish.

Occurrence, associations, and uses: Native copper is most commonly associated with basic extrusive igneous rocks where it has formed by reaction between copper-bearing solutions and iron minerals. In this type of deposit, which has been extensively mined on the Keweenaw Peninsula of Michigan, native copper is associated with chalcocite, bornite, epidote, calcite, prehnite, datolite, chlorite, zeolites, and small amounts of native silver. Copper and other minerals fill cavities and fissures in amygdaloidal lavas and interstices in interbedded sandstones and conglomerates. The copper also replaces, either partly or completely, the grains and pebbles. Single masses of copper weighing many tons have been found in the Keweenaw district. Production from these deposits, including that involving underground mining, dates back to prehistoric tribes.

Copper was one of the first metals used by man; it is thought that in the eastern Mediterranean countries, stone implements gave way to those made of bronze about 7000 years ago. Since then, the outstanding characteristics of ductility, malleability, resistance to corrosion, and high electrical conductivity have continually increased the many uses and thus the demand for copper. Indeed,

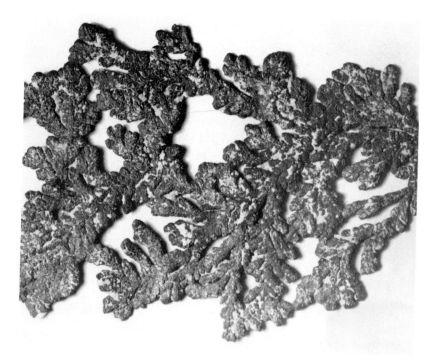

Figure 8-7. Dendritic native copper, 15 cm wide. (Courtesy of Queen's University.)

copper is one of the elements that must be considered essential to modern industry; for example, a large percentage of electrical wire is copper, and copper is an ingredient of numerous important alloys.

PLATINUM GROUP

The platinum group includes the two metals platinum and palladium, plus natural alloys of platinum-iridium and iridium-osmium. Native platinum commonly contains iron in solution. Platinum, palladium, and iridium are isometric, with the same structure as minerals of the gold group. Whereas all those minerals have a face-centered cubic closest packing structure, osmium and the alloys of iridium and osmium have a hexagonal closest packing structure (Figures 8-2b, 8-3b, and 8-8).

Platinum, Pt

Crystal system, class, and space group: Isometric; $4/m\bar{3}2/m$; $Fm3m$.
Cell dimensions and content: $a = 3.9231$; $Z = 4$.
Habit: Crystals rare; typically as grains, scales, or nuggets.
Cleavage and fracture: None. Hackly.
Hardness: $4-4\frac{1}{2}$, increases with iron content.

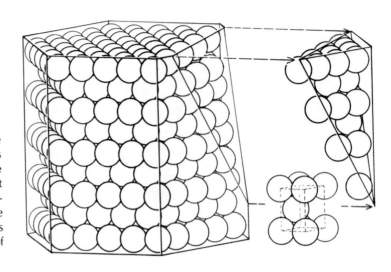

Figure 8-8. Hexagonal close packing showing the atoms as spheres in their true size relative to the space. The simplest unit cell is a 60° rhombic prism containing two atoms. This structure is typical of magnesium and is found in some native metals of the platinum group.

Tenacity: Malleable and ductile.

Specific gravity: 14–19 (21.46 for pure Pt).

Color and streak: Whitish steel-gray to dark gray.

Luster and light transmission: Metallic. Opaque.

Chemistry: Native platinum contains iron, up to about 28 percent; thus, it is commonly distinctly magnetic and typically exhibits polarity when iron-rich. It can also contain palladium, rhodium, iridium, and copper.

Diagnostic features: Gray color, high specific gravity, and the fact that it can be dissolved only by hot aqua regia. It does not tarnish as silver does.

Occurrence, associations, and uses: Platinum occurs in basic and ultrabasic igneous rocks, associated with olivine, pyroxene, chromite, and magnetite. It also occurs as grains or nuggets in river gravels derived from areas containing such ultrabasic rocks. Platinum and a few platinum-bearing minerals are mined from magmatic concentrations in dunite in, for example, the Bushveld Igneous Complex in the Transvaal, South Africa. Large quantities are recovered from river placers along both flanks of the Ural mountains in the USSR, and minor quantities are recovered from placers in many other parts of the world.

Platinum, because of its resistance to corrosion by nearly all chemicals, finds its greatest use in the chemical industries—for example, as a catalyst in the production of both organic and inorganic chemicals. [It is used in catalytic converters in automobiles.] In addition, because of its good electrical conductivity, and its high melting temperature, it has many other important applications.

IRON GROUP

The iron group includes two minerals: (1) iron, Fe [body-centered cubic closest packing (Figure 8-1b), containing minor nickel; $a = 2.874$, Disko Island Greenland] and (2) nickel-iron, Ni-Fe (face-centered cubic closest packing, 77–24 percent nickel; $a = 3.560$, 75 percent Ni, Josephine County, Oregon). Both are extremely rare as terrestrial material, but they make up the bulk of the metallic constituents of meteorites as kamacite and taenite, respectively. Iron is a rare constituent of basalts, with perhaps the most notable occurrence at Disko Island, Greenland, where it occurs as both large masses and small embedded grains. Nickel-iron has been found as small plates and grains in serpentinite, as an alteration product of Ni-bearing olivine, and in a number of placers derived from those rocks. These minerals, which are magnetic, are distinguished from magnetite by their lower hardnesses and their malleability.

ARSENIC GROUP

The arsenic group includes native arsenic, As; antimony, Sb; and bismuth, Bi. These semimetals crystallize with a rhombohedral lattice. In accordance with their threefold valence and their position in group V of the periodic table, each of these atoms is bonded to six of its neighbors, but more closely to three than to the other three. The shorter of the bonds result in a puckered sheet of atoms, which is connected to the next sheet by the longer bonds. The sheets extend in the plane (0001), and thus perfect basal cleavage can occur without breaking of the shorter bonds (Figure 8-9). In arsenic, the shorter bonds are largely homopolar in character; in antimony and bismuth, they become increasingly metallic in character, accompanied by decreases in hardness and melting point and increases in atomic weight and specific gravity.

Crystal system, class and space group: Hexagonal (trigonal); $\bar{3}2/m$; $R\bar{3}m$.

	As	Sb	Bi
Cell dimensions:	$a = 3.760$	$a = 4.307$	$a = 4.546$
	$c = 10.548$	$c = 11.273$	$c = 11.860$
Cell content:	$Z = 6$	$Z = 6$	$Z = 6$
Twinning:	$\{10\bar{1}4\}$	$\{10\bar{1}4\}$	$\{10\bar{1}4\}$
Cleavage:	$\{0001\}$ perfect	$\{0001\}$ perfect	$\{0001\}$ perfect
Hardness:	$3\frac{1}{2}$	$3–3\frac{1}{2}$	$2–2\frac{1}{2}$
Specific gravity:	5.7	6.7	9.7–9.8
Color:	Tin-white, tarnishing to gray-black	Tin-white	Silver-white, reddish hue
Streak:	Tin-white to gray	Gray	Silver-white

240   Figure 8-9. Structure of antimony. Solid lines represent short bonds of essentially covalent character; dashed lines represent longer bonds. The cleavage is horizontal.

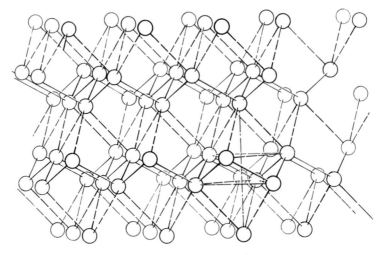

Rotation 1:3   Tilt 1:18¼

Arsenic, As

See also the preceding tabulated data.

Habit: Natural crystals extremely rare; commonly massive, in concentric layers; less commonly reniform or stalactitic.

Chemistry: Most arsenic contains some antimony; some arsenic also contains minor amounts of iron, nickel, silver, or sulfur. Although arsenic and antimony form an unbroken solid-solution series at high temperature in the laboratory, equilibration at lower temperatures results in exsolution of an intermetallic compound, AsSb, over a wide range of compositions. The name *stibarsen* is applied to naturally occurring specimens of this intermediate compound, and the term *allemontite* is applied to intergrowths of stibarsen with either arsenic or antimony.

Diagnostic features: Habit and hardness. Upon heating, arsenic is volatile without fusion, giving a garlic odor and white fumes of As_2O_3.

Occurrence, associations, and uses: Arsenic occurs in hydrothermal veins with silver, cobalt, or nickel ores. It can also be associated with barite, cinnabar, realgar, orpiment, stibnite, and galena.

Arsenic is used widely in herbicides and insecticides and for alloying with lead for shot. Adequate supplies of arsenic for these and other commercial needs are recovered from smelter fumes given off during the treatment of metallic ores that contain arsenic in the form of arsenopyrite, enargite, or tennantite.

Antimony, Sb

See also the data given in the introduction to this group of minerals.

Habit: Crystals extremely rare; typically massive; lamellar and distinctly cleavable; also radiated, botryoidal, or reniform.

Diagnostic features: Melts readily at 630°C, leaving a brittle metallic globule commonly coated with acicular needles of a white oxide.

Occurrence, associations, and uses: Antimony of hydrothermal origin occurs in veins with silver ores, commonly associated with stibnite and less commonly with sphalerite, pyrite, galena, and quartz.

Antimony has several different uses, most of which relate to the fact that the metal expands upon consolidation from its melt. Alloyed with lead, antimony is used for such diverse things as type metal, storage battery plates, shrapnel balls, and collapsible tubes (e.g., for toothpaste). Also, with other or additional elements, it is included in the white metal alloys such as pewter (Pb-Sn-Sb). Antimony and the antimony oxides contribute somewhat to commercial sources of antimony, but most antimony is obtained from stibnite and antimonial lead ores.

Bismuth, Bi

See also the data given in the introduction to this group of minerals.

Habit: Natural crystals rare and indistinct; usually in reticulated or arborescent shapes; foliated, granular.

Color: The silver-white color, commonly with a reddish hue, tends to darken with exposure and to develop an iridescent tarnish.

Diagnostic features: Color, hardness, and specific gravity. It melts readily at 270°C, forming a brittle globule that is soluble in nitric acid.

Occurrence, associations, and uses: Bismuth occurs in hydrothermal veins with cobalt, nickel, silver, and tin minerals and in pegmatites.

Bismuth has many noteworthy uses. Among them are those associated with its presence in certain compounds that have curative and smoothness qualities (e.g., in Pepto-Bismol and cosmetics) and those that depend on the fact that bismuth lowers, in some cases quite markedly, the melting points of alloys that may, as a consequence, be used for such things as sprinklers in fire control systems and electrical fuses. It also is used in many glazes. Most of the bismuth used in commerce is obtained from deposits of tin, copper, and silver, and as a refinery by-product from the treatment of lead ores.

SULFUR GROUP

The sulfur group comprises the mineral sulfur, which is naturally occurring α sulfur or rhombic sulfur, together with the natural forms of β and γ sulfur, both of which are monoclinic. The latter two minerals are rare. In orthorhombic sulfur, the atoms are bonded into puckered rings of eight atoms (Figure 8-10d) that can be regarded as S_8 molecules. Each atom is bonded to two neighbors in its ring by homopolar bonds; the distance between S atoms within different rings, even adjacent rings, is much greater.

Sulfur, S

Crystal system, class, and space group: Orthorhombic; 2/m2/m2/m; Fddd.
Cell dimensions and content: a = 10.45, b = 12.84, c = 24.46; Z = 128.
Habit: Crystals common; typically dipyramidal; thick tabular on {001}

Figure 8-10.   Sulfur (rhombic or alpha-sulfur) crystals. Forms: pinacoids c{001}, b{010}; rhombic prisms n{011}, e{101}; rhombic dipyramids s{113}, p{111}. (c) A twin crystal with twin plane {011}. (d) An S_8 ring (each sulfur atom has two closest neighbors); 16 such rings are contained in the face-centered orthorhombic unit cell.

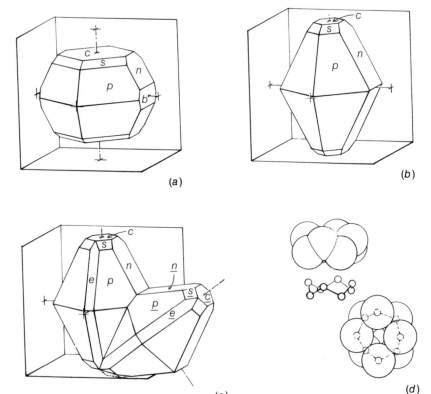

(a)

(b)

(c)

(d)

(Figures 8-10 and 8-11); also massive, in spherical or reniform shapes, incrusting.

Twinning: Rare on {011} (Figure 8-10c).

Common forms and angles:

$(001) \wedge (011) = 62°18'$ $(001) \wedge (113) = 45°11'$

$(001) \wedge (101) = 66°52'$ $(111) \wedge (11\bar{1}) = 36°39'$

$(010) \wedge (111) = 53°13'$ $(111) \wedge (1\bar{1}1) = 73°34'$

Cleavage and fracture: {001}, {110} indistinct. Conchoidal to uneven.

Hardness: $1\frac{1}{2}$–$2\frac{1}{2}$.

Tenacity: Brittle to slightly sectile.

Specific gravity: 2.07.

Color and streak: Yellow to yellowish brown. White.

Luster: Adamantine, resinous to greasy.

Optical properties:

anisotropic, biaxial (+) $2V \cong 69°$

refractive indices: α, 1.958; β, 2.038; γ, 2.245.

weakly pleochroic

Diagnostic features: Color, low hardness, brittleness, melts at 113°C, burns at 270°C in air with a blue flame, yielding sulfur dioxide; insoluble in water and unaffected by most acids, soluble in carbon disulfide and some oils.

Occurrence, associations, and uses: Native sulfur occurs in regions of recent volcanic activity where it has been deposited as a direct sublimation product from volcanic gases, has been formed by incomplete oxidation of hydrogen sulfide from volcanic sources, or has resulted from the decomposition of hydrogen sulfide in thermal spring waters. It also occurs in sedimentary sequences that contain sulfates along with organic materials (e.g., in bituminous limestone), as a product of anaerobic bacterial sulfur-forming activity. Substantial quantities of such sulfur occur in the cap rocks of some of the salt domes of the Texas-Louisiana gulf coastal region; this sulfur generally occurs in a cavernous calcite cap rock and extends into the layer of anhydrite that covers the main body of the salt.

Sulfur and chemical compounds of sulfur are used widely in industry and agriculture. Among their almost innumerable uses are those for drugs, manufacture of explosives, fertilizers, synthetic fibers, herbicides and insecticides, metallurgical processing, paper and plastic manufacture, and refining of petroleum.

The low melting point of sulfur enables its recovery from sedimentary beds at depth by the Frasch process: hot water is pumped under pressure into a well penetrating the sulfur-bearing formation; this procedure causes molten sulfur to collect at the bottom; it is then forced by hot compressed air up through a small inner pipe, as red liquid sulfur, and is piped to a stockpile. Large quantities of sulfur are also recovered as sulfur dioxide from smelter gases and by-product pyrite and as elemental sulfur from pyrite and sour (H_2S-bearing) natural gas.

Figure 8-11. Sulfur. The largest crystal, ~3 cm across, is oriented with *b* axis vertical; the large pinacoid c{001} is beveled by the rhombic pyramid p{111}. (Caltanisetta, Sicily.) (Courtesy the Royal Ontario Museum.)

CARBON GROUP

The carbon group includes two natural forms of native carbon—diamond and graphite.* These minerals present the greatest contrast in structure and properties to be found in any pair of polymorphic substances. Diamond is isometric ($a = 3.567$ Å), with each carbon atom linked tetrahedrally to four neighboring carbon atoms (Figure 8-12). The close three-dimensional covalent linking completing the outer electron shell for each atom by sharing of electrons is manifested in the great hardness of diamond. Graphite is hexagonal, with the carbon atoms

*Two additional polymorphs, *lonsdaleite* and *chaoite*, have been recognized in meteorites.

Figure 8-12. Diamond structure. (a) Atoms in the face-centered cubic unit cell and the tetrahedral bonds from each atom to its four closest neighbors. (b) Atoms in nearly true size relative to space and showing the relation of the octahedral {111} cleavage to the structure.

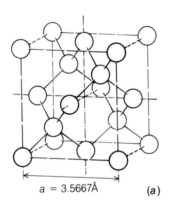

$a = 3.5667$Å   (a)

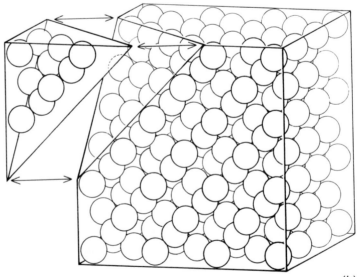

(b)

linked together in sheets that are widely spaced along the c axis (Figure 8-13). This wide spacing of the sheets of carbon atoms accounts for the perfect and easy cleavage of graphite.

	Diamond	Graphite
Crystal system:	Isometric	Hexagonal*
Crystal class:	$4/m\,\overline{3}2/m$	$6/m\,2/m\,2/m$
Space group:	$Fd3m$	$P6_3/mmc$
Cell dimensions:	$a = 3.567$	$a = 2.464$
		$c = 6.736$
Cell content:	$Z = 8$	$Z = 4$
Common forms:	{111}, {110}, {100}	{0001}, {10$\overline{1}$0}, {10$\overline{1}$1}, {10$\overline{1}$2}
Twinning:	{111}	—
Cleavage:	{111} perfect	{0001} perfect, easy
Fracture:	Conchoidal	None
Tenacity:	Brittle	Flexible, not elastic, greasy feel
Hardness:	10	1–2
Specific gravity:	3.50	2.09–2.23
Color:	Colorless and varied	Black
Streak:	White	Black
Luster:	Adamantine	Metallic to dull
Light transmission:	Transparent to translucent	Opaque

*A second polytype has been reported as rhombohedral, $\overline{3}m$, $R\overline{3}m$, $c = 10.06$, $Z = 6$. The two may be referred to as 2H and 3R graphites.

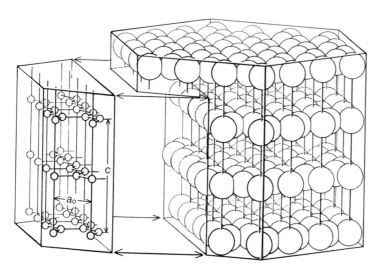

Figure 8-13. Graphite structure. Atoms are closely linked in sheets parallel to {0001} and widely spaced along c, resulting in the perfect basal cleavage {0001}.

$a_0 = 2.464Å$ $c = 6.736Å$

246

Diamond, C

See also the preceding tabulated data.

Habit: Crystals common; typically octahedral, also dodecahedral, cubic, or tetrahedral; faces commonly curved, flattened on {111}.

Twinning: Common on {111}, giving simple and multiple twins (Figure 8-14).

Color: Pale yellow, brown, white to blue-white; also orange, pink, blue, green, red, and black.

Optical properites:
isotropic
refractive index: *n*, 2.4195

Diagnostic features: Hardness and luster.

Occurrence and uses: Diamond occurs as scattered crystals in an ultrabasic olivine-rich or olivine- and phlogopite-rich porphyry called kimberlite. The kimberlite occurs in roughly carrot-shaped intrusive bodies with diameters up to 950 meters and depths of 1200 meters or more. Many kimberlite pipes occur in South Africa, but only a

Figure 8-14. Diamond, ~16 mm long, exhibiting a spinel {111} twin (see Figure 2-68), from Mir Pipe, Mironovo, Yakutia, Siberia. (Photograph by S. C. Chamberlain. By permission, from S. C. Chamberlain and V. T. King, 1981. "The William W. Pinch Collection." *Rocks and Minerals*, **56,** 49–66.)

few contain sufficient diamond to warrant mining; similar pipes occur in the Yakutia Region in Siberia, and some of these are reported to be large producers.

The exposed kimberlite of these masses has been altered by surface weathering to "yellow ground," whereas at greater depths only the olivine is altered (to serpentine) and the rock is called "blue ground." The diamonds can be recovered from the friable yellow ground by passing it, along with water, over greased bronze tables; the diamonds adhere to the grease. The harder blue ground, however, must be crushed before tabling. The content of the profitable pipes ranges from about 0.1 to 0.35 carats per ton (1 carat = 0.2 gram). The Premier Mine, near Pretoria, South Africa, has produced about $5\frac{1}{2}$ tons of diamonds from 100 million tons of rock, which is about 0.0000055 percent. This mine also yielded the largest known diamond, the Cullinan, which weighed 3106 carats (~ 0.6 kg) before it was cut.

Diamonds are also found in river and beach gravels and, at present, more than 90 percent of the world's diamond production is obtained from such deposits. The color and transparency of diamond range greatly and have a considerable bearing on its value. The colorless and water-clear material continues to be the most desirable of all substances for faceted gems. Water-clear diamond in other colors is also used for gem cutting, but it is usually not valued as highly. Diamond not of gem quality, usually called bort, has many specialized industrial uses, most of which depend on its being the hardest known substance. These uses include dies for drawing very fine wire, cutting tools for special machine work, diamond-drill bits that enable the recovery of a cylindrical sample of rock, abrasives for grinding and cutting wheels and for the fine grinding or polishing of diamond itself and of many other hard substances. The term bort, in its older and more restricted usage, applies to granular and cryptocrystalline diamond, gray to black in color due to impurities and inclusions. This material shows no distinct cleavage and, for this reason, is tougher and less brittle than other diamond. Another variety, which is highly prized for industrial cutting, is called carbonado; it is a black or gray-black bort, massive, with a density less than that of other diamond. Of the annual production, about half, by weight, is bort for industrial uses. Only about 5 percent is suitable for cut stones of 1 carat or larger.

From early times until about 1730, India was the only source of diamonds. Brazil became the chief producer from that time until the discovery of the South African kimberlite pipes in 1867. Since then, the pipes and river and beach gravels in adjoining parts of Africa have been the source of most commercial diamond. A few diamonds have been found in alluvium in Australia and in glacio-fluvial deposits at scattered points in North America. Also, one occurrence of kimberlite in Arkansas has yielded several thousand small diamonds of good color.

Graphite, C

See also the data given in the introduction to this group of minerals.

Habit: Crystals uncommon, tabular on {0001} with I-order hexagonal prism and dipyramids {$10\bar{1}1$}, {$10\bar{1}2$}; commonly as embedded foliated masses, isolated tabular grains, and scales; also columnar, radiated, or earthy.

Diagnostic features: Extreme softness, greasy feel, low specific gravity, black color, marks on paper readily; it is distinguished from molybdenite by its black color, its black streak on glazed porcelain, and specific gravity.

Occurrence and uses: Graphite is of common occurrence in rocks produced by regional or contact metamorphism. It occurs in marble, gneiss, schist, quartzite, and metamorphosed coal beds. In thoroughly recrystallized limestone (marble), it is present as distinct micaceous flakes. Graphite also occurs in igneous rocks, pegmatite dikes, and veins. In some places, these rocks occur in contact with metamorphosed graphite sediments, and it has been suggested that the carbon was probably introduced from the sedimentary rocks. In other places, the graphite is thought to have formed from primary constituents of the magma. Graphite in shear zones, contact metamorphic deposits, and hydrothermal veins may also have had its origin in magmatic vapors or from adjacent sediments.

The graphite resulting from regional metamorphism of sedimentary rocks probably formed by crystallization of carbon from organic remains. Although there is no direct evidence for the nature of such a process, graphite—especially that in rocks of Precambrian age—may have originated from the reduction of calcium carbonate by some inorganic process.

Graphite's high melting temperature (3000°C) and insolubility in acid result in its having many uses other than as the lead in pencils. For example, it is used in foundry facings, crucibles, lubricants, paints, electrodes, and generator brushes.

9 Class II: Sulfides and Sulfosalts

The sulfides include a large group of minerals, predominantly metallic in character, with the general formula A_mX_p, in which X, the larger atom, is sulfur or, to a lesser extent, arsenic, antimony, bismuth, selenium, or tellurium, and the smaller atom is one or more of the metals. In a few cases, sulfur and arsenic or antimony are present in about equal amounts.

The sulfosalts, also included in this chapter, have the general formula $A_mB_nX_p$, which can be written in most cases as a double sulfide $A_mX_q \cdot B_nX_{(p-q)}$. The common elements in the sulfosalts are: Ag, Cu, and Pb as A; As, Sb, Bi, and Sn as B; and S as X. Iron, Ni, and Hg are rare; Tl, V, and Ge each occur as major constituents in one or two rare sulfosalt minerals.

In atomic bonding, the sulfide minerals range from typical ionic compounds in some of the simple sulfides, through the homopolar character of pure ZnS, to other sulfides that are essentially alloys with distinct metallic characteristics. The sulfur atoms (and also As and Sb) are separate in some minerals, whereas they are linked in pairs by homopolar bonds in others, such as pyrite. There is considerable structural similarity between sulfides and sulfosalts. Many of these minerals are opaque or nearly so; those that are not typically have very high indices of refraction and, in many cases, transmit only red light. They range in hardness from about 1 for molybdenite to 6 or more for pyrite and sperrylite.

On the following list, the sulfides are arranged according to the decreasing $A:X$ ratio, and then the sulfosalts are arranged according to the decreasing $(A + B):X$ ratio. [Only a few of the sulfosalt minerals, which as a group are relatively rare, are described in this book.]

A_2X Type
    Argentite, Ag_2S
    Chalcocite, Cu_2S

A_3X_2 Type
    Bornite, Cu_5FeS_4

AX Type
    Galena, PbS
    Sphalerite, Wurtzite, $(Zn,Fe)S$
    Chalcopyrite, $CuFeS_2$
    Nickeline Group
        Pyrrhotite, $Fe_{1-x}S$
        Nickeline, $NiAs$
        Breithauptite, $NiSb$
    Millerite, NiS
    Pentlandite, $(Fe,Ni)_9S_8$
    Covellite, CuS
    Cinnabar, HgS
    Realgar, AsS
    Orpiment, As_2S_3
    Stibnite, Sb_2S_3

AX_2 Type
    Pyrite Group
        Pyrite, FeS_2
        Sperrylite, $PtAs_2$
    Cobaltite, $CoAsS$
    Marcasite, FeS_2
    Arsenopyrite, $FeAsS$
    Molybdenite, MoS_2
    Krennerite Group
        Krennerite, $(Au,Ag)Te_2$
        Calaverite, $AuTe_2$
        Sylvanite, $(Au,Ag)Te_2$

AX_3 Type
    Skutterudite Series $(Co,Ni)As_{2-3}$

A_3BX_3 Type
    Pyrargyrite, Ag_3SbS_3
    Proustite, Ag_3AsS_3
    Tetrahedrite, $(Cu,Fe)_{12}Sb_4S_{13}$

Tennantite $(Cu,Fe)_{12}As_4S_{13}$

A_3BX_4 Type

Enargite, Cu_3AsS_4

A_2BX_3 Type

Bournonite, $PbCuSbS_3$

ABX_2 Type

Boulangerite, $Pb_5Sb_4S_{11}$

Argentite, Ag_2S

Crystal system, class and space group: Isometric; $4/m\bar{3}2/m$ (179°– 586°C); *Im3m.*

Cell dimensions and content: $a = 4.89$; $Z = 2$.

Habit: Crystals uncommon; cubic, octahedral, rarely dodecahedral (see remarks under Chemistry); commonly in groups of parallel individuals; also arborescent, filiform, massive, or as a coating.

Cleavage and fracture: {001}, {011} indistinct. Subconchoidal.

Hardness: $2–2\frac{1}{2}$.

Tenacity: Very sectile.

Specific gravity: 7.2–7.4.

Color and streak: Black. Black and shining.

Luster and light transmission: Metallic. Opaque.

Chemistry: The cubic modification of Ag_2S, called argentite, is stable only above 179°C. The cubic crystals are therefore paramorphs; they possess the monoclinic internal crystal structure of the dimorphic modification called *acanthite.*

Diagnostic features: Color, sectility; lacks perfect cleavage of galena, which it resembles; upon heating, emits sulfurous fumes and yields silver.

Occurrence, associations, and uses: "Argentite" typically occurs with the ruby silvers (pyrargyrite and proustite) and native silver. It also occurs as microscopic inclusions in galena, the material being called *argentiferous galena.* Probably the most important primary silver mineral, "argentite" occurs widely in hydrothermal sulfide ore deposits.

Chalcocite, Cu_2S

Crystal system, class, and space group: Monoclinic; $2/m$; $P2/c$.

Cell dimensions and content: $a = 15.23$, $b = 11.88$, $c = 13.50$; $\beta = 116° 16'$; $Z = 48$. [A distinct pseudocell with $a' = b/3 = 3.96$, $b' = a/2 \sin \beta = 6.83$, $c' = c/2 = 6.75$, $Z = 4$ is also pseudo-hexagonal with a (hexagonal) $= 3.96$, $c = 6.75$; $Z = 2$.]

Habit: Crystals rare; prismatic along c or b, or thick tabular on $\{\bar{1}02\}$

252

({001} in pseudocell) (Figure 9-1); also compact, massive.

Twinning: Very common on {230} ({110} in pseudocell, giving pseudo-hexagonal forms) and microscopic lamellar twinning.

Cleavage and fracture: {110} indistinct. Conchoidal.

Hardness: $2\frac{1}{2}$–3.

Tenacity: Brittle to somewhat sectile.

Specific gravity: 5.5–5.8.

Color and streak: Dark gray to black.

Luster and light transmission: Metallic. Opaque.

Diagnostic features: Hardness, black color, slight sectility, and association with other primary and secondary copper minerals; soluble in nitric acid. (Hexagonal Cu_2S, dimorphous with chalcocite, is stable above 105°C.) There are, however, a number of closely related minerals that are not readily distinguished from chalcocite; included are *digenite* (Cu_9S_5), *djurleite* ($Cu_{31}S_{16}$), and *anilite* (Cu_7S_4). In fact, recent investigations have led to the suggestion that about half of the "chalcocite specimens" in university and museum collections are probably djurleite.

Occurrence, associations, and uses: Chalcocite occurs most commonly as fine-grained, massive material, showing alteration to co-

Figure 9-1.   Chalcocite crystals. Forms: pinacoids   c{001}, b{010}, a{100}; rhombic prisms l{130}, n{230}, m{110}, f{012}, d{021};   rhombic   dipyramids z{113}, v{112}, p{111}. (d) Twin crystal, twin plane {110} or {130}, composition plane {130}.

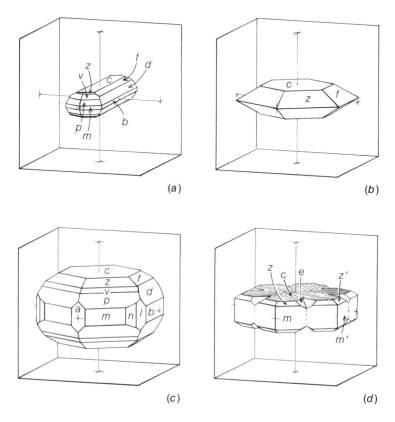

(a)

(b)

(c)

(d)

vellite, malachite, or azurite. It can also be associated with native copper or cuprite.

The principal occurrence of chalcocite is in the supergene enriched zone of sulfide deposits. In arid and semi-arid climates, oxidizing surface waters attack the primary sulfides—chalcopyrite, bornite, pyrite, enargite, and others—forming soluble sulfates. These soluble sulfates react at greater depth with the primary sulfides, and secondary chalcocite is deposited as a blanketlike deposit at the water table level. This chalcocite blanket contains a considerably higher content of copper than does the unaltered primary ore, and it has yielded important amounts of copper at Rio Tinto in Spain, Ely in Nevada, Morenci in Arizona, Bingham Canyon in Utah, and at several other localities.

Chalcocite also occurs in hydrothermal sulfide veins with bornite, enargite, chalcopyrite, tennantite-tetrahedrite, covellite, pyrite, and quartz, as at Butte, Montana. These veins are concentrated in a faulted and fractured area at the margin of the Boulder bathylith granodiorite. Chalcocite has been one of the most important sources of copper.

Bornite, Cu_5FeS_4

Crystal system, class, and space group: Isometric; $4/m\bar{3}2/m$; $Fd3m$ above ~228°C. Tetragonal; $\bar{4}2m$; $P\bar{4}2_1c$ below ~228°C.

Cell dimensions and content: $a = 5.47$; $Z = 1$—cell representing the high-temperature disordered structure; $a = 10.94$, $c = 21.88$; $Z = 16$ cell for $P\bar{4}2_1c$. Bornite crystals with an ordered structure with either isometric or tetragonal structures have been reported with unit cells of 8, 64, 216, or 320 times the volume of the disordered unit cell.

Habit: Crystals rare; cubic or dodecahedral, with faces typically rough or curved; generally massive.

Twinning: On {111}.

Cleavage and fracture: {111} traces; conchoidal to uneven.

Hardness: 3.

Specific gravity: 5.06–5.08.

Color and streak: Copper-red to golden brown or bronzelike on fresh fracture surfaces, quickly tarnishing to purplish iridescent. Grayish brown.

Luster and light transmission: Metallic. Opaque.

Alteration: Bornite may alter to chalcocite, chalcopyrite, covellite, cuprite, chrysocolla, malachite, or azurite.

Diagnostic features: Color usually enables identification of bornite and its distinction from chalcocite and chalcopyrite. Soluble in nitric acid with separation of sulfur and resulting in formation of a bluish solution.

Occurrence, associations, and uses: Bornite, a common and wide-spread copper mineral, occurs in many of the important copper deposits of the world. It is present in dikes, in basic intrusives, disseminated in basic rocks, in contact metamorphic deposits, in pegmatites, and in quartz veins. Crystals of bornite have been occasionally found as druses in vugs and veins, for example, at Butte, Montana, and Bristol, Connecticut.

Galena, PbS

Crystal system, class and space group: Isometric; $4/m\bar{3}2/m$; Fm3m.

Cell dimensions and content: a = 5.936; Z = 4.

Crystal structure: The atomic arrangement (Figure 9-2) is the same as that of halite (Figure 11-1). However, galena and the isostructural minerals PbSe (*clausthalite*) and PbTe (*altaite*), which are much rarer, display semi-metallic bonding in place of the ionic bonding of halite. In galena, each lead atom is bonded to six sulfur atoms, and each sulfur atom is, in turn, bonded to six lead atoms.

Habit: Crystals common; cubic or cubo-octahedral (Figures 9-3 and 9-4), also octahedral, many large crystals consisting of subparallel segments; commonly massive, coarse to very fine granular.

Twinning: Twin plane {111}, penetration or contact twins; twin plane {114} with lamellae giving rise to diagonal striations on cleavage faces.

Cleavage: {001} highly perfect and easy; the cleavage occurs between atomic planes of greatest spacing in the structure and perpendicular to the Pb-S bonds.

Figure 9-2.   (a) Galena structure showing packing of the atoms in actual relative sizes and in an open view with atoms shown considerably smaller. (b) Galena structure, one unit cube with a cube diagonal [111], a threefold axis, vertical; one coordination octahedron of six Pb around one S is shown.

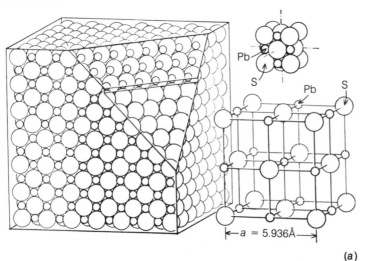

(a)

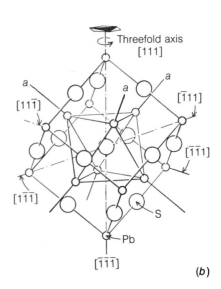

(b)

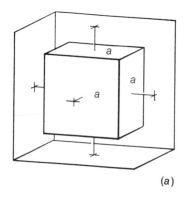

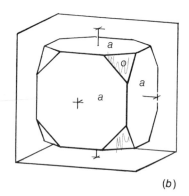

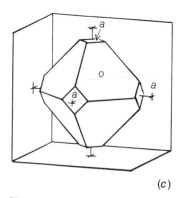

(a)

(b)

(c)

Figure 9-3. Galena crystals. Forms: a{100}, o{111}.

Hardness: 2$\frac{1}{2}$.

Specific gravity: 7.58.

Color and streak: Lead-gray.

Luster and light transmission: Metallic. Opaque.

Chemistry and alteration: Most galena is very nearly pure PbS. The silver, arsenic, and antimony reported in chemical analyses are largely due to inclusions of "argentite" or tetrahedrite, small amounts of which are difficult to detect macroscopically.

In the zone of weathering, galena is readily oxidized to secondary lead minerals, including cerussite, anglesite, pyromorphite, or mimetite.

Diagnostic features: Readily distinguishable by its perfect cleavage, metallic luster, color, and hardness. Its cubic cleavage, greater specific gravity, and darker color distinguish it from stibnite, with which it is most easily confused. The perfect cleavage is obscured in the extremely fine-grained material. Upon heating, galena emits sulfurous fumes and yields a globule of metallic lead.

Occurrence, associations, and uses: The most important lead mineral in the earth's crust, galena is also one of the most common sulfide minerals. It occurs in many types of deposits: in sedimentary rocks, in hydrothermal veins (etc.), and also in pegmatites. Extensive deposits of galena and sphalerite occur as irregular masses in solution cavities and in brecciated zones in limestone. In the typical deposits of this last type (for example, in the Tri-State district), coarsely crystallized galena and sphalerite occur with chert, marcasite, pyrite, chalcopyrite, dolomite, and calcite.

Extensive hydrothermal vein deposits that contain galena also commonly contain significant silver values. In veins generally thought to be of intermediate temperature origin, galena—along with sphalerite, chalcopyrite, pyrite, tetrahedrite, and silver minerals—occurs with quartz and siderite (e.g., Coeur d'Alene, Idaho), with barite and siderite, or with fluorite and barite. In veins and replacement deposits thought to be of high-temperature origin, the sulfides are found with feldspar, garnet, and rhodonite (Broken

Figure 9-4. Galena cubes on gray chert from Joplin, Missouri. Field pictured is 5 cm wide. (Courtesy of Queen's University.)

Hill, New South Wales), or with garnet, diopside, actinolite, and biotite (Sullivan Mine, Kimberley, British Columbia). In these deposits, the galena and sphalerite are typically fine grained. Large deposits of galena with rich silver values occur as replacements of limestone or dolostone in Mexico. Galena also occurs in deposits of contact metamorphic origin. Galena is the most important source mineral for lead, although anglesite and cerussite are important ore minerals in oxidized ore deposits, particularly in Mexico.

Lead finds its main uses because of its resistance to corrosion. Among its uses are those for plates in storage batteries, electric cable sheathing, lead pipe, and solder. The use of lead as a basic ingredient of "lead pigment" in paints and as an antiknock additive for gasolines is decreasing because of environmental considerations.

Sphalerite, Wurtzite, (Zn,Fe)S

These minerals are dimorphic forms (actually polytypes) of ZnS, sphalerite having face-centered cubic stacking of the ZnS tetrahedral layers and wurtzite having hexagonal stacking (Figure 9-5). Polytypes with more complex stacking sequences have been observed both in nature and in synthetic preparations. Pure Zn-sphalerite changes to wurtzite when heated above 1020°C; the temperature is lower for materials higher in iron. Sphalerite occurs much more commonly in nature, although many natural specimens apparently contain domains of both kinds of stacking.

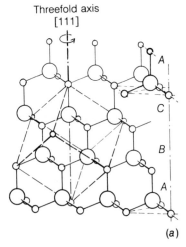

Threefold axis [111]

(a)

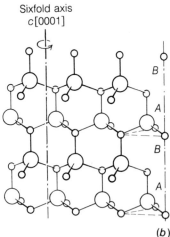

Sixfold axis c[0001]

(b)

Figure 9-5. The stacking of tetrahedral layers in sphalerite and wurtzite. The vertical axis of wurtzite is sixfold, but in the structure it is in reality a sixfold screw axis involving a translation of one-half c for each 60° of rotation. Large circles, S; small circles, Zn.

	Sphalerite	Wurtzite
Crystal system:	Isometric	Hexagonal
Crystal class:	$\bar{4}3m$	$6mm$
Space group:	$F\bar{4}3m$	$P6_3mc$
Cell dimensions (pure ZnS):	$a = 5.4060$	$a = 3.820$ $c = 6.260$
Cell content:	$Z = 4$	$Z = 2$
Habit:	Tetrahedral	Pyramidal
Twinning:	[111] twin axis	—
Common forms:	{001}, {111}, {1$\bar{1}$1}, {211}	{000$\bar{1}$}, {10$\bar{1}$0}, {10$\bar{1}$1}, {10$\bar{1}\bar{1}$}
Cleavage:	{110} perfect	{11$\bar{2}$0} perfect, {0001} indistinct
Hardness:	$3\frac{1}{2}$–4	$3\frac{1}{2}$–4
Specific gravity (pure ZnS):	4.096	4.089
Color (common):	Brown to yellow	Brownish black
Streak:	Brown to light yellow	Brown
Luster:	Resinous to nearly metallic	Resinous

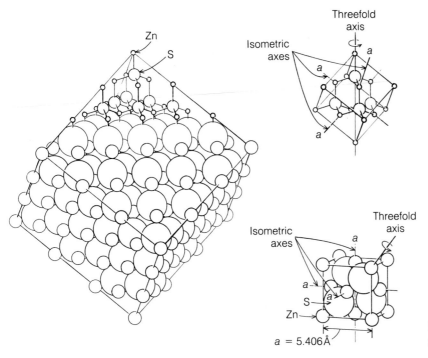

Figure 9-6.   Atomic packing in sphalerite.

Sphalerite, (Zn,Fe)S

See also the preceding tabulated data.

Crystal structure: The structure of sphalerite (Figures 9-6 and 9-7) is analogous to that of diamond (Figure 8-12), with four zinc atoms at the points of a face-centered cubic lattice—comparable to four of the carbon atoms in diamond—and four sulfur atoms at the points of a face-centered cubic lattice that is displaced one-quarter along the body diagonal of the first cube—comparable in position to the other four carbon atoms of diamond. Each zinc atom is coordinated with four sulfur atoms, and each sulfur atom is coordinated with four zinc atoms. These tetrahedra of SZn_4 are all oriented the same way, with a triangular face of the tetrahedron parallel to (111), and the ZnS_4 tetrahedra are all oriented in the opposite way, also with a triangular face parallel to (111). As a consequence, the whole structure shows tetrahedral symmetry, as do the crystals. Each {111} face on a crystal is not identical with its parallel and opposite face; as shown in Figure 9-7, there are alternate layers of all Zn and all S atoms in the {111} planes. That is to say, along the axis [111], normal to (111), the sequence is ZnS-ZnS, etc. to the left, and SZn-SZn, etc. to the right; thus, the [[111]] axes are polar in character and show pyroelectricity. It has also been noted that the faces of $+o\{111\}$ and $-o\{1\bar{1}1\}$ are affected quite differently by solvents. In

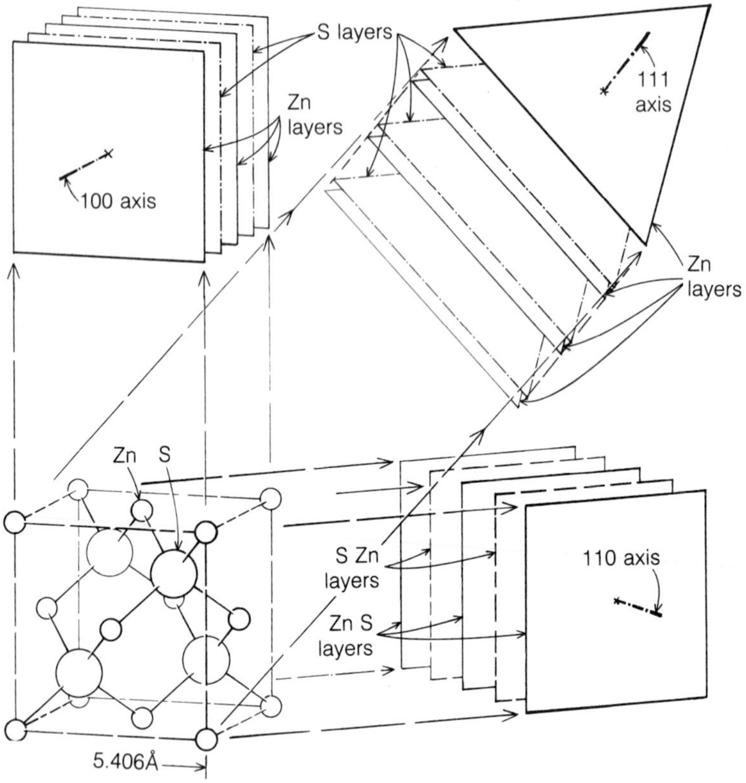

Figure 9-7. Sphalerite structure, showing equal spacing of atomic layers parallel to (100) and (110) and unequal spacing of Zn and S layers parallel to (111), resulting in the nonequivalence of {111} and {$\bar{1}\bar{1}\bar{1}$} (the positive and negative tetrahedra).

sphalerite, the tetrahedral layers parallel to {111} are stacked in the sequence *ABC, ABC, ABC, ...* This sequence results in the cubic structure, as in the case of the face-centered cubic metals, because this sequence results in four identical threefold axes. On the other hand, in wurtzite, the stacking sequence is *AB, AB, . . . ,* which results in a hexagonal structure.

The cleavage, parallel to {110}, breaks relatively few bonds; it occurs between planes containing equal numbers of Zn and S atoms. The spacing between Zn planes and S planes in {111} is larger but does not result in a {111} cleavage.

Habit: Crystals common; tetrahedral (Figure 9-8) or dodecahedral, typically distorted and complex; rough, curved faces are common. Also as cleavable masses, coarse-to-fine granular, fibrous, concretionary, and botryoidal.

Twinning: Common on [111], as simple or multiple contact or penetration; also lamellar gliding twins, due to stress, typically on {111} as composition plane. Directed pressure causes glide twinning on {111}.

Color and streak: Commonly yellow, yellow-brown, or dark brown;

less commonly red, green, white, or nearly colorless (when free of iron). Brown to light yellow or white.

Luster and light transmission: Resinous to adamantine, almost metallic in high iron varieties. Transparent to translucent.

Optical properties:

  isotropic

  refractive index: *n,* 2.37–2.50

  pale yellow to pale brown

Chemistry and alteration: The sphalerite structure is characteristic of a large number of *AX* compounds, of which zinc, cadmium, and mercury with sulfur, and also mercury selenide and mercury telluride, are known in nature. Only zinc sulfide, however, is common. Solid solution is common in this group. Although there is some substitution of cadmium for zinc in sphalerite (up to 1.66 percent), the most common metals that substitute for zinc are iron and, to a lesser extent, manganese. Thus, it seems odd that iron and manganese do not form sulfides that are isostructural with sphalerite; instead, they normally combine with sulfur in sixfold coordination, with either the hexagonal nickeline structure or the cubic halite structure.

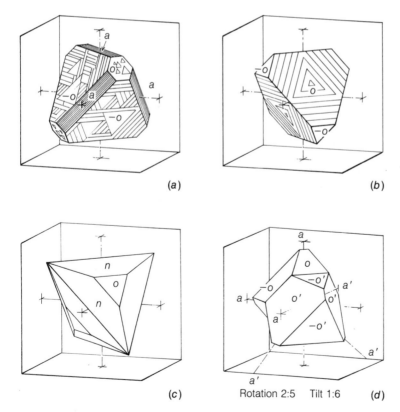

(a)   (b)

(c)   Rotation 2:5   Tilt 1:6   (d)

Figure 9-8. Sphalerite crystals. Forms: cube a{100}, positive tetrahedron o{111}, negative tetrahedron −o{1$\bar{1}$1}, positive tristetrahedron n{211}. (d) Contact twin on {111}.

With increased substitution of iron, sphalerite is darker in color, has a slightly increased cell edge, and has a lower density. Such iron content, as recorded in analyses, ranges up to about 26 percent, which corresponds to nearly 50 mol percent of FeS. This solubility of FeS in ZnS increases with temperature, so the darker, high-iron varieties are of high-temperature origin. In fact, where sphalerite and an iron sulfide have formed together, the iron content of the sphalerite gives an indication of the temperature of deposition.

Under many conditions, sphalerite weathers, leaving a limonite residue. In dry regions, the hydrous zinc sulfate, goslarite, may be formed instead. Sphalerite may also be altered to hemimorphite or smithsonite; actually, some specimens of these minerals are pseudomorphic after sphalerite.

Diagnostic features: The identification of sphalerite is often difficult because of its extremely variable color and luster. Cleavage and hardness are the most dependable macroscopic properties. Yellow, yellow-brown, brown, and dark brown are the most common colors. Sphalerite with other colors is not uncommon and may deceive even the expert mineralogist. The resinous luster is most common. Sphalerite disolves in hydrochloric acid with the evolution of hydrogen sulfide. It is pyroelectric, and some of it exhibits triboluminescence.

Occurrence, associations, and uses: Sphalerite, the most important zinc mineral, occurs intimately associated with galena in most of the important Pb-Zn deposits. In some replacement bodies, it is associated with chalcopyrite, pyrite, pyrrhotite, and magnetite; in these cases, galena is generally not present. The Tri-State district referred to under galena has produced tremendous quantities of zinc from sphalerite. The masses of sulfide, dominantly sphalerite, occur along solution channels in flat-lying limestone; most of the sphalerite is associated with chert layers. Sphalerite also occurs widely in hydrothermal veins and replacement bodies with galena and silver minerals—for example, at Coeur d'Alene, Idaho.

Sphalerite provides, by far, the bulk of the world's zinc, although smithsonite and hemimorphite are local ore minerals of zinc, especially in Mexico, and zincite, willemite, and franklinite are the ore minerals in the deposits at Franklin and Sterling Hill, New Jersey. The world's cadmium production is also obtained from zinc minerals as a by-product of zinc smelting.

Wurtzite, (Zn,Fe)S

See also the data given in the introduction to this group of minerals.

Crystal structure: The wurtzite structure can be described as a stacking of tetrahedral sheets of ZnS in a sequence *AB, AB,* . . . , parallel to

(0001). The Zn_4S tetrahedra all point in the same direction, with a triangular face parallel to (0001) (Figure 9-5b). The structure is therefore hemimorphic and distinct from sphalerite, in which the ZnS tetrahedral sheets are stacked in a sequence ABC, ABC, ...

Habit: Crystals uncommon; hemimorphic pyramidal, short prismatic to tabular on {0001}; typically striated on {10$\bar{1}$0} and {10$\bar{1}$1}; the large basal pedion {000$\bar{1}$} generally dominant; commonly fibrous, columnar, or as concentrically banded crusts.

Chemistry: Most wurtzite, like sphalerite, contains Fe and may also contain Cd and Mn in substitution for Zn; CdS occurs with the wurtzite structure as the mineral greenockite.

Optical properties:
anisotropic, uniaxial (+)
refractive indices: ω, 2.356; ϵ, 2.378
may appear opaque or nearly so

Diagnostic features: Habit. Massive wurtzite is macroscopically indistinguishable from dark-colored sphalerite.

Occurrence: Wurtzite is found to a minor extent in various sulfide ores.

Chalcopyrite, $CuFeS_2$

Crystal system, class, and space group: Tetragonal; $\bar{4}2m$; $I\bar{4}2d$.
Cell dimensions and content: $a = 5.28$, $c = 10.41$; $Z = 4$.
Habit: Crystals common; typically tetrahedral in aspect with {112} the dominant form; the disphenoid {112} approaches closely an isometric tetrahedron, since $c/2 = 0.9858$; the faces of {112} are commonly large, dull, striated, whereas those of {1$\bar{1}$2} are small, brilliant, and not striated (Figure 9-9); commonly massive compact, sometimes botryoidal.

Twinning: On {112}, contact, penetration, or lamellar.

Crystal structure: Though tetragonal in symmetry, the structural arrangement in chalcopyrite is very similar to that in sphalerite. The unit cell is very close in dimension to two cubes. In each pseudocubic half-cell, the metal atoms occupy face-centered positions, and the sulfur atoms occupy the same positions as the sulfur in sphalerite (Figure 9-10). The arrangement of metals around sulfur is again tetrahedral, the forms {111} and {112} on chalcopyrite are disphenoids, and {112} approximates very closely the isometric tetrahedron. The special distribution of copper and iron atoms, as shown in Figure 9-10, results in a tetragonal unit cell with c double the pseudocube edge. The unit cell then contains 4[$CuFeS_2$]. Analyses of chalcopyrite reveal very little divergence from this formula, but selenium may replace sulfur to a minor extent.

Cleavage and fracture: {011} generally distinct. Uneven.
Hardness: $3\frac{1}{2}$–4.

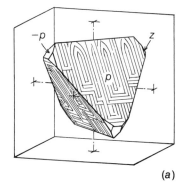

(a)

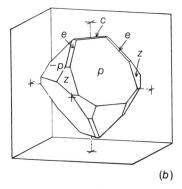

(b)

Figure 9-9. Chalcopyrite crystals. Forms: positive tetragonal disphenoid p{112}, negative tetragonal disphenoid −p{1$\bar{1}$2}; tetragonal dipyramids z{011}, e{012}, pinacoid c{001}. Natural etching on p faces is consistent with the tetragonal symmetry.

262

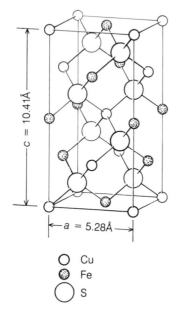

Cu
Fe
S

Figure 9-10. Structure of chalcopyrite. Note similarity of a half-cell to the cubic cell of sphalerite (Figure 9-7).

Specific gravity: 4.1–4.3.

Color and streak: Brass-yellow, commonly tarnished slightly iridescent. Greenish black.

Luster and light transmission: Metallic. Opaque.

Alteration: Chalcopyrite alters to chalcocite, covellite, chrysocolla, malachite, and iron oxides. Under the action of weathering, chalcopyrite, along with other sulfides, is oxidized to sulfates, which are carried down into the ore body leaving much of the iron behind as a gossan. The copper-bearing solutions react with primary sulfides to produce supergene sulfides, including chalcocite and covellite, which partially or completely replace chalcopyrite and other primary sulfides in a layer at the level of the water table. This zone is termed the *zone of supergene enrichment.*

Diagnostic features: Crystals distinctive. Chalcopyrite is readily distinguished from pyrite by its lower hardness, from gold by its brittle character, from pyrrhotite by its color and nonmagnetic character, and from bornite by its color. It is soluble in nitric acid, with separation of sulfur and coloring of solution greenish.

Occurrence, associations, and uses: Chalcopyrite is the most widespread copper mineral and one of the most important sources of that metal. It has been formed under a great variety of conditions. Most sulfide ore deposits contain some chalcopyrite; a few carry important amounts of the mineral.

In vuggy and drusy cavities and in veins, chalcopyrite may occur as small crystals on sphalerite, galena, or dolomite. In veins of moderate- to high-temperature origin, it usually occurs as irregular blebs and masses. In a few deposits, chalcopyrite and pyrite occur with tourmaline or quartz as gangue minerals. It is associated with cassiterite at Cornwall, England. In the Rouyn district of Quebec, chalcopyrite is found in massive, hydrothermal replacement bodies of pyrite and pyrrhotite with more-or-less sphalerite and magnetite; these ores also contain substantial values in gold. It also occurs in sulfide replacement lenses with pyrite, pyrrhotite, and minor sphalerite, bornite, hematite, magnetite, quartz, chlorite, and other silicates in metamorphic schists at Ducktown, Tennessee.

Chalcopyrite is the important primary copper mineral in most of the porphyry-copper deposits. Chalcopyrite, pyrite, and bornite, along with minor sphalerite and molybdenite, occur disseminated and in closely spaced veinlets with quartz in the upper fractured zones of igneous intrusions of monzonite, quartz monzonite, or diorite porphyry and in the surrounding country rocks. Deposits of this kind, which are important copper producers in Utah, Arizona, Nevada, New Mexico, and Chile, have been deeply weathered, thus leaving a leached capping and an enriched zone or blanket of secondary sulfides. In some of these deposits, the primary chalcopyrite and/or other primary copper minerals have been altered to the carbonate, malachite (Ajo, Arizona), or to basic sulfates or chlorides (antlerite, atacamite, *etc.* at Chuquicamata, Chile). In others, the primary sulfides are present in sufficient quantity to constitute ores. Molybdenite, unaffected by weathering or enrichment, is recovered as an important by-product.

In some deposits, chalcopyrite is associated with large masses of pyrite, as at Mt. Lyell, Tasmania and at Rio Tinto, Spain. The latter deposits include the largest known pyritic bodies in the world. They have been mined for 3000 years, first for gold, later for copper and sulfur with minor lead, zinc, gold, silver, nickel, cobalt, and other metals.

Small amounts of chalcopyrite, pyrite, and other sulfides occur in sedimentary rocks. A few contact metamorphic deposits in limestone contain chalcopyrite with garnet, tremolite, and other lime silicates—for example, at Bisbee, Morenci, and Silver Bell, Arizona, and Seven Devils, Idaho. The massive sulfide deposits at Sudbury, Ontario contain important amounts of chalcopyrite along with pyrrhotite and pentlandite.

NICKELINE GROUP

The nickeline group comprises the minerals pyrrhotite, nickeline, breithauptite, and a few rarer minerals; all may have the nickel arsenide structure (Figure 9-12) possessed by many metallic compounds of AX type. Most pyrrhotite shows a deficiency of iron leading to the general formula $Fe_{1-x}S$, and well-ordered material has a true unit cell that is a multiple of the simple unit cell listed in the following tabulation. However, several different superstructures, based on diverse ordered distributions of vacancies, have been reported for pyrrhotite—for example, the 4C structure of Fe_7S_8 and the 6C structure of $Fe_{11}S_{12}$—and, in some cases, different superstructures have been found to constitute intimately associated domains of individual crystals. The mineral with composition close to the ideal FeS is known as *troilite*; it has been found in many meteorites.

Crystal system, class, and space group: Hexagonal; $6/m2/m2/m$; $P6_3/mmc$.

	Pyrrhotite	Nickeline	Breithauptite
Cell dimensions:	$a = 3.452$	$a = 3.609$	$a = 3.93$
	$c = 5.762$	$c = 5.019$	$c = 5.13$
Cell content:	$Z = 2$	$Z = 2$	$Z = 2$
Cleavage:	None	None	None
Fracture:	Uneven	Uneven	Uneven
Hardness:	$3\frac{1}{2}-4\frac{1}{2}$	$5-5\frac{1}{2}$	$5\frac{1}{2}$
Specific gravity:	4.58–4.65	7.78	8.23

	Pyrrhotite	Nickeline	Breithauptite
Luster:	Metallic	Metallic	Metallic
Color:	Bronze-yellow	Pale copper-red	Copper-red
Streak:	Gray-black	Brownish black	Reddish brown
Light transmission:	Opaque	Opaque	Opaque
Magnetism:	Magnetic, variable	Nonmagnetic	Nonmagnetic

Pyrrhotite, $Fe_{1-x}S$

See also the preceding tabulated data.

Crystal structure: The structure given in the tabulation is one that apparently obtains above ~ 300°C. Below ~ 250°C, pyrrhotite appears to assume diverse orthorhombic and monoclinic structures. Typical crystals, however exhibit hexagonal forms. (The tabulated hexagonal structure is described under nickeline.)

Habit: Crystals uncommon; typically tabular to platy on {0001} (Figure 9-11); commonly massive, granular.

Twinning: Rare on {10$\bar{1}$2} as twin plane.

Color: Bronze-yellow, tarnishing darker on exposure.

Chemistry: Pyrrhotite analyses show a range in sulfur content from 50 to 55.5 atomic percent. Although this content was at first interpreted as excess sulfur, it was soon recognized that such is not possible because there is not sufficient space for additional sulfur atoms. Instead, as we have already noted, it reflects deficiencies of iron that are related to vacancies; the vacancies appear to be disordered in the high-temperature hexagonal structure and ordered according to different schemes in the diverse low-temperature structures. The variation in composition is, therefore, generally expressed by the formula $Fe_{1-x}S$, where x can range up to 0.2.

Small amounts of nickel, copper, and cobalt are often reported in analyses, but these metals are probably present in minute inclusions of the commonly associated minerals pentlandite and chalcopyrite rather than in substitution for iron.

Diagnostic features: Crystals rare. Pyrrhotite is distinguished from chalcopyrite by color and magnetism, albeit variable, and from pyrite by color and hardness. It can be distinguished from pentlandite only with difficulty. It is decomposed by hydrochloric acid with evolution of hydrogen sulfide.

Occurrence, associations, and uses: Pyrrhotite occurs principally in basic igneous rocks, either disseminated as minute flects and blebs or segregated in large masses, as at Sudbury, Ontario. Chalcopyrite, pentlandite, and other nickel sulfides are commonly associated with pyrrhotite.

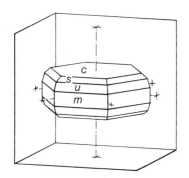

Figure 9-11.   Pyrrhotite crystal. Forms: pinacoid c{0001}; hexagonal prism m{10$\bar{1}$0}; hexagonal dipyramids s{10$\bar{1}$2}, u{20$\bar{2}$1}.

Pyrrhotite also occurs in pegmatites, in contact metamorphic deposits, and as early formed masses with other sulfides in vein and replacement deposits. It occurs as nodules (troilite) in iron meteorites.

Until the 1950s, pyrrhotite was not an economically important mineral. Since 1955, pyrrhotite concentrates from the Sudbury, Ontario ores have been converted into a high-grade iron ore, the sulfur being collected for production of sulfuric acid. Selenium, which probably substitutes for sulfur in pyrrhotite, is also recovered during the smelting process.

Nickeline, NiAs

See also the data given in the introduction to this group of minerals.

Crystal structure: The metal atoms are in closely packed hexagonal layers with interleaved closely packed layers of arsenic (or sulfur, as in pyrrhotite) atoms with each arsenic resting between three metal atoms below and three above (Figure 9-12). Similarly, each metal rests on three arsenics with three others above. All metal layers have their atoms in rows parallel to *c*, whereas the arsenic layers are staggered with repetition along *c* on alternate layers. The unit cell contains 2[NiAs].

Habit: Crystals rare; tabular on {0001} or pyramidal on {10$\bar{1}$1}; typically massive or reniform with columnar structure.

Alteration: Nickeline alters readily to pale green annabergite. If intergrown with cobalt arsenides, the alteration product is white, becoming pink with higher cobalt content (erythrite).

Diagnostic features: Hardness and color; when heated, it emits arsenical fumes with a garlic odor; when dissolved in HNO_3, it gives a greenish solution.

The rarer mineral, *breithauptite* (NiSb), which is isostructural with nickeline, can be distinguished by its darker, distinctly violet-red color.[It has been found to occur in veins with nickeline (etc.) at, for example, Cobalt, Ontario.]

Occurrence, associations, and uses: Nickeline, not a common mineral, occurs with pyrrhotite, chalcopyrite, and other nickel sulfides in basic igneous rocks and ore deposits derived from them, and also in hydrothermal vein deposits along with cobalt arsenides and silver. It has been reported to occur as spheroids with concentric shells in the Natsume Nickel Mine, Japan and at several localities in Germany and France; as massive material in veins at Cobalt, Ontario; and to a minor extent in the nickel ores at Sudbury, Ontario.

Name: Nickeline is often called *niccolite*.

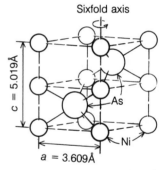

Figure 9-12. Structure of nickeline. Here, as in wurtzite, the vertical axis c, sixfold in crystals, is a sixfold screw axis in the structure.

266

Millerite, NiS

Crystal system, class, and space group: Hexagonal (trigonal); *3m*; *R3m*.

Cell dimensions and content: a = 9.62, c = 3.16; Z = 9.

Habit: Crystals common; typically very slender to capillary along *c*, commonly in radiating, fibrous groups (Figure 9-13); rarely granular.

Cleavage: $\{10\bar{1}1\}$ and $\{01\bar{1}2\}$ perfect.

Hardness: $3–3\frac{1}{2}$.

Tenacity: Brittle, but capillary crystals are elastic.

Specific gravity: 5.5.

Color and streak: Pale brass-yellow. Greenish black.

Luster and light transmission: Metallic. Opaque.

Diagnostic features: Similar to pyrite in color, but readily distinguished by its crystal form, cleavage, and hardness.

Occurrence, associations, and uses: Normally a mineral of low-temperature origin, millerite occurs rather commonly as tufts of capillary crystals in cavities in limestone or dolostone or in carbonate veins. It also occurs as an alteration of other nickel minerals and as a late mineral in veins that contain other sulfides and nickel minerals.

Millerite is of only minor importance as a source of nickel.

Pentlandite, (Fe,Ni)$_9$S$_8$

Crystal system, class, and space group: Isometric; $4/m\bar{3}2/m$; *Fm3m*.

Cell dimensions and content: a = 10.04; Z = 4.

Habit: Crystals nonexistent (?); massive, typically in granular aggregates; most large grains exhibit well-developed octahedral parting planes.

Cleavage, parting, and fracture: None. {111}. Conchoidal.

Hardness: $3\frac{1}{2}–4$.

Specific gravity: 4.6–5.0.

Color and streak: Light bronze-yellow. Light bronze-brown.

Luster and light transmission: Metallic. Opaque.

Chemistry: Most pentlandite analyses show an iron-to-nickel ratio close to 1:1. Slight differences may be due to admixed pyrrhotite, which is extremely difficult to avoid completely in preparing a sample for analysis. Cobalt is often recorded to the extent of about 1 percent; this cobalt, which can substitute for iron or nickel, probably accounts for a small cobalt production at Sudbury, Ontario.

Diagnostic features: Nearly all pentlandite is intimately associated with pyrrhotite, from which it can be distinguished megascopically only by its marked octahedral parting. It is remarkably similar to pyrrhotite in hardness, density, color, and streak. Although it is

(a)

(b)

Figure 9-13. Millerite. (a) Straight filiform tufts in a vug from the Sterling Mine, Antwerp, New York. Longest needles are ~ 2 cm long. (Photograph by S. C. Chamberlain. By permission, from S. C. Chamberlain, 1980, "The George Robinson Collection." *Rocks and Minerals*, **55,** 70–74.) (b) Curved filiform needles in a geode from Halls Gap, Kentucky. Specimen is ~ 6 cm across. (Photograph by John Medici. By permission, from cover of *Rocks and Minerals*, **56,** 3 (1981).)

nonmagnetic, the variable magnetism of pyrrhotite and the prevalence of intergrowths of the two minerals render this feature of little value in identification.

Occurrence, associations, and uses: Pentlandite occurs in basic igneous rocks along with iron and nickel sulfides and arsenides that probably have accumulated by some process of magmatic segregation. The best known occurrence of pentlandite is at Sudbury, where it is the chief source of nickel in the ore.

Covellite, CuS

Crystal system, class, and space group: Hexagonal; 6/m2/m2/m; P6₃/mmc.

Cell dimensions and content: a = 3.792, c = 16.344; Z = 6 (Figure 9-15a).

Habit: Crystals uncommon; typically in hexagonal plates with {0001} showing hexagonal striations (Figure 9-14), bounded by steep pyramid faces, {10$\bar{1}$1} and others, which are horizontally striated; commonly massive. [The thin platy habit in a mineral with the lattice dimensions of covellite affords a fine example of the Law of Bravais (page 34)].

Cleavage: {0001} perfect giving thin flexible leaves.

Hardness: 1½–2.

Specific gravity: 4.6–4.76.

Color and streak: Indigo-blue or darker, commonly iridescent in brass-yellow and dark-red. Lead-gray to black.

Luster and light transmission: Metallic. Opaque except in very thin plates.

Optical properties:
anisotropic, uniaxial (+)
transmits light in thin (< 0.0005 mm) plates only

Chemistry: Although covellite very closely approaches the ideal formula CuS, traces of iron are occasionally reported.

Diagnostic features: Color and perfect cleavage distinguish covellite from chalcocite and bornite; the opacity distinguishes it from other common blue minerals; flakes ignite easily and burn with a blue flame.

Occurrence and associations: Commonly associated with other copper sulfides, chalcopyrite, chalcocite, and bornite, most covellite occurs in the zones of secondary enrichment where it has been formed by alteration of the primary sulfides. Fine crystals of primary covellite occurred in hydrothermal veins at Butte, Montana.

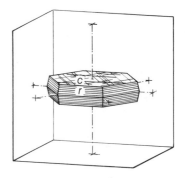

Figure 9-14. Covellite crystal. Typical tabular crystal with large pinacoid c{0001} beveled by narrow faces of the hexagonal dipyramid r{10$\bar{1}$1}.

Cinnabar, HgS

Crystal system, class, and space group: Hexagonal (trigonal); 32; P3₁21 or P3₂21.

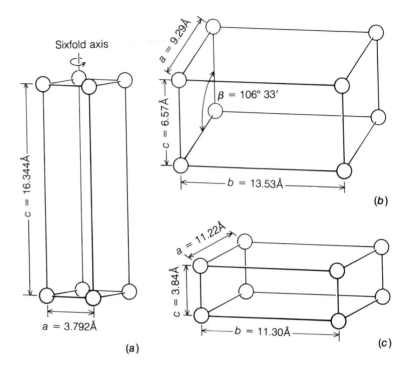

Figure 9-15. (a) Covellite, unit cell of structure. In contrast to the usual tabular habit of covellite crystals, the unit cell has c much longer than a. (b) Realgar, unit cell of structure, showing the monoclinic character. (c) Stibnite, unit cell of structure. In contrast to the markedly acicular habit of stibnite, the unit cell has c much shorter than a or b.

Cell dimensions and content; a = 4.14, c = 9.49; Z = 3.

Habit: Crystals common; rhombohedral or thick tabular on {0001} to short prismatic (Figure 9-16); common in crystalline incrustations; granular, massive, or as an earthy coating.

Twinning: Common with {0001} as twin plane and c axis as the twin axis; simple and penetration.

Cleavage: {10$\bar{1}$0} perfect.

Hardness: 2–2$\frac{1}{2}$.

Tenacity: Somewhat sectile.

Specific gravity: 8.09.

Color and streak: Scarlet-red to brownish red. Scarlet.

Luster and light transmission: Adamantine to submetallic, inclining to metallic when dark colored, or dull in friable varieties. Transparent.

Optical properties:
anisotropic, uniaxial (+)
refractive indices: ω, 2.905; ϵ, 3.256.
deep red

Alteration: In the zone of weathering, cinnabar may be altered to native mercury, mercury oxide (*montroydite*), and/or mercurous chloride (*calomel*). Mercury sulfide is also found as *metacinnabar,* which is black isometric HgS, which is isomorphous with sphalerite.

Diagnostic features: Color, cleavage, and specific gravity; when heated, it is volatile (the fumes are toxic!).

Occurrence, associations, and uses: Cinnabar is the most important mercury mineral. Most of it occurs in veins or impregnations formed at relatively low temperatures near recent volcanic rocks or in hot-spring areas. It occurs in fractures, impregnating and replacing quartz in sandstone or quartzite. It is commonly associated with pyrite, marcasite, and stibnite; with opal, chalcedony, quartz, calcite, and/or dolomite; or with carbonaceous material in shales and slates.

    The most important mercury deposit in the world is at Almaden, Spain, where cinnabar impregnates and replaces quartzite that is interbedded with bituminous shales. These mines have been operating since Roman times, when Pliny recorded that 10,000 pounds of mercury a year were brought to Rome. They are still in operation and have known reserves capable of supplying the world for 100 years.

Realgar, AsS

Crystal system, class, and space group: Monoclinic; $2/m$; $P2_1/n$.

Cell dimensions* and content: $a = 9.29$, $b = 13.53$, $c = 6.57$; $\beta = 106° 53'$; $Z = 16$ (Figure 9-15b).

Habit: Crystals uncommon; generally short prismatic and striated parallel to c (Figure 9-17); also coarse-to-fine granular, compact, or as an incrustation.

Cleavage: {010} good; {$\bar{1}$01}, {100}, and {120} indistinct.

Hardness: $1\frac{1}{2}$–2.

Tenacity: Sectile.

Specific gravity: 3.56.

Color and streak: Aurora-red to orange-yellow. Orange-red.

Luster and light transmission: Resinous to greasy. Transparent when fresh.

Optical properties:
    anisotropic, biaxial (−): $2V = 39°$
    refractive indices: α, 2.538; β, 2.684; γ, 2.704
    slightly pleochroic

Alteration: Disintegrates on long exposure to light to a red-yellow orpiment-bearing powder.

Diagnostic features: Color and hardness; on heating, it emits a garlic odor.

Occurrence, associations, and uses: Realgar occurs as a minor constit-

---

*In monoclinic and triclinic mineral descriptions, the interaxial angles are given as an essential part of the cell dimensions.

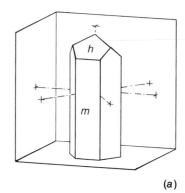

269

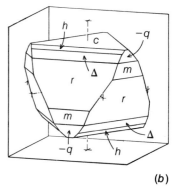

(a)

(b)

Figure 9-16. Cinnabar crystals. Forms: pinacoid $c\{0001\}$, hexagonal prism $m\{10\bar{1}0\}$; positive rhombohedrons $h\{10\bar{1}3\}$, $\Delta\{10\bar{1}2\}$, $r\{10\bar{1}1\}$; negative rhombohedron $-q\{02\bar{2}1\}$.

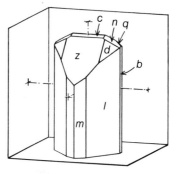

Figure 9-17. Realgar crystal. Forms: pinacoids $c\{001\}$, $b\{010\}$, $z\{101\}$; rhombic prisms $l\{120\}$, $m\{110\}$, $n\{011\}$, $d\{111\}$, $q\{\bar{1}21\}$.

uent in hydrothermal sulfide veins with orpiment and other arsenic minerals, with stibnite, or with lead, silver, or gold ores. It contributes to the arsenic content of such ores, which yield arsenic oxide on smelting. It occurs sporadically in limestones, dolostones, and claystones, as a sublimation product from volcanic emanations, and also in hot-spring deposits.

Orpiment, As_2S_3

Crystal system, class, and space group: Monoclinic; $2/m$; $P2_1/n$.
Lattice, cell dimensions, and content: $a = 11.49$, $b = 9.59$, $c = 4.25$; $\beta = 90° 27'$; $Z = 4$.
Habit: Crystals uncommon; typically small, short prismatic on c, with pseudo-orthorhombic appearance or obviously monoclinic with a prominent zone parallel to [103]; commonly in foliated, columnar, or fibrous masses.
Cleavage: {010} perfect, lamellae flexible but not elastic.
Hardness: $1\frac{1}{2}$–2.
Tenacity: Sectile.
Specific gravity: 3.49.
Color and streak: Lemon-yellow to brownish yellow. Pale yellow.
Luster and light transmission: Pearly on cleavage surfaces, elsewhere resinous. Transparent.
Optical properties:
   anisotropic, biaxial (−): $2V = 76°$
   refractive indices: α, 2.4; β, 2.81; γ, 3.02
   slightly pleochroic
Diagnostic features: Its yellow color and excellent cleavage.
Occurrence and associations: Orpiment, typically a very low-temperature hydrothermal mineral, occurs in veins and in hot-spring deposits or as a common alteration product of realgar or other arsenic minerals. It occurs with stibnite, realgar, arsenic, calcite, barite, or gypsum.

Stibnite, Sb_2S_3

Crystal system, class, and space group: Orthorhombic; $2/m2/m2/m$; Pbnm.
Cell dimensions and content: $a = 11.22$, $b = 11.30$, $c = 3.84$; $Z = 4$ (Figure 9-15c).
Habit: Crystals common; stout-to-slender prismatic along c (Figure 9-18), typically striated or grooved parallel to c; most prisms appear bent or twisted as a consequence of translation gliding; commonly in complex aggregates of acicular crystals (Figure 9-19); also in radiating or columnar masses, or granular.

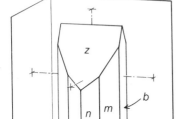

(a)

(b)

Figure 9-18. Stibnite crystals. Forms: pinacoid b{010}; rhombic prisms m{110}, n{210}, z{501}; rhombic dipyramid s{111}. Drawings show only the termination of prismatic crystals.

Cleavage: {010} perfect and easy.

Hardness: 2.

Tenacity: Flexible, not elastic, slightly sectile.

Specific gravity: 4.63.

Color and streak: Lead-gray, commonly with a blackish to iridescent tarnish. Lead-gray.

Luster and light transmission: Metallic, brilliant on cleavage and fresh surfaces. Opaque.

Chemistry: Stibnite varies very little in composition from the ideal Sb_2S_3, although small amounts of Fe, Pb, Cu, and other metals are reported in analyses.

Diagnostic features: Distinguished from galena by cleavage, habit, and lower density; soluble in hydrochloric acid. The rarer mineral, bismuthinite, Bi_2S_3, which is isostructural with stibnite, has similar physical properties but a much higher specific gravity (6.78).

Occurrence, associations, and uses: Stibnite occurs most commonly in hydrothermal vein and replacement deposits of apparently low-temperature origin and in hot-spring deposits. It is generally associated with realgar, orpiment, galena, lead sulfantimonites, marcasite, pyrite, cinnabar, calcite, ankerite, barite, chalcedony, or quartz. The minerals are usually found filling fissures, joints, and rock pores, or as irregular or lenslike replacement bodies. Under surface and near-surface conditions, stibnite may oxidize to white or yellowish stibiconite [$Sb_3O_6(OH)$]; in some cases, this oxidation has resulted in the formation of stibiconite pseudomorphs after stibnite.

Stibnite is the most important ore mineral of antimony. Extensive deposits in sandstones have been mined in Hunan Province, China, and in limestones at several localities in Mexico. A large amount of antimony metal is, however, obtained as a by-product from the smelting of lead ores both in the United States and in Canada (British Columbia).

Figure 9-19. Stibnite aggregate of acicular crystals, Manhattan, Nevada. Field is ~ 6 cm wide. (Courtesy of Queen's University.)

PYRITE GROUP

The AX_2 type of sulfide minerals can be classified into several groups. The isometric minerals of the pyrite and cobaltite groups, which are closely related structurally and chemically, are examples. Although a large number of substances, in which X may be S, Se, Te, As, or Sb, crystallize with the pyrite structure, most of the naturally occurring compounds are sulfides. Pyrite is by far the most common; sperrylite is relatively uncommon but economically important; MnS_2 (hauerite), RuS_2 (laurite), $AuSb_2$ (aurostibite), and others are very rare.

In the pyrite structure (Figure 9-20), the metal atoms occupy the face-centered cubic lattice positions, and the sulfur atoms lie in pairs

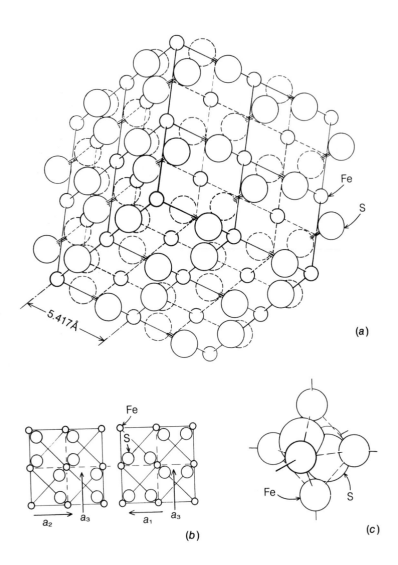

Fe

S

5.417Å

(a)

Figure 9-20. Structure of pyrite. (a) The iron atoms are in a face-centered cubic arrangement with sulfur atoms located on the cube diagonals. (b) In the structure projected on (100) and (010), the sulfur pairs are parallel to a_3 on (100) and to a_1 on (010), corresponding to the direction of striations on (100) and (010) on the cube in Figure 9-21a. (c) The octahedral grouping of iron around one pair of sulfur atoms.

Fe

S

a_2 a_3 a_1 a_3

(b)

Fe

Fe   S

(c)

along the trigonal axes of the lattice, with each S located about three-eighths of the length of the diagonal from an iron atom. The sulfur atoms are in sixfold coordination about each iron atom, with the sulfur atoms in S_2 pairs separated by about the same distance as the S atoms in the covalently bonded S_8 rings in native sulfur. The sulfur pairs are, in turn, coordinated to six iron atoms. The pyrite group includes some of the hardest of the sulfide minerals, with pyrite $6-6\frac{1}{2}$, sperrylite 6–7, and laurite $7\frac{1}{2}$. The compund FeS_2 also occurs as an orthorhombic polymorph, marcasite.

Crystal system, and class; and space group: Isometric; $2/m\bar{3}$; Pa3.

	Pyrite	Sperrylite
Cell dimension:	$a = 5.417$	$a = 5.967$
Cell content:	$Z = 4$	$Z = 4$
Common forms:	{100}, {111}, {210}, {321}	{100}, {111}, {210}
Cleavage:	{001} indistinct	{001} indistinct
Fracture:	Conchoidal	Conchoidal
Hardness:	$6-6\frac{1}{2}$	$6-7$
Specific gravity:	5.01	10.58
Luster:	Metallic	Metallic
Color:	Brass-yellow	Tin-white
Magnetism:	Paramagnetic	—

Pyrite, FeS_2

See also the preceding tabulated data.

Habit: Crystals common, typically cubic, also pyritohedral or octahedral (Figures 9-21 and 9-22); the pyritohedral and cubic faces are generally striated parallel to the edge (a) between them (Figures 9-21a, b, and c); these striae are due to oscillatory growth of the two forms, which tends to produce rounded faces; many crystals are abnormally developed; also massive, granular, in some cases radiated, reniform, or globular.

Twinning: . Twin axis [110], interpenetrating, known as the Iron Cross Law (Figures 9-21f and 9-23).

Color and streak: Pale brass-yellow, iridescent where tarnished. Greenish black or brownish black.

Chemistry and alteration: In most occurrences, pyrite is very close to the ideal composition FeS_2. Uncommonly, nickel, cobalt, or both substitute for iron; these varieties are referred to as nickelian or cobaltian pyrite. The name *bravoite* is used for those rare occurrences of $(Ni,Fe)S_2$, in which iron is about 50 mol percent or less. Two minerals with the pyrite structure, in which nickel (*vaesite*) or cobalt (*cattierite*) is the major metal component, have been described from the Katanga District of Zaire. Theoretically, it is possible that any composition of $(Fe,Ni,Co)S_2$ may occur with the pyrite structure. The substitution of Ni for Fe, for example, is accompanied by an increase in the edge length of the cubic unit cell (Figure 9-24), by a decrease in hardness, and by a change in color toward silver-white or gray. The oxidation of pyrite to limonite may take place in such a way as to preserve the crystal form of pyrite as a limonite pseudomorph.

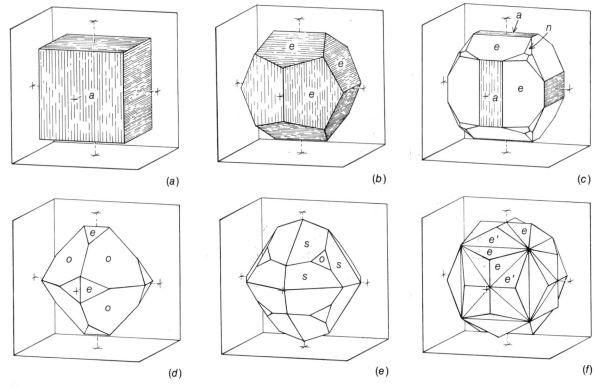

Figure 9-21. Pyrite crystals. Forms: cube a{100}, pyritohedron e{210}, octahedron o{111}, trapezohedron n{112}, diploid s{321}. [(a), (b), and (c) show the common striations on cube and pyritohedral faces.] (f) Pyritohedron e{210} in twinned position e', twin axis [110], Iron Cross Law.

Diagnostic features: Distinguished from most other yellowish metallic minerals by its superior hardness. It is distinguished macroscopically from marcasite with certainty only by its crystal form, although marcasite is typically paler in color and much of it exhibits alteration. Pyrite is insoluble in hydrochloric acid but, in fine powder, it is completely dissolved in strong nitric acid.

Occurrence, associations, and uses: Pyrite is by far the most widespread and commonly occurring sulfide mineral. It can be found in almost any type of geological environment: as an accessory mineral in igneous rocks, both acidic and basic types; as a magmatic segregation from igneous rocks; in pegmatites; in contact metamorphic deposits; in hydrothermal sulfide veins and replacement deposits; as a sublimation product; and in sedimentary and metamorphic rocks.

Most massive hydrothermal sulfide bodies contain pyrite, and in some, as at Rio Tinto, Spain and at Noranda, Quebec, it is the most important sulfide. Pyrite is rarely of economic importance itself, but its presence directs attention to vein or replacement deposits that may contain chalcopyrite, as in many massive sulfide deposits, or gold, as in some of the Precambrian gold deposits where much of the gold is present as minute grains in and with pyrite. Pyrite changes on oxidation to iron sulfates that break down, leaving limonite; thus, the presence of extensive pyrite in a rock at some depth can be revealed by a capping of limonite, gossan, at the surface.

Pyrite yields some commercial sulfur; most such pyrite is, however, mined primarily because it is present in ores containing valuable metals. Nonetheless, in a few cases, local demands for sulfuric acid have resulted in the mining of pyrite as a source of sulfur.

Sperrylite, PtAs$_2$

See also the data given in the introduction to this group of minerals.

Occurrence, associations, and uses: Distinct crystals occur in the pyrrhotite-pentlandite-chalcopyrite ores at Sudbury, Ontario, where the mineral is responsible for important platinum production. It also occurs in igneous rocks of the Bushveld Complex of South Africa and as a detrital mineral in certain river gravels.

Cobaltite, CoAsS

Crystal system, class, and space group: Orthorhombic; *mm2*; *Pca2$_1$*.
Cell dimensions and content: $a \cong b \cong c \cong 5.57$; $Z = 4$.
Crystal structure: As might be expected from the cell dimensions and composition, the structure of cobaltite, along with that of the rarer minerals *gersdorffite* (NiAsS) and *ullmannite* (NiSbS), was long cor-

Figure 9-22. Pyrite crystal: striated pyritohedron e{210}. (Courtesy of Smithsonian Institution, National Museum of Natural History.)

Figure 9-23. Twinned pyrite crystal, Iron Cross Law, twin axis [110], combining two pyritohedral crystals as in Figure 9-21f. (Courtesy of Queen's University.)

Figure 9-24. Pyrite, variation in edge length of cubic unit cell with nickel content.

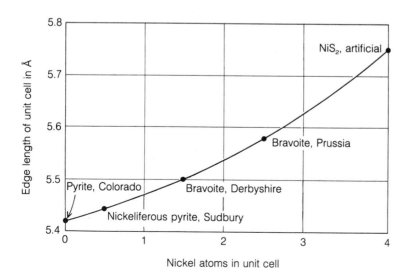

related with that of pyrite. Group relations require additional clarifi-
cation.

Habit: Pseudoisometric; crystals uncommon; typically in cubes {100},
or pyritohedra {210}, or as combinations of these forms, generally
with faces striated as in pyrite; octahedral; also granular to compact.

Cleavage and fracture: {001} perfect. Uneven.

Hardness: $5\frac{1}{2}$.

Specific gravity: 6.33.

Color and streak: Silver-white inclined to red, also steel-gray with a
violet tinge. Gray-black.

Luster and light transmission: Metallic. Opaque.

Diagnostic features: Softer ($H = 5\frac{1}{2}$) than pyrite ($H = 6-6\frac{1}{2}$) and with
perfect {001} cleavage. The reddish tinge can help to distinguish
cobaltite from smaltite. Heated, it gives a characteristic garlic odor.
Powdered, it dissolves in warm nitric acid, coloring the solution
pink.

Occurrence, associations, and uses: Cobaltite occurs in vein deposits
with other cobalt and nickel sulfides or arsenides (at Cobalt, Ontario
and elsewhere) and is also disseminated in certain metamorphic
rocks. Gersdorffite is a minor nickel mineral in the pyrrhotite-
pentlandite-chalcopyrite ores at Sudbury, Ontario.

Marcasite, FeS$_2$

Crystal system, class, and space group: Orthorhombic; 2/m2/m2/m;
Pnnm.

Cell dimensions and content: $a = 4.445$, $b = 5.425$, $c = 3.388$;
$Z = 2$.

Crystal structure: In marcasite, which is dimorphous with pyrite, the
metal atoms occupy the lattice points of an orthorhombic body-
centered lattice, and the sulfur atoms surround each iron in sixfold
coordination (Figure 9-25b). The FeS$_6$ octahedra are all oriented
with two edges parallel to c and two edges perpendicular to c; the
latter edges are shared, resulting in chains of octahedra parallel to
c. The chains centered on the corners of the unit cell have their
shared edges inclined to a and b, whereas the chain at body center
of the cell is displaced one-half along c with the shared edges also
inclined to a and b but in a reverse relation to the corner chains. This
inclination is such that the sulfurs of the center chain approach those
in the corner chains in S$_2$ pairs, with the same separation as in the
S$_2$ pairs in pyrite. The S$_2$ pairs have their long axis in the plane (001)
and are located between iron atoms along the long axis b. The
coordination of S to Fe is identical to that in pyrite. The structure also
shows a close similarity to that of rutile, in which the octahedra have
their shared edges at 45° to the a axes, the oxygens do not occur as
O$_2$ pairs, and the symmetry is tetragonal.

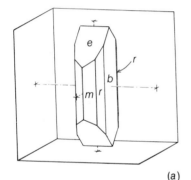

Figure 9-25.   (a) Marcasite crystal. Forms: pinacoid b{010},
rhombic prisms m{110}, r{140},
e{101}. (b) Structure of marcasite, showing iron atoms on a
body-centered orthorhombic lattice with pairs of sulfur atoms;
each iron atom is octahedrally
surrounded by six sulfurs.

Habit: Crystals common; typically tabular {010}, also pyramidal (Figure 9-25); faces commonly curved; also stalactitic, globular, or reniform (Figure 9-26b) with radiating internal structure and exterior covered with projecting crystals.

Twinning: Extremely common on {100} as the well-known swallowtail contact twins (Figure 9-26a); also common on {101}, generally repeated.

Cleavage and fracture: {101}, distinct. Uneven.

Hardness: $6–6\frac{1}{2}$.

Specific gravity: 4.89.

Color and streak: Pale brass-yellow, tending toward light green on exposure, tin-white on fresh fracture. Grayish or brownish black.

Luster and light transmission: Metallic. Opaque.

Chemistry and alteration: Marcasite, dimorphous with pyrite, shows very little variation from the composition FeS_2. It decomposes more readily than pyrite; under atomspheric conditions, it alters to ferrous sulfate and sulfuric acid. In addition, marcasite may invert, forming pyrite paramorphs, or it may be altered to form limonite pseudomorphs.

Diagnostic features: Crystal form and hardness; also, when fine powder is treated first with cold nitric acid and then boiled after the initial vigorous reaction has ceased, marcasite is decomposed with separation of sulfur whereas, when similarly treated, pyrite is dissolved completely.

Occurrence and associations: Most marcasite occurs in near-surface deposits where it has been formed at low temperatures from acid solutions; pyrite, the more stable form, is deposited under conditions of higher temperature and alkalinity or low acidity. Most marcasite is of supergene origin, but some has been deposited from ascending low-temperature vein solutions. Consequently, it is most frequently found in sedimentary rocks, limestone, clays, or lignite, in many cases as concretions or as replacement-type fossils.

Marcasite is quite common with the low-temperature galena and sphalerite ores in Mississippi Valley districts, generally as individual crystals or as cockscomb groups.

Marcasite is of no economic value.

(a)

(b)

Figure 9-26. Marcasite. (a) Crystals in fine gray chalk, Folkestone, England, main crystal ~ 3 cm long. (Courtesy of Royal Ontario Museum.) (b) Radiating nodule of crystals, Joplin, Missouri, 5 cm wide. (Courtesy of Queen's University.)

Arsenopyrite, FeAsS

Crystal system, class, and space group: Triclinic (morphologically orthorhombic); $\bar{1}$; $P\bar{1}$.

Cell dimensions and content: $a = 5.744$, $b = 5.675$, $c = 5.785$; $\alpha = 90° 00'$, $\beta = 112° 10'$, $\gamma = 90° 00'$; $Z = 4$.

Habit: Crystals common, prismatic with elongation on [010] or [101], generally as a combination of two pseudo-orthorhombic prism zones [101] with [$\bar{1}$01] (Figures 9-27 and 9-28), faces striated

278

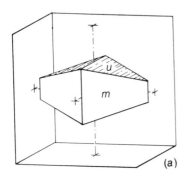

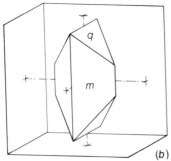

(a)

(b)

Figure 9-27.   Arsenopyrite crystals. Forms: $m\{100\}$ and $\{001\}$, pinacoids; $u\{14\bar{1}\}$ and $q\{11\bar{1}\}$, prisms.

Figure 9-28.   Arsenopyrite crystals. (a) In fine-grained quartz, showing striated faces $\{14\bar{1}\}$ with pinacoids $\{100\}$ and $\{001\}$ (see Figure 9-27). (b) and (c) Twinned crystals. Forms $\{1\bar{1}1\}$, $\{11\bar{1}\}$, and $\{12\bar{1}\}$ with $\{001\}$ twin plane. Specimens from Deloro, Ontario; all crystals ~ 2 cm across. (Courtesy of Queen's University.)

parallel to [101] (bisector of obtuse angle β), columnar exhibiting distinct rhombic cross sections; also granular or compact.

Twinning: Common on $\{001\}$ as contact or penetration twins (Figure 9-28), also on $\{1\bar{1}1\}$ as cruciform twins; pseudo-orthorhombic crystals result because of twinning on $\{10\bar{1}\}$ and $\{101\}$.

Cleavage and fracture: $\{100\}$ and $\{001\}$ distinct. Uneven.

Hardness: $5\frac{1}{2}$–6.

Specific gravity: 6.07.

Color and streak: Silver-white to steel-gray. Dark grayish black.

Luster and light transmission: Metallic. Opaque.

Chemistry: Arsenopyrite, essentially FeAsS, commonly contains some cobalt in substitution for iron; thus, the composition may grade into glaucodot, (Co,Fe)AsS, in which cobalt predominates.

Diagnostic features: The crystal form with rhombic cross sections serves to distinguish it from smaltite. It emits a garliclike odor when ground or hammered. It is decomposed by nitric acid with separation of sulfur.

Occurrence, associations, and uses: Arsenopyrite, the most abundant arsenic mineral, typically occurs as an early formed mineral apparently representing moderate- to high-temperature conditions. It occurs most commonly in high-temperature gold-quartz veins, as in South Dakota and Quebec; in high-temperature cassiterite veins, as in Cornwall, England; and with scheelite or in some contact metamorphic deposits with gold and other sulfides. It is less common in apparently lower-temperature veins with gold and quartz, as in the Mother Lode, California or in nickel-cobalt-silver veins as at Cobalt, Ontario.

Arsenopyrite is of widespread occurrence in many parts of the world. It is not mined for its arsenic content because most commercial arsenic is recovered from smelter fumes resulting from treatment of ores of other materials that contain some arsenopyrite. This source more than satisfies the market for arsenic, most of which is used in insecticides, pesticides, and herbicides.

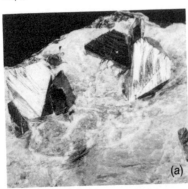

(a)

(b)

(c)

Molybdenite, MoS$_2$

Crystal system, class, and space group: Hexagonal; $6/m2/m2/m$; $P6_3/mmc$.

Cell dimensions and content: $a = 3.16$, $c = 12.32$; $Z = 2$.

Crystal structure: MoS$_2$ has three polymorphs—molybdenite-2H (hexagonal); molybdenite-3R (trigonal); and jordisite (amorphous). When molybdenite is referred to with no suffix, it refers to molybdenite-2H. In this polymorph, MoS$_6$ octahedra are linked together by sharing edges to form sheets parallel to {0001}. The sheets are linked by weak bonds between sulfurs of adjacent sheets, which results in the perfect basal cleavage (Figure 9-29).

Habit: Crystals uncommon; hexagonal, thin-to-thick tabular on {0001} with prism {10$\bar{1}$0} and dipyramids {10$\bar{1}$1} and {10$\bar{1}$2} as narrow horizontally striated faces; commonly foliated, massive, or in scales.

Cleavage: {0001} perfect, laminae flexible but not elastic.

Hardness: $1–1\frac{1}{2}$.

Tenacity: Sectile.

Specific gravity: 4.62–4.73.

Color and streak: Lead-gray with bluish tinge. Bluish gray on paper, greenish on glazed porcelain.

Luster and light transmission: Metallic. Opaque.

Diagnostic features: Perfect basal cleavage and hardness similar to that of graphite. Molybdenite has a distinct steel-blue color, as compared with the lead-gray of graphite, and it gives a greenish streak on glazed porcelain. It has a greasy feel. Molybdenite and graphite may occur in similar environments.

Occurrence, associations, and uses: Molybdenite, the most common molybdenum mineral, occurs as an accessory in some granites and pegmatites, in a few cases in commercially important quantities. Molybdenite also occurs in deep-seated veins with scheelite, wolframite, topaz, and fluorite, and in contact metamorphic deposits with lime silicates, scheelite, or chalcopyrite.

Molybdenite is the principal source of molybdenum, which has many applications in special steel alloys. In recent years, the major sources of molybdenite have been quartz veinlets in granite at Climax, Colorado, and the copper mines of Utah, New Mexico, and Arizona.

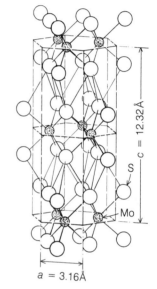

Figure 9-29. Structure of molybdenite, MoS$_2$. Octahedral layers of Mo parallel to {0001} are loosely bonded to adjacent layers, resulting in basal cleavage similar to that in graphite.

KRENNERITE GROUP

This group comprises three minerals: *krennerite*, (Au,Ag)Te$_2$; *calaverite*, AuTe$_2$; and *sylvanite*, (Au,Ag)Te$_2$. These three minerals are the most important naturally occurring compounds of gold and are also three of a relatively small group of minerals that contain tellurium as a major constituent. All three occur in hydrothermal veins. All three minerals have been identified in gold mines of the Cripple Creek district, Colorado and the Rouyn district, Quebec. Calaverite and sylvanite, more widespread in occurrence than krennerite, also occur in the Mother Lode, California (including Calaveras County, for which calaverite is named), and in the Kirkland Lake gold district of Ontario, in which they are associated with quartz, pyrite, native gold, and rare tellurides of bismuth, lead, mercury, and nickel.

	Krennerite	Calaverite	Sylvanite
Crystal system:	Orthorhombic	Monoclinic	Monoclinic
Crystal class:	$mm2$	$2/m$	$2/m$
Space group:	$Pma2$	$C2/m$	$P2/c$
Cell dimensions:	$a = 16.54$	$a = 7.19$	$a = 8.96$
	$b = 8.82$	$b = 4.41$	$b = 4.49$
	$c = 4.46$	$c = 5.08$	$c = 14.62$
		$\beta = 90° \pm 30'$	$\beta = 145°\,26'$
Cell content:	$Z = 8$	$Z = 2$	$Z = 2$
Cleavage:	{001} perfect	None	{010} perfect
Hardness:	2–3	$2\frac{1}{2}$–3	$1\frac{1}{2}$–2
Specific gravity:	8.62	9.24	8.16

Calaverite, $AuTe_2$

See also the preceding tabulated data.

Habit: Crystals rare; bladed or lathlike, short prisms typically elongated and striated parallel to b; also massive, granular.
Twinning: Common {101}, {310}, or {111}.
Color and streak: Brass-yellow to silver-white. Yellowish to greenish gray.
Luster and light transmission: Metallic. Opaque.
Diagnostic features: Gives gold globule on heating; reacts with acids similar to the way sylvanite does; lacks good cleavage of sylvanite.

Sylvanite, $(Au,Ag)Te_2$

See also the data in the introduction to this group of minerals.

Habit: Crystals rare; short prismatic along c or b, also tabular on {100} or {010}; often skeletal or bladed, also granular.
Twinning: Common on {100} as simple contact, lamellar, or penetration twins.
Color and streak: Steel-gray to silver-white.

Luster and light transmission: Brilliant metallic. Opaque.

Diagnostic features: Crystal habit and cleavage; decomposed by nitric acid yields rusty colored gold and a solution that, upon addition of hydrochloric acid, gives a $AgCl_2$ precipitate; powdered, in heated H_2SO_4, gives a violet-red solution.

Occurrence, associations, and uses: See the introductory remarks.

Skutterudite Series

Strictly speaking, this series includes only two minerals, *skutterudite* ($CoAs_{2-3}$) and *nickel–skutterudite* ($NiAs_{2-3}$). In practice, however, some mineralogists also include the isostructural, arsenic-deficient minerals *smaltite* and *chloanthite*.

The portions of the series are not sharply separated or readily distinguishable. In general, however, the high-cobalt members yield pink erythrite [$Co_3(AsO_4)_2 \cdot 8H_2O$] on weathering, whereas the rarer high-nickel members yield green annabergite [$Ni_3(AsO_4)_2 \cdot 8H_2O$]. [These minerals, which also form a series, belong to the vivianite group (p. 000).]

In addition to the diverse cobalt-nickel contents, iron may substitute up to about 12 percent; the series with Fe:(Co + Ni) greater than 1:1, however, is not represented among minerals. The arsenic-to-metal ratio ranges from a maximum of 3:1 in the skutterudites to about 2:1 in some smaltite and chloanthite.

Crystal system, class, and space group: Isometric; $2/m\overline{3}$; Im3.

Cell dimensions and content: $a = 8.19–8.29$; $Z = 8$.

Habit: Crystals rare; typically cubic, cubo-octahedral, or octahedral, rarely modified by dodecahedron or pyritohedron; commonly massive, dense fine granular; also colloform.

Cleavage and fracture: {100}, {111}, distinct, but variable and not characteristic. Conchoidal to uneven.

Hardness: $5\frac{1}{2}$–6.

Specific gravity: 6.5.

Color and streak: Tin-white to silver-gray, in some cases tarnished iridescent. Black.

Luster and light transmission: Metallic, brilliant. Opaque.

Diagnostic features: Distinguished from arsenopyrite by crystal form; when heated, these minerals give a strong garliclike odor. Some specimens are macroscopically indistinguishable from some cobaltite.

Occurrence, associations, and uses: The skutterudite minerals typically occur in veins with other cobalt and nickel minerals, especially cobaltite and nickeline, but also arsenopyrite, silver, and bismuth. The minerals have yielded important amounts of cobalt

282

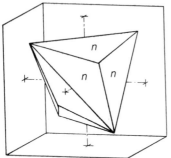

(a)

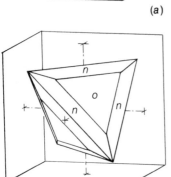

(b)

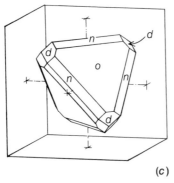

(c)

Figure 9-30. Tetrahedrite crystals. Forms: dodecahedron $d\{110\}$, positive tetrahedron $o\{111\}$, positive tristetrahedron $n\{211\}$.

(and the by-product arsenic) from the ores of Cobalt, Ontario, where they have been a by-product of silver mining since 1904.

Pyrargyrite, Ag_3SbS_3, and Proustite, Ag_3AsS_3

Crystal system, class, and space group: Hexagonal (trigonal); $3m$; $R3c$.

	Pyrargyrite	Proustite
Cell dimensions:	$a = 11.06$	$a = 10.79$
	$c = 8.74$	$c = 8.69$
Cell content:	$Z = 6$	$Z = 6$
Hardness:	$2\frac{1}{2}$	$2\text{–}2\frac{1}{2}$
Specific gravity:	5.85	5.57
Luster:	Adamantine (metallic)	Adamantine (metallic)
Color:	Deep red	Scarlet-vermilion
Streak:	Red	Vermilion
Light transmission:	Translucent	Translucent
Optical properties:	uniaxial $(-)$	uniaxial $(-)$
	ω, 3.084; ϵ, 2.881	ω, 3.0877; ϵ, 2.7924
	deep red	blood red

Habit: Crystals rare; typically prismatic with hexagonal form due to $\{11\bar{2}0\}$, commonly with hemimorphic development with trigonal pyramids $\{01\bar{1}2\}$, $\{10\bar{1}1\}$, or $\{10\bar{1}\bar{1}\}$; also massive compact.

Twinning: Common on $\{10\bar{1}4\}$ and $\{10\bar{1}1\}$.

Cleavage: $\{10\bar{1}1\}$ distinct.

Chemistry: The two minerals are isostructural, but analyses record only minor amounts of As substituting for Sb in pyragyrite, and the same is true of Sb in proustite. There is no evidence for an extended solid-solution series.

Occurrence, associations, and uses: These minerals are commonly called ruby-silvers. They occur in apparently low-temperature silver veins, typically as late-forming primary minerals, but they may also have formed as a result of processes of secondary enrichment.

Pyrargyrite is more common than proustite, and it often constitutes an important ore of silver. It occurs with proustite, argentite, tetrahedrite, silver, calcite, dolomite, or quartz. Fine crystals have been found in the silver veins of Saxony and Bohemia, and important quantities are found in many silver mines the world over. Proustite commonly occurs in the same veins as pyrargyrite and silver.

Tetrahedrite, $(Cu,Fe)_{12}Sb_4S_{13}$, and Tennantite, $(Cu,Fe)_{12}As_4S_{13}$

Crystal system, class, and space group: Isometric; $\bar{4}3m$; $I\bar{4}3m$.

Cell dimensions and content: $a = 10.21$ (tennantite); $a = 10.34$–10.48 (tetrahedrite), greater with higher silver contents; $Z = 2$.

Habit: Crystals uncommon; typically tetrahedral (Figures 9-30 and 9-31); also massive, granular to compact.

Twinning: Twin axis [111] as contact or penetration twins.

Fracture: Subconchoidal to uneven.

Hardness: 3–$4\frac{1}{2}$ (tennantite harder).

Specific gravity: 4.6–5.1 (greater with higher antimony and silver contents).

Color and streak: Flint-gray to iron-black. Black to brown.

Luster and light transmission: Metallic. Opaque.

Chemistry: These minerals are the end members of a continuous solid-solution series; the name *tetrahedrite* is used if Sb predominates, and the name *tennantite* is used if As predominates. Although copper is the most important metal in all specimens, iron is ever present in substitution from about 1 to 5 percent and some mineralogists include it in the formulas of these minerals [$(Cu,Fe)_{12} \ldots$]. Zinc, reported up to about 9 percent in many analyses, may be due in part to admixed sphalerite. Silver is commonly present and, in rare cases, is the dominant metal; this silver-bearing species is called *freibergite*. The presence of mercury in tetrahedrite has also been recorded.

Tetrahedrite commonly alters to malachite, azurite, and antimony oxides.

Diagnostic features: Crystal form and lack of cleavage. Decomposed by nitric acid.

Occurrence, associations, and uses: Tetrahedrite is probably the most widespread and economically important sulfosalt mineral. It is an important copper ore mineral, and the argentian varieties add appreciably to the silver values. It is also an important antimony-bearing mineral in these ores.

These minerals commonly occur in hydrothermal veins of copper, lead, zinc, and silver minerals; they occur rarely in contact metamorphic deposits. Tennantite occurs less commonly than tetrahedrite.

Figure 9-31. Tetrahedrite crystals showing $o\{111\}$, with negative tetrahedron $- o\{\bar{1}11\}$, and $d\{110\}$, in calcite, locality unknown. Specimen is 7 cm wide. (Courtesy of Queen's University.)

Enargite, Cu_3AsS_4

Crystal system, class, and space group: Orthorhombic; $2mm$; Pnm.

Cell dimensions and content: $a = 6.47$, $b = 7.44$, $c = 6.19$; $Z = 2$.

Habit: Crystals common; tabular on {001} or prismatic along c and typically striated parallel to c; commonly massive, granular, or prismatic.

Twinning: Common on {320}, in some cases as star-shaped cyclic trillings.

Cleavage and fracture: {110} perfect; {100}, {010} distinct. Uneven.

Hardness: 3.

Specific gravity: 4.45.

Color and streak: Grayish black to iron-black. Black.

Luster and light transmission: Metallic, tarnishing dull. Opaque.

Chemistry and alteration: Enargite is copper arsenic sulfide (Cu_3AsS_4) in which Sb may substitute for As to about 6 percent by weight, and minor iron may substitute for copper. *Famatinite* is the antimony analog of enargite and can occur with it. Enargite may alter to tennantite or, on oxidation, to a variety of copper arsenate, as at Tintic, Utah.

Diagnostic features: Striated crystals are distinctive; cleavage.

Occurrence, associations, and uses: Enargite occurs in vein and replacement deposits, apparently formed at moderate temperatures, at some places in sufficient quantity to be important as an ore of copper. It can also occur as a late-formed mineral in some low-temperature deposits. Enargite is typically associated with pyrite, sphalerite, bornite, galena, tetrahedrite, covellite, chalcocite, barite, and quartz.

Bournonite, $PbCuSbS_3$

Crystal system, class, and space group: Orthorhombic; $2/m2/m2/m$; $Pn2_1m$.

Cell dimensions and content: $a = 8.15$, $b = 8.70$, $c = 7.80$; $Z = 4$.

Habit: Crystals uncommon; typically short prismatic along c to tabular on {001}; $(hk0)$ faces striated parallel to c and $(h0l)$ faces striated parallel to b; commonly as subparallel aggregates; also massive, granular.

Twinning: Very common on {110}, in many cases repeated, thus forming cruciform or wheellike aggregates.

Cleavage and fracture: {010} indistinct. Subconchoidal.

Hardness: $2\frac{1}{2}$–3.

Specific gravity: 5.83.

Color and streak: Steel-gray to iron-black.

Luster and light transmission: Metallic, brilliant to dull. Opaque.

Chemistry and alteration: Bournonite may contain up to about 3 percent As; there is no evidence for extended solid solution to the rarer mineral $PbCuAsS_3$ (*seligmannite*). Bournonite alters to antimony oxides, cerussite, malachite, or azurite.

Diagnostic features: Alternating brilliant and dull faces on some crystals; rather common twinning; decomposed by HNO_3, yielding a pale blue-green solution.

Occurrence, associations, and uses: Bournonite is one of the more common sulfosalts, occurring widely in apparently moderate temperature hydrothermal veins, associated with galena, tetrahedrite, sphalerite, chalcopyrite, pyrite, siderite, quartz, and rarely with stibnite, antimony sulfosalts, rhodochrosite, dolomite, and barite. It contributes to the lead, copper, and antimony recovery from some complex sulfide vein deposits. Well-crystalized material has been found in certain sulfide veins of the old mining districts in, for example, Cornwall, England. Large crystals have been found with siderite and sphalerite at Park City, Utah, in Arizona (Boggs Mine), in California (Cerro Gordo Mine), and at Austin, Nevada.

Boulangerite, $Pb_5Sb_4S_{11}$

Crystal system, class, and space group: Monoclinic; $2/m$; $P2_1/a$.
Cell dimensions and content: $a = 21.56$, $b = 23.51$, $c = 8.09$; $\beta = 100°48'$; $Z = 8$.
Habit: Crystals uncommon; long prismatic to acicular and deeply striated parallel to c, rarely terminated by well-developed faces; also plumose, fibrous, compact fibrous masses.
Cleavage: {100} good.
Hardness: $2\frac{1}{2}$–3.
Tenacity: Brittle, thin fibers flexible.
Specific gravity: 6.23.
Color and streak: Bluish lead-gray. Brownish gray.
Luster and light transmission: Metallic. Opaque.
Diagnostic features: Boulangerite is similar in physical properties to stibnite and a number of other lead-antimony sulfosalts from which it cannot be distinguished, with certainty, without quantitative chemical analysis or X-ray study.
Occurrence, associations, and uses: Boulangerite is common in vein deposits presumed to have formed at low or moderate temperatures; it is associated with galena, stibnite, sphalerite, pyrite, quartz, siderite, and other lead sulfosalts. It occurs in lead-zinc veins at several localities in Europe and western North America.

10 Class III: Oxides and Hydroxides

Class III, the oxides and hydroxides, includes compounds in which atoms or cations, typically of one or more metals, are combined with oxygen. In some cases, hydrogen is one of the cations and is present as hydroxyl or as water of hydration. For the most part, the oxide minerals exhibit ionic bonding. The oxide class does not include those compounds that have discrete anionic radicals in their structures—for example, the carbonates and the sulfates. On the other hand, multiple oxides such as minerals of the columbite series $[(Fe,Mn)Nb_2O_6]$ are included on the basis of their crystal chemistry. They are included despite the fact that, on chemical grounds alone, they are sometimes referred to as, for this example, niobates.

The oxides are classified conveniently on the basis of the $A:X$ ratio; they are followed by the hydroxides.

Oxides
A_2X Type
Cuprite, Cu_2O

AX Type
Periclase, MgO
Zincite, $(Zn,Mn)O$

AB₂X₄ Type
- Spinel Group
    - Spinel, $MgAl_2O_4$
    - Magnetite, $FeFe_2O_4$
    - Chromite, $FeCr_2O_4$
    - Franklinite, $(Zn,Mn,Fe)(Fe,Mn)_2O_4$
  - Hausmannite, $MnMn_2O_4$
  - Chrysoberyl, $BeAl_2O_4$

A₂X₃ Type
- Hematite Group
    - Corundum, Al_2O_3
    - Hematite, Fe_2O_3
    - Ilmenite, $FeTiO_3$
  - Braunite, $3Mn_2O_3 \cdot MnSiO_3$
  - Pyrochlore-Microlite Series:
    - $(Na,Ca)_2Nb_2O_6(OH,F)–(Na,Ca)_2Ta_2O_6(O,OH,F)$
  - Romanechite, $BaMn_9O_{16}(OH)_4$

AX₂ Type
- Rutile Group
    - Rutile, TiO_2
    - Cassiterite, SnO_2
    - Pyrolusite, MnO_2
  - Anatase, TiO_2
  - Brookite, TiO_2

Columbite-Tantalite, $(Fe,Mn)(Nb,Ta)_2O_6$

Uraninite, UO_2

Hydroxides
- Brucite, $Mg(OH)_2$
- Gibbsite, $Al(OH)_3$
- Boehmite, $AlO(OH)$
- Manganite, $MnO(OH)$
- Diaspore, $AlO(OH)$
- Goethite, $FeO(OH)$

OXIDES

Cuprite, Cu_2O

Crystal system, class, and space group: Isometric; $4/m\bar{3}2/m$; $Pn3m$.
Cell dimensions and content: $a = 4.2696$; $Z = 2$.
Crystal structure: As noted by R. C. Evans (1964, *An Introduction to Crystal Chemistry*, Cambridge University Press, p. 164), the structure of cuprite "is unique among inorganic compounds in that it consists of two identical interpenetrating frameworks which are not

directly bonded together; starting from any one atom it is possible to reach half, but only half, of the remainder by traveling along Cu-O bonds."

The symmetry of cuprite presents an unusual problem. Many crystals have faces of the general forms {hkl} that display gyroidal symmetry, and this class has long been accepted as the correct crystal class. But, the structure—one of the first analyzed by Bragg—apparently possesses the full isometric symmetry $4/m\overline{3}2/m$, and etch figures also indicate the higher symmetry.

Habit: Crystals common; typically octahedral or cubic, less commonly dodecahedral; in some cases, elongated along *a* into capillary fibers, and then known as *chalcotrichite;* also massive, granular, or earthy.

Cleavage and fracture: {111} distinct, interrupted. Conchoidal.

Hardness: $3\frac{1}{2}$–4.

Specific gravity: 6.10.

Color and streak: Red, nearly black. Brownish red.

Luster and light transmission: Submetallic to adamantine or earthy. Red by transmitted light, yellow in thinner splinters.

Optical properties:
 isotropic
 refractive index: n, 2.849
 orange-yellow to red

Alteration: Crystals of cuprite are commonly altered to malachite pseudomorphs. Cuprite may also alter to copper or other secondary copper minerals.

Diagnostic features: Softer than hematite but harder than cinnabar and proustite; the latter two also have a lighter streak. Cuprite is soluble in concentrated hydrochloric acid, coloring the solution blue, which on cooling and diluting with cold water, yields a heavy, white precipitate of cuprous chloride.

Occurrence, associations, and uses: Cuprite commonly occurs in the oxidized zone of copper deposits. It is typically associated with copper, malachite, azurite, iron oxides, clays, and the black copper oxide *tenorite* (CuO). Crystals of cuprite are generally closely associated with native copper. Cuprite is widespread in its occurrence, especially in those parts of the world where the results of supergene alteration on primary copper deposits have not been destroyed by glaciation. It has sometimes been an important ore of copper, as at Bisbee, Arizona. Especially fine specimens have been found in the tin and copper mines of Cornwall, England; at Chessy near Lyon, France; and in the Onganja Mine, Namibia.

Periclase, MgO

Crystal system, class, and space group: Isometric; $4/m\overline{3}2/m$; Fm3m.

Cell dimensions and content: a = 4.213; Z = 4.

Habit: Crystals rare; octahedral; typically in irregular or rounded grains.

Cleavage: {100} perfect.

Hardness: $5\frac{1}{2}$.

Specific gravity: 3.58.

Color and streak: Colorless to grayish white, also yellow to brown or black due to iron or foreign inclusions. White.

Luster and diaphaneity: Vitreous. Transparent.

Optical properties:
> isotropic
> refractive index: n, 1.735–1.745
> colorless

Chemistry and alteration: Periclase has been found to contain up to about 8 percent Fe substituting for Mg. Artificial MgO forms a complete solid-solution series with FeO.

> Periclase alters readily to brucite or to hydromagnesite (a hydrated magnesium carbonate).

Diagnostic features: Cleavage, hardness, and common alteration to brucite; also easily soluble in dilute hydrochloric or nitric acid.

Occurrence and associations: Periclase is a rare mineral, occurring as disseminated grains and clusters of grains in marbles. It is formed by dissociation of dolomitic limestones during high-temperature metamorphism. Periclase has been recognized in limestone blocks ejected from Monte Somma, Vesuvius, Italy and in metamorphosed impure limestones in, for example, Crestmore, California, where it is associated with forsterite, magnesite, brucite, chondrodite, and spinel.

Zincite, (Zn,Mn)O

Crystal system, class, and space group: Hexagonal; 6mm; $P6_3mc$.

Cell dimensions and content: a = 3.249, c = 5.205; Z = 2.

Crystal structure: Zincite has a hemimorphic hexagonal structure, with Zn in fourfold coordination, which is closely similar to that of the hexagonal zinc sulfide, wurtzite.

Habit: Crystals rare; hemimorphic (Figure 10-1); typically massive, foliated, also compact, in rounded masses.

Cleavage, parting, and fracture: {10$\bar{1}$0} perfect, but commonly difficult. {0001} common. Conchoidal.

Hardness: 4.

Specific gravity: 5.68 (pure).

Color and streak: Orange-yellow to deep red, probably due to its manganese content, pure ZnO being white. Orange-yellow.

Luster and light transmission: Subadamantine. Translucent in thin slivers.

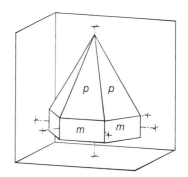

Figure 10-1. Zincite crystal. Forms: hexagonal prism $m\{10\bar{1}0\}$; hexagonal pyramid $p\{10\bar{1}1\}$; pedion $- c\{000\bar{1}\}$ (not visible).

Optical properties:
    anisotropic, uniaxial (+)
    refractive indices: ω, 2.013; ϵ, 2.029
    deep red to yellow
Diagnostic features: Color and streak.
Occurrence and associations: Zincite is a rare mineral except in the
    zinc deposits of Franklin and Sterling Hill, New Jersey, where it
    amounted to about 1 percent of the ore. It occurs with willemite and
    franklinite in calcite, largely in granular form.

SPINEL GROUP

The term *spinel* is widely applied to a large number of oxides that occur
in nature, in artificial laboratory preparations, and in slags resulting from
metallurgical operations. The structures are usually described as double
oxides AB_2X_4 in which A is one or more divalent metals (Mg, Fe, Zn, Mn,
Ni), B is one or more trivalent metals (Al, Fe, Cr, Mn) or Ti^{+4}, and X is
oxygen. Most of the natural spinels fall into three series—spinel, in
which B is mainly Al^{+3}; magnetite, in which B is mainly Fe^{+3}; and
chromite, in which B is mainly Cr^{+3}. In nature, extensive solid solution
occurs within each series, whereas it occurs only to a lesser extent
between members of different series. Much broader solid-solution variations have been found in artificial preparations.

The face-centered cubic unit cell of spinel, with $a = 8.0$–8.5, contains $8[AB_2X_4]$. In the structure (Figure 10-2), the oxygen atoms lie in
approximately cubic closest packing. The 8 A-type metal atoms, which
have the same arrangement as the carbon atoms in diamond (Figure
8-12), are in fourfold coordination between a tetrahedral group of oxygen atoms; the 16 B atoms are in sixfold coordination between octahedral groups of oxygen. In turn, each oxygen is linked to one A and
three B atoms. This ideal spinel structure presents an anomaly since, in
most minerals, the B metal ions, which are smaller than the A ions,
occur in higher coordination with oxygen. Consequently, the spinel
structure is not in conformity with the general rules for determining
coordination numbers. In the case of some spinels—$MgFe_2O_4$ (mag-
nesioferrite), $Ti(Fe,Mg)_2O_4$ (ulvöspinel), and some artificial spinels—the
fourfold coordinated A positions are occupied by 8 of the smaller trivalent atoms, leaving, in $MgFe_2O_4$, 8 Mg^{+2} and 8 Fe^{+3} distributed at
random over the 16 sixfold coordinated B positions. In ulvöspinel, Mg^{+2}
and Fe^{+2} fill the A positions, and Ti^{+4} with Fe^{+3} and some Al^{+3} probably
fill the B positions. Many spinels have not been studied in sufficient
detail to indicate the type to which they belong.

The occurrence of natural intergrowths of two spinel minerals suggests that solid solution in the spinels is, in many instances, limited by
the temperature of formation, although the solid solution may be more

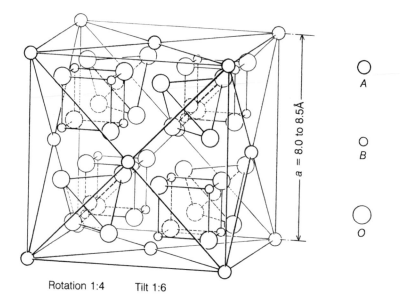

Rotation 1:4   Tilt 1:6

$a = 8.0$ to 8.5Å

A

B

O

Figure 10-2. Structural arrangement in spinel. Of the medium circles representing the eight A atoms, four are shown in tetrahedral coordination with oxygen (larger circles) as four tetrahedra within the cube. The small circles representing the sixteen B atoms are in octahedral coordination with oxygen. The B atoms are shown at the alternate corners of four small cubes with edge length $a/4$.

extended or complete at higher temperatures. Therefore, such intergrowths can be used as geothermometers. Intergrowths of magnetite and ilmenite, which is not a spinel, indicate that Ti does not readily enter into the spinel structure at normal temperatures of formation.

Spinels may also vary in the $A:B$ ratio—for example, in some synthetic spinels with $MgO \cdot nAl_2O_3$, in which n may be greater than unity. This is in agreement with the fact that Al_2O_3 and Fe_2O_3 possess the spinel structure in their gamma modifications. The structure still includes 32 oxygen atoms per unit cell, but some of the A and B metal sites are vacant, giving an overall composition of B_2O_3; this is an example of a defect structure. The substance $\gamma\text{-}Fe_2O_3$ occurs in nature as the mineral maghemite.

Crystal system, class, and space group: Isometric; $4/m\bar{3}2/m$; $Fd3m$.

	Spinel	Magnetite	Chromite	Franklinite
Cell dimension:	$a = 8.080$ (pure)	$a = 8.391$	$a = 8.36$	$a = 8.466$
Content:	$Z = 8$	$Z = 8$	$Z = 8$	$Z = 8$
Hardness:	$7\frac{1}{2}$–8	$5\frac{1}{2}$–$6\frac{1}{2}$	$5\frac{1}{2}$–6	$5\frac{1}{2}$–$6\frac{1}{2}$
Specific gravity:	3.581 (pure)	5.17 [5.20 (pure)]	4.5–4.8 [5.09 (pure)]	5.07–5.22

Spinel, MgAl$_2$O$_4$

See also the preceding tabulated data.

Habit: Crystals common; typically octahedral, commonly modified by the cube or dodecahedron; also massive, coarse granular to compact, as irregular or rounded grains.

Twinning: Common on {111} (Spinel Law, Figure 2-80), in many cases flattened on the composition plane (111).

Cleavage, parting, and fracture: None. Indistinct parting on {111}. Conchoidal.

Specific gravity: 3.581 (pure), higher with the substitution of iron, zinc, or manganese for magnesium.

Color and streak: Ranges from red (ruby spinel) to blue, green, brown, or nearly colorless; dark green to brown or black for impure varieties. White to gray, green, or brown.

Luster and light transmission: Vitreous. Transparent to subtranslucent.

Optical properties:
  isotropic
  refractive index: *n*, 1.719 (pure)

Chemistry: In spinel, Fe^{+2}, Zn, and less commonly Mn^{+2} substitute for Mg in all proportions. Al is the dominant trivalent cation, but Fe^{+3}, Cr, Mn^{+3}, V, and Ti may substitute for Al to a considerable extent.

Diagnostic features: Crystal shape, hardness; insoluble in acids except concentrated sulfuric acid.

Occurrence and uses: The spinels, typically high-temperature minerals, occur as accessory minerals in basic igneous rocks, in highly aluminous metamorphic rocks, in contact-metamorphic limestone, and rarely in ore veins and pegmatites. Because of their resistance to both chemical and physical weathering and erosion processes, spinels also occur in gravels and sands.

  Spinels of gem quality, including the rose-red balas ruby, spinel-ruby, and spinels of some other colors, have been found in contact-metamorphic limestones and alluvium derived from them, mainly in Sri Lanka, Burma, India, and Afghanistan.

Magnetite, FeFe$_2$O$_4$

See also the data given in the introduction to this group of minerals.

Habit: Crystals common; typically octahedral, less commonly dodecahedral; also massive, coarse or fine granular.

Twinning: Common on {111} (Spinel Law).

Cleavage and parting: None, but commonly shows distinct parting on {111}.

Color and streak: Black.

Luster and light transmission: Splendent to dull metallic. Opaque.

Optical properties (opaque except in thin splinters):
  isotropic
  refractive index: 2.42

Chemistry: Magnetite has the general formula AB_2O_4, in which A is chiefly Fe^{+2}, but Mg, Zn, Mn, and less commonly Ti and Ni, may substitute for it, and B is essentially Fe^{+3} with small amounts of Al, Cr, Mn^{+3}, and V substituting for it. Whereas in most natural spinels the $A:B$ ratio is close to $1:2$, in magnetite, the ratio of $Fe^{+2}:Fe^{+3}$ may decrease toward γ-Fe_2O_3, maghemite, which is isostructural with magnetite with the same number of oxygen atoms but with some iron positions vacant.

Diagnostic features: Black color and streak and strong magnetic character distinguish magnetite from other minerals high in iron; magnetite is harder than native iron, which is malleable; it dissolves slowly in hydrochloric acid. Some magnetite specimens show polarity; they are called lodestone (Figure 10-3).

Occurrence, associations, and uses: Magnetite is one of the most widespread of the oxide minerals. It is common (1) as a minor accessory mineral in igneous rocks; (2) as a magmatic segregation deposit with apatite and pyroxene, of which important deposits are mined at Kiruna in Sweden; (3) in contact-metamorphic deposits, such as limestones, with garnet, diopside, olivine, pyrite, hematite, and chalcopyrite as at Iron Springs, Utah, and at Cornwall, Pennsylvania; (4) as crystals in chlorite schists; (5) as replacement deposits associated with biotite, amphiboles, epidote, and feldspars; (6) in some high-temperature sulfide veins; and (7) as a detrital mineral in beach and river sands.

Magnetite occurs in several deposits large enough for commercial recovery as iron ore. Some of the deposits contain P as apatite, Ti as ilmenite, and S in sulfides, which are also recovered.

Figure 10-3. Magnetite, var. lodestone, locality unknown.

Chromite, $FeCr_2O_4$

See also the data given in the introduction to this group of minerals.

Habit: Crystals rare, octahedral; typically massive, fine granular to compact.

Cleavage and fracture: None. Uneven.

Color and streak: Black. Brown.

Luster and light transmission: Metallic. Almost opaque.

Optical properties (opaque or nearly so except in thin splinters):
  isotropic
  refractive index: 2.08–2.16
  brown to brownish black

Chemistry: Chromite forms a series with *magnesiochromite* ($MgCr_2O_4$) and *hercynite* ($FeAl_2O_4$). Many natural chromites approximate the composition $(Mg,Fe)Cr_2O_4$, with 8 formula units per unit cell (Figure 10-4). In the majority of analyses, Mg ranges from 4 to 6, Fe^{+2} from 2 to 4, Cr from 7 to 14, Al from 2 to 9, and Fe^{+3} from 0 to 2 atoms per unit cell. The Cr_2O_3 content of most analyzed chromites extends from 30 percent to 61 percent (maximum 67.91 percent) in chromite proper, but it is known to reach 79.04 percent in magnesiochromite. Metallurgical grade chromite must have a minimum ratio of Cr : Fe of 2.5 : 1, which corresponds roughly to 42 percent Cr_2O_3. Chromite ore with from 33 to 48 percent Cr_2O_3 and from 12 to 30 percent Al_2O_3 is suitable for chrome refractory brick.

Diagnostic features: Weak magnetism and brown streak distinguish chromite from magnetite; also, chromite is commonly associated with olivine and/or serpentine.

Occurrence, associations, and uses: Most chromite occurs as an accessory mineral in ultrabasic igneous rocks, such as peridotites, and in serpentinites derived from them. The chromite occurs as segregated masses, lenses, or as disseminated grains, locally in sufficient quantities to serve as an ore. In a few localities, chromite has been concentrated in detrital sands. Olivine, serpentine, pyroxene, chro-

Figure 10-4.   Triangular prism diagram, showing the compositional range of natural terrestrial chromites in terms of six end-member components. Observed compositions are represented as points at the extremities of each horizontal line. As can be seen, these points lie in a blanketlike space that extends from the $MgO \cdot Al_2O_3$—$MgO \cdot Cr_2O_3$ join, along the upper left of the diagram, to the $FeO \cdot Fe_2O_3$ corner at the lower right. Most of the points fall toward the upper left end of the blanket, in the area that corresponds to the range of composition indicated in the description.

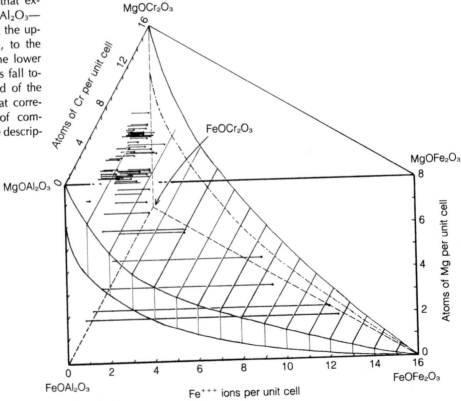

mian spinel, chromium garnet (uvarovite), chromian chlorites, magnetite, and pyrrhotite are common associates.

Chromite is the only ore mineral of chromium, which is of great value as a ferro-alloy metal, especially in stainless steels. The mineral is also used for making refractories.

Franklinite, $(Zn,Mn,Fe)(Fe,Mn)_2O_4$

See also the data given in the introduction to this group of minerals.

Habit: Crystals uncommon; octahedral; also massive.
Cleavage: None, but parting on {111}.
Color and streak: Black to brownish black. Reddish brown.
Luster: Metallic to semimetallic.
Chemistry: Most franklinite is dominantly $ZnFe_2O_4$, but it also has Mn^{+2} and Fe^{+2} substituting for Zn, and Mn^{+3} substituting for Fe^{+3}.
Diagnostic features: Resembles magnetite but is only weakly magnetic.
Occurrence, associations, and uses: Franklinite is the dominant ore mineral of the zinc deposits formerly worked at Franklin and Sterling Hill, New Jersey. It constituted thick beds in crystalline limestone and was associated with zincite, willemite, and rhodonite.

Hausmannite, $MnMn_2O_4$

Crystal system, class, and space group: Tetragonal; $4/m2/m2/m$; $I4_1/amd$.
Cell dimensions and content: $a = 5.76$, $c = 9.44$; $Z = 4$.
Crystal structure: Hausmannite has a distorted spinel structure; the degree of distortion decreases on heating and, above about 1160°, Mn_3O_4 is isometric.
Habit: Crystals uncommon; typically pseudo-octahedral with dipyramid {101}; also massive granular.
Twinning: Common on {112}, repeated.
Cleavage and fracture: {001} perfect. Uneven.
Hardness: $5\frac{1}{2}$–6.
Specific gravity: 4.84.
Color and streak: Brownish black. Chestnut-brown.
Luster and light transmission: Submetallic. Almost opaque.
Optical properties (nearly opaque except in thin splinters):
anisotropic, uniaxial (−)
refractive indices: ω, 2.46; ϵ, 2.15
Diagnostic features: Color, streak; soluble in hot hydrochloric acid with evolution of chlorine.
Occurrence, associations, and uses: Most hausmannite occurs in

apparently high-temperature hydrothermal veins, in contact-metamorphic deposits, and as a product of recrystallization of sedimentary or residual manganese deposits. It is probably an important constituent of manganese deposits formed by circulating meteoric waters and also in many residual deposits.

Hausmannite occurs at numerous manganese-mineral localities —for example, along with pyrolusite and psilomelane in the Batesville District of Arkansas.

Chrysoberyl, $BeAl_2O_4$

Crystal system, class, and space group: Orthorhombic; $2/m2/m2/m$; $Pmcn$.

Cell dimensions and content: $a = 5.48$, $b = 4.43$, $c = 9.41$; $Z = 4$.

Crystal structure: The formula of chrysoberyl is similar to that of spinel, although the symmetry is orthorhombic. Whereas its structure is similar to that of spinel (in that Be is in fourfold coordination with O and Al is in sixfold coordination with O), it is isostructural with olivine (Be and Si, and Al and Mg occupy corresponding positions in the two structures). The hexagonal close packed oxygen arrangement results in a pseudo-hexagonal lattice with $a:b$ close to $1:\sqrt{3}$ and in common twinning on {130}.

Habit: Crystals common; tabular on {001}.

Twinning: Common on {130} as twin plane (contact or penetration) commonly repeated, forming pseudo-hexagonal "sixlings" (Figure 10-5).

Cleavage: {110} distinct; {010} indistinct.

Hardness: $8\frac{1}{2}$.

Specific Gravity: 3.75.

Color and streak: Green, greenish white, and yellowish green. White.

Luster and light transmission: Vitreous. Transparent.

Optical properties:
anisotropic, biaxial (+): $2V = 10°-70°$
refractive indices: α, 1.732–1.747; β, 1.734–1.749; γ, 1.741–1.758
may be slightly pleochroic

Diagnostic features: Color, hardness, twinning; insoluble in acids.

Occurrence, associations, and uses: Most chrysoberyl occurs in granite pegmatite and aplite; it also occurs in mica schists and as a detrital mineral in placers with diamond, corundum, garnet, and cassiterite. Chrysoberyl that is transparent and of good color is used as a gem. The variety *alexandrite*, which is red by natural sunlight and green by artificial light, is especially prized. Another gem variety, which exhibits a chatoyant effect due to particularly oriented needlelike inclusions, is called *cat's eye*.

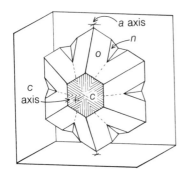

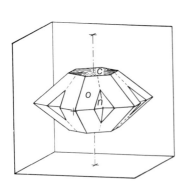

Figure 10-5.   Chrysoberyl, twinned crystal. Forms: pinacoid c{001}; rhombic dipyramids o{111}, n{121}; cyclic twin on {130} as twin plane and composition plane.

HEMATITE GROUP

This group includes corundum, hematite, and the ilmenite series. All are rhombohedral and virtually isostructural, the first two with symmetry $\bar{3}2/m$ and the members of the ilmenite series [$FeTiO_3$ (ilmenite), $MgTiO_3$, and $MnTiO_3$] with the lower symmetry $\bar{3}$.

The so-called corundum structure contains $2[Al_2O_3]$ in the primitive rhombohedral unit. The arrangement of the oxygen atoms is approximately hexagonal closest packing with the close packed layers parallel to {0001}. Octahedrally coordinated cations (i.e., each cation coordinated to six oxygen atoms) may lie between these layers. If all such spaces were filled, an AX composition would result; in corundum, however, only two-thirds of these positions are occupied, each in octahedral coordination with three oxygen atoms above (along c) and three below. Groups of three oxygen atoms form a triangular face parallel to {0001} and common to two octahedra. Thus, three oxygen atoms with an aluminum above and below constitute Al_2O_3 groups with the form of a trigonal dipyramid. There are two orientations of these groups in the structure, which are related by the vertical symmetry planes (glide planes, in this case).

In the ilmenite minerals, where the cation positions in corundum and hematite are occupied in an ordered fashion by two cations (Fe and Ti, Mg and Ti, or Mn and Ti), the symmetry plane parallel to {$11\bar{2}0$} is not present and the symmetry is $\bar{3}$. Both cations are in octahedral coordination with oxygen.

Corundum and hematite show very little solid solution, whereas ilmenite forms a series in which Mg or Mn may substitute for Fe in all proportions. The high magnesium and manganese minerals of the ilmenite series are, however, of rare occurrence and are not discussed further here.

	Corundum	Hematite	Ilmenite
Crystal system:	Hexagonal (trigonal)	Hexagonal (trigonal)	Hexagonal (trigonal)
Symmetry:	$\bar{3}2/m$	$\bar{3}2/m$	$\bar{3}$
Space group:	$R\bar{3}c$	$R\bar{3}c$	$R\bar{3}$
Cell dimensions:	a = 4.758	a = 5.039	a = 5.093
	c = 12.991	c = 13.76	c = 14.06
Content:	Z = 6	Z = 6	Z = 6
Cleavage:	None	None	None
Parting:	{0001}{$01\bar{1}2$}	{0001}{$01\bar{1}2$}	{0001}{$01\bar{1}2$}
Hardness:	9	5–6	5–6
Specific gravity:	4.0–4.1;3.98 (pure)	5.26; 5.256 (pure)	4.72;4.79 (pure)

298

Corundum, Al_2O_3

See also the preceding tabulated data.

Habit: Crystals common; typically tabular on {0001} to short prismatic along c with {11$\bar{2}$0} (Figure 10-6), also steep pyramidal with {11$\bar{2}$3}, {11$\bar{2}$1}, and others; commonly rough, rounded, barrel-shaped crystals (Figure 10-6c and 10-7a): rarely rhombohedral; prism and pyramid faces horizontally striated due to oscillatory combination of faces with different slope; massive granular (in emery), in rounded grains.

Twinning: Common on {01$\bar{1}$2} and {0001}, in many cases lamellar, giving a lamellar structure and striations on {0001} and {01$\bar{1}$2} (Figure 10-7b).

Common angles and forms:
(0001) $\wedge$ (11$\bar{2}$3) = 61°11'   (0001) $\wedge$ (01$\bar{1}$2) = 57°35'
(0001) $\wedge$ (11$\bar{2}$1) = 79°37'   (0001) $\wedge$ (10$\bar{1}$1) = 72°23'

Parting: Common on {0001} and {01$\bar{1}$2}, with marked striations on the parting planes.

Tenacity: Brittle, but very tough when compact.

Color and streak: Blue (var. *sapphire*), pink to blood-red (var. *ruby*); also yellow, yellow-brown, green, purple to violet; some crystals are color-zoned (Figure 10-7b); pure Al_2O_3 is white. White.

Luster and light transmission: Submetallic-splendent to vitreous. Transparent to translucent.

Optical properties:
anisotropic, uniaxial (−)
refractive indices: ω, 1.767–1.772; ϵ, 1.759–1.762
may be color-zoned

Diagnostic features: Hardness; crystal form (barrel shape is characteristic), parting; insolubility.

Occurrence, associations, and uses: Most corundum occurs in rocks that have a lower silica and higher alumina content than that required for feldspar. Most such rocks are syenites that contain feldspar, nepheline, and sodalite, but no quartz. The nepheline syenites and nepheline-feldspar pegmatites at Bancroft and Craigmont, Ontario, are examples. Corundum has also resulted from recrystallization and contact metamorphism of highly aluminous rocks, such as bauxite. The finest gem corundum occurs in recrystallized limestone and in its deeply weathered surface zone in Burma. Placer deposits are another source; those in Sri Lanka have yielded most of the near-gem material used in industry.

As might be expected, corundum has long been recognized as an important mineral because of its value both as an abrasive and as a gemstone. Among natural minerals, corundum is next in hardness to diamond. Corundum is also widely used as an abrasive in the form of *emery* although emery is more properly the name for a black

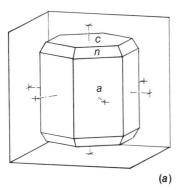

(a)

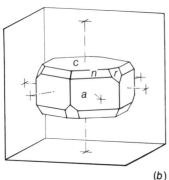

(b)

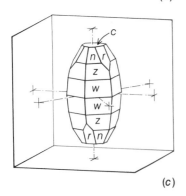

(c)

Figure 10-6. Corundum crystals. Forms: pinacoid c{0001}; hexagonal prisms a{11$\bar{2}$0}; negative rhombohedron r{01$\bar{1}$2} hexagonal dipyramids n{11$\bar{2}$3}, z{11$\bar{2}$1}, ω\{7·7·$\overline{14}$·3\}.

(a)

(b)

Figure 10-7. Corundum. (a) Rough, greenish brown, barrel-shaped crystal (~ 8 cm high) showing a hexagonal dipyramid close to $\{44\bar{8}3\}$ (locality unknown). (b) Basal parting plane (0001) showing striations due to the trace of parting planes $\{01\bar{1}2\}$ as horizontal lines and of parting planes $\{1\bar{1}02\}$ at 30° to right of vertical; the parting surface $\{01\bar{1}2\}$ shows at the bottom of the picture; vertical lines and those inclined 60° to left of vertical are lines of differing growth color due to zones parallel to the prism $\{11\bar{2}0\}$ or a hexagonal dipyramid $\{h \cdot h \cdot \overline{2h} \cdot 1\}$. Color bronze-brown. Specimen is 6 cm across. Craigmont, Raglan Township, Ontario. (Courtesy of Queen's University.)

or gray-black rock that is a granular mixture of corundum with magnetite and/or hematite and spinel.

Corundum of gem quality occurs in a wide variety of colors: red (ruby), blue (sapphire), yellow (oriental topaz), green (oriental emerald), and purple (oriental amethyst). The color of ruby is apparently due to a small Cr content, and the color of sapphire to Fe or Ti. Most varieties prized for gem purposes are transparent and free of the parting so common in some corundum. The varieties called star sapphire and star ruby are highly prized for the six-rayed star formed by internal reflection in a strong light. Good quality corundum lacking the standards of color and transparency required for gems has been widely used for jewel bearings in watches and other instruments.

Emery has long been obtained commercially from Naxos, Samos, and other Greek islands; it has also been mined at Chester, Massachusetts.

Large amounts of artificial corundum (alundum) for abrasive purposes are produced by melting bauxite in an electric furnace. This material and artificial silicon carbide have displaced natural corundum for many abrasive uses. Emery, however, is still preferred by many for the fine-grinding of glass in the manufacture of precision optical components, and natural corundum is preferred for some abrasive-wheel applications.

Rubies and sapphires can also be synthesized by melting and recrystallizing alumina in an oxy-hydrogen flame. Gems cut from carefully synthesized boules colored by chromium, cobalt, or titanium salts are not easily distinguished from natural stones.

Hematite, Fe_2O_3

See also the data given in the introduction to this group of minerals.

Habit: Crystals uncommon; typically thick-to-thin tabular on {0001}, commonly as subparallel growths on {0001}, and also rhombohedral (Figure 10-8); micaceous to platy, compact columnar, fibrous and radiating, in reniform masses with smooth fracture and in botryoidal or stalactitic shapes; commonly earthy and frequently admixed with clay and other impurities; also granular, friable to compact, concretionary, or oolitic.

Twinning: On {0001} as penetration twins and on {01$\bar{1}$2}, typically lamellar.

Parting and fracture: On {0001} and {01$\bar{1}$2} due to twinning. Uneven.

Hardness: 5–6, but apparently much softer in earthy varieties.

Tenacity: Crystals are brittle; thin laminae are elastic.

Color and streak: Dull to bright red (earthy and compact material); steel-gray (crystals and *specularite*), rarely iridescent. Red-brown.

Luster and light transmission: Metallic (crystals and specularite) to submetallic to dull in other varieties. Thin splinters deep blood-red by transmitted light.

Optical properties:
anisotropic, uniaxial (−)
refractive indices: ω, 3.15–3.22; ϵ, 2.87–2.94
pleochroic in yellow- and red-browns

Chemistry: Hematite differs but little from the ideal composition, with 69.94 percent Fe. Some Ti may be present, however, and up to several percent of water is present in some of the fibrous or ocherous varieties.

Diagnostic features: Characteristically shows a red-brown streak; the color in massive and earthy varieties is also red to red-brown, whereas in the variety called specular hematite or specularite, the color is medium to dark gray and the luster metallic and splendent. The specular material is commonly foliated or micaceous and very brittle; in this form, the red-brown color may be difficult to obtain on the streak plate; it can, however, be seen by pulverizing the material. Hematite is soluble in concentrated hydrochloric acid.

Occurrence, associations, and uses: Hematite is the most important and widely used source mineral for iron; even though it is exceeded in iron content by magnetite, hematite is much more abundant and, in most cases, is more readily accessible for surface mining.

As a primary mineral, hematite occurs as an uncommon accessory in igneous rocks. It also occurs as a minor constituent in high-temperature hydrothermal veins and in contact-metamorphic de-

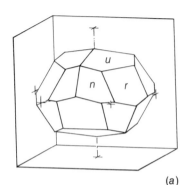

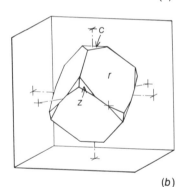

Figure 10-8.   Hematite crystals. Forms: pinacoid c{0001}; negative rhombohedra   r{01$\bar{1}$2}, u{01$\bar{1}$8}; hexagonal dipyramids n{11$\bar{2}$3}, z{11$\bar{2}$1}.

posits, commonly associated with magnetite, and it occurs in small amounts, coloring many sedimentary rocks and soils red. The huge deposits of hematite worked as iron ore are mainly of sedimentary origin, several of which have undergone subsequent concentration or enrichment by meteoric waters and/or by hydrothermal solutions. The extensive Clinton iron beds of Upper Silurian age that occur over a strike distance of about 700 miles from central New York to Alabama are sedimentary beds of oolitic hematite, associated with clay and limonite; they are probably of marine shallow-water deposition. Similar beds of Ordovician age crop out in the Wabana Basin on the east coast of Newfoundland, and they extend, with a low-angle dip, beneath the ocean bottom for several miles.

To the south and west of Lake Superior, extensive iron deposits consist of hematite ranging from hard specular hematite to earthy types containing minor limonite and magnetite; they have resulted from enrichment of ferruginous siliceous sediments of Precambrian age. The original iron-bearing sediments, which have been regionally metamorphosed and are generally known as taconite or banded iron-formation, are now composed of silica as chert, and iron as hematite, magnetite, siderite, or greenalite. The iron-enrichment is thought to have resulted from removal of silica and carbonates by meteoric and/or hydrothermal waters, leaving a high-grade iron ore as a residue. Similar high-grade hematite deposits also occur in Australia, Brazil, Venezuela, Liberia, China, and Canada.

Another kind of extensive hematite deposit has been formed as a result of surface weathering. Examples include leaching of limestones containing siderite lenses in North Africa and leaching of serpentine in Cuba.

The importance of iron and steel in modern life is evident to everyone, since these substances are essential in nearly all forms of construction, transport, communications, and manufacturing. A continued supply of iron ore, available at a reasonable cost, is essential for the maintenance of present-day industry.

Ilmenite, $FeTiO_3$

See also the data given in the introduction to this group of minerals.

Habit: Crystals uncommon; thick tabular on {0001} (Figure 10-9), with hexagonal prisms {10$\bar{1}$0}, {11$\bar{2}$0}, and one or more rhombohedra; commonly compact massive, as embedded grains or loose in sand.

Twinning: Common on {0001}, also lamellar on {01$\bar{1}$2}.

Parting and fracture: On {0001} and {01$\bar{1}$2}. Conchoidal.

Color and streak: Iron-black. Black.

Luster and light transmission: Metallic to submetallic. Nearly opaque.

302

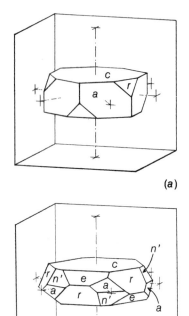

(a)

(b)

Figure 10-9.   Ilmenite crystals. Forms: pinacoid c{0001}; hexagonal prism a{11$\bar{2}$0}; negative rhombohedron r{01$\bar{1}$2}; positive rhombohedron e{10$\bar{1}$4}; negative rhombohedron n'{11$\bar{2}$3}.

Optical properties (opaque except in thin splinters):
  anisotropic, uniaxial (−)
  refractive indices: ω, ~ 2.7

Chemistry and alteration: Ilmenite is typically close to $FeTiO_3$ in composition, but it may contain up to about 6 percent Fe_2O_3 in solid solution. Minor substitution of Mg for Fe (1 percent or less MgO) is not uncommon; high-Mg ilmenite (MgO ranging between 5 and 15 percent) is a characteristic mineral of kimberlites, and the identification of such ilmenite in alluvial deposits has been used as a prospecting guide for kimberlites and hence for possible diamond deposits. These so-called kimberlitic ilmenites grade into *geikielite* ($MgTiO_3$), in which Mg is dominant over Fe. The manganese analog of ilmenite occurs as the rare mineral *pyrophanite* ($MnTiO_3$).

Ilmenite may alter to a gray-white material known as *leucoxene*, which is a mixture of titanium minerals, such as rutile, anatase, and titanite (sphene).

Diagnostic features: Readily distinguished from hematite by its black streak and from magnetite by its nonmagnetic character. Some ilmenite, however, appears slightly magnetic due to intergrown magnetite.

Occurrence, associations, and uses: Most ilmenite occurs in close association with gabbros, diorites, and anorthosites as veins, disseminated deposits, or large masses. It also occurs as an accessory mineral in igneous rocks, in pegmatites, and in quartz veins with chalcopyrite and hematite. In addition, some heavy black beach sands in, for example, Florida contain important amounts of ilmenite.

Ilmenite is used as a source of titanium, most of which is used as paint pigment, but some of which is used in the production of titanium metal and alloys.

Braunite, $3Mn_2O_3 \cdot MnSiO_3$

Crystal system, class, and space group: Tetragonal; 4/m2/m2/m; I4/acd.

Cell dimensions and content: a = 9.38, c = 18.67; Z = 8.

Habit: Crystals rare; pyramidal with {101} and {311}; also granular massive.

Twinning: On {112}, contact twins.

Cleavage and fracture: {112}, perfect. Uneven.

Hardness: 6–6$\frac{1}{2}$.

Specific gravity: 4.72–4.83.

Color and streak: Dark brownish black to steel-gray.

Luster and light transmission: Submetallic. Opaque.

Diagnostic features: Cleavage form is distinct from that of other manganese minerals, such as hausmannite and pyrolusite; soluble in

HCl with evolution of Cl and leaving a residue of gelatinous silica.

Occurrence, associations, and uses: Braunite is commonly associated with other manganese oxide minerals and barite. It occurs primarily in veins and lenses resulting from the metamorphism of manganese oxides and silicates and as a secondary mineral formed under weathering conditions, generally with pyrolusite and psilomelane.

Braunite occurs at numerous localities, and good crystals have been found at many localities in North America and elsewhere. In some localities it is an ore mineral for manganese.

Pyrochlore–Microlite Series:
$(Na,Ca)_2Nb_2O_6(OH,F)$-$(Na,Ca)_2Ta_2O_6(O,OH,F)$

Crystal system, class, and space group: Isometric; $4/m\bar{3}2/m$; $Fd3m$.

Cell dimensions and content: $a = 10.37$–10.41; $Z = 8$.

Habit: Crystals uncommon; typically octahedral; also in irregular masses or embedded grains.

Cleavage or parting, and fracture: {111} generally distinct. Subconchoidal to uneven.

Hardness: 5–$5\frac{1}{2}$.

Specific gravity: 4.2–6.4, greater with higher tantalum content.

Color and streak: Brown with yellowish or reddish shades to black (pyrochlore); pale yellow to brown, red, olive-buff or green (microlite). White.

Luster and light transmission: Vitreous to resinous. Dark varieties nearly opaque.

Optical properties:
isotropic
index of refraction: 1.93–2.02 (microlite); 1.96–2.01 (pyrochlore)

Chemistry: Most specimens of these minerals are metamict, but an isometric structure like that of the nonmetamict material may be restored by heating. A large number of elements, including rare earths, uranium, and thorium, occur in substitution for Na and Ca. Nb predominates in pyrochlore and Ta in microlite; Ti is generally present in rather minor amounts.

Diagnostic features: Crystal form; commonly radioactive.

Occurrence, associations, and uses: Most pyrochlore occurs associated with zircon, apatite, and other rare-earth minerals in pegmatites derived from alkalic rocks, particularly in Norway and in Hastings County, Ontario. It also occurs as an accessory mineral in nepheline syenite (and other alkalic dike rocks) and in carbonatites associated with alkalic intrusives, as in the Alnö Region of Sweden, in East Africa, and at Fen in Norway. It has been mined as a source of niobium.

Most microlite occurs with tantalite or columbite in the albitized parts of some granite pegmatites, particularly in Norway, in

Varuträsk (Sweden), and in the northeastern United States. It is also found in placers in Western Australia. Microlite was mined as a source of tantalum from the Harding pegmatite in New Mexico during World War II.

Romanechite, $BaMn_9O_{16}(OH)_4$

Crystal system, class, and space group: Orthorhombic; 222; P222.
Cell dimensions and content: a = 9.45, b = 13.90, c = 5.72; Z = 2.
Habit: Massive as botryoidal, reniform, or mamillary crusts; stalactitic; also earthy.
Hardness: 5–6.
Specific gravity: 4.7.
Color and streak: Iron-black to dark steel-gray; brownish black to black.
Luster and light transmission: Submetallic. Opaque.
Diagnostic tests: Hardness, botryoidal form, and streak.
Occurrence, associations, and uses: Romanechite is typically a secondary mineral formed under surface conditions of temperature and pressure. It is generally associated with pyrolusite, goethite, limonite, and hausmannite. It occurs as large residual deposits resulting from the weathering of manganous carbonates or silicates and also as concretionary masses in clays and in lake and swamp deposits. Romanechite is one of a number of hard, hydrous manganese oxides that occur as fine-grained, massive, or botryoidal material (Figure 10-10). The name *psilomelane* was formerly applied to material of this sort from many localities, but modern methods of investigation have identified a number of distinct mineral species in these hard

Figure 10-10. Colloform "psilomelane" from Crimora, Virginia. Length of specimen is 63 cm. (Photograph by Wards Natural Science Establishment, Inc. Courtesy of C. A. Michael.)

ores. It appears that much of the material that was formerly called psilomelane is romanechite. Other common constituents include cryptomelane (KMn_8O_{16}), hollandite ($BaMn_8O_{16}$), coronadite ($PbMn_8O_{16}$), manjiroite [$(Na,K)Mn_8O_{16} \cdot nH_2O$], and todorokite [$(Mn,Ca,Mg)Mn_3O_7 \cdot H_2O$]. See also *wad* (p. 308).

RUTILE GROUP

The rutile group includes rutile (TiO_2), pyrolusite (MnO_2), cassiterite (SnO_2), and the rare mineral plattnerite (PbO_2). Although the members of the group are isostructural, they show almost no tendency to form series between the species, probably because of the rather large differences in the radii of the metal ions. Rutile and cassiterite occur in granitic rocks, particularly greisens, probably formed at high temperature; pyrolusite forms at much lower temperatures.

The structure of rutile (Figure 10-11) is typical of the structure of this group. This structure is tetragonal with 2[TiO_2] per unit cell. The metal atoms are located at the points of a body-centered tetragonal lattice. Each metal is coordinated to six oxygens, two along [110] in line with the body-centered metal atom, two above and two below along [1$\bar{1}$0] in line with the corner metal atoms. The octahedra lie with two edges parallel to c and two parallel to [110] or [1$\bar{1}$0]; they form chains, parallel to c, that are linked by shared edges. The chains containing the corner metal atoms have the shared edges parallel to [110], whereas those containing the body-centered metal atoms have their shared edges parallel to [1$\bar{1}$0]. The chains are linked by sharing corner oxygen atoms. This structure is similar to, but more symmetric than, that of marcasite.

Some analyses of rutile show a minor tin content, but many more show substantial amounts of iron (up to about 11 percent Fe_2O_3 or 15 percent FeO), niobium (up to 32 percent Nb_2O_5), or tantalum (up to 35 percent Ta_2O_5). The Fe^{+2} occurs with Nb or Ta. This content of Fe, Nb, and Ta, present in solid solution in all cases, leads to a general formula $Fe_x(Nb,Ta)_{2x}Ti_{1-3x}O_2$, where x reaches a maximum of 0.2.

Some studies of *ilmenorutile* [$(Ti,Nb,Fe)_3O_6$] have shown however, that at least several specimens are intergrowths of rutile and columbite. Therefore, some chemical analyses recorded in the literature represent a mixture. On the other hand, there is an iron-tantalum mineral from Ross Lake, Northwest Territories, that does have the rutile structure.

Artificial $FeTa_2O_6$ and the mineral *tapiolite*, $Fe(Ta,Nb)_2O_6$, have the ordered trirutile structure in which the c length is three times the c length of rutile. This structure is dimorphous with columbite-tantalite (Fe,Mn)-(Nb,Ta)$_2O_6$. Tapiolite, in which Ta $>$ Nb, has been identified in pegmatites in Custer County, South Dakota, and elsewhere. In rutile containing substantial iron and niobium or tantalum, the structure is typically a partially disordered form of the trirutile structure.

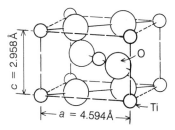

Figure 10-11. Structure of rutile. Large oxygen atoms surround smaller Ti atoms in octahedral (sixfold) coordination. Each oxygen is in threefold coordination. Top of cell is square in outline.

Crystal system, class, and space group: Tetragonal; 4/m2/m2/m; P4₂/mnm.

	Rutile	Pyrolusite	Cassiterite	Plattnerite	Tapiolite
Cell					
dimensions:	a = 4.594	a = 4.39	a = 4.738	a = 4.941	a = 4.754
	c = 2.958	c = 2.86	c = 3.188	c = 3.374	c = 9.228
Cell content:	Z = 2	Z = 2	Z = 2	Z = 2	Z = 2
Cleavage:	{110}	{110}	{100}	None	None
	distinct	perfect	indistinct		
Hardness:	$6-6\frac{1}{2}$	$<5\frac{1}{2}$	6–7	$5\frac{1}{2}$	$6-6\frac{1}{2}$
Specific					
gravity:	4.25 (calc.)	5.06	6.99 (calc.)	9.42	8.17 (calc.)
Common					
color:	Brown	Black	Brown	Black	Black

Rutile, TiO₂

See also the preceding tabulated data.

Habit: Crystals common; typically prismatic to slender or acicular; prism faces vertically striated, commonly terminated by dipyramids {101} or {111} (Figures 10-12 and 10-13); also granular massive.

Twinning: Very common on {101}; simple contact (Figures 10-12a and 10-13b); repeated contact twins giving geniculate forms (Figures 10-12b and 10-13d) or cyclic forms with six or eight individuals (Figure 2-79 and 10-13c); also polysynthetic.

Common forms and angles:
(110) ∧ (111) = 47° 41'   (110) ∧ (101) = 57° 13'
(011) ∧ (0$\bar{1}$1) = 65° 34'   (101) ∧ (130) = 18° 26'

Cleavage and fracture: {110} distinct, {100} less distinct. Conchoidal to uneven.

Color and streak: Reddish brown, rarely yellowish, bluish violet or black; pure TiO_2 is white; artificial rutile crystals used as gemstones under the name *titania* are very pale yellow. Pale brown or grayish black; white for pure TiO_2.

Luster and light transmission: Adamantine to metallic. Transparent in thin pieces.

Optical properties:
anisotropic, uniaxial (+)
indices of refraction: ω, 2.065–2.616; ϵ, 2.890–2.903
may be nearly opaque

Diagnostic features: Habit, including twinning; luster; insoluble in acids. Some rutile exhibits triboluminescence.

Occurrence, associations, and uses: Rutile is by far the most common of the three polymorphic forms of titanium dioxide. It is often

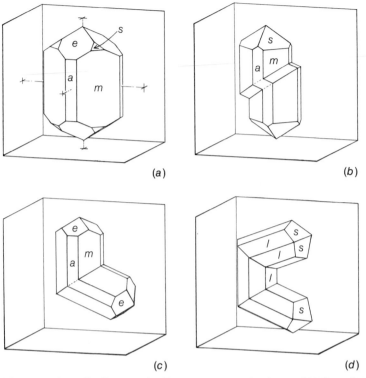

Figure 10-12. Rutile crystals. Forms: tetragonal prisms a{100}, m{110}; ditetragonal prism *l*{310}; tetragonal dipyramids e{101}; s{111}.

formed as an alteration product of other titanium minerals, especially titanite and ilmenite. It is widespread as a minor accessory mineral in igneous rocks, gneisses, and schists. It also occurs in greisens, pegmatites, veins, and in several metamorphic rocks. Minute needles of rutile are included in some quartz (so-called rutilated quartz) and mica crystals. Rutile is also a constituent of some beach sands (e.g., in Australia).

The bulk of rutile and ilmenite that is produced is used as a source of titania for paint pigments, but metal production appears to be rather steadily increasing. Although rutile is preferred for metal production, supplies may not be adequate, so ilmenite will also be used as a source of titanium metal.

Pyrolusite, MnO_2

See also the data given in the introduction to this group of minerals.

Habit: Crystals uncommon as well-developed prisms; typically massive, columnar, or fibrous and divergent; also in reniform coatings and in concretionary forms, granular to powdery, and commonly

Figure 10-13. Rutile. (a) Single crystal with tetragonal prisms {110}, {100}, striated and tetragonal dipyramid {101}. (b) Simple twin on (101) as twin plane. (c) Cyclic twin with eight individuals with {101} as twin plane. (d) Geniculate twinning on {101}; striations parallel to [001] are vertical in the largest individual, locality unknown. (a) and (b): greatest dimension ~ 1cm; (c): ~ 2 cm across; (d): greatest dimension ~ 5 cm. (Courtesy of Queen's University.)

308

reported to be the Mn-oxide constituting dendritic growths on fracture surfaces of rocks (Figure 10-14). It is common as pseudomorphs, with orthorhombic form, after manganite.

Cleavage and fracture: {110} perfect. Uneven.

Hardness: 6–6$\frac{1}{2}$ (crystals), 2–6 (massive material); fibrous or granular material soils fingers easily and marks paper.

Specific gravity: 5.06 (crystals), ranging down to 4.4 (massive material).

Color and streak: Light to dark steel-gray to iron-gray, commonly bluish. Black.

Luster and light transmission: Metallic. Opaque.

Diagnostic features: Usually distinguished from other manganese minerals by its black streak and low hardness (it soils the fingers).

Occurrence, associations, and uses: Pyrolusite, one of the most common manganese minerals, is formed under highly oxidizing conditions. It occurs in bogs, lakes, and shallow marine deposits; in the oxidized zone of ore deposits or rocks that contain manganese; and as deposits formed by circulating meteoric waters. It is generally associated with other manganese and iron oxides and hydroxides. At many localities, it replaces other manganese minerals such as rhodonite, alabandite (MnS), hausmannite, and rhodochrosite. Pyrolusite is one of the important ore minerals of manganese. A large percentage of the manganese that is produced from these ores is used in steel alloys; the remainder is used in dry batteries and chemicals.

The term *wad* is frequently used as a field term for substances consisting mainly of manganese oxides, in much the same sense as

Figure 10-14. Mn-oxide (formerly assumed widely to be pyrolusite) dendritic growth on fracture surface of sandstone, locality unknown. Specimen is ~ 15 cm across. (Courtesy of Royal Ontario Museum.)

limonite is used for hydrous iron oxides and bauxite is used for hydrous aluminum oxides. That is to say, wad is typically a fine-grained mixture, chiefly pyrolusite and romanechite and/or one or more of the other complex hydrous manganese oxides identifiable only by detailed laboratory study. Hence, the different wads typically contain other metallic oxides such as BaO, CuO, K_2O, Na_2O, CoO, Fe_2O_3, PbO_2, and Li_2O, at least some of which are present as constituents of the component manganese minerals.

Wad may occur as masses that are earthy or compact, reniform, concretionary, incrusting, or as stains. Typically it is without internal structure although, in some cases, it is fibrous, banded, or scaly. It is typically soft; much of it soils one's fingers. Its specific gravity, 2.8–4.4, is usually difficult to determine because of its loose and porous quality. It is dull, opaque, and may be black or bluish or brownish black in color and has a brown-to-black streak. Most wad occurs in bog and lake deposits, in clays, in shallow marine sediments, in the oxidized zone of ore deposits, and as a residual product of weathering in areas of manganese-bearing rocks. It is formed under oxidizing conditions at normal temperatues and pressures as a direct precipitate due to chemical or biogenic action. It is of widespread occurrence in many parts of the world, especially in the manganese ore-producing areas.

Cassiterite, SnO_2

See also the data given in the introduction to this group of minerals.

Habit: Crystals common; short prismatic, less commonly pyramidal, with prisms {100}, {110}, {210} and/or dipyramids {111}, {101}, {321} (Figure 10-15a); also massive, in radially fibrous botryoidal crusts or concretionary masses (wood-tin); also as brown rounded pebbles with a concretionary structure and as fine sandlike grains (stream tin).

Twinning: Very common on {101}; contact (Figure 10-15b) or penetration twins, generally repeated.

Common forms and angles:
(110) ∧ (111) = 46° 25′   (100) ∧ (101) = 56° 04′
(321) ∧ (231) = 20° 57′   (101) ∧ (011) = 46° 30′

Cleavage, parting, and fracture: {100} distinct, {110} indistinct. {111} parting common. Subconchoidal to uneven.

Color and streak: Yellow or red-brown to brown-black; color may be unevenly distributed within crystals and in growth bands of concretionary types (white, pure SnO_2). White, grayish, or brownish.

Luster and light transmission: Adamantine and splendant. Light-colored material transparent.

Optical properties:
anisotropic, uniaxial (+)

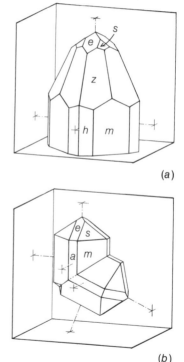

(a)

(b)

Figure 10-15. Cassiterite crystals. Forms: tetragonal prisms a{100}, m{110}; ditetragonal prism h{210}; tetragonal dipyramids e{101}, s{111}; ditetragonal dipyramid z{321}. (b) Twinned on (011) as twin plane and composition plane.

indices of refraction: ω, 1.990–2.010; ϵ, 2.091–2.100
may be pleochroic and/or color-zoned

Chemistry: Well-crystallized cassiterite is close to SnO_2 in composition, but some has up to 3 percent Fe^{+3} substituting for Sn, and less commonly minor niobium and tantalum. The material widely called *wood-tin*—showing botryoidal and reniform shapes, concentric and radially fibrous internal structure, brown color, and lower specific gravity—commonly contains inclusions of hematite and silica.

Diagnostic features: Often difficult to identify; best distinguished by its color, specific gravity, hardness, and crystal form. It has the highest specific gravity among the common light-colored, nonmetallic minerals. Specimens of wolframite have similar color and specific gravity but are readily scratched by steel. It is slowly attacked by acids; a fragment, placed in dilute hydrochloric acid with a little metallic zinc, becomes coated with a dull gray deposit of metallic tin, which turns bright when rubbed.

Occurrence, associations, and uses: Cassiterite, the most important ore of tin, is one of a very few tin minerals. It typically occurs in high-temperature hydrothermal veins or metasomatic deposits that are genetically associated with siliceous igneous rocks. Wolframite, tourmaline, topaz, quartz, fluorite, arsenopyrite, muscovite, Li-micas, bismuthinite, bismuth, and molybdenite are commonly associated minerals. Greisen, the coarse-grained rock containing these minerals, is formed by hydrothermal alteration of granitic rocks. Cassiterite also occurs in pegmatite dikes and in the unusual Bi-Pb-Ag veins in Bolivia. Cassiterite is also a common alluvial mineral in areas of many greisens.

Tin was known to the ancients and was used in making bronze at least as early as 3200 BC. The Phoenicians brought tin from Spain and Cornwall for trade in Mediterranean countries as early as 1100 BC. Vein deposits containing cassiterite were mined in Cornwall and other areas in Europe from the early fifteenth century. The Cornish deposits became the world's most important source of tin during the eighteenth and nineteenth centuries. The ancient sources of tin have dwindled, and today Malaya, Indonesia, Bolivia, Zaire, Thailand, and Nigeria produce most of the world's tin ore. Except for the vein deposits of Bolivia, nearly all of this production is from alluvial deposits.

Currently, the main uses of tin include its use for tinplating, in diverse solders, as a component in several alloys (e.g., bronze and pewter), and in the chemical industry.

Anatase, TiO_2, and Brookite, TiO_2

These two minerals are crystalline modifications of TiO_2 polymorphous with rutile. Anatase is converted to rutile on heating to about 915°C; brookite is converted to rutile at about 700°C.

	Anatase	Brookite
Crystal system:	Tetragonal	Orthorhombic
Crystal class:	4/m2/m2/m	2/m2/m2/m
Space group:	I4₁/amd	Pbca
Cell dimensions:	a = 3.783	a = 5.449
	c = 9.51	b = 9.184
		c = 5.145
Cell content:	Z = 4	Z = 8
Cleavage:	{001}, {101} perfect	{120} indistinct
Hardness:	5½–6	5½–6
Specific gravity:	3.90	4.14
Habit:	Crystals uncommon; pyramidal {101}, or tabular {001}	Crystals uncommon; tabular {010}, prismatic along c with {120}
Luster:	Adamantine	Metallic adamantine
Color:	Yellowish and reddish brown to black; also colorless, greenish, lilac, gray	Brown yellowish to reddish brown; dark brown to black
Streak:	Colorless to pale yellow	Colorless to grayish
Optical properties	Uniaxial (−)	Biaxial (+): 2V = ~0–30°
refractive indices:	ω, 2.561; ε, 2.488	α, 2.583; β, 2.584; γ, 2.700
color:	may be pleochroic in browns.	may be pleochroic in browns.

Occurrence and uses: Anatase and brookite are rarer than rutile; they typically occur in vein or crevice deposits formed from leaching of gneisses or schists by hydrothermal solutions. Anatase also occurs, though uncommonly, in pegmatites; brookite occurs rarely in pegmatites; also, both minerals occur as detrital grains in sediments and sedimentary rocks.

Although neither is of economic value as a source of titanium, in some industrial uses, the anatase form of titanium dioxide is preferred to the rutile form.

Columbite–Tantalite Series, $(Fe,Mn)(Nb,Ta)_2O_6$

Crystal system, class, and space group: Orthorhombic; 2/m2/m2/m; Pbcn.

Cell dimensions and content: a = 5.74, b = 14.27, c = 5.09 (17 per-

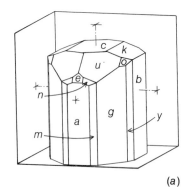

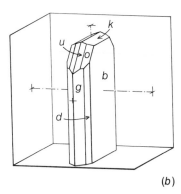

(a)

(b)

Figure 10-16. Columbite-tantalite crystals. Forms: pinacoids $c\{001\}$, $b\{010\}$, $a\{100\}$; rhombic prisms $m\{110\}$, $g\{130\}$, $y\{160\}$, $d\{170\}$, $k\{011\}$; rhombic dipyramids $u\{111\}$, $o\{131\}$.

Figure 10-17. Columbite-tantalite crystal from the Mitchell Pegmatite near Bedford, Virginia. Dimensions of the crystal are ~ 37 x 25 x 9 mm.

cent Ta_2O_5); $a = 5.77$, $b = 14.46$, $c = 5.06$ (red manganotantalite); $Z = 4$.

Habit: Crystals common; short prismatic along c, thin tabular on $\{010\}$ (Figures 10-16 and 10-17) or thick tabular on $\{100\}$; commonly in large groups of parallel or subparallel crystals; massive.

Twinning: Common on $\{201\}$, as contact twins; also as penetration or repeated twins.

Cleavage and fracture: $\{010\}$ distinct, $\{100\}$ less distinct. Subconchoidal.

Hardness: 6 (columbite) to $6–6\frac{1}{2}$ (tantalite).

Specific gravity: 5.2 (columbite) to 7.95 (tantalite).

Color and streak: Iron-black to brownish black (manganotantalite may be dark red); commonly tarnished iridescent. Dark red to black.

Luster and light transmission: Submetallic, often brilliant, subresinous. Transparent in thin splinters.

Optical properties (both minerals are nearly opaque except in thin splinters):
anisotropic; biaxial (−)—columbite; biaxial (+)—tantalite; $2V$, large
indices of refraction:

	Columbite	Tantalite
α	2.20–2.30	2.15–2.20
β	2.30–2.40	2.17–2.25
γ	2.35–2.45	2.25–2.35

Chemistry: These minerals form an almost continuous series of solid solutions within the range shown in the formula. Fe or Mn may predominate in either mineral, and Mg may be a major component of columbite. The specific gravity increases linearly with increase in Ta and thus is useful in estimating Ta content. (Contrariwise, the Fe:Mn ratio has only a slight effect on the specific gravity.)

Currently accepted names for members of the series, including Mg-rich columbite, are as follows: *ferrocolumbite* (FeNb$_2$O$_6$), *manganocolumbite* [(Mn,Fe)(Nb,Ta)$_2$O$_6$], *tantalite* [(Fe,Mn)(Ta,Nb)$_2$O$_6$], *ferrotantalite* (FeTa$_2$O$_6$), *manganotantalite* (MnTa$_2$O$_6$), and *magnocolumbite* [(Mg,Fe,Mn)(Nb,Ta)$_2$O$_6$].

Diagnostic features: Hardness, color, and common tarnish; habit.

Occurrence, associations, and uses: These minerals occur in granite pegmatites, particularly in those that contain albite, Li silicates, and Li-Mn-Fe phosphate minerals; that is, the commonly associated minerals are albite, microcline, beryl, lepidolite, muscovite, tourmaline, spodumene, amblygonite, apatite, microlite, and cassiterite. Some reported occurrences of tantalite have been shown to be tapiolite (tetragonal FeTa$_2$O$_6$), which is dimorphous with ferrotantalite. Columbite-tantalite also occurs as detrital grains.

This series includes the most abundant niobium-tantalum minerals and constitutes the most important source for these metals.

Uraninite, UO$_2$

Crystal system, class, and space group: Isometric; $4/m\bar{3}2/m$; Fm3m.

Cell dimensions and content: $a = 5.4682$ (pure); $Z = 4$. The cell edge decreases (to about 5.4) with oxidation of U^{+4} to U^{+6}; it is greater with substitution of Th for U ($a = 5.5997$, pure ThO$_2$).

Habit: Crystals common; typically cubic octahedral, or cubo-octahedral; also massive (pitchblende); dense botryoidal, or reniform, with a banded structure; in dendritic aggregates of small crystals.

Fracture: Uneven to conchoidal.

Hardness: 5–6.

Specific gravity: 10.95, decreasing with oxidation of U^{+4} to U^{+6}, also with substitution of thorium or cerium for uranium; crystals 8–10; massive 6.5–8.5.

Color and streak: Steely black, brownish black to black. Brownish black or grayish.

Luster and light transmission: Submetallic to pitchlike, also dull. Nearly opaque.

Optical properties (nearly opaque except in thin splinters):
isotropic
greenish, yellowish, or dark brown

Chemistry: Whereas uraninite is essentially UO$_2$ with the fluorite structure, essentially all of the natural material is more or less oxidized and grades toward U$_3$O$_8$ in actual composition. The crystallized material, called *uraninite proper* by some mineralogists, contains 1 percent or more of thorium and rare-earth metals. The substitution of Th for U, to the extent of more than 40 percent ThO$_2$

in nature, may be complete in artificial preparations and extends to pure ThO_2, which occurs in nature as the mineral *thorianite*. Rare-earth elements substitute for U to at least 10 percent. Pegmatitic uraninites are almost all of this type, and the specific gravity generally falls in the range between 8 and 10. The isostructural substance $(Ce,Th)O_2$ (*cerianite*) also occurs as a mineral.

Massive or colloform uraninite, commonly called *pitchblende*, typically contains less than 1 percent of Th and rare-earth elements substituting for U. The specific gravity is generally in the range from 6.5 to 8.5.

Lead is always present, since it is the stable end product of the radioactive disintegration of uranium and thorium; helium is also present for the same reason.

Diagnostic features: Specific gravity, color, pitchy luster, and radioactivity; it has the highest density of any oxide mineral when pure.

Occurrence: The principal occurrences of uraninite can be grouped under several main types:

1. Uraninite, generally containing more than 1 percent Th and rare-earth elements, is found in granite and syenite pegmatites. It is associated with zircon, tourmaline, monazite, mica, and feldspar, and also with complex oxides of the rare-earth elements along with Nb-Ta-Ti. The mineral occurs either with distinct crystal form or massive (ordinarily without colloform structure). Typical occurrences are numerous, for example, at Bancroft, Ontario, and in some of the pegmatites in New England, in southern Norway, and in East Africa.

2. Colloform crusts (pitchblende) occur in high-temperature hydrothermal tin veins, along with cassiterite, pyrite, chalcopyrite, and arsenopyrite—for example, in Cornwall, England.

3. Colloform masses occur in moderate-temperature hydrothermal veins with pyrite, chalcopyrite, galena, carbonates, barite, fluorite, and bismuth, silver, and Co-Ni-Ag minerals. This type of association is important in, for example, the deposits of Great Bear Lake, Canada.

4. Masses occur in veins similar to those mentioned in number 3, but without the Co-Ni-Ag minerals, at, for example, Gilpin County, Colorado, and South Alligator River, Northern Territory, Australia.

5. Minute grains occur in quartz-pebble conglomerates, for example, in the Transvaal gold-bearing conglomerates and in the Blind River district of Ontario. These low-grade ores contain large reserves of uranium dioxide. Although their origin is a subject of considerable speculation, the distribution of the uraninite indicates an origin approximately contemporaneous with the formation of the conglomerate. Other minerals containing U, Th, and rare-earth ele-

ments, commonly as detrital grains, also occur in the conglomerates.

HYDROXIDES

Brucite, Mg(OH)$_2$

Crystal system, class, and space group: Hexagonal (trigonal); $\bar{3}2/m$; $P\bar{3}m1$.

Cell dimensions and content: $a = 3.147$, $c = 4.769$; $Z = 1$.

Crystal structure: Brucite has a layer structure in which each layer, parallel to {0001}, consists of two sheets of OH's in hexagonal close packing, with a sheet of Mg atoms between them (Figure 10-18a). Each Mg atom is in sixfold coordination with OH's. Each OH fits into trigonal depressions of the adjacent layer. The layers are held together by weak secondary forces between adjacent OH sheets.

Habit: Crystals rare; typically broadly tabular on {0001}; as subparallel aggregates of plates; foliated massive; fibrous with fibers separable and elastic.

Cleavage: {0001} perfect, foliae flexible and sectile.

Hardness: $2\frac{1}{2}$.

Specific gravity: 2.39.

Color and streak: White to pale green, light brown, gray, or blue. White.

Luster and light transmission: Pearly on cleavage surfaces, elsewhere waxy to vitreous. Transparent.

Optical properties:
anisotropic, uniaxial (+)
indices of refraction: ω, 1.559–1.590; ϵ, 1.580–1.600

Chemistry: Manganese and iron commonly substitute for Mg to the extent of a few percent. [Mn(OH)$_2$ (*pyrochroite*) and Ca(OH)$_2$ (*portlandite*) are isostructural with brucite.]

Diagnostic features: Occurrence in three distinct forms: the platy or foliate type; the fibrous material elongated on an a axis; and the massive material with a soapy appearance. Brucite, in foliate form, is harder than talc and gypsum, is not greasy to feel as talc, and is not elastic like muscovite. The fibers are not as silky or as fine as those of chrysotile. In addition, brucite is easily soluble in hydrochloric acid.

Occurrence, associations, and uses: Brucite, generally associated with calcite, aragonite, talc, magnesite, and other rarer minerals, is typically a low-temperature, hydrothermal vein mineral. It occurs in serpentinite and in chloritic and dolomitic schists, as well as an alteration product of periclase in crystalline limestones.

Fibrous brucite occurs in serpentinite in New Jersey and is common in the asbestos mines of Asbestos and Black Lake, Quebec.

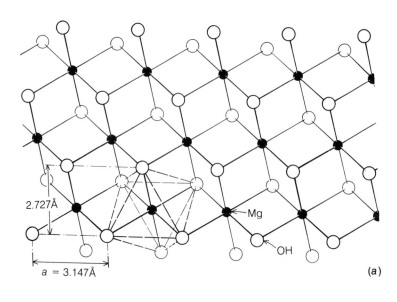

(a)

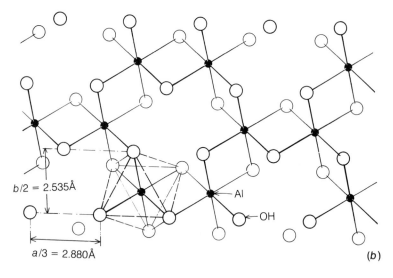

(b)

Figure 10-18. (a) Single layer of the brucite structure with Mg in octahedral coordination with (OH); the hexagonal *a* axis is horizontal. (b) Single layer of the gibbsite structure with Al in octahedral coordination with (OH); only two-thirds of the possible octahedral sites are occupied. The *a* axis of the monoclinic lattice is horizontal.

Granular brucite, pseudomorphous after periclase, occurs in crystalline limestone at Crestmore, California. It is recovered commercially from crystalline limestone at Wakefield, Quebec, and from lenses in a magnesite rock in the Paradise Range, Nye County, Nevada, for use in the production of specialty refractories.

Gibbsite, Al(OH)₃

Crystal system, class, and space group: Monoclinic; $2/m$; $P2_1/n$.

Cell dimensions and content: a=8.641, b=5.070, c=9.719; β=94° 34'; Z=8.

Crystal structure: Gibbsite has a layer structure (Figure 10-18*b*) that is similar in some respects to that of brucite. Each layer, one-half of the c axis period in thickness, is composed of two sheets of OH ions approximately in the hexagonal closest packing arrangement with Al lying between the two sheets. Each Al is in octahedral coordination with six OH's, but only two-thirds of the possible octahedral sites are occupied, in contrast with brucite in which all the sites are occupied. In gibbsite, the layers are packed together so that the OH's in adjoining layers are in line parallel to c instead of nesting into the trigonal depressions as in brucite. This arrangement requires a c length slightly more than double that of brucite. The relatively weak attraction between OH in adjoining layers results in the perfect cleavage {001}.

Habit: Crystals rare; tabular on {001} with {100} and {110}; radiating spheroidal concretions; also stalactitic and encrusting with a smooth surface; compact earthy.

Twinning: Common with [130] as twin axis or {001} as twin plane.

Cleavage: {001} perfect.

Hardness: $2\frac{1}{2}$–$3\frac{1}{2}$.

Tenacity: Tough.

Specific gravity: 2.4.

Color and streak: White; grayish, greenish, or reddish white. White.

Luster and light transmission: Pearly on cleavage surfaces, vitreous on others. Transparent.

Optical properties:
anisotropic, biaxial (+): 2V = ~0°–40°
indices of refraction: α, 1.568–1.580; β, 1.568–1.580; γ, 1.587–1.600

Diagnostic features: Gives strong clay odor when breathed upon; when crystalline, it is softer than diaspore; when fine grained, it is not distinguishable from boehmite.

Occurrence, associations, and uses: Gibbsite is one of the important constituent minerals of bauxite. It occurs with boehmite and diaspore in earthy, oolitic, and pisolitic bauxites, constituting the chief mineral at some occurrences. It also occurs as a low-temperature hydrothermal mineral in veins or cavities in some alkalic and aluminous igneous rocks.

Boehmite, AlO(OH)

Crystal system, class, and space group: Orthorhombic; 2/m2/m2/m; Amam.

Cell dimensions and content: a = 2.868, b = 12.227, c = 3.700; Z = 4.

Habit: Crystals rare; microscopic crystals tabular on {001}; commonly disseminated or in pisolitic aggregates.

Cleavage: {010}, seldom visible.

Specific gravity: 3.0.

Color and streak: White.

Optical properties:
anisostropic, biaxial (+): 2V = ~80°
indices of refraction: α, 1.64–1.65; β, 1.65–1.66; γ, 1.65–1.67

Chemistry: Boehmite is isostructural with the rare iron mineral *lepidocrocite*, FeO(OH), but not with manganite, MnO(OH). Boehmite and lepidocrocite are polymorphous with diaspore, AlO(OH), and goethite, FeO(OH), respectively. Boehmite and lepidocrocite possess a sheet structure that gives rise to the distinct {010} cleavage.

Diagnostic features: Essentially indistinguishable from other bauxite constituents by macroscopic means.

Occurrence, associations, and uses: Boehmite is widely distributed as a major constituent of most *bauxite*. As such, it constitutes an important ore mineral of aluminum. It and the other minerals of bauxite (e.g., diaspore and gibbsite) and hydrous iron oxides result from weathering, under tropical conditions, of aluminum silicate rocks low in free quartz.

Manganite, MnO(OH)

Crystal system, class, and space group: Monoclinic; 2/m (pseudo-orthorhombic); $B2_1/d$.

Cell dimensions and content: a = 8.94, b = 5.28, c = 5.74; β = 90° 00'; Z = 8.

Habit: Crystals uncommon; striated and short, or long prismatic parallel to c (Figure 10-19); also columnar to fibrous.

Twinning: Twin plane {011} as contact or penetration twins, commonly repeated.

Cleavage and fracture: {010} perfect; {110} and {001} good. Uneven.

Hardness: 4.

Specific gravity: 4.33.

Color and streak: Dark steel-gray to iron-black. Reddish brown to black.

Luster and light transmission: Submetallic. Almost opaque.

Optical properties:
anisotropic, biaxial (+)
indices of refraction: α, 2.25; β, 2.25; γ, 2.35
may be pleochroic

Diagnostic features: Habit; soluble in concentrated hydrochloric acid with evolution of chlorine.

Occurrence, associations, and uses: Manganite occurs in low-

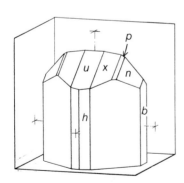

Figure 10-19.   Manganite crystal. Forms: monoclinic pinacoids b{010}, u{101}; prisms m{210}, h{810}, x{818}, p{212}, n{111}. The crystals usually develop with pseudo-orthorhombic form.

temperature hydrothermal veins with barite, calcite, siderite, and hausmannite; in deposits formed by meteoric waters associated with pyrolusite, goethite, and romanechite; also with other manganese minerals in residual clays. It is commonly altered to pyrolusite and less commonly to other manganese oxides. Manganite, however, is seldom an important constituent of manganese ores.

Diaspore, AlO(OH)

Crystal system, class, and space group: Orthorhombic; $2/m2/m2/m$; *Pbnm.*

Cell dimensions and content: $a = 4.396$, $b = 9.426$, $c = 2.844$; $Z = 4$.

Habit: Crystals rare; thin platy or tabular on {010}, elongated to acicular crystals parallel to c; foliated massive and in thin scales; also stalactitic; disseminated.

Cleavage and fracture: {010} perfect, {110} less perfect. Conchoidal.

Hardness: $6\frac{1}{2}$–7.

Tenacity: Very brittle.

Specific gravity: 3.3–3.5 (3.380, calc.).

Color and streak: White, grayish white, colorless, rarely greenish brown, lilac, or pink. White.

Luster and light transmission: Brilliant vitreous, pearly on cleavage faces. Transparent to subtranslucent.

Optical properties:
anisotropic, biaxial (+)
indices of refraction: α, 1.682–1.706; β, 1.705–1.725; γ, 1.730–1.752

Diagnostic features: Distinguished from axinite, which it resembles, by cleavage and hardness. Decrepitates rather violently; luster is distinctive for coarsely crystalline material. Diaspore is not macroscopically discernible in bauxite.

Occurrence, associations, and uses: Diaspore is of widespread occurrence as massive fine-granular material with boehmite and gibbsite in bauxites (Figure 10-20). It also occurs with corundum in emery rock and in certain crystalline limestones; as a late hydrothermal mineral in some alkalic pegmatites; and as a hydrothermal alteration of other aluminous minerals.

Goethite, FeO(OH)

Crystal system, class, and space group: Orthorhombic; $2/m2/m2/m$; *Pbnm.*

Cell dimensions and content: $a = 4.596$, $b = 9.957$, $c = 3.021$; $Z = 4$.

Figure 10-20. Bauxite from Lightner Prospect, near Spottswood, Virginia, exhibiting typical concretionary pisolites of gibbsite with minor amounts of diaspore and traces of kaolin.

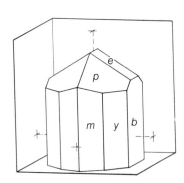

Figure 10-21. Goethite crystal. Forms: pinacoid $b\{010\}$; rhombic prisms $y\{120\}$, $m\{110\}$, $e\{021\}$; rhombic dipyramid $p\{121\}$.

Habit: Crystals uncommon; prismatic and striated parallel to c (Figure 10-21); also thin tabular and scaly parallel to $\{010\}$; aggregates of capillary to acicular crystals; typically massive; reniform, botryoidal, or stalactitic masses (Figure 10-22) with internal radiating fibrous or concentric structure; bladed or columnar; compact; earthy; pisolitic and oolitic; loose and porous; also as pseudomorphs after pyrite.

Cleavage and fracture: $\{010\}$ perfect, $\{100\}$ good. Uneven.

Hardness: $5–5\frac{1}{2}$.

Specific gravity: 3.3 to 4.3.

Color and streak: Blackish brown to yellow or reddish brown to brownish yellow in massive and earthy forms. Brownish yellow, orange-yellow.

Luster and light transmission: Crystals adamantine; metallic to dull; fibrous material silky. Transparent in thin splinters.

Optical properties:
anisotropic, biaxial (−): $2V = \sim 0°–27°$
indices of refraction: α, 2.260–2.275; β, 2.393–2.409; γ, 2.398–2.515
pleochroic

Chemistry: Pure goethite is usually close to $FeO(OH)$ in composition.

Fine-grained and earthy goethite commonly contains adsorbed water and/or siliceous and/or clayey impurities. *Lepidocrocite* is a relatively rare polymorph of goethite.

Diagnostic features: Distinguished from romanechite and colloform hematite by streak and color; upon dissolving in HCl, colors the solution yellow.

Occurrence, associations, and uses: Most of the common yellow-brown and brown ferric oxides usually known as *limonite* properly belong to this species. Most goethite has been formed as the result of the chemical weathering of iron minerals, such as siderite, pyrite, magnetite, and glauconite, under oxidizing conditions at ordinary temperatures. It also has been formed by direct precipitation from marine or meteoric waters in bogs and lagoons. It occurs as crystals, or botryoidal or stalactitic forms, in near-surface fissures and pockets in many kinds of rocks. Goethite, as the main constituent of limonite, occurs as a gossan or weathered capping on veins or replacement deposits that are rich in iron-bearing sulfides. Residual limonite cappings are also found over rocks containing ferrous-iron minerals where rock decay has progressed sufficiently; these rock products are called *laterites*. Large laterite bodies, resulting from the weathering of serpentine, occur in Cuba. Oolitic limonite, partly to wholly goethite, is largely responsible for the iron content of the extensive sedimentary "Minette" ores of eastern France. Large tonnages of residual limonite, also largely goethite, are now being mined in the Labrador Trough located along the Quebec-Labrador boundary.

Figure 10-22. Goethite stalactitic mass showing internal radial fibrous structure and lustrous black surface, Hörhausen, Germany. Specimen is ~ 12 cm across. (Courtesy of Royal Ontario Museum.)

11 Class IV: Halides

The halides comprise those minerals in which a halogen element—F, Cl, Br, I—is the sole or principal anion. They include simple structures (e.g., that of halite) and structures that contain anion groups, such as $(AlF_6)^{-3}$ (e.g., cryolite). Some halides contain water of crystallization; others are oxyhalides or hydroxyhalides. In addition, it is noteworthy here that some minerals classified as silicates, phosphates, etc. also contain minor amounts of halogen elements, particularly fluorine and chlorine.

Chlorine and fluorine are far more abundant (130 and 625 parts per million, respectively) in the earth's crust than bromine and iodine (2.5 and 0.5 parts per million, respectively). From a mineralogical point of view, this abundance is reflected by the fact that even though bromine is more concentrated in the ocean waters (65 parts per million), it is very rare in minerals.

Fluorine is of strikingly different geological occurrence from the other halogens. It occurs almost exclusively in minerals of igneous rocks, pegmatites, hydrothermal veins, or rocks altered by pyrometasomatic action. On the other hand, the great bulk of the chlorine of the earth's crust is dissolved in sea water or occurs in solid stratiform deposits of soluble salts that were formed by the evaporation of sea water at various periods in geological time, whereas it occurs as only a minor element in a few minerals of igneous or metamorphic origin—for example, in apatite, sodalite, scapolite.

The important halide minerals are:

Halite   NaCl
Sylvite   KCl
Fluorite CaF_2
Carnallite $KMgCl_3 \cdot 6H_2O$
Cryolite Na_3AlF_6

In addition, chlorides, bromides, and iodides (in decreasing order of importance) occur in the oxidized zone of many ore deposits, especially in the arid regions of the earth. Examples are the basic copper chloride—atacamite—an important constituent of the ore in the Chilean copper deposits; the insoluble halides of silver—*chlorargyrite* (AgCl), *bromargyrite* (AgBr), and *iodargyrite* (AgI)—that have constituted an important part of some ore, for example, in the early days of development in the silver mines of Saxony, Mexico, New South Wales, and the southwestern United States; and halides of lead, mercury, and bismuth that are also fairly common at a few localities.

HALITE GROUP

The crystal structure of halite (Figure 11-1) was one of the first to be analyzed by X-ray diffraction methods. The halite structure type is present in a large number of *AX* compounds with radius ratios between 0.41 and 0.73; minerals of the halite group, the galena group, and the periclase group are included. These minerals are isostructural. [In addition, halite is isotypic with, for example, calcite.] The important members of the halite group are halite (NaCl) and sylvite (KCl). At ordinary temperatures there is virtually no substitution of Na by K in halite. In the halite structure, one kind of ion occurs at the points of a face-centered cubic lattice and the other kind of ion occurs midway between each pair of the first kind along the [100] directions. Each ion is in octahedral coordination with six ions of the opposite kind, rather than linked to one ion in particular. In fact, the analysis of this structure first established the

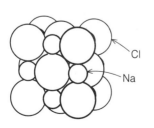

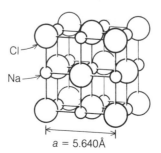

$a = 5.640Å$

Figure 11-1. Structure of halite. Na and Cl are each octahedrally surrounded by six atoms of Cl or Na.

important point in the crystal chemistry of typical inorganic compounds and minerals that atoms are not grouped into discrete molecules.

Crystal system, class, and space group: Isometric; $4/m\bar{3}2/m$; Fm3m.

	Halite	Sylvite
Cell edge:	$a = 5.6402$	$a = 6.2931$
Cell content:	$Z = 4$	$Z = 4$
Cleavage:	{100} perfect	{100} perfect
Hardness:	$2\frac{1}{2}$	$2\frac{1}{2}$
Specific gravity:	2.164 (calc.)	1.9865 (calc.)

Halite, NaCl

See also the preceding tabulated data.

Habit: Crystals common; typically as cubes, commonly with cavernous and stepped faces, hopper crystals (Figure 11-2); also massive, coarsely granular to compact; rarely columnar.

Color and streak: Colorless, white, gray, yellow, red, rarely blue or purple. White.

Luster and light transmission: Vitreous. Transparent to translucent.

Optical properties:
   isotropic
   index of refraction: 1.544

Diagnostic features: Distinguished by its salty taste, low hardness, and cubic cleavage. Some halite is fluorescent; some exhibits triboluminescence.

Occurrence, associations, and uses: Halite is by far the most common water-soluble mineral, and it occurs as extensive beds formed by evaporation of sea water. The beds occur interstratified with beds of shale, limestone, dolostone, and rock-gypsum or rock-anhydrite. Halite also occurs as a sublimation in volcanic regions and as an efflorescence in arid areas. And, along with other salts, it occurs in playa deposits in dried inland lake basins such as on the Great Salt Lake Desert in Utah.

Stratiform deposits of halite, of widespread occurrence throughout the world, range in age from Paleozoic to Recent. Deformation of thick salt beds, generally at considerable depths, has resulted in the local extrusion of plug-like or chimney-like masses, ranging from a few hundred feet to several miles in diameter, into overlying sediments, forming masses called salt domes or salt plugs. In the Gulf Coast region of the United States, approximately 250 of these masses have been discovered in the search for oil and gas.

Halite finds its greatest use in the production of sodium com-

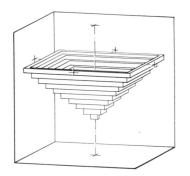

Figure 11-2. Halite, hopper crystal with cube faces. This form often results from growth of crystals floating on the surface of natural brines.

pounds, chlorine, and as an ice control agent for highways and walkways. Other uses include those for culinary purposes and for several diverse uses in the chemical and metallurgical industries.

Sylvite, KCl

See also the data given in the introduction to this group of minerals.

Habit: Crystals common; typically as cubes; massive; coarsely granular to compact.

Tenacity: Not as brittle as halite; distinctly sectile, as compared with halite, when scratched with a knife point.

Color and streak: Colorless to white, also grayish, bluish, yellowish red, or red; the red tints are due to included particles of hematite. White.

Luster and light transmission: Vitreous. Transparent to translucent.

Optical properties:
   isotropic
   index of refraction: n,1.4903

Diagnostic features: Similar in appearance to halite (and they occur together in some evaporite salt beds), sylvite can be best distinguished from halite by its bitter taste and its tendency to sectility. On scratching a smooth surface, a knife point raises a distinct powder with halite, but very little with sylvite. This latter test is dependable and easier to apply than tasting, especially if one is examining many feet of drill core. Also, in a given deposit, sylvite can have a slightly different transparency than that of halite, and it can be seen, in many cases, to fill in around euhedral cubes of halite, clearly indicating the later formation of sylvite.

Occurrence, associations, and uses: Sylvite occurs principally in some of the bedded deposits of halite and gypsum, but it is much less common than halite. The most famous locality for sylvite and other potash salts is at Stassfurt, in Germany, where about 30 different saline minerals have been recognized. Important quantities of sylvite, however, have been mined from bedded salt deposits in New Mexico, Texas, and Saskatchewan. A large percentage of the world production of sylvite and other potassium salts is used in fertilizers. The remainder has diverse uses, especially in the manufacture of potassium hydroxide, which finds broad use in many processes, such as the making of dyes and soaps.

Fluorite, CaF$_2$

Crystal system, class, and space group: Isometric; $4/m\bar{3}2/m$; Fm3m.

Cell dimensions and content: $a = 5.4626$; $Z = 4$.

326

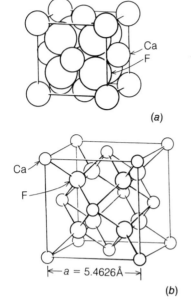

(a)

Ca
F

←— $a = 5.4626Å$ —→

(b)

Figure 11-3.   Structure of fluorite. Each fluorine is coordinated
to four calcium atoms, and each
calcium to eight fluorine atoms.

Crystal structure: The structure of fluorite (Figure 11-3) is unique among natural halides, although it is also the structure of uraninite, UO_2, and of the related oxides, ThO_2 and CeO_2. The Ca atoms occupy face-centered cubic positions; the F atoms occupy the body-center of each cubelet, the length of which is one-half the cube-edge length of the unit cell. The unit cell contains $4[CaF_2]$. Each Ca atom is in eightfold coordination with eight F atoms at the corners of a cube with $a/2$, and each F atom is in fourfold coordination with four Ca atoms at the corners of a tetrahedron.

Habit: Crystals common (Figure 11-4); typically cubes, less commonly dodecahedra, octahedra, cubes, or octahedra modified by {110}, {310}, {311}, or {421}; cube faces smooth and lustrous, octahedron faces rough and dull; commonly composite with many small crystals in parallel aggregation; also massive, coarse-to-fine granular, or compact.

Twinning: Interpenetrating cubes with [111] as twin axis (Figure 11-4).

Cleavage: {111} perfect.

Hardness: 4.

Specific gravity: 3.18.

Color and streak: Colorless and transparent when pure, commonly wine-yellow, green, greenish blue, violet-blue, also white-gray, sky-blue, purple, bluish black, or brown; the color is commonly distributed in zones parallel to crystal faces; massive material may also exhibit zones of different color. White.

Luster and light transmission: Vitreous. Transparent to translucent.

Optical properties:
   isotropic
   index of refraction: 1.433–1.435

Chemistry: Although most fluorite is fairly pure, Y, Ce, and traces of other rare-earth elements may be present substituting for Ca.

Diagnostic features: Distinguished by octahedral cleavage; softer than feldspars; harder than calcite and does not effervesce with acid; commonly fluorescent; less commonly triboluminnescent.

Occurrence, associations, and uses: Fluorite occurs in mineral deposits of widely different character. It occurs most commonly as a vein mineral—in some cases as the principal constituent, in others as a gangue mineral—with lead and silver ores, and generally associated with quartz, calcite, dolomite, and barite. It also occurs in cavities in dolostones and limestones, and is commonly associated with minerals such as celestite, anhydrite, gypsum, dolomite, sulfur, and millerite.

   Fluorite is also characteristic (1) of some pneumatolytic deposits such as greisens; (2) of high-temperature cassiterite veins associated with topaz, tourmaline, lepidolite, apatite, or quartz; (3) as a late hydrothermal mineral in cavities and joints in granite; and (4) in pegmatites.

   The occurrences of fluorite are widespread. The principal producing countries are the United States, Germany, Mexico, the Soviet Union, Canada, England, and Italy.

Large quantities of fluorite are used in metallurgical operations as a flux (especially in the manufacture of open-hearth steel and in the smelting of aluminum) and in the manufacture of chemicals. A small amount of colorless, transparent fluorite is used for optical purposes.

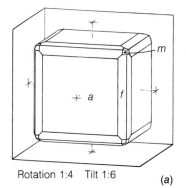

Rotation 1:4   Tilt 1:6   (a)

Carnallite, KMgCl$_3 \cdot$6H$_2$O

Crystal system, class, and space group: Orthorhombic; 2/m2/m2/m; Pbnn.

Cell dimensions and content: a = 9.56, b = 16.05, c = 22.56; Z = 12.

Habit: Crystals uncommon: pseudo-hexagonal due to nearly equal development of {hhl} and {0kl} forms, tabular on {001}; typically massive, granular.

Cleavage and fracture: None. Conchoidal.

Hardness: 2$\frac{1}{2}$.

Specific gravity: 1.60.

Color and streak: Colorless to milk-white, commonly reddish due to enclosed oriented scales of hematite. White.

Luster and light transmission: Greasy, dull. Transparent to translucent.

Taste: Bitter.

Optical properties:

anisotropic, biaxial (+): 2V = ~70°

indices of refraction: α, 1.465–1.467; β, 1.472–1.475; γ, 1.494–1.497

Diagnostic features: Distinguished from other salts by its lack of cleavage, its conchoidal fracture, and its deliquescent nature. [Specimens must be kept in sealed bottles because they will deliquesce in a humid atmosphere.] It is easily soluble in water and has a bitter taste.

Occurrence, associations, and uses: Carnallite occurs in the upper layers of some saline evaporite deposits, in association with halite, sylvite, and *kieserite* (MgSO$_4 \cdot$H$_2$O). Carnallite and kieserite contribute to the magnesium content of natural brines, from which some magnesium is produced, and carnallite also provides some potash. The principal known occurrence is in the salt deposits of Stassfurt, Germany.

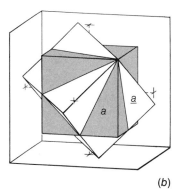

(b)

Figure 11-4. Fluorite crystals. Forms: cube a{100}; tetrahexahedron f{310}; trapezohedron m{311}. (b) Interpenetrating cubes twinned on [111] as twin axis.

Cryolite, Na$_3$AlF$_6$

Crystal system, class, and space group: Monoclinic; 2/m; P2$_1$/n.

Cell dimensions and content: a = 5.40, b = 5.60, c = 7.78; β = 90° 11'; Z = 2.

Habit: Crystals uncommon; cuboidal (pseudo-cubic) with {001} and {110}; {101}, {$\bar{1}$01}, and {011} together simulate an octahedron; also massive, coarsely granular.

Twinning: Very common with several twin laws occurring together; [110], [021], [$\bar{1}$11], [100] as twin axes as well as others, repeated or polysynthetic, reflecting the pseudo-isometric character of the cell bounded by {001} and {110}.

Cleavage and parting: None. Parting on {001} and {110} giving cuboidal forms.

Hardness: $2\frac{1}{2}$.

Specific gravity: 2.97.

Color and streak: Colorless to white; also brownish or reddish. White.

Luster and light transmission: Vitreous to greasy. Translucent.

Optical properties:
anisotropic, biaxial (+): $2V = \sim43°$
indices of refraction: α,1.338; β,1.338; γ,1.339

Diagnostic features: Characterized by the cuboidal form of the parting fragments, the white color, and the greasy luster.

Occurrence, associations, and uses: Cryolite, not a common mineral, occurs in important quantities at only two localities—Ivigtut in West Greenland and Miask in the Soviet Union. At Ivigtut, it occurs in a pegmatitic body associated with a small granite stock. The associated minerals are microcline, quartz, siderite, fluorite, and minor amounts of sulfides. It is commonly altered to other rare fluorides.

The importance of this unusual mineral derives from its use, in molten form, as the electrolyte in the Hall–Héroult process for the electrolytic reduction of aluminum ores to metallic aluminum. Much of the cryolite used in this operation, however, is now manufactured from fluorite by a process developed when supplies of cryolite were subject to interruption during World War II.

12 Class V: Carbonates, Nitrates, Borates

The carbonates include some very common and widespread minerals. The fundamental anionic unit is $(CO_3)^{-2}$, which is a planar group with C at the center of an equilateral triangle with an O atom at each of its apices. The nitrates, which occur rarely as minerals, include some minerals that are isostructural with carbonates—for example, nitratite with calcite. The borates may be based on the simple anionic group $(BO_3)^{-3}$ or on anionic complexes that include both BO_3 triangles and BO_4 tetrahedra (e.g., in borax and colemanite). In these complexes, some of the oxygen positions are generally occupied by (OH).

Some carbonate minerals form extensive rock masses in sedimentary and metamorphosed sedimentary rock sequences. Thus, calcite and dolomite are the principal constituents of limestone and calcitic marble, and of dolostone and dolomitic marble. Other carbonates also occur in some evaporite deposits. Carbonates are also common in hydrothermal veins and replacement zones.

Nitrates, being very soluble minerals, occur only in very arid, virtually rainless, regions of the earth. Borates occur for the most part in dry lake basins in mountainous regions and in lacustrine formations of Tertiary age.

The most common carbonate minerals fall into the calcite, dolomite, and aragonite groups as follows:

Calcite, CaCO₃	Dolomite, CaMg(CO₃)₂	Aragonite, CaCO₃

Magnesite, MgCO₃
Siderite, FeCO₃
Rhodochrosite,
  MnCO₃
Smithsonite, ZnCO₃
	Ankerite, CaFe(CO₃)₂
	Kutnahorite,
	  CaMn(CO₃)₂
		Witherite, BaCO₃
		Strontianite, SrCO₃
		Cerussite, PbCO₃

Each group includes three or more isostructural substances. $CaCO_3$ is polymorphous as calcite and aragonite. The other members of the calcite group with metal ions of smaller radius than Ca^{+2}, however, do not crystallize with the aragonite structure, and the members of the aragonite group with larger metal ions than Ca^{+2} do not crystallize with the calcite structure.

Dolomite is closely related to calcite; in fact, they are almost isostructural, the rhombohedral lattices in both having similar dimensions and the minerals of both groups having similar rhombohedral cleavage with interfacial angles of 73–75° and 105–107°. In calcite, however, the Ca ions are all structurally equivalent, and they occur in rows parallel to the threefold axis (Figure 12-1), whereas in dolomite, Ca and Mg ions occupy these same positions alternately. Consequently, the dolomite structure differs from the calcite structure in not possessing twofold axes parallel to the hexagonal a axes and the symmetry planes perpendicular to them.

The many chemical analyses of minerals of these groups show that they enter a wide range of solid solutions. However, since the conditions of temperature, pressure, and availability of ions at the time of formation are not known for most mineral specimens, it is not possible to outline in detail the limits of substitution of one ion for another in any of these structures or the conditions of temperature or pressure that are favorable to such substitution. Laboratory investigations of artificial materials are slowly providing information of this sort. The extent of substitution occurring in the minerals is indicated under each species.

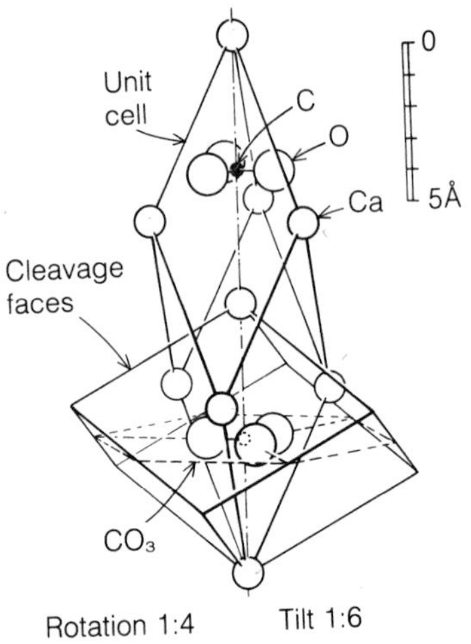

Figure 12-1. Structure of calcite. The axial ratio given in the text corresponds to the cleavage rhombohedron as the unit cell; that rhombohedron is face centered. The simpler structural (rhombohedral) unit cell containing 2[CaCO₃] is also shown.

CALCITE GROUP

The structure of calcite (Figure 12-1) is analogous to that of halite (Figure 11-1) if we consider the unit cube of halite to be shortened along one trigonal axis to give interaxial angles of 101° 55' (the interedge angles in the calcite cleavage rhombohedron) instead of the 90° angles of halite. The Ca atoms are situated at the points of a face-centered rhombohedron (corresponding to face-centered cubic positions in halite, Cl in Figure 11-1). The CO₃ ions are situated midway between Ca atoms along rows parallel to the rhombohedral cleavage edges (corresponding to the Na positions in Figure 11-1); thus, one CO₃ group is situated at the body-center of the rhombohedron. The CO_3 ions are triangular in form, planar parallel to {0001}, and each group is turned 60° with respect to the next one (note the orientation of CO_3 groups along c in Figure 12-1). The simplest unit cell of the structure is a rhombohedron with interedge angle of 46° 4' corresponding to hexagonal dimensions $a = 4.989$, and $c = 17.062$. This unit cell corresponds to the rhombohedron {40$\bar{4}$1} in the conventional axial ratio for calcite based on the cleavage rhombohedron as {10$\bar{1}$1}. In the following mineral descriptions, all indices apply to the axial ratio in which the cleavage is {10$\bar{1}$1}. In addition to the minerals described here, the group includes *gaspéite* [(Ni,Mg,Fe)CO_3], *sphaerocobaltite* ($CoCO_3$), and *otavite* ($CdCO_3$).

Crystal system, class, and space group: Hexagonal (trigonal); $\bar{3}2/m$; $R\bar{3}c$.

Cell content: $Z = 2$ (for simplest rhombohedral cell), $Z = 6$ (hexagonal unit cell).

	Calcite	Magnesite	Siderite	Rhodochrosite	Smithsonite
Cell dimensions:	$a = 4.989$	$a = 4.633$	$a = 4.72$	$a = 4.777$	$a = 4.66$
	$c = 17.062$	$c = 15.015$	$c = 15.46$	$c = 15.67$	$c = 14.98$
Cell content:	$Z = 6$	$Z = 6$	$Z = 6$	$Z = 6$	$Z = 6$
Cleavage:	{10$\bar{1}$1}	{10$\bar{1}$1}	{10$\bar{1}$1}	{10$\bar{1}$1}	{10$\bar{1}$1}
Hardness:	3	4	4	4	$4–4\frac{1}{2}$
Specific gravity:	2.711 (calc.)	3.009 (calc.)	3.96 (pure)	3.69 (calc.)	4.43 (pure)
Optical properties:	uniaxial ($-$)	uniaxial ($-$)	uniaxial ($-$)	uniaxial ($-$)	uniaxial ($-$)
refractive indices: ω	1.658	1.700	1.875	1.816	1.850
ϵ	1.468	1.509	1.633	1.597	1.625

Calcite, $CaCO_3$

See also the preceding tabulated data.

Habit: Crystals common; extremely diverse in development, prismatic along c, tabular on {0001}, rhombohedral {01$\bar{1}$2}, {02$\bar{2}$1}, {40$\bar{4}$1}, scalenohedral {21$\bar{3}$1} (Figures 12-2, 12-3, and 2-51), parallel and other aggregates (Figure 12-4); massive, coarse-to-very-fine granular; stalactitic (Figure 12-5), nodular, tuberose, and coralloidal shapes; oolitic or pisolitic, typically concentrically banded and internally radiating; also as paramorphs after aragonite.

Twinning: Two twin laws are common on calcite: (1) Twin plane and composition plane {0001} (see Figures 12-2e, 12-6, and 2-77a); re-entrant angles are commonly present about the equator of the twinned crystal, except those bounded by {10$\bar{1}$0}. (2) Twin plane and composition plane {01$\bar{1}$2}, as simple contact twins (Figure 2-77); also lamellar (Figure 12-7), produced by pressure (Figure 2-73), either during metamorphism or artifically, and commonly

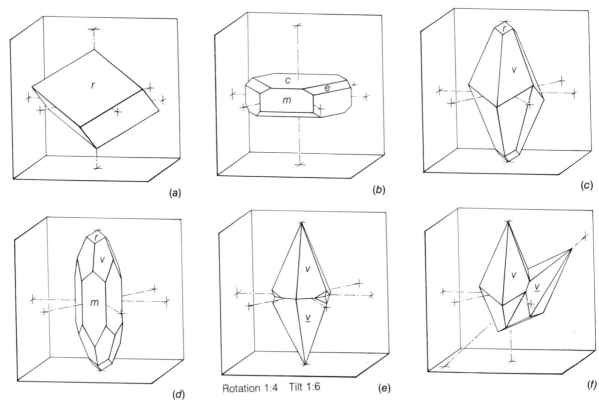

(a)   (b)   (c)

(d)

Rotation 1:4   Tilt 1:6   (e)   (f)

Figure 12-2. Calcite crystals. Forms: pinacoid c{0001}; hexagonal prism m{10$\bar{1}$0}; positive rhombohedron r{10$\bar{1}$1}; negative rhombohedron e{01$\bar{1}$2}; trigonal scalenohedron v{21$\bar{3}$1}. (e) Twinned on (0001) as twin plane and composition plane. (f) Twinned on ($\bar{2}$0$\bar{2}$1) as twin plane and composition plane.

manifested by visible striations on the {10$\bar{1}$1} cleavage planes of grains in marble (Figure 12-7b). Less common twin crystals have the twin plane {10$\bar{1}$1} (Figure 12-8) or {02$\bar{2}$1} (Figure 12-2f).

Common forms and angles:

(0001) $\wedge$ (10$\bar{1}$1) = 44°38'   (01$\bar{1}$0) $\wedge$ (01$\bar{1}$2) = 63°44'
(10$\bar{1}$1) $\wedge$ ($\bar{1}$101) = 74°57'   (21$\bar{3}$1) $\wedge$ (3$\bar{1}$$\bar{2}$1) = 35°35'
(40$\bar{4}$1) $\wedge$ ($\bar{4}$401) = 114°10'   (0001) $\wedge$ (02$\bar{2}$1) = 63°08'

Specific gravity: 2.71 (pure), higher with substitution of Fe, Mn, or Zn for Ca.

Color and streak: Colorless and transparent or white when pure; differs widely with substitution or with mechanically included material: yellow, brown, pink, blue, lavender, greenish, gray, black, greenish due to chlorite, reddish brown due to hematite. White to grayish.

Luster and light transmission: Vitreous. Transparent to translucent.

Chemistry: Calcite is the stable form of $CaCO_3$ at most temperatures and pressures. Calcite may be pure $CaCO_3$, or it may contain other metals (e.g., Fe, Mg, and/or Mn) in substitution for Ca. In general, only minor substitution of Mg occurs in calcite, but the tendency to form the ordered phase dolomite is very great, in keeping with the large difference in ionic radii between Ca and Mg. Fe^{+2} substitutes for calcium to a somewhat greater extent; the ordered phase ankerite is less common. Mn^{+2} substitutes for Ca extensively; nonetheless, the ordered phase kutnahorite has been identified at several localities. [At Franklin, New Jersey, kutnahorite, calcian rhodochrosite, and manganoan calcite have all been identified, and solid solution extends from pure $CaCO_3$ to 40 weight percent $MnCO_3$, and from pure $MnCO_3$ to about 25 weight percent $CaCO_3$.] Zn and Co have also been found substituting for Ca to a minor extent.

Diagnostic features: Colorless calcite exhibits excellent double refraction. All calcite effervesces freely in cold dilute hydrochloric acid. It is also characterized by its hardness (3), perfect rhombohedral cleavage with angles between adjacent planes of 75° or 105°, and its vitreous luster. It is distinguished from dolomite by the ready effervescence of fragments (as well as powder) in acid, as just noted, and by the direction of twin striations (Figures 12-7a and 12-10); it is distinguished from aragonite by its cleavage and lower specific gravity. Some calcite is fluorescent; some is triboluminescent.

Occurrence, associations, and uses: Calcite is a very common and widely distributed mineral in the earth's crust. It is an important rock-forming mineral in sedimentary and metamorphosed sedimentary rocks. It occurs in nearly pure form in great thicknesses of chalk and limestone and as a cementing material in other sedimentary rocks. Marble, sometimes termed *crystalline limestone*, can range from almost pure calcite, as in the pure white statuary marble, to diverse assemblages of minerals including calcite, diopside, tremolite, and many others, depending on the composition of the original limestone, the conditions of metamorphism, and the materials added from adjacent rock masses or igneous sources. Calcite is a

Figure 12-3. Two crystals of honey-colored calcite in random intergrowth, with forms $r\{10\bar{1}1\}$ and $v\{21\bar{3}1\}$, Joplin, Missouri, 6 cm across. (Courtesy of Queen's University.)

Figure 12-4. Calcite; crystal mass, ~ 7 cm in greatest dimension, from a cave in Virginia. (Courtesy of R. J. Holden.)

334

Figure 12-5. Calcite; stalactitic group, ~ 7 cm in greatest dimension, from a cave in Virginia. (Courtesy of R. J. Holden.)

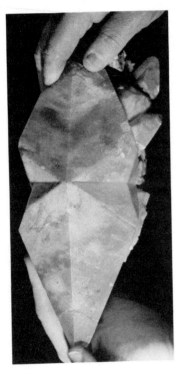

Figure 12-6. Calcite twinned on (0001) as both twin plane and composition plane (see Figure 12-2e). Specimen from Lone Jack Quarry, Rockbridge County, Virginia. (Courtesy of R. J. Holden.)

common constituent of altered basic igneous rocks, where it has developed by alteration of calcium silicates. It also has been formed as a late hydrothermal deposit in veins, both of meteoric and juvenile origin; in these deposits, it is typically associated with sulfides, quartz, barite, fluorite, dolomite, and siderite. Crystals of calcite occur lining vugs and fractures in limestone. Calcite stalactites and stalagmites occur in larger cavities, such as caves and solution channels. Calcite is deposited as travertine, tufa, or calc-sinter from springs and streams. Calcite of optical quality has been obtained from a cavernous zone in a basalt in Iceland (hence, the common name *Iceland spar* for optical material), as well as from New Mexico and elsewhere. In the United States, exceptional crystals have been found in, for example, the Michigan copper deposits and in the low-temperature lead-zinc veins of the central states.

Calcite, in its wide variety of occurrences, has many uses. The clear transparent material has long been used for optical purposes, particularly in the construction of polarizing prisms. From very early times, calcite or limestone has been burned to quicklime (CaO), slaked to hydrated lime Ca(OH)$_2$, and mixed with sand to make mortar. It also finds wide use in chemical industries and as a fertilizer. Portland cement, now widely used in concrete for building purposes, is made by burning a finely ground mixture of 75 percent CaCO$_3$ and 25 percent clayey minerals. Limestone is the raw material for some rock-wool insulating material. Limestone and lime are used in metallurgical processes for smelting both iron and nonferrous metals; they are most used as a flux to help remove impurities as slag. Limestones and marbles of many types are widely used in building: as dimension stone for actual construction; as a facing or veneer on concrete walls; as interior ornamental finish stone; and for decorative carvings and statuary. Crushed limestone is widely used as the coarse aggregate in concrete and as railroad ballast, and it is mixed with asphalt for road surfacing.

Magnesite, MgCO$_3$

See also the data given in the introduction to this group of minerals.

Habit: Crystals uncommon; rhombohedral $\{10\bar{1}1\}$ or prismatic along c; commonly massive, coarse-to-fine granular, or compact, earthy to chalky, or in lamellar or coarsely fibrous aggregates.

Specific gravity: 3.009 (pure), higher with substitution of Fe for Mg.

Color and streak: Colorless, white, grayish white, yellowish to brown. White.

Luster and light transmission: Vitreous. Transparent to translucent.

Chemistry: In MgCO$_3$, Fe^{+2} can substitute for Mg and a complete series extends to siderite. Mn and Ca substitute for Mg to a small extent.

Diagnostic features: Scarcely affected by cold hydrochloric acid, but dissolves with effervescence in hot hydrochloric acid. The compact massive forms of magnesite may resemble chert in appearance, but they are inferior in hardness; this hardness difference, however, is often difficult to discern because much magnesite is intimately intergrown with ultrafine-grained silica.

Occurrence: Magnesite, much less common than calcite, occurs only rarely as a sedimentary rock. It has also been formed by alteration of rocks that consist largely of magnesium silicates (e.g., serpentine, olivine, and pyroxene) by carbonated waters; magnesite of this origin is typically cryptocrystalline. In addition, it occurs as crystalline stratiform beds of metamorphic origin, commonly associated with talc-chlorite and/or mica schists and as a replacement of calcite-rich rocks by magnesium-bearing solutions.

Most of the replacement deposits are distinctly crystalline and occur associated with, or in, dolostone, which appears to represent an intermediate product of the replacement process. Extensive deposits of this type are mined in, for example, the state of Washington and Quebec. Veins and masses resulting from alteration of magnesian silicates are mined extensively in, for example, California. Magnesite is calcined for the manufacture of refractory bricks, cements, and flooring. It is also used for making magnesium metal.

(a)

(b)

Figure 12-7. Lamellar twinning in calcite. (*a*) Twin plane (01$\bar{1}$2) gives striations parallel to the long diagonal of the rhombic cleavage plane. (*b*) Lamellar twinning on (01$\bar{1}$2) manifested as the horizontal light lines on a salmon-pink calcite from Frontenac County, Ontario, 8 cm across. (Photograph courtesy of Queen's University.)

Siderite, $FeCO_3$

See also the data given in the introduction to this group of minerals.

Habit: Crystals common; typically rhombohedral {10$\bar{1}$1}, less commonly {01$\bar{1}$2}, {02$\bar{2}$1}, {40$\bar{4}$1}; also thin-to-thick tabular on {0001}; most crystal faces are curved or composite; massive, coarse-to-fine granular, botryoidal or globular, oolitic, earthy, or stony.

Common forms and angles:
(10$\bar{1}$1) $\wedge$ ($\bar{1}$101) = 73°00'   (0001) $\wedge$ (05$\bar{5}$1) = 78°03'

Specific gravity: 3.96, less with substitution of Mn^{+2}, Mg, or Ca for Fe^{+2}.

Color and streak: Yellowish brown, and grayish brown, to brown and reddish brown, also gray, yellowish gray, and greenish gray. (The dark colors are due to partial oxidation of Fe^{+2}.) White.

Luster and light transmission: Vitreous. Translucent.

Chemistry: Mg may substitute for Fe^{+2} to form a complete series to magnesite; Mn^{+2} may also substitute for Fe^{+2}. Ca may substitute for Fe^{+2} to a minor extent, but the difference in radius between these ions precludes extended solid solution and the ordered phase ankerite forms instead.

Diagnostic features: Becomes magnetic upon heating; slowly soluble with effervescence in cold HCl and rapidly so in hot HCl. Siderite

336

Figure 12-8. Calcite. Twin crystal on {10$\bar{1}$1} as twin plane, with prism m{10$\bar{1}$0} and negative rhombohedron e{01$\bar{1}$2}. Alternate faces of {10$\bar{1}$0} are horizontally striated, due to narrow {h0$\bar{h}$l} faces. (Sudbury, Ontario.) (Courtesy of Queen's University.)

can be distinguished from other carbonates by its color, specific gravity, and its alteration to iron oxides; it is distinguished from sphalerite by its rhombohedral cleavage.

Occurrence, associations, and uses: Siderite is widespread as a bedded deposit in sedimentary rock sequences consisting largely of shale and/or coal seams. It is typically massive and fine grained, and, in some places, concretionary; impure siderite-rich rocks include clay-ironstones and bituminous or black-band ores. Such deposits have been used as iron ores in Poland (Radom), in England (Somerset, Durham, Yorkshire), and in Pennsylvania. Metamorphosed rocks that had these rocks as their precursors occur in Austria (Erzberg, Hüttenberg) and Ontario (Michipicoten).

Siderite is present in many hydrothermal veins as a gangue mineral of primary origin—for example, in the silver-lead veins of the Coeur D'Alene district, Idaho, and the Slocan district, British Columbia. It is the predominant mineral in some veins. Siderite also occurs as a hydrothermal replacement of limestone by iron-bearing solutions. Deposits of this type constitute important iron ores at Bilbao, Spain, and in Algeria. Siderite also occurs rarely in pegmatites, as in Greenland with cryolite.

Siderite alters readily to hematite or limonite. Important residual bodies of these minerals that have been formed in this way occur, for example, in Algeria.

Rhodochrosite, $MnCO_3$

See also the data given in the introduction to this group of minerals.

Habit: Distinct crystals uncommon; rhombohedral {10$\bar{1}$1}; most are rounded or composite; massive, coarsely granular to compact; columnar, incrusting; also globular and botryoidal.

Color and streak: Various shades of pink, rose, and rose-red, also yellowish gray to brown. White.

Luster and light transmission: Vitreous. Translucent.

Chemistry: Much rhodochrosite contains small amounts of Fe^{+2} substituting for Mn^{+2}. A complete series may extend to siderite. Ca commonly substitutes for Mn^{+2} and a series extends to about 25 weight percent $CaCO_3$. With larger amounts of $CaCO_3$, the ordered phase kutnahorite, with the dolomite structure, is the stable mineral.

Diagnostic features: Soluble in warm hydrochloric acid with effervescence; turns black but remains nonmagnetic when heated. It is distinguished by its pink color, rhombohedral cleavage, and hardness of 4. The pink manganese silicate, rhodonite, has a hardness of 6.

Occurrence, associations, and uses: Rhodochrosite occurs as a primary gangue mineral in many apparently low-to-moderate-temperature hydrothermal veins. It occurs with silver, lead, zinc, and copper sulfide ores at Butte, Montana, and at other localities in the western United States and in Europe; it is associated with calcite, siderite, dolomite, fluorite, barite, quartz, manganite, alabandite (MnS), tetrahedrite, and sphalerite. Rhodocrosite also occurs with rhodonite, garnet, hausmannite, and other minerals in some high-temperature metasomatic deposits and as a secondary mineral in many residual or sedimentary manganese oxide deposits.

A substantial proportion of the high-grade manganese ore that has been produced in the United States was rhodochrosite obtained from veins in the Butte district of Montana.

Smithsonite, $ZnCO_3$

See also the data given in the introduction to this group of minerals.

Habit: Distinct crystals uncommon; rhombohedral $\{10\bar{1}1\}$ or $\{02\bar{2}1\}$; faces curved and rough, or composite; typically botryoidal, reniform, or stalactitic; as crystalline incrustations; coarsely granular to compact; earthy, or as porous or cavernous masses.

Color and streak: Grayish white to dark gray, greenish brownish white, green to apple-green, bluish green, blue, yellow, brown, and white. White.

Luster and light transmission: Vitreous. Translucent.

Chemistry: Fe^{+2}, Ca, Co, Cu, Mn, Cd, and Mg may substitute for Zn, but they are rarely present in excess of a few percent. Smithsonite is isostructural with calcite, and thus the metal ion, Zn, is in octahedral coordination, which is unusual for zinc since it normally occurs in tetrahedral coordination with oxygen (hemimorphite, willemite, zincite) or with sulfur (sphalerite, wurtzite).

Diagnostic features: Distinguished by its effervescence in warm hydrochloric acid; distinguished from other carbonates by its hardness of 4 and its high specific gravity.

Occurrence, associations, and uses: Most smithsonite is secondary and occurs in the oxidized zone of ore deposits, apparently derived by alteration from primary zinc minerals. It also occurs replacing calcareous rocks adjacent to such ore deposits. It is typically associated with hemimorphite and secondary lead and copper minerals such as cerussite, malachite, anglesite, and pyromorphite.

Although smithsonite occurs in the surficial oxidized zone of many lead-zinc deposits and has provided some of the zinc produced from those deposits, it has only rarely served as an important ore of zinc as, for example, in the mining districts of northern Mexico.

DOLOMITE GROUP

The dolomite structure results from the regular substitution of alternate atoms of Ca in calcite by another divalent metal, generally Mg, Fe, or

Mn. An ordered structure results because of the difference in radii between Ca and the other metals. Dolomite [CaMg(CO$_3$)$_2$] is of common occurrence, consistent with the large difference in radii between Ca and Mg and the limited solid solution between calcite and magnesite. Ankerite [Ca(Fe,Mg,Mn)(CO$_3$)$_2$] is less common, while kutnahorite [Ca(Mn,Mg,Fe)(CO$_3$)$_2$] is rare. A fourth member of the group is norsethite [BaMg(CO$_3$)$_2$]. Also, nordenskiöldine [CaSn(BO$_3$)$_2$] is isostructural with dolomite. The inequivalence of Ca with the second metal in these structures results in a lower symmetry for this structure than for the calcite structure.

Crystal system, class, and space group: Hexagonal (trigonal); $\bar{3}$; $R\bar{3}$.
Cell content: Z = 1 for simplest rhombohedral cell, Z = 3 for hexagonal unit cell.

	Dolomite	Ankerite	Kutnahorite
Cell dimensions:	a = 4.842	a = 4.832*	a = 4.85
	c = 15.95	c = 16.14	c = 16.34
Cell content:	Z = 3	Z = 3	Z = 3
Cleavage:	{10$\bar{1}$1}	{10$\bar{1}$1}	{10$\bar{1}$1}
Hardness:	3$\frac{1}{2}$–4	3$\frac{1}{2}$–4	3$\frac{1}{2}$–4
Specific gravity:	2.85	3.01	3.12
Optical properties:	uniaxial (−)	uniaxial (−)	uniaxial (−)
refractive indices: ω	1.679	1.690–1.750	1.727
ϵ	1.500	1.510–1.548	1.535

* Dimensions for mineral with Fe : Mg = 1 : 1.1.

Dolomite, CaMg(CO$_3$)$_2$

See also the preceding tabulated data.

Habit: Crystals common; typically rhombohedral {10$\bar{1}$1} or {40$\bar{4}$1}, also prismatic with {11$\bar{2}$0} or tabular on {0001}; {10$\bar{1}$1} commonly striated horizontally, also curved or made up of subparallel individuals that merge into saddle-shaped forms (Figure 12-9); massive, coarse-to-fine granular, columnar; as a rock-forming mineral typically compact (massive like ordinary limestone) or granular.

Twinning: Common on {0001}, also on {10$\bar{1}$0} and {11$\bar{2}$0}; lamellar twinning on {02$\bar{2}$1} (Figure 12-10), common in grains of dolomite in marble.

Common forms and angles:
  {10$\bar{1}$1} $\wedge$ {$\bar{1}$101} = 73°16'   {0001} $\wedge$ {40$\bar{4}$1} = 75°16'

Specific gravity: 2.85 for pure dolomite, about 3.02 for a mineral with Mg:Fe = 1:1.

Color and streak: Colorless, white, gray, or greenish; yellowish brown or brown with high Fe^{+2} content; also pink or rose; ferroan dolomite

Figure 12-9. Saddle-shaped groups of dolomite crystals, Florence Mine, Beckermet, Cumbria, United Kingdom. Largest group is ~2 cm in greatest dimension. (Photograph courtesy of S. C. Chamberlain.)

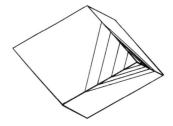

Rotation 1:3   Tilt 1:3

Figure 12-10. Lamellar twinning in dolomite, with $(02\bar{2}1)$ as twin plane; striations appear parallel to short diagonal of the rhombic cleavage plane. Compare this diagram with Figure 12-7a.

and ankerite turn dark brown or reddish on weathering, as does siderite. White.

Luster and light transmission: Vitreous to pearly. Transparent to translucent.

Chemistry: Fe^{+2} commonly substitutes for Mg in dolomite, and a complete series probably extends to ankerite. Mn, on the other hand, occurs to the extent of a few percent and, in nearly all cases, it occurs along with iron. In addition, Co and Zn may also substitute for Mg, but only to a minor extent.

Diagnostic features: Some dolomite is triboluminescent. Dolomite is only slightly attacked by cold, dilute hydrochloric acid, in contrast to calcite, but it does dissolve readily in warm acids. In coarsely crystalline granular material, the direction of twinning lamellae distinguishes dolomite from calcite. There are also stain tests (see, for example, *American Geological Institute Data Sheet,* new series #52).

Occurrence, associations, and uses: Dolomite is of widespread occurrence in sedimentary strata. Although dolomite, or protodolomite, can be formed by direct precipitation from seawater, most dolomite is believed to have originated through transformation (dolomitization) of sedimentary $CaCO_3$ by magnesium-bearing solutions. Most dolomite rocks are mixtures of dolomite and calcite. These dolostones can be recrystallized by metamorphism, and siliceous dolostones can develop diopside, tremolite, and other silicates in this process. Dolomite also occurs in hydrothermal veins along with fluorite, barite, calcite, siderite, quartz, and metallic ore minerals; in cavities and veins in limestone or dolomitic rocks; and as veins or crystals in serpentine, talcose rocks, and altered basic igneous

340

cf: Slake.
Combine (lime) with water
to produce Calcium Hydroxide

rocks. It occurs as a common gangue mineral in the lead-zinc deposits of Missouri and adjoining areas; in sulfide veins of many mining districts of the western United States; in the metalliferous veins of Cornwall, England; as a gangue mineral in gold-quartz veins; and as a replacement mineral in the adjoining rocks in the Mother Lode district of California and in the Porcupine and Larder Lake gold camps of Ontario and Quebec. In these last occurrences, it is associated with bright green chromian muscovite (or sericite). Rock-forming dolomite is typically low in iron. Iron-rich dolomite is common with iron ores.

Dolomite is used as a source of magnesium or calcium metal, of magnesia for refractory bricks or in blast furnace fluxes, and as dolostone for a flux in blast furnaces. The slag produced from dolostone does not slake; consequently, it can be used for such things as light-weight aggregate, whereas slag from limestone cannot be used for such purposes.

ARAGONITE GROUP

As already noted, aragonite is polymorphous with calcite, but other members of this group—witherite, strontianite, and cerussite—do not have polymorphous forms with the calcite structure that occur as minerals. The aragonite structure is stable for calcium and the larger metal ions, whereas the calcite structure is stable for calcium and smaller ions. A parallel is found in the nitrates, wherein nitratite ($NaNO_3$) is isostructural with calcite, and niter (KNO_3) is isostructural with aragonite.

The aragonite structure includes planar triangular CO_3 groups parallel to {001}. The calcium atoms lie approximately in the positions of hexagonal closest packing; this arrangement gives aragonite a pseudohexagonal character that is responsible for its great tendency to form twin crystals (Figure 2-69). Each CO_3 group lies between six Ca atoms arranged so that each O is linked to three Ca atoms with the calcium in

Figure 12-11.* Aragonite crystals. Forms: pinacoid $b\{010\}$; rhombic prisms $m\{110\}$, $k\{011\}$, $\lambda\{091\}$; rhombic dipyramid $\sigma\{991\}$. (b) Twinned on {110} as twin plane and composition plane.

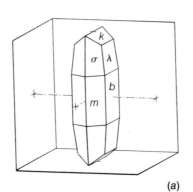

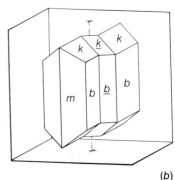

(a)

(b)

ninefold coordination with oxygen. [In calcite, each O is linked to two Ca atoms with the calcium in octahedral (sixfold) coordination with oxygen.]

Cyrstal system, class; and space group: Orthorhombic; *2/m2/m2/m; Pmcn.*

	Aragonite	Witherite	Strontianite	Cerussite
Cell dimensions:	$a = 4.959$	$a = 5.313$	$a = 5.107$	$a = 5.195$
	$b = 7.968$	$b = 8.904$	$b = 8.414$	$b = 8.436$
	$c = 5.741$	$c = 6.430$	$c = 6.029$	$c = 6.152$
Cell content:	$Z = 4$	$Z = 4$	$Z = 4$	$Z = 4$
Hardness:	$3\frac{1}{2}$–4	3–$3\frac{1}{2}$	$3\frac{1}{2}$	3–$3\frac{1}{2}$
Specific gravity:	2.930 (calc.)	4.308 (calc.)	3.785 (calc.)	6.582 (calc.)
Optical properties:	biaxial (−)	biaxial (−)	biaxial (−)	biaxial (−)
~2V	18°	16°	7°	9°
refractive indices: α	1.530	1.529	1.516	1.803
β	1.680	1.676	1.644	2.074
γ	1.685	1.677	1.666	2.076

Aragonite, $CaCO_3$

See also the preceding tabulated data.

Habit: Crystals common; most are twinned; short-to-long prismatic along c, also acicular or chisel-shaped with a steep prism {091} and pyramids {991} (Figure 12-11a); pseudo-hexagonal symmetry may be marked; also as columnar aggregates or crusts; as radiating groups of acicular crystals; in coralloidal, reniform, or globular shapes, pisolitic with a radially fibrous and concentrically zoned structure; stalactitic.

Twinning: Twin plane {110}, very common, generally repeated (Figure 12-11b), giving pseudo-hexagonal aggregates of both contact and penetration types (Figure 12-12); also as thin polysynthetic lamellae producing fine striations on {001} or parallel to c.

Cleavage and fracture: {010} distinct. Subconchoidal.

Tenacity: Brittle.

Color and streak: Colorless to white; also gray, yellowish, blue, green, pale-to-deep violet, or rose-red. White.

Luster and light transmission: Vitreous. Transparent to translucent.

Chemistry: Aragonite is polymorphous with calcite, which is the stable form, and it may change spontaneously into calcite. Aragonite will begin to invert to calcite upon heating to 400°C in dry air or at lower temperatures in contact with water. Sr and Pb are found to substitute for Ca to the extent of a few percent. Aragonite is formed under a much narrower range of conditions than calcite and is much less

Figure 12-12. Aragonite. Pseudo-hexagonal twinned crystal due to cyclic interpenetration twinning on {110} as twin plane, Molina de Aragon, Spain, ~ 4 cm across. (Photograph courtesy of S. C. Chamberlain.)

widespread, being found chiefly in low-temperature, near-surface deposits. There are no geologically old aragonites.

Diagnostic features: Specific gravity; soluble with effervescence in cold dilute hydrochloric acid. May be distinguished from calcite by cleavage and higher specific gravity; most columnar aragonite has cleavage parallel to elongation, whereas calcite shows rhombohedral cleavage transecting the columnar crystals. Some aragonite exhibits triboluminescence.

Occurrence, associations, and uses: The principal modes of occurrence are: (1) as crystals, pisolites, or sinter deposits from hot springs and geysers; (2) as disseminated crystals or masses in gypsum or clay; (3) with limonite, calcite, malachite, smithsonite, and other secondary minerals in oxidized zones of ore deposits; (4) in veins and cavities—along with calcite, dolomite, and diverse magnesium minerals—in serpentine, altered basic igneous rocks, and basalt; (5) with celestite and sulfur, as in Sicily; (6) in the hard parts of fossil or living organisms, such as oysters, and in pearls; (7) as stalactites and other speleothems in caves; and (8) constituting marble, apparently formed under high pressure, that occurs with glaucophane schist.

Witherite, BaCO$_3$

See also the data given in the introduction to this group of minerals.

Habit: Crystals uncommon; typically repeatedly twinned on {110} giving pseudo-hexagonal dipyramids; also short prisms parallel to c; faces are typically rough and horizontally striated; also in globular, tuberose, and botryoidal forms; columnar, granular, and coarsely fibrous.

Twinning: Universal on {110}.

Cleavage: {010} distinct.

Color and streak: Colorless to milky, white, or grayish; also weakly tinted yellow, brown, or green. White.

Luster and light transmission: Vitreous. Transparent to translucent.

Chemistry: In nature, witherite ranges only slightly in composition, although a complete series with SrCO$_3$ does exist in artificial materials.

Diagnostic features: High specific gravity. Soluble with effervescence in dilute hydrochloric acid; the resulting solution, even when very dilute, gives a precipitate upon the addition of sulfuric acid.

Occurrence, associations, and uses: Witherite is not a common mineral, although, as a barium mineral, it is next in importance to barite (BaSO$_4$). Much of it occurs in low-temperature hydrothermal veins with fluorite, barite, and/or galena. It also has been found with sanbornite (BaSi$_2$O$_5$), gillespite (BaFeSi$_4$O$_{10}$), and other Ba silicates in several deposits in the western United States. It is of minor importance commercially as a source for barium metal and hence for other industrially produced barium compounds.

Strontianite, $SrCO_3$

See also the data given in the introduction to this group of minerals.

Habit: Crystals uncommon; short or long prismatic parallel to c, commonly pseudo-hexagonal due to equal development of {110} and {010}, with {111} and {021}; also massive, columnar to fibrous, granular, and in rounded masses.

Twinning: Very common on {110} as contact twins, typically repeated.

Cleavage: {110} nearly perfect.

Specific gravity: 3.785 (calc. for pure $SrCO_3$), 3.72 (3.3 percent CaO), and less with higher Ca contents.

Color and streak: White to gray; yellowish, or greenish. White.

Luster and light transmission: Vitreous. Translucent.

Chemistry: Nearly all strontianite includes some Ca in substitution for Sr.

Diagnostic features: Dissolves with effervescence in dilute hydrochloric acid; even a moderately dilute resulting solution gives a precipitate of strontium sulfate on addition of sulfuric acid. On intense heating in a flame, it swells up, throws out fine sprouts, and colors the flame crimson (strontium). Characterized by its high specific gravity and its effervescence in acid, it is distinguished from witherite and aragonite by its flame color and from celestite by its poor cleavage and solubility in acid.

Occurrence, associations, and uses: Strontianite is typically a low-temperature hydrothermal mineral, commonly associated with barite, celestite, and calcite in veins in limestone and less commonly with metallic minerals. It also occurs as concretionary masses in limestone or clay and with celestite in the cap rock of some of the salt domes in Texas and Louisiana.

Strontianite is less common than celestite, but it is preferred as an ore of strontium. Commercial quantities have been obtained in the vicinity of Münster, Germany, where it occurs as small veins in marl deposits. Large deposits occur in limestone in the Strontium Hills, San Bernadino County, California.

Cerussite, $PbCO_3$

See also the data given in the introduction to this group of minerals.

Habit: Crystals common; tabular on {010} or dipyramidal and pseudo-hexagonal (Figure 12-13); stellate and reticular aggregates due to

344

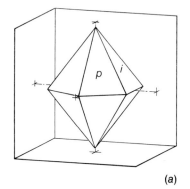
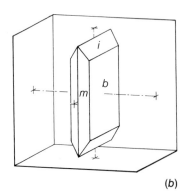
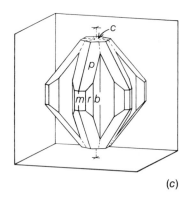

(a)

(b)

(c)

Figure 12-13. Cerussite crystals. Forms: pinacoids $c\{001\}$, $b\{010\}$; rhombic prisms $m\{110\}$, $r\{130\}$, $i\{021\}$; rhombic dipyramid $p\{111\}$. (c) Cyclic twin on $\{110\}$ as twin plane.

twinning (Figure 12-14); also massive, granular to dense and compact.

Twinning: Almost universal; most common on $\{110\}$, repeated contact types producing stellate aggregates (Figure 12-13c) or as twin lamellae; also on $\{130\}$.

Cleavage and fracture: $\{110\}$, $\{021\}$ distinct. Conchoidal.

Color and streak: Colorless to white and gray or smoky; darker colors due to impurities. White.

Luster and light transmission: Adamantine. Translucent.

Diagnostic features: Soluble in warm dilute nitric acid with effervescence. It can be distinguished by its high specific gravity, white color, brittleness, and adamantine luster; the effervescence in

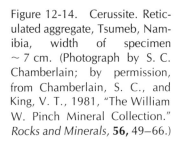

Figure 12-14. Cerussite. Reticulated aggregate, Tsumeb, Namibia, width of specimen ~ 7 cm. (Photograph by S. C. Chamberlain; by permission, from Chamberlain, S. C., and King, V. T., 1981, "The William W. Pinch Mineral Collection." *Rocks and Minerals,* **56,** 49–66.)

HNO$_3$ distinguishes it from anglesite. It decrepitates upon heating.

Occurrence, associations, and uses: Cerussite is a common mineral in the upper oxidized zone of ore deposits that contain galena. It occurs as crystalline crusts or dense masses, concentrically banded, commonly with a core of unaltered galena. Anglesite, PbSO$_4$, may have formed as a first step in the oxidation of, and later alteration to, cerussite. Most cerussite is associated with anglesite, limonite, malachite, smithsonite, and other secondary minerals. Fine crystalline material, typically twinned or in reticulate aggregates, occurs in many lead-mining districts, for example, at Tsumeb, Namibia, and in the southwestern United States.

Malachite, Cu$_2$(CO$_3$)(OH)$_2$

Crystal system, class, and space group: Monoclinic; 2/m; P2$_1$/a.

Cell dimensions and content: a = 9.51, b = 12.02, c = 3.25; β = 98° 42'; Z = 4.

Habit: Distinct crystals rare; characteristically prismatic to fine acicular, parallel to c, and grouped in tufts or rosettes; commonly massive or incrusting, with mammillary surface, botryoidal (Figure 12-15), or tuberose; internally divergent and fibrous and color-banded; also as pseudomorphs after, for example, azurite and cuprite.

Twinning: Twin plane {100} very common.

Cleavage and fracture: {$\overline{2}$01} perfect, {010} distinct. Uneven on massive material.

Hardness: 3$\frac{1}{2}$–4.

Specific gravity: 4.05 (down to 3.6 for massive material).

Color and streak: Bright green. Pale green.

Luster and light transmission: Silky, velvety, or dull. Translucent.

Optical properties:

anisotropic, biaxial (−): 2V ≅ 43°

refractive indices: α, 1.655; β, 1.875; γ, 1.909

pleochroic

Diagnostic features: Can be distinguished by its green color, typically banded, its effervescence in cold dilute HCl, and its common botryoidal habit.

Occurrence, associations, and uses: Malachite is widespread as a secondary copper mineral, commonly associated with the less common mineral azurite. It may occur as pseudomorphs after azurite or cuprite. It occurs in the upper oxidized zones of copper deposits, especially in regions where limestone is present; it is commonly associated with limonite, calcite, chalcedony, and chrysocolla, and less commonly with other secondary copper, lead, and zinc minerals.

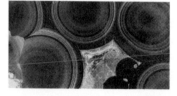

Figure 12-15. Malachite. Section transecting botryoidal material and exhibiting light and dark green, concentric banding. Width of field is 5 cm. (Courtesy of Queen's University.)

It has found wide use for decorative purposes and costume jewelry. Important amounts have been recovered at Mednorudiansk, near the town of Nizhni Tagil, Siberia, and in copper-mining districts of Africa, Australia, and the southwestern United States.

Azurite, $Cu_3(CO_3)_2(OH)_2$

Crystal system, class, and space group: Monoclinic; $2/m$; $P2_1/c$.
Cell dimensions and content: $a = 4.97$, $b = 5.84$, $c = 10.29$; $\beta = 92° 25'$; $Z = 2$.
Habit: Crystals common; of diverse habits, for example, tabular on {001} or {102}, short prismatic along c or b (Figure 12-16); massive or stalactitic with a columnar or radial structure; also earthy.
Cleavage and fracture: {011} perfect, {100} distinct. Conchoidal.
Hardness: $3\frac{1}{2}$–4.
Specific gravity: 3.77.
Color and streak: Azure-blue to very dark blue in crystals, lighter in massive or earthy types. Light blue.
Luster and light transmission: Vitreous. Transparent.
Optical properties:
   anisotropic, biaxial (+); $2V \cong 68°$
   refractive indices: α, 1.730; β, 1.754–1.758; γ, 1.835–1.838
   pleochroic
Diagnostic features: Can be distinguished by its hardness, its blue color, its effervescence in cold dilute HCl, and its common occurrence with malachite.
Occurrence and associations: Azurite forms in the upper oxidized zone of copper deposits by reaction between carbonated waters and copper minerals or between copper sulfate solutions and limestone. Nearly all azurite is associated with malachite; it is also commonly associated with limonite, calcite, chalcocite, chrysocolla, copper oxides, and other secondary copper minerals. Although much azurite occurs interbanded with malachite in botryoidal material, it is common as distinct crystals implanted on malachite or other secondary minerals. Many azurite crystals have been altered to malachite pseudomorphs. Notable localities have been Chessy, France; Tsumeb, South West Africa; Broken Hill, New South Wales; and Bisbee, Arizona.

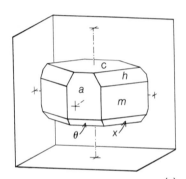

(a)

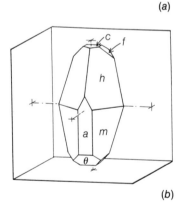

(b)

Figure 12-16.   Azurite crystals. Forms: pinacoids   c{001}, a{100}, $\theta\{\bar{1}02\}$; prisms $m\{110\}$, $f\{012\}$, $h\{111\}$, $x\{\bar{1}12\}$.

Nitratite, $NaNO_3$

Crystal system, class, and space group: Hexagonal (trigonal); $\bar{3}2/m$; $R\bar{3}c$.
Cell dimensions and content: $a = 5.0696$, $c = 16.829$; $Z = 6$.

Crystal structure: Nitratite is isostructural with calcite, but no substitution of either Ca or CO_3 occurs.

Habit: Crystals uncommon; rhombohedral $\{10\bar{1}1\}$; typically massive and granular or as incrustations.

Twinning: Common on $\{01\bar{1}2\}$, $\{0001\}$, or $\{02\bar{2}1\}$.

Cleavage: $\{10\bar{1}1\}$ perfect.

Hardness: $1\frac{1}{2}$–2.

Tenacity: Somewhat sectile.

Specific gravity: 2.25.

Color and streak: Colorless, white, or tinted brown or gray by impurities. White.

Luster and light transmission: Vitreous. Transparent.

Optical properties:
   anisotropic, uniaxial (−)
   refractive indices: ω, 1.5874; ϵ, 1.3361

Diagnostic features: Water soluble giving a cooling taste. Nitratite exhibits the same twin gliding as calcite (see Figures 2-73 and 2-74) but, in nitratite, because of weaker bonding, it can be induced by only the pressure of a finger.

Occurrence, associations, and uses: This water-soluble salt commonly occurs as a surface impregnation or efflorescence in arid areas. It is typically associated with gypsum, halite, and other nitrates and sulfates. The only significant deposits of nitratite occur in a narrow belt about 450 miles long in the deserts of northern Chile, along the virtually rainless eastern slopes of the coast range. These deposits, which consist of a near-surface layer from a few inches to a few feet in thickness, and which contain numerous other salts, have supplied important amounts of nitrate for the world's fertilizer demands. The nitrate rock also contains iodates that have provided the only mineral source of iodine. In recent years, artificial nitrates prepared by fixation of atmospheric nitrogen have cut seriously into the market for Chilean nitrate.

Name: In most former literature, this mineral is referred to as *soda niter*.

Borax, $Na_2B_4O_7 \cdot 10H_2O$

Crystal system, class, and space group: Monoclinic; $2/m$; $C2/c$.

Cell dimensions and content: $a = 11.858$, $b = 10.674$, $c = 12.197$; $\beta = 106° 35'$; $Z = 4$.

Habit: Crystals common; typically short prismatic parallel to c, somewhat tabular on $\{100\}$ (Figure 12-17).

Cleavage and fracture: $\{100\}$, $\{110\}$ perfect. Conchoidal.

Hardness: 2–$2\frac{1}{2}$.

Specific gravity: 1.71.

Color and streak: Colorless to white. White.

348

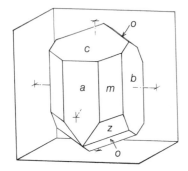

Figure 12-17. Borax crystal. Forms: pinacoids c{001}, b{010}, a{100}; prisms m{110}, o{$\bar{1}$12}, z{$\bar{1}$11}.

Luster and light transmission: Vitreous to dull [Borax, per se, is clear; the dull white material usually found is tincalconite ($Na_2B_4O_7 \cdot 5H_2O$).]. Translucent to opaque.

Optical properties:
anisotropic, biaxial (−): $2V \cong 40°$
refractive indices: α, 1.447; β, 1.469; γ, 1.472

Alteration: Commonly loses water and alters to tincalconite.

Diagnostic features: Characterized by crystal form, brittleness, and solubility in water.

Occurrence, associations, and uses: Borax is an evaporite mineral associated with halite, sulfates, carbonates, other borates, and muds in dried-up lakes and playas. Large deposits of borax and borate brines are worked as a source of borates in California. The best-known deposits are at Kramer and at Searles Lake. The latter is a pan-shaped desert basin, about 8 km by 16 km in area, that contains many substances besides borates. The Kramer deposit of borax, chiefly *kernite* ($Na_2B_4O_7 \cdot 4H_2O$) and other borate minerals, is the largest known reserve of boron compounds in the world.

Colemanite, $Ca_2B_6O_{11} \cdot 5H_2O$

Crystal system, class, and space group: Monoclinic; 2/m; $P2_1/a$.

Cell dimensions and content: a = 8.743, b = 11.264, c = 6.102; $\beta = 110° 07'$; Z = 2.

Habit: Crystals uncommon; typically equant, though rather diverse (Figure 12-18); common forms {110}, {001}, {100}, {$\bar{2}$01}, {$\bar{1}$21}, {011}, {021}, {111}, {$\bar{1}$11}, {$\bar{2}$21}; also massive, cleavable to granular or compact.

Cleavage and fracture: {010} perfect, {001} distinct. Uneven.

Hardness: $4\frac{1}{2}$.

Specific gravity: 2.42.

Color and streak: Colorless, white, yellowish, or gray. White.

Luster and light transmission: Vitreous to brilliant. Transparent to translucent.

Optical properties:
anisotropic, biaxial (+): $2V \cong 55°$
refractive indices: α, 1.586; β, 1.592; γ, 1.614

Diagnostic features: Characteristic crystal form, one perfect cleavage, and its exfoliation on heating.

Occurrence: Colemanite is present in many of the California borax deposits, where it occurs principally in geodes in the sedimentary rocks. It has apparently been formed by the action of meteoric waters on borax and other borates that were originally precipitated in a playa deposit and later buried by sedimentation. Prior to the discovery of the Kramer kernite deposit, colemanite was the principal source of the borates produced in California.

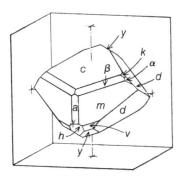

Figure 12-18. Colemanite crystal. Forms: pinacoids c{001}, a{100}, h{$\bar{2}$01}; prisms m{110}, k{011}, α{021}, β{111}, y{$\bar{1}$11}, v{$\bar{2}$21}, d{$\bar{1}$21}.

13

Class VI: Sulfates, Chromates, Molybdates, Tungstates

This group of minerals includes those anisodesmic salts with anionic groups of the general type $(XO_4)^{-n}$, in which X is a hexavalent ion in tetrahedral coordination with oxygen, yielding $(XO_4)^{-2}$. In the sulfates, chromates, and a few very rare selenates and tellurates, the anion groups are symmetric tetrahedra $(XO_4)^{-2}$.

The sulfates comprise a large number of minerals: some are of primary origin in veins; others form as evaporites; still others are of secondary origin in the oxidized zone of ore deposits. The sulfates can be subdivided into anhydrous, hydrated, and those containing hydroxyl or a halogen. Chromates, rare as minerals, include one well-known species, crocoite ($PbCrO_4$); they generally occur as secondary minerals in ore deposits. Chromates of potassium, however, occur with iodates in some nitrate deposits. Selenates have long been recognized as very rare secondary minerals; they occur in the oxidized zones of some ore deposits, particularly those that contain the rare copper selenide minerals. A few tellurates are also known.

The molybdates and tungstates are $A_m(XO_4)_n$ compounds in which the $(XO_4)^{-2}$ anionic group is a distorted tetrahedron, and X is Mo or W. The X atom in these structures is appreciably larger in radius than S, so there is no substantial substitution of S for Mo or W. Partial or complete series exist between some molybdates and tungstates.

Anhydrous Sulfates
        Type AXO_4
                Barite, $BaSO_4$
                Celestite, $SrSO_4$
                Anglesite, $PbSO_4$
                Anhydrite, $CaSO_4$
Hydrated Sulfates
        Type $AXO_4 \cdot xH_2O$
                Gypsum, $CaSO_4 \cdot 2H_2O$
                Chalcanthite, $CuSO_4 \cdot 5H_2O$
                Melanterite, $FeSO_4 \cdot 7H_2O$
                Epsomite, $MgSO_4 \cdot 7H_2O$
Anhydrous Sulfates Containing Hydroxyl
        Type $A_m(XO_4)_pZ_q$
                Brochantite, $Cu_4(SO_4)(OH)_6$
                Antlerite, $Cu_3(SO_4)(OH)_4$
        Type $A_2(XO_4)Z_q$
                Alunite, $KAl_3(SO_4)_2(OH)_6$
                Jarosite, $KFe_3(SO_4)_2(OH)_6$
Anhydrous Chromates
                Crocoite, $PbCrO_4$
Molybdates, Tungstates
        Type AXO_4
                Wolframite, $(Fe,Mn)WO_4$
                Scheelite, $CaWO_4$
                Wulfenite, $PbMoO_4$

ANHYDROUS SULFATES

The principal anhydrous sulfates that occur in nature are the members of the isostructural barite group—barite, celestite, and anglesite—plus the mineral anhydrite, which is not so isostructural with barite. In the barite structure, S is in tetrahedral coordination with oxygen, and the cations Ba, Sr, or Pb are in twelvefold coordination with oxygen. Anhydrite, with its calcium ion of relatively small ionic radius, has a different structure in which calcium is in eightfold coordination with oxygen.

Crystal system and class: Orthorhombic; $2/m2/m2/m$.

	Barite	Celestite	Anglesite	Anhydrite
Space group:	*Pnma*	*Pnma*	*Pnma*	*Cmcm*
Cell	$a = 8.87$	$a = 8.38$	$a = 8.47$	$a = 6.22$
dimensions:	$b = 5.45$	$b = 5.37$	$b = 5.39$	$b = 6.97$
	$c = 7.14$	$c = 6.85$	$c = 6.94$	$c = 6.96$
Cell content (z):	4	4	4	4

	Barite	Celestite	Anglesite	Anhydrite
Cleavage:	{001}, {210} perfect	{001} perfect, {210} good	{001} good, {210} distinct	{010} perfect, {001} nearly perfect, {100} good
Hardness:	$3–3\frac{1}{2}$	$3–3\frac{1}{2}$	$2\frac{1}{2}–3$	$3\frac{1}{2}$
Specific gravity:	4.48 (calc.)	3.971 (calc.)	6.323 (calc.)	2.963 (calc.)
Optical properties:	biaxial (+)	biaxial (+)	biaxial (+)	biaxial (+)
~2V	36–40	50	60–75	43
refractive indices: α	1.634–1.637	1.621–1.622	1.878	1.570
β	1.636–1.639	1.623–1.624	1.883	1.576
γ	1.646–1.649	1.630–1.633	1.895	1.614

Barite, $BaSO_4$

See also the preceding tabulated data.

Habit: Crystals common; tabular on {001}, also prismatic and elongated parallel to c, a, or b (Figures 13-1 and 13-2).

Common forms and angles:
    (001) $\wedge$ (101) = 38°51′   (210) $\wedge$ (2$\bar{1}$0) = 78°20′
    (001) $\wedge$ (011) = 52°42′   (001 $\wedge$ (211) = 64°18′

Color and streak: Colorless to white, also yellow, brown, reddish, gray, greenish, or blue: color may be distributed in growth zones. White.

Luster and light transmission: Vitreous. Transparent to translucent.

Chemistry: A complete solid-solution series probably exists between barite and celestite. In most mineral occurrences, however, the composition is close to one or the other end-member composition.

Diagnostic features: Its cleavage with angles of 90°, 78°20′, and 101°40′, its crystal form, its specific gravity, and its insolubility in acids. In addition, barite is commonly platy in divergent groups; whereas celestite tends to be fibrous and radiating.

Occurrence, associations, and uses: Barite, the most common barium mineral, occurs principally as a gangue mineral in hydrothermal metalliferous veins that have formed at moderate or low temperatures; in rare cases, it is the major constituent of a vein. It is generally associated with fluorite, calcite, siderite, dolomite, quartz, galena, manganite, and stibnite. It is widely distributed as veins, lenses, cavity fillings, or replacement deposits in limestones where it was apparently formed by either hypogene or groundwater solu-

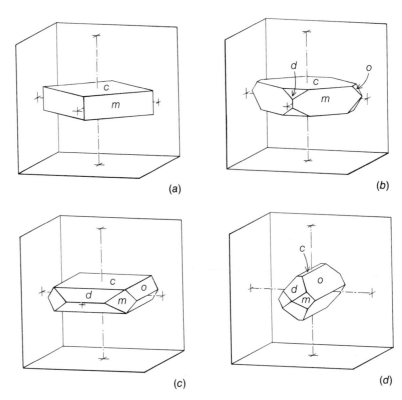

Figure 13-1. Barite crystals. Forms: pinacoids c{001}; rhombic prisms m{210}, o{011}, d{101}.

Figure 13-2. Barite crystals. Tabular on {001} with {210} (see Figure 13-1a). Crystals, along with the saddle-shaped mass of dolomite (left of center), are in a predominantly calcite-lined vug from Lone Jack Quarry, Rockbridge County, Virginia. Prism faces of individual crystals are up to 5 mm long.

tions. Because of its insolubility, it is relatively common in residual clay deposits formed by the weathering of limestones. It also occurs as concretions and "desert roses" in, for example, central Oklahoma.

Barite has many diverse uses, most of which relate to its being inert, white, and relatively heavy. The largest use in the United States is for drilling mud. Other uses include its inclusion in paint, coated papers, glazes and enamels, linoleum, rubber, and several plastics and resins.

Name: Barite is often spelled *baryte* in many places other than North America.

Celestite, SrSO$_4$

See also the data given in the introduction to this group of minerals.

Habit: Crystals common; typically tabular on {001}, also elongated along a, b, or c (Figure 13-3).

Common forms and angles:

$(001) \wedge (101) = 39°24'$ $(210) \wedge (2\bar{1}0) = 75°58'$

$(001) \wedge (011) = 52°04'$ $(001) \wedge (211) = 64°22'$

Color and streak: Colorless to pale blue, also white, reddish, or greenish. White.

Luster and light transmission: Vitreous. Transparent to translucent.

Chemistry: As noted, celestite is isostructural with barite and commonly contains small amounts of Ba substituting for Sr. Substantial substitution (up to 39.85 percent $BaSO_4$) has been reported for a celestite from vugs in Paleozoic limestone near Lansdowne in eastern Ontario.

Diagnostic features: Distinguished from barite by its lower specific gravity, by the fine fibrous and radiating character of some celestite, and by its characteristic (though not universal) pale blue color. It dissolves slowly in most concentrated acids.

Occurrence, associations, and uses: Celestite occurs chiefly in sedimentary stratiform deposits of gypsum, anhydrite, or halite, commonly in association with sulfur and in cavities and veins and as disseminated grains in limestone or dolostone, generally along with fluorite and/or gypsum. It may have been deposited directly from sea water or precipitated in favorable horizons by meteoric waters carrying strontium in solution. Celestite also occurs as a primary mineral in hydrothermal veins. In the United States and Canada, celestite is relatively common in cavities in Paleozoic limestones and dolostones in, for example, New York, Ohio, and Ontario. It also occurs in California with gypsum and halite in lakebed deposits at several localities in San Bernadino County.

　　Celestite is the principal source of strontium, although the less abundant strontianite is also used since it is less costly to convert to a useful strontium salt. Strontium salts find many minor uses, as in the beet-sugar industry, for fireworks, in electrical batteries, paints, rubber, and glass.

Name: Celestine has also been widely applied to this mineral.

Figure 13-3. Celestite crystals. Forms: pinacoid $c\{001\}$; rhombic prisms $m\{210\}$, $o\{011\}$, $l\{102\}$, $d\{101\}$; rhombic dipyramid $z\{211\}$.

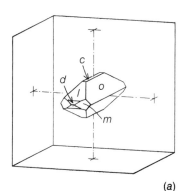

(a)

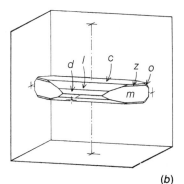

(b)

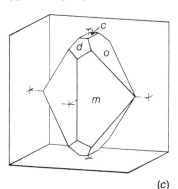

(c)

Anglesite, PbSO₄

See also the data given in the introduction to this group of minerals.

Habit: Crystals uncommon; generally well developed, tabular on {001}, or prismatic and elongate parallel to *c*, *a*, or *b* (Figure 13-4); commonly massive, granular to compact, nodular; also massive with concentric banding and enclosing a core of galena.

Fracture: Conchoidal.

Color and streak: Colorless to white, commonly tinged gray, yellow, or green. White.

Luster and light transmission: Adamantine. Transparent to opaque.

Diagnostic features: High specific gravity and adamantine luster.

Occurrence and associations: Anglesite is a secondary mineral found in the oxidized zone of ore deposits that contain galena. Usually formed by oxidation of galena, it is associated with cerussite and other secondary lead minerals, gypsum, and silver halides.

Anhydrite, CaSO₄

See also the data given in the introduction to this group of minerals.

Habit: Crystals uncommon; typically massive, fine granular, fibrous, and radiated or plumose.

Cleavage: The cleavages {010}, {100}, and {001} (though of distinctly different quality) simulate cubic cleavage in the massive material.

Color and streak: Colorless to bluish or violet, gray to dark gray. White to grayish white.

Luster and light transmission: Vitreous to pearly; vitreous on {001} and {100}, pearly on {010}. Transparent to translucent.

Chemistry: Much anhydrite includes minor amounts of Sr and/or Ba substituting for Ca.

Diagnostic features: Three cleavages at right angles to each other. It is

Figure 13-4.   Anglesite crystals. Forms: pinacoids   *c*{001}, *b*{010}; rhombic prisms *m*{210}, *o*{011}, *d*{101}; rhombic dipyramids   *y*{111},   *z*{211}, *s*{232}.

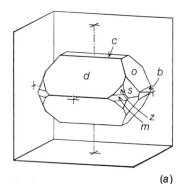

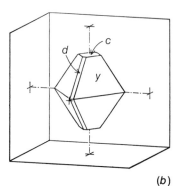

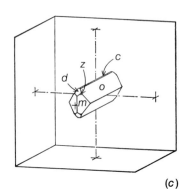

(a)

(b)

(c)

harder than gypsum and has a higher specific gravity than calcite and dolomite. The cleavages differ markedly from those of barite. Anhydrite is soluble, without effervescence, in hot hydrochloric acid, and addition of the resulting solution to barium chloride solution gives a white precipitate of barium sulfate.

Occurrence and associations: Anhydrite is an important rock-forming mineral, occurring as extensive beds interstratified with gyprock, limestone, dolostone, and salt. Anhydrite may be deposited directly from evaporating sea water, at temperatures of 42°C or higher, or at lower temperatures with increased salinity. (At lower temperatures or lower salinity, gypsum is deposited.) Beds of anhydrite may have been formed, in part, by dehydration of preexisting beds of gypsum but, on the other hand, anhydrite may be converted to gypsum by the action of meteoric waters. Anhydrite is also found as an accessory mineral in sedimentary rocks and, to a minor extent, as a gangue mineral in metalliferous veins. In addition, it is an important mineral in the cap rock of many salt domes.

HYDRATED SULFATES

Gypsum, $CaSO_4 \cdot 2H_2O$

Crystal system, class, and space group: Monoclinic; $2/m$; $A2/n$.
Cell dimensions and content: $a = 5.68$, $b = 15.18$, $c = 6.29$; $\beta = 113° 50'$; $Z = 4$.
Habit: Crystals common; typically simple in habit, tabular on {010} (Figure 13-5); many crystals have warped or curved surfaces, and long prismatic crystals are commonly bent or curled because of translation gliding (Figure 13-6); also granular, massive, coarse to very fine, foliated, or as fibrous veinlets; also as pseudomorphs after anhydrite.

Figure 13-5. Gypsum crystals. Forms: pinacoids b{010}, e{$\bar{1}$03}; prisms f{120}, n{011}, l{$\bar{1}$11}.

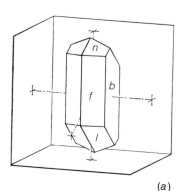

(a)

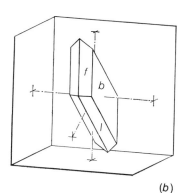

(b)

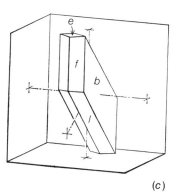

(c)

356

Figure 13-6.   Gypsum "flower," reported by some mineralogists to appear curled because of translation gliding, from Mammoth Cave, Kentucky. (Photograph by Mark Elliott; by permission, from Currens, J. C., 1981. "Kentucky Caves." *Rocks and Minerals*, **56,** 93–97.)

Twinning: Contact twinning on {100} or {$\bar{1}$01} is common, giving swallow-tail twins; also as cruciform penetration twins (Figures 2-75*b* and 13-7).

Common forms and angles:
(010) $\wedge$ (120) = 55°37′   (010) $\wedge$ (011) = 69°16′
(010) $\wedge$ ($\bar{1}$11) = 71°50′   (100) $\wedge$ (001) = 66°10′

Cleavage: {010} good, yielding thin foliae; also {100} and {011} distinct, giving rhombic fragments with angles of 66° and 114°. The fragments are flexible but not elastic because of translation gliding, which is irreversible.

Hardness: 2.

Specific gravity: 2.32.

Color and streak: Colorless; also white, gray, yellowish, or orange-brown in massive varieties. White.

Luster and light transmission: Vitreous, pearly on {010} cleavage; some varieties are silky. Transparent to translucent.

Optical properties:
anisotropic, biaxial (+): $2V \cong 58°$
refractive indices: α, 1.519–1.521; β, 1.522–1.526; γ, 1.529–1.531

Chemistry and alteration: Gypsum shows almost no variation in composition. The coloration is due to the presence of iron oxide or clay impurities. Gypsum converts slowly to the hemihydrate in air at about 70°C, and more rapidly above 90°C; commercially, the hemihydrate, known as plaster of Paris, is made by heating gypsum at

(a) (b)

Figure 13-7. Gypsum (a) Simple twin with {$\bar{1}01$} as twin plane, [101] vertical; rough faces on top are {120}; trace of {001} cleavage visible at upper left. (b) Cleavage fragment from a swallow-tail twin, twin plane {100} vertical; inclined lines are fibrous fractures parallel to a. In each photograph, {010} is parallel to page, localities unknown. (a) is 15 cm high; (b) is 2 cm wide. (Courtesy of Queen's University.)

190–200°C; at higher temperatures, anhydrite is produced (see Figure 13-8).

Varieties: Several varieties of gypsum are recognized: the coarsely crystallized material, typically colorless and transparent, is called *selenite;* aggregate material with a parallel fibrous structure is called *satin spar;* and the fine-grained massive material is commonly called *alabaster.*

Diagnostic features: Low hardness (2) and the perfect cleavage with two poorer cleavages. The presence of much water distinguishes it from anhydrite. Soluble in hot dilute hydrochloric acid and the resulting solution gives, with the addition of barium chloride, a white precipitate of barium sulfate.

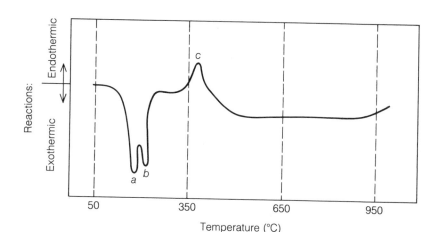

Figure 13-8. Differential thermal analysis graph for gypsum. On the diagram, a is caused by a loss of water ($CaSO_4 \cdot 2H_2O \rightarrow CaSO_4 \cdot \frac{1}{2}H_2O$); b is caused by loss of remaining water ($CaSO_4 \cdot \frac{1}{2}H_2O \rightarrow CaSO_4$); c is caused by a structural inversion ($CaSO_4 \rightarrow \beta CaSO_4$).

Occurrence, associations, and uses: Gypsum occurs as extensive sedimentary deposits interbedded with limestone, red shales, claystone, and rock salt. It is normally the first salt deposited in the evaporation of sea water, followed by anhydrite and halite as the salinity increases, which are followed, in rare cases, by more soluble sulfates and other salts of Mg and K. Gypsum also occurs in many other ways: as saline lake deposits; with native sulfur around volcanic fumaroles; as an efflorescence on soils or in limestone caves; in the cap rock of salt domes; and in gossans over pyritic mineral deposits in limestone areas.

Gypsum is mined extensively in many parts of the world for use in the construction industry, especially for manufacture of plasters, gypsum wallboard, and roof tiles; in cements; and as a filler in paper and paints. It is also often used as a soil conditioner (landplaster) and fertilizer.

Chalcanthite, $CuSO_4 \cdot 5H_2O$

Crystal system, class, and space group: Triclinic; $\bar{1}$; $P\bar{1}$.

Cell dimensions and content: $a = 6.122$, $b = 10.695$, $c = 5.96$; $\alpha = 97° 35'$, $\beta = 107° 10'$, $\gamma = 77° 33'$; $Z = 2$.

Habit: Crystals uncommon; short prismatic parallel to [001] (Figure 13-9); also massive and granular, stalactitic, and reniform.

Cleavage and fracture: $\{1\bar{1}0\}$ indistinct. Conchoidal.

Hardness: $2\frac{1}{2}$.

Specific gravity: 2.28.

Color and streak: Sky-blue of different shades. White.

Luster and light transmission: Vitreous. Subtransparent.

Optical properties:

anisotropic, biaxial (−): $2V = 56°$

refractive indices: α, 1.516; β, 1.539; γ, 1.546

pale blue

Diagnostic features: Color; water solubility; loss of water on heating.

Occurrence, associations, and uses: This soluble mineral, often called *bluestone*, occurs with other hydrated sulfates of copper and iron in the oxidized near-surface zones of copper sulfide ore deposits. Copper and iron sulfates result from the oxidation of chalcopyrite and other sulfides by meteoric waters; locally, in arid regions, these salts form as substantial deposits in the oxide zone. Some of these deposits are of commercial importance, particularly in the Chilean copper deposits (e.g., Chuquicamata and Quetena). Copper and sulfate ions are commonly present in mine waters, from which chalcanthite may crystallize as stalactites or crusts in mines; in some mines, the copper is recovered from such waters by precipitating it with metallic iron.

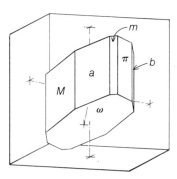

Figure 13-9. Chalcanthite crystal. Forms: pinacoids $b\{010\}$, $a\{100\}$, $m\{110\}$, $\pi\{130\}$, $M\{1\bar{1}0\}$, $\omega\{\bar{1}\bar{1}1\}$.

Melanterite, $FeSO_4 \cdot 7H_2O$

Crystal system, class, and space group: Monoclinic; $2/m$; $P2_1/c$.

Cell dimensions and content: $a = 14.11$, $b = 6.51$, $c = 11.02$; $\beta = 105° 15'$; $Z = 4$.

Habit: Crystals rare; equant to short prismatic along c with {110} and {001}; generally in stalactitic or concretionary forms; also as fibrous to capillary aggregates and crusts or massive and pulverulent.

Cleavage and fracture: {001} perfect, {110} distinct. Conchoidal.

Hardness: 2.

Specific gravity: 1.898 (pure).

Color and streak: Green of various shades; greenish blue and blue in specimens with substitution of Cu for Fe; also greenish white. White.

Luster and light transmission: Vitreous. Translucent to subtranslucent.

Optical properties:
anisotropic, biaxial (+): $2V = 85°$
refractive indices: α 1.471; β, 1.478; γ, 1.486
colorless to pale green

Chemistry and alteration: Cu commonly substitutes for Fe^{+2}, and a series extends to about Fe:Cu = 1:1.9. Mg also substitutes for Fe, and a partial series extends to $MgSO_4 \cdot 7H_2O$, the monoclinic polymorph of epsomite. Less commonly, Zn, Co, Ni, and Mn are also present in substitution for Fe. In many occurrences, several metal ions substitute for Fe together. Melanterite dehydrates to the pentahydrate or to lower hydrates at room temperature, depending on the relative humidity.

Diagnostic features: Solubility in water; sweetish, astringent taste; yields water and SO_2 on strong heating.

Occurrence and associations: Melanterite and its compositional varieties are secondary minerals formed by the oxidation of pyrite, marcasite, and copper-bearing pyritic ores. It occurs widely as an efflorescence on the walls and timbers of mine workings in the oxidized zone of pyritic ore bodies, especially in arid regions. Melanterite is commonly associated with epsomite, chalcanthite, gypsum, and other hydrous or basic sulfates.

Epsomite, $MgSO_4 \cdot 7H_2O$

Crystal system, class, and space group: Orthorhombic; 222; $P2_12_12_1$.

Cell dimensions and content: $a = 11.86$, $b = 11.99$, $c = 6.858$; $Z = 4$.

Habit: Artificial crystals short prismatic along c; natural crystals rare; typically fibrous to hairlike or acicular crusts with fiber axis c; also as efflorescence, botryoidal or reniform masses.

Cleavage: {010} perfect, {101} distinct.
Hardness: 2–2½
Specific gravity: 1.678 (calc.).
Color and streak: Colorless in crystals; white for massive material. White.
Luster and light transmission: Vitreous, silky to earthy on fibrous material. Translucent to subtranslucent.
Optical properties:
   anisotropic, biaxial (−): $2V \cong 50°$
   refractive indices: α, 1.430–1.440; β, 1.452–1.462; γ, 1.457–1.469
Chemistry and alteration: There is minor substitution of Zn, Fe, Co, Ni, and/or Mn for Mg in epsomite. The substitution may be more extensive in artificial crystals, and complete series are known to extend to $NiSO_4 \cdot 7H_2O$ and to $ZnSO_4 \cdot 7H_2O$. In dry air, at ordinary temperatures, epsomite loses up to $1H_2O$, changing to *hexahydrite.*
Diagnostic features: Very soluble in water; bitter-to-metallic taste; on heating, it will dissolve in its own water of crystallization.
Occurrence and associations: Epsomite occurs as crusts or fibrous efflorescences in mine workings in coal and metal deposits; in limestone caves; in sheltered places on outcrops of magnesian rocks, particularly in the presence of pyrite or pyrrhotite; in oxidized zones of pyritic deposits in arid regions; and in mineral spring deposits, salt lake deposits, and oceanic salt deposits. It has long been known as a deposit from mineral waters at Epsom, Surrey, England, and at Sedlitz, Bohemia.

ANHYDROUS SULFATES CONTAINING HYDROXYL

Brochantite, $Cu_4SO_4(OH)_6$

Crystal system, class, and space group: Monoclinic; $2/m$; $P2_1/a$.
Cell dimensions and content: $a = 13.08$, $b = 9.85$, $c = 6.02$; $\beta = 103° 22'$; $Z = 4$.
Habit: Crystals uncommon; typically prismatic to acicular parallel to c (Figure 13-10); commonly as loosely coherent aggregates of acicular crystals; also in groups of drusy crusts; massive, granular.
Twinning: Twin plane {100}, common as contact twin.
Cleavage: {100} perfect.
Hardness: 3½–4.
Specific gravity: 3.97.
Color and streak: Emerald-green to blackish green. Pale green.
Luster and light transmission: Vitreous. Transparent to translucent.
Optical properties:
   anisotropic, biaxial (−): $2V \cong 77°$

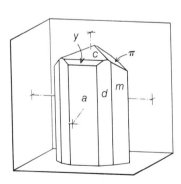

Figure 13-10. Brochantite crystal. Forms: pinacoids c{001}, a{100}, y{201}; prisms m{110}, d{210}; π{$\bar{1}$11}.

refractive indices: α, 1.728; β, 1.771; γ, 1.800
weakly pleochroic

Diagnostic features: Loses water on heating. Brochantite is similar to antlerite and atacamite in color. All three minerals occur in the oxidized zones of copper deposits and are difficult to distinguish from one another macroscopically, except by careful study of the crystals and cleavage, combined with laboratory tests.

Occurrence and associations: Brochantite occurs associated with malachite, limonite, cuprite, and chrysocolla. It occurs abundantly in some of the copper deposits of Chile and in the southwestern United States.

Antlerite, $Cu_3SO_4(OH)_4$

Crystal system, class, and space group: Orthorhombic; *2/m2/m2/m; Pnam.*

Cell dimensions and content: a = 8.24, b = 11.99, c = 6.03; Z = 4.

Habit: Crystals uncommon; typically thick tabular on {010} or short prismatic along c (Figure 13-11); also as cross-fiber veinlets, friable aggregates, and granular masses.

Cleavage: {010} perfect.

Hardness: $3\frac{1}{2}$.

Specific gravity: 3.88.

Color and streak: Emerald-green to blackish green. Pale green.

Luster and light transmission: Vitreous. Translucent.

Optical properties:
anisotropic, biaxial (+): 2V = 53°
refractive indices: α, 1.726; β, 1.738; γ, 1.789
pleochroic in greens

Diagnostic features: Extremely difficult to distinguish from other copper sulfates.

Occurrence, associations, and uses: Similar to and commonly associated with brochantite and atacamite, antlerite occurs in many copper deposits in arid regions, also with chalcanthite and gypsum.
Antlerite is the principal ore mineral of copper at the important mine at Chuquicamata, Chile.

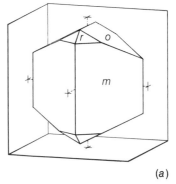

(a)

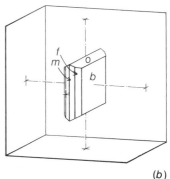

(b)

Figure 13-11. Antlerite crystals. Forms: pinacoids b{010}; rhombic prisms m{110}, f{130}, o{011}; rhombic dipyramid r{111}.

Alunite, $KAl_3(SO_4)_2(OH)_6$

Crystal system, class, and space group: Trigonal; *3m; R3m.*

Cell dimensions and content: a = 6.97, c = 17.38; Z = 3.

Crystal structure: Alunite has been found to be strongly pyroelectric and piezoelectric, and it is for this reason that it is assigned to the class *3m*. Some investigators have been unable to confirm this and assign it to the class $\overline{3}2/m$.

Habit: Crystals rare; generally small with {0001} and the rhombohedron {01$\overline{1}$2} and pseudocubic in appearance; in most occurrences, massive (granular to dense).

Cleavage and fracture: {0001} distinct. Uneven to conchoidal.

Hardness: $3\frac{1}{2}$–4.

Specific gravity: 2.82.

Color and streak: White, also grayish, yellowish, or reddish. White.

Luster and light transmission: Vitreous, pearly on {0001}. Transparent to translucent.

Optical properties:
anisotropic, uniaxial (+)
refractive indices: ω, 1.568–1.585; ϵ, 1.590–1.601

Chemistry: Alunite is a basic sulfate of aluminum and potassium in which Na is often found in substitution for K and may extend to Na:K = 7:4. Where Na exceeds K, the mineral is called *natroalunite*. The alunite group also includes jarosite and the several compositional variants of jarosite.

Diagnostic features: Decrepitates when heated; soluble in sulfuric acid. In the massive form, it is difficult to distinguish macroscopically from limestone, dolostone, anhydrock, or rock magnesite without chemical tests. A simple chemical check is that alunite dispersed in water gives an acid solution.

Occurrence: Alunite is of widespread occurrence in near-surface rocks that have been altered by volcanic solutions containing sulfuric acid. This process of alteration, called *alunitization*, is a widespread feature of the altered or mineralized volcanic rocks of the western United States, for example, in the Goldfield District, Nye County, Nevada, and at Marysvale, Utah.

Jarosite, $KFe_3(SO_4)_2(OH)_6$

Crystal system, class and space group: Trigonal; *3m*; R3m.

Cell dimensions and content: a = 7.21, c = 17.03; Z = 3.

Habit: Crystals rare; minute, tabular on {0001} or pseudocubic with {01$\overline{1}$2}, as crusts or coatings of minute crystals; granular massive, fibrous, nodular; also earthy and pulverulant.

Cleavage: {0001} distinct.

Hardness: $2\frac{1}{2}$–$3\frac{1}{2}$.

Density: 3.25.

Color and streak: Ocherous, amber-yellow to dark brown. Pale yellow.

Luster and light transmission: Vitreous to resinous. Translucent to subtranslucent.

Optical properties:
  anisotropic, uniaxial $(-)$
  refractive indices: ω, 1.815–1.820; ϵ, 1.713–1.715
  pleochroic in yellows
Chemistry: Jarosite is isostructural with alunite. Ordinarily, there is no appreciable substitution of Al for Fe^{+3} or vice versa; there are a few occurrences, however, whereby the possibility of a continuous series between alunite and jarosite is indicated. Na commonly substitutes for K in jarosite, and a series extends toward natrojarosite. Other members of this group have NH_4, Ag, Pb, or H_2O in place of K.
Diagnostic features: Soluble in hydrochloric acid; surface alters quickly to limonite minerals.
Occurrence: Jarosite is a widespread secondary mineral that occurs in the form of crusts or coatings on pyrite-bearing ores and adjoining rocks. It is a common constituent of limonitic gossans and some of it is difficult to distinguish from earthy limonite.

ANHYDROUS CHROMATES

Crocoite, $PbCrO_4$

Crystal system, class, and space group: Monoclinic; $2/m$; $P2_1/n$.
Cell dimensions and content: $a = 7.11$, $b = 7.41$, $c = 6.77$; $\beta = 102°\ 27'$; $Z = 4$.
Crystal structure: Crocoite is isostructural with monazite.
Habit: Crystals uncommon; typically prismatic and striated along c; also massive, columnar to granular.
Cleavage: {110} distinct.
Hardness: $2\frac{1}{2}$–3.
Tenacity: Sectile.
Specific gravity: 6.10.
Color and streak: Hyacinth-red, deep orange-red, or orange. Orange-yellow.
Luster and light transmission: Adamantine to vitreous. Transparent.
Optical properties:
  anisotropic, biaxial $(+)$: $2V = 57°$
  refractive indices: α, 2.29; β, 2.36; γ, 2.66
  pleochroic in reds
Diagnostic features: Color, luster, crystal habit, and specific gravity.
Occurrence and associations: Crocoite is a secondary mineral of rather limited occurrence; it occurs in gossans of ore deposits, commonly associated with pyromorphite, cerussite, and other second-

ary lead minerals and also in lead ores spatially associated with chromite-bearing ultramafic rocks (e.g., in Tasmania).

MOLYBDATES, TUNGSTATES

Wolframite, (Fe,Mn)WO₄

Crystal system, class, and space group: Monoclinic; $2/m$; $P2/c$.

Cell dimensions and content: $a = 4.81$, $b = 5.73$, $c = 4.97$; $\beta = 90°$ 49'; $Z = 2$; dimensions are slightly larger for $MnWO_4$ and slightly smaller with $\beta = 90°$ for $FeWO_4$.

Habit: Crystals common; typically short prismatic along c, characteristically flattened on {100} (Figure 13-12); faces are striated parallel to c; also in subparallel groups, lamellar or massive granular; huebnerite ($MnWO_4$) is commonly prismatic to long prismatic along c; ferberite ($FeWO_4$) is commonly elongated along b. Wolframite also occurs as pseudomorphs after scheelite.

Twinning: Twin plane {100}, common as simple contact twins.

Cleavage: {010} perfect.

Hardness: $4-4\frac{1}{2}$.

Specific gravity: 7.1–7.5, higher with greater iron content.

Color and streak: Brownish black to iron-black. Reddish brown to brownish black.

Luster and light transmission: Submetallic. Subtranslucent to opaque.

Optical properties:

anisotropic, biaxial (+): $2V = 73°-79°$

refractive indices (Li): α, 2.17–2.26; β, 2.22–2.32; γ, 2.32–2.42

pleochroic

Chemistry: The wolframite minerals form a complete series between ferberite ($FeWO_4$) and huebnerite ($MnWO_4$); nearly all specimens show intermediate composition, and wolframite with 20–80 atomic percent iron (and 80–20 atomic percent manganese) is the mineral that occurs most widely.

Diagnostic features: Color, one direction of perfect cleavage, and high specific gravity. Huebnerite is reddish brown, with a resinous luster, and is translucent. Ferberite is black, with a nearly black streak and an almost metallic luster; is virtually opaque and is weakly magnetic.

Occurrence, associations, and uses: The minerals of this series constitute the principal ores of tungsten. Most wolframite occurs:

1. In greisens, quartz-rich veins, or complex pegmatitic masses associated with granitic intrusive rocks. There is generally evidence of pneumatolytic origin, and the commonly associated minerals are topaz, cassiterite, arsenopyrite, lithia micas, and tourmaline.

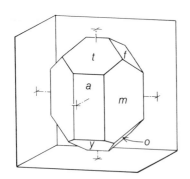

Figure 13-12.   Wolframite crystal. Forms: pinacoids a{100}, t{102}, y{$\bar{1}$02}; prisms m{110}, f{011}.

2. In high-temperature hydrothermal veins associated with pyrrhotite, pyrite, chalcopyrite, bismuthinite, and some of the minerals listed in number 1. At least some of these veins may be genetically associated with the deposits of 1; in both types of occurrences, wall-rock alteration, accompanied by the development of tourmaline, mica, topaz, fluorite, and chlorite, tends to be well marked.

The wolframite group minerals also occur in some moderate- and low-temperature veins along with sulfides, cassiterite, scheelite, bismuth, quartz, and/or siderite; in contact metamorphic deposits adjacent to granitic intrusives; and in alluvial deposits.

Scheelite, CaWO₄

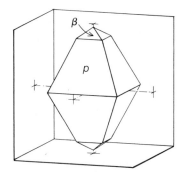

Figure 13-13. Scheelite crystal. Forms: tetragonal dipyramids $p\{101\}$, $\beta\{103\}$.

Crystal system, class, and space group: Tetragonal; 4/m; 14₁/a.
Cell dimensions and content: a = 5.242, c = 11.372; Z = 4.
Habit: Crystals common; typically dipyramidal with {101} and {112} predominant (Figure 13-13); commonly massive, granular.
Twinning: Common on {110}, in most occurrences as penetration twins.
Cleavage and fracture: {101} distinct. Uneven.
Hardness: 4½–5.
Specific gravity: 6.12 (pure), lower with substitution of Mo for W.
Color and streak: Colorless to white, pale yellow, or brownish; also greenish, gray, or reddish; fluoresces bright bluish white in short-wave, ultraviolet radiation. White.
Luster and light transmission: Vitreous. Transparent.
Optical properties:
anisotropic, uniaxial (+)
refractive indices: ω, 1.9208; ϵ, 1.9375
Chemistry: Mo may substitute for W, and a partial series extends toward *powellite* (CaMoO₄). Most scheelite, however, contains only minor amounts of Mo. The color of the fluorescence of scheelite varies with Mo content; whereas with X-rays, shortwave ultraviolet, and cathode rays, more-or-less pure scheelite fluoresces bright bluish white; with increasing Mo content, it tends to fluoresce white to yellowish white.
Diagnostic features: Light color, high specific gravity, crystal form, and fluorescence. A test for tungsten, however, may be necessary. Scheelite is thermoluminescent.
Occurrences, associations, and uses: Scheelite commonly occurs with wolframite in deposits of apparently high-temperature origin. It occurs in contact metamorphic deposits (which include most economic occurrences); in high-temperature hydrothermal veins with wolframite and other associated minerals listed for that mineral; as

a minor mineral in some gold-quartz veins; and in some pegmatites and moderate-temperature hydrothermal veins.

Scheelite is an important source of tungsten.

Wulfenite, PbMoO$_4$

Figure 13-14.   Wulfenite crystal. Forms: pinacoid c{001} (or pedions c{001} and −c{00$\bar{1}$}), tetragonal dipyramid n{101} (or tetragonal pyramids n{101} and −n{10$\bar{1}$}).

Crystal system, class, and space group: Tetragonal; 4/m; I4$_1$/a.

Cell dimensions and content: a = 5.435, c = 12.11; Z = 4.

Habit: Crystals common; square tabular on {001}, bounded by the prism {100} and pyramids {101} and {10$\bar{1}$} (Figure 13-14); also massive, course-to-fine granular; some crystals are distinctly hemimorphic indicating pyramidal symmetry (4).

Cleavage: {101} distinct.

Hardness: 3.

Tenacity: Extremely brittle.

Specific gravity: 6.5–7.0 (6.815 calc.), lower with substitution of Ca for Pb, or higher with substitution of W for Mo.

Color and streak: Orange-yellow to wax-yellow, also yellowish gray, olive-green, or brown. White.

Luster and light transmission: Resinous. Transparent or translucent.

Optical properties:

anisotropic, uniaxial (−)

refractive indices: ω, 2.4053; ε, 2.2826

pleochroic in yellows and/or oranges

Chemistry: W substitutes for Mo to a small extent, but evidence of a series extending to PbWO$_4$ is lacking. Ca substitutes for Pb, indicating a partial series to powellite (CaMoO$_4$).

Diagnostic features: Square tabular crystals, color, and luster. It is decomposed by concentrated acids. It decrepitates in a flame.

Occurrence, associations, and uses: Wulfenite is a secondary mineral formed in the oxidized zone of ore deposits that contain lead and molybdenum minerals. It is commonly associated with pyromorphite, vanadinite, cerussite, limonite, calcite, and other secondary minerals.

Next to molybdenite, wulfenite is the most common molybdenum mineral. It is a minor source of molybdenum.

14 Class VII: Phosphates, Arsenates, Vanadates

The phosphate, arsenate, and vanadate group of minerals includes a large number of naturally occurring oxysalts with anionic groups of the type $(XO_4)^{-n}$ in which X is P, As, or V, and n is 3. Extensive substitution between P and As, and between As and V, are common. These minerals are usually classified on the basis of the $A{:}(XO_4)$ ratio for anhydrous, acid, and hydrated salts. Relatively common members of this class that are described here are listed next.

Anhydrous Normal Phosphates
　　Type $A(XO_4)$
　　　　Xenotime, YPO_4
　　　　Monazite, $(Ce,La,Nd,Th)PO_4$

Hydrated Normal Phosphates
　　Type $A_3(XO_4)_2 \cdot 8H_2O$
　　　　Vivianite, $Fe_3(PO_4)_2 \cdot 8H_2O$
　　　　Erythrite, $Co_3(AsO_4)_2 \cdot 8H_2O$

Anhydrous Phosphates with Hydroxyl or Halogen
　　Type $AB(XO_4)Z_q$
　　　　Amblygonite Series: $(Li,Na)Al(PO_4)(F,OH)$

Type $A_5(XO_4)_3Z_q$
  Apatite Group
    Apatite Series: $Ca_5(PO_4)_3(F,Cl,OH)$
    Pyromorphite Series: $Pb_5[(PO_4),(AsO_4),(VO_4)]_3Cl$

Hydrated Phosphates Containing Hydroxyl
  Turquoise, $CuAl_6(PO_4)_4(OH)_8 \cdot 5H_2O$

Uranyl Phosphates
  Autunite Group
    Autunite, $Ca(UO_2)_2(PO_4)_2 \cdot 10-12H_2O$
    Torbernite, $Cu(UO_2)_2(PO_4)_2 \cdot 8-12H_2O$
  Meta-autunite Group
    Meta-autunite, $Ca(UO_2)_2(PO_4)_2 \cdot 2-6H_2O$
    Metatorbernite, $Cu(UO_2)_2(PO_4)_2 \cdot 8H_2O$
Vanadium Oxysalts
  Carnotite: $K_2(UO_2)_2(VO_4)_2 \cdot 3H_2O$

ANHYDROUS NORMAL PHOSPHATES

Xenotime, YPO_4

Crystal system, class, and space group: $4/m2/m2/m$; $14_1/amd$.
Cell dimensions and content: $a = 6.89$, $c = 6.03$; $Z = 4$.
Crystal structure: Isostructural with zircon.
Habit: Crystals uncommon; short to long prismatic along c; also equant, pyramidal; closely resembles zircon in habit; as radial aggregates of coarse crystals and as rosettes.
Cleavage: {100} perfect.
Hardness: 4–5.
Specific gravity: 4.3–4.7 (4.25, calc. for pure YPO_4).
Color and streak: Commonly yellowish brown to reddish brown, also medium brown, flesh-red, grayish white, yellow, or greenish. Pale brown, yellowish, or reddish.
Luster and light transmission: Vitreous to resinous. Translucent to opaque.
Optical properties:
  anisotropic, uniaxial (+)
  refractive indices: ω, 1.720–1.724; ϵ, 1.816–1.827
  slightly pleochroic
Chemistry: Xenotime is essentially YPO_4; rare-earth elements, particularly erbium and cerium, may substitute for Y. Th, U, Zr, and Ca also substitute for Y in small amounts. The substitution of Zr^{+4} and U^{+4} for Y^{+3} apparently involves concomitant substitution of SiO_4 for PO_4.

Diagnostic features: Easily confused with zircon; can be distinguished by its inferior hardness and perfect cleavage.

Occurrence: Xenotime is of widespread occurrence as a minor accessory mineral in acidic and alkalic igneous rocks and as larger crystals in the associated pegmatites. It is also a common detrital mineral and has been recorded as occurring in mica- and quartz-rich gneisses.

Monazite, (Ce,La,Nd,Th)PO$_4$

Crystal system, class, and space group: Monoclinic; $2/m$; $P2_1/n$.

Cell dimensions and content: $a = 6.79$, $b = 7.01$, $c = 6.46$; $\beta = 103°\,38'$; $Z = 4$.

Habit: Crystals uncommon; most are small and typically equant, but may be flattened on {100} or somewhat elongated on b (Figure 14-1). The crystal faces are commonly rough, striated, or uneven.

Twinning: Common on {100} as contact or cruciform twins.

Cleavage and parting: {100} distinct. {001} parting.

Hardness: $5–5\frac{1}{2}$.

Specific gravity: 4.6–5.4.

Color and streak: Yellowish or reddish brown to brown. Nearly white.

Luster and light transmission: Resinous or waxy. Subtranslucent.

Optical properties:
anisotropic, biaxial (+): $2V = 6°–19°$
refractive indices: α, 1.774–1.800; β, 1.777–1.801; γ, 1.828–1.851
slightly pleochroic

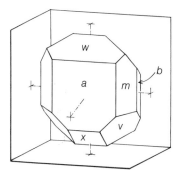

Figure 14-1. Monazite crystal. Forms: pinacoids $b\{010\}$, $a\{100\}$, $w\{101\}$, $x\{10\bar{1}\}$; rhombic prisms $m\{110\}$, $v\{111\}$.

Chemistry: Monazite is an anhydrous phosphate, principally of cerium and/or lanthanum, with lesser amounts of the heavier rare earths. Some Th is generally present, commonly to the extent of 4–12 percent; Nd and/or Ca may substitute in small amounts for (Ce,La); U has been reported for a few specimens. The corresponding phosphate of yttrium and the heavier rare-earth elements is xenotime, which is isostructural with zircon.

Diagnostic features: Crystal habit and hardness; for example, monazite is softer than zircon and harder than sphene.

Occurrence, associations, and uses: Monazite is widespread as an accessory mineral in granite, syenite, and certain gneisses, and as relatively large crystals in some pegmatite masses. In some regions, detrital sands derived from these rocks contain commercial quantities of monazite. It is commonly associated with zircon, xenotime, magnetite, apatite, columbite, and rare-earth niobate-tantalates. It is important as a source of cerium and thorium.

HYDRATED NORMAL PHOSPHATES

Vivianite, $Fe_3(PO_4)_2 \cdot 8H_2O$

Crystal system, class, and space group: Monoclinic; 2/m; C2/m.
Cell dimensions and content: $a = 10.06$, $b = 13.41$, $c = 4.696$; $\beta = 104°\ 18'$; $Z = 2$.
Habit: Crystals common; typically prismatic along c with pinacoids {010} and {100} dominant; as reniform or tubular masses or concretions, or incrusting with a divergent bladed or fibrous structure; also earthy or pulverulent.
Cleavage: {010} perfect; thin laminae are flexible.
Hardness: $1\frac{1}{2}$–2.
Tenacity: Somewhat sectile.
Specific gravity: 2.68.
Color and streak: Colorless and transparent when fresh, rapidly becoming pale to dark blue when exposed in air, due to oxidation. White to bluish or brownish.
Luster and light transmission: Vitreous. Transparent to translucent.
Optical properties:
   anisotropic, biaxial (+): $2V = 63°$–$83°$
   refractive indices: $\alpha,1.5788$–1.616; $\beta,1.6024$–1.656; $\gamma,1.6294$–1.675
   pleochroic in greens and blue
Chemistry and alteration: The hydrated arsenates of Co, Ni, and Zn with the composition $A_3(XO_4)_2 \cdot 8H_2O$ are isostructural with vivianite and comprise the vivianite group, but vivianite shows only a slight tendency to form a solid solution with the other members. Mn^{+2}, Mg, and Ca may substitute for Fe^{+2} to a limited extent. The principal variations in vivianite, however, are due to the partial oxidation of Fe^{+2} to Fe^{+3}; with this change, the mineral becomes dark blue to nearly black.
Diagnostic features: Soft; typically altered to a blue or green color; flexible cleavage lamellae are distinctive; yields water on heating; readily soluble in acids.
Occurrence and associations: Vivianite, a mineral of secondary origin, occurs commonly in gossans of metallic ore deposits and in the surface zones of pegmatites that contain iron-manganese phosphates. It also occurs as concretions in clays and as crystal clusters in recent sedimentary deposits where it is associated with bone or other organic phosphatic remains.

Erythrite, $Co_3(AsO_4)_2 \cdot 8H_2O$

Crystal system and class: Monoclinic; 2/m; C2/m.
Cell dimensions and content: $a = 10.20$, $b = 13.37$, $c = 4.74$; $\beta = 105°\ 01'$; $Z = 2$.

Habit: Crystals rare; typically prismatic to acicular along *c*, flattened on {010}, and striated parallel to *c*; commonly as radial groups or reniform shapes with a drusy surface and columnar or fibrous structure; also earthy and pulverulent; and as thin crusts.

Cleavage: {010} perfect.

Hardness: $1\frac{1}{2}$–$2\frac{1}{2}$.

Tenacity: Somewhat sectile.

Specific gravity: 3.06.

Color and streak: Crimson-red, paler with noteworthy Ni content and white or gray with Ni > Co; annabergite, the nickel analogue, is apple-green. Paler than color.

Luster and light transmission: Adamantine to dull. Transparent to subtranslucent.

Optical properties:
anisotropic, biaxial (±); $2V \cong 90°$
refractive indices: α,1.622–1.629; β,1.658–1.663; γ,1.681–1.701
pleochroic in reds and violet

Chemistry: Erythrite, in contrast to vivianite, shows rather wide variations in composition. Ni commonly substitutes for Co, and a complete series extends to *annabergite,* $Ni_3(AsO_4)_2 \cdot 8H_2O$. Ca, Zn, Mg, and Fe^{+2} also substitute for Co or Ni, but only Zn forms an isostructural mineral. [A complete series may extend to the mineral *koettigite* $Zn_3(AsO_4)_2 \cdot 8H_2O$.]

Diagnostic features: Color and the association with other cobalt minerals. Erythrite yields water and gives the odor of arsenic when heated.

Occurrence and associations: Erythrite and annabergite are secondary minerals formed by surface oxidation of cobalt and nickel arsenides. Their occurrence as pulverulent coatings on the primary minerals led to the common terms *cobalt-bloom* (pink) and *nickel-bloom* (green). Where the primary mineral contains both nickel and cobalt in substantial quantities, the resulting bloom is white or gray. The presence of pink cobalt-bloom in the zone of surface weathering is a distinctive indication of the presence of cobalt arsenides; its presence led prospectors to the cobalt-silver veins of Cobalt, Ontario and to the cobalt-silver-uranium veins of Great Bear Lake, Canada.

ANHYDROUS PHOSPHATES WITH HYDROXYL OR HALOGEN

Amblygonite, $(Li,Na)Al(PO_4)(F,OH)$

Crystal system, class, and space group: Triclinic; $\bar{1}$; $P\bar{1}$.

Cell dimensions and content: $a = 5.19$, $b = 7.12$, $c = 5.04$; $\alpha = 112° 02'$, $\beta = 97° 49'$, $\beta = 68° 07'$; $Z = 2$.

Habit: Crystals uncommon; typically rough, especially when large; most small crystals are equant; also in large cleavable masses, columnar, or compact.

Twinning: Common on $\{\bar{1}\bar{1}1\}$; lamellar on $\{111\}$.

Common forms and angles:

$(100) \wedge (110) = 45°06'$ $(100) \wedge (0\bar{1}1) = 75°11'$

$(100) \wedge (001) = 90°15'$ $(010) \wedge (021) = 25°31'$

Cleavage and fracture: $\{100\}$ perfect, $\{110\}$ good, $\{0\bar{1}1\}$ distinct. Uneven.

Hardness: $5\frac{1}{2}$–6.

Specific gravity: 2.98–3.11.

Color and streak: White to milky or creamy, also yellowish, pinkish, greenish, bluish, and gray; rarely colorless. White.

Luster and light transmission: Vitreous to greasy, pearly on well-developed cleavages. Transparent to translucent.

Optical properties:

anisotropic, biaxial (−): $2V = 50°$–$90°$

refractive indices: $\alpha, 1.575$–1.595; $\beta, 1.587$–1.610; $\gamma, 1.588$–1.612

Chemistry: Amblygonite ranges considerably in composition because of the substitution of Na for Li and the substitution of OH for F. Li is normally present much in excess of Na, and F in excess of OH as in amblygonite proper. When OH exceeds F, either the name *natromontebrasite* or *montebrasite* applies; known occurrences where Na is in excess of Li are rare.

Diagnostic features: Easily mistaken for white albite, with which it is generally associated, but the different cleavage angles and the less perfect quality of the cleavage, together with the higher density of amblygonite, provide the means for distinguishing between the two minerals.

Occurrence, associations, and uses: The members of this series commonly occur as large crystals (up to a few meters long) in granitic pegmatites that are rich in Li and PO_4. Amblygonite is typically associated with spodumene, apatite, lepidolite, tourmaline, and other Li minerals. It also occurs in high-temperature veins and greisens, generally with cassiterite, topaz, and mica. It is sometimes used as an ore of lithium.

APATITE GROUP

The members of the apatite group described in this book fall into two series, the apatite series and the pyromorphite series. The group also includes the svabite series and the "silicate-apatite" series, both of which are represented by only rare minerals not described in this book. Whereas most minerals of the apatite group crystallize in the hexagonal system, chlorapatite, for example, may also crystallize in the monoclinic system (space group $P2_1/a$; $a = 19.210$, $b = 6.785$, $c = 9.605$; $\beta = 120°$).

The general formula widely applied to these minerals is $A_5(XO_4)_3(F,Cl,OH)$ in which A = Ba, Ca, Ce, K, Na, Pb, and Sr, and X = As, C, P, S, Si, and V. This formula, however, is somewhat misleading in that structural requirements preclude site-by-site substitution of, for example, Cl for F in the apatite series. Nonetheless, substitutional solid solution is, in general, extensive among members of individual series but is rather limited among members of different series. The limitation is easily understood when it is noted, for example, that the unit cell volume of minerals of the apatite series is approximately 20 percent to 30 percent smaller than that of minerals of the pyromorphite series.

The apatite series includes:

Fluorapatite,	$Ca_5(PO_4)_3F$
Chlorapatite,	$Ca_5(PO_4)_3Cl$
Hydroxylapatite,	$Ca_5(PO_4)_3(OH)$
Carbonate-fluorapatite,	$Ca_5(PO_4,CO_3)_3F$
Carbonate-hydroxylapatite,	$Ca_5(PO_4,CO_3)_3(OH)$

Carbonate-fluorapatite has been given the alternative formula $Ca_5(P,C)_3(O,F)_{13}$ with F and CO_2 each > 1 weight percent and it has been referred to widely as *francolite*. Carbonate-hydroxylapatite has been given the alternative formula $Ca_5(P,C)_3(O,OH)_{13}$ with $CO_2 > 1$ weight percent, and it has been referred to widely as *dahllite*. Sr, Mn, Ce, rare-earth elements, and, to a lesser extent, Na, may substitute in part for Ca; S and Si can substitute in part for P.

The pyromorphite series includes:

Pyromorphite,	$Pb_5(PO_4)_3Cl$
Mimetite,	$Pb_5(AsO_4)_3Cl$
Vanadinite,	$Pb_5(VO_4)_3Cl$

Substitution of V, As, and P for one another in minerals of this series is common. Also, as the formulas indicate, Cl, which is least common in the apatite series, is the only significant halogen in the pyromorphite series.

Apatite

(The data presented are for fluorapatite, the member of the series most likely to be discernable macroscopically, except as noted otherwise.)

Crystal system, class, and space group: Hexagonal; $6/m$; $P6_3/m$.
Cell dimensions and content: $a = 9.37$, $c = 6.88$; $Z = 2$.
Habit: Crystals common; typically thick tabular on {0001} with {10$\bar{1}$0} and dipyramid forms; the hexagonal prism {41$\bar{5}$0} and/or dipyramids {21$\bar{3}$1}, {31$\bar{4}$1} occur rarely as small faces; also as long prisms, typically {10$\bar{1}$0} and {10$\bar{1}$1}, terminated by dipyramids

374

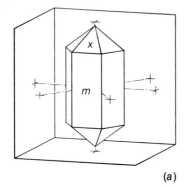

(a)

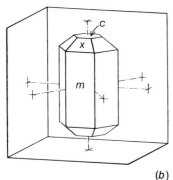

(b)

Figure 14-2. Apatite crystals. Forms: pinacoid c{0001}; hexagonal prism m{10$\bar{1}$0}; hexagonal dipyramid x{10$\bar{1}$1}.

(Figure 14-2). Crystals, especially in marbles, commonly appear to have had their edges dissolved (Figure 14-3). [The carbonate apatites are commonly massive, coarse granular to cryptocrystalline and compact, in some cases globular, stalactitic, oolitic, or earthy.]

Twinning: Relatively common on {21$\bar{3}$1} and {12$\bar{3}$1}.

Common forms and angles:
   (10$\bar{1}$0) $\wedge$ (10$\bar{1}$1) = 49°41′
   (0001) $\wedge$ (11$\bar{2}$1) = 55°46′

Cleavage and fracture: {0001} and {10$\bar{1}$0} indistinct. Conchoidal to uneven.

Hardness: 5.

Specific gravity: 3.17–3.23.

Color and streak: Greenish yellow, bluish green, grayish green, pale pink, violet; also blue, orange, colorless, brown, flesh to brownish red. White.

Luster and light transmission: Vitreous to greasy. Transparent to subtranslucent.

Optical properties:
   anisotropic, uniaxial (−): [Chlorapatite, some of which is monoclinic, may be uniaxial (−) or biaxial (−) with a 2V $\cong$ 10°; the carbonate apatites, which may be pseudo-hexagonal, may also give biaxial influence figures.]
   refractive indices: ω, 1.632; ϵ, 1.628. [Values for other apatites are: chlorapatite (ω,1.668; ϵ,1.665); hydroxylapatite (ω,1.651; ϵ,1.647); carbonate-fluorapatite (mean index, ~ 1.630); carbonate-hydroxylapatite (mean index, 1.520–1.610).] Low birefringence is characteristic.

Chemistry: See the statements in the introductory remarks.

Diagnostic features: Crystal form, color, luster, and hardness; for example, apatite is distinguished from beryl by its inferior hardness and from quartz by color and hardness. Massive, sugary varieties may resemble diopside or olivine but, again, the apatite is of inferior hardness and typically of darker color. Apatite is soluble in acids. The presence of phosphorus may be confirmed by putting a drop of HNO_3 on a surface and then adding some ammonium-molybdate, which yields a yellow ammonium phosphomolybdate precipitate.

Occurrence, associations, and uses: The minerals of the apatite series are, by far, the most abundant phosphorus-bearing minerals in the earth's crust. D. McConnell (1973, *Apatite. . .* New York: Springer-Verlag, 111 pp.) includes in a data summary the following occurrences for the diverse apatites:

1. Fluorapatite: in igneous rocks of essentially all compositions, in metamorphic rocks, in pegmatites, and as detrital grains in some sediments.

Figure 14-3. Fluorapatite; crystal, in Precambrian marble, with crystal edges rounded, as if dissolved or melted off, Mulvaney Property, Pitcairn, New York. Length of crystal is ~6 cm. (Photograph courtesy of S. C. Chamberlain.)

2. Chlorapatite: in veins associated with mafic igneous rocks, in certain iron deposits, and in silicified marbles.

3. Hydroxylapatite (typically, if not exclusively, containing F and/or Cl and, in some cases CO_2): in diverse metamorphic rocks.

4. Carbonate-fluorapatite: as the principal component of phosphorites and other phosphatic rocks, in hydrothermal veins, and as a biochemically formed component of some brachiopods and conodonts.

5. Carbonate-hydroxylapatite: as insular rock phosphates—that is, those formed as a result of interactions between calcareous materials, such as coral and bird excrement; as biological hard parts (e.g., bones and teeth); and as secondary deposits, typically crusts atop preexisting apatite crystals, in cavities in diverse rocks.

The apatite in igneous rocks occurs typically as microscopic crystals in only accessory amounts. Exceptions include apatite-rich segregations in alkalic rocks of the Kola Peninsula of the Soviet Union, in the rutile (and/or ilmenite)-apatite rock called nelsonite of Nelson County, Virginia, and also some apatite-rich carbonatite dikes in southern and eastern Africa.

Some of the apatites found in pegmatites are relatively large (up to about 250 mm along c). Many of these are Mn-bearing and fluoresce under all wavelengths in the ultraviolet range.

Massive, fine-grained or cryptocrystalline apatite—generally consisting largely or wholly of one of the carbonate apatites—is widely referred to as *collophane*, a rather useful field designation somewhat analogous to limonite, bauxite, and wad. Much of this material is horn-like, with a dense, layered, or colloform structure; some of it is concretionary; it may be pulverulent. It is typically light gray to nearly white, yellowish, or brown; has a dull luster; has an apparent hardness of only 3–4; and has a specific gravity between 2.5 and 2.9.

Apatite, as a source of phosphorus, has many uses. By far the greatest volume of it is used for fertilizers and is derived from phosphorite and other diagenetic, sedimentary, and residual deposits such as the "pebble phosphate" deposits of Florida. Other major uses include the production of phosphoric acid and detergents (e.g., tripolyphosphate).

Pyromorphite Series, $Pb_5[(PO_4),(AsO_4),(VO_4)]_3Cl$

Crystal system, class, and space group: Hexagonal; $6/m$; $P6_3m$.

	Pyromorphite	Mimetite	Vanadinite
Cell dimensions:	$a = 9.97$	$a = 10.26$	$a = 10.331$
	$c = 7.32$	$c = 7.44$	$c = 7.343$
Cell content:	$Z = 2$	$Z = 2$	$Z = 2$
Hardness:	$3\frac{1}{2}$–4	$3\frac{1}{2}$–4	$2\frac{1}{2}$–3
Specific gravity:	7.04	7.24	6.86
Optical properties:	uniaxial $(-)$	uniaxial $(-)$	uniaxial $(-)$
refractive indices: ω	2.058	2.124–2.263	2.416
ϵ	2.048	2.106–2.239	2.350

Pyromorphite, $Pb_5(PO_4)_3Cl$, and Mimetite, $Pb_5(AsO_4)_3Cl$

See also the preceding tabulated data.

Habit: Crystals common; typically prismatic and simple in habit with the commonest forms $\{10\bar{1}0\}$, $\{0001\}$, and $\{10\bar{1}1\}$; also in rounded barrel-shaped forms and as subparallel groups of prismatic crystals; commonly globular, reniform or wartlike; also granular or fibrous.

Cleavage: $\{10\bar{1}1\}$ indistinct.

Fracture: Uneven.

Color and streak: Green, yellow, or brown of various shades for pyromorphite, generally pale yellow to yellow-brown or colorless for mimetite. Nearly white.

Luster and light transmission: Resinous. Transparent to translucent.

Chemistry: These two minerals form a complete solid-solution series; the name *pyromorphite* applies to that half of the series with PO_4 predominant; the name *mimetite* applies to the half with AsO_4 predominant. Ca substitutes to a minor extent for Pb; V substitutes for (As,P) to a small extent in mimetite.

Diagnostic features: Crystal form, high luster, and high specific gravity. It is soluble in HNO_3. It is difficult, however, to distinguish pyromorphite from mimetite without chemical tests.

Occurrence and associations: These minerals are of secondary origin and occur in the oxidized zones of ore deposits that contain galena. In most cases, pyromorphite is more common than mimetite. Generally associated minerals are limonite and cerussite and, to a lesser extent, smithsonite, hemimorphite, anglesite, malachite, vanadinite, and wulfenite.

Vanadinite, $Pb_5(VO_4)_3Cl$

See also the data given in the introduction to this group of minerals.

Habit: Crystals common; typically short to long prismatic along c with the most common forms being $\{10\bar{1}0\}$, $\{0001\}$, $\{10\bar{1}1\}$ and $\{20\bar{2}1\}$; some crystals are cavernous (i.e., hollow prisms); also in rounded forms, in subparallel groupings, and in globules.

Fracture: Uneven to conchoidal.

Specific gravity: 6.88.

Color and streak: Orange-red to brownish red. White to yellowish.

Luster and light transmission: Subresinous. Nearly opaque.

Chemistry: In vanadinite, P may substitute for V to a small extent, and As may substitute for V up to about As:V = 1:1.

Diagnostic features: Distinguished by its luster, high specific gravity, crystal form, and, from pyromorphite-mimetite, by its color. Upon dissolving in HNO_3, it turns the solution yellow; upon dissolving in HCl, it turns the solution green.

Occurrence, associations, and uses: Vanadinite is a secondary mineral that occurs in the surface oxidized zone of ore deposits that contain galena and other sulfides. It is commonly associated with pyromorphite, wulfenite, cerussite, anglesite, and limonite. The vanadium in the mineral probably represents an enrichment from the sparse vanadium content of primary sulfide and gangue minerals.

Vanadinite is a minor source of vanadium, particularly from the oxidized ores of Broken Hill, Zambia, and the Otavi District of South West Africa.

HYDRATED PHOSPHATES CONTAINING HYDROXYL

Turquoise, $CuAl_6(PO_4)_4(OH)_8 \cdot 5H_2O$

Crystal system, class, and space group: Triclinic; $\bar{1}$; $P\bar{1}$.

Cell dimensions and content: $a = 7.48$, $b = 9.95$, $c = 7.68$; $\alpha = 111°39'$, $\beta = 115°23'$, $\gamma = 69°26'$; $Z = 1$.

Habit: Crystals very rare; typically massive, dense, and cryptocrystalline to fine granular; also as veinlets or crusts and concretionary.

Cleavage and fracture: {001} perfect; {010} good but rare. Conchoidal to smooth in massive material.

Hardness: 5–6.

Specific gravity: 2.6–2.8.

Color and streak: Sky-blue, bluish green to apple-green. White or greenish to pale green.

Luster and light transmission: Waxy, vitreous in crystals. Transparent to subtranslucent.

Optical properties:
anisotropic, biaxial (+): $2V \cong 40°$
refractive indices: α,1.61; β,1.62; γ,1.65
slightly pleochroic

Chemistry: Turquoise is a member of a mineral series that, through substitution of Fe for Al, probably extends to the relatively rare mineral *chalcosiderite*. Most turquoise falls near the high Al end of the series.

Diagnostic features: Its distinctive blue color. Also, after pulverization and heating, it is soluble in HCl.

Occurrence, associations, and use: Turquoise is a secondary mineral that occurs with limonite, chalcedony, and kaolin, and is formed by the action of surface waters on aluminous igneous, sedimentary, or metamorphic rocks. It is found chiefly in arid regions.

The fine blue cryptocrystalline material has been valued as an ornamental or semiprecious stone since ancient times. Fine quality turquoise has been obtained from numerous localities in the southwestern United States.

URANYL PHOSPHATES

Autunite Group
    Autunite, $Ca(UO_2)_2(PO_4)_2 \cdot 10\text{–}12H_2O$
    Torbernite, $Cu(UO_2)_2(PO_4)_2 \cdot 8\text{–}12H_2O$
Meta-autunite Group
    Meta-autunite, $Ca(UO_2)_2(PO_4)_2 \cdot 2\text{–}6H_2O$
    Metatorbernite, $Cu(UO_2)_2(PO_4)_2 \cdot 8H_2O$

The autunite group of minerals, which consists of some 13 recognized species, conforms to the general formula $A(UO_2)_2(XO_4)_2 \cdot 8–12H_2O$, in which A may be Ba, Ca, Cu, Fe, $\frac{1}{2}$(HAl), Mg, Mn, or Na$_2$, and X may be As, P, or V.

The meta-autunite group, which includes some 15 recognized species, conforms to the general formula $A(UO_2)_2(XO_4)2 \cdot 4–8H_2O$, in which A may be Ba, Ca, Co, Cu, Fe, K$_2$, Mg, (NH$_4$)$_2$, or Zn, and X may be As or P.

The two groups have similar layer structures. Minerals of the autunite group are common in natural surroundings, whereas those of the meta-autunite group are more likely to be found in collections. The transition temperatures between the pairs, however, differ with the identities of A and X and also with vapor pressure; some of these changes are reversible and, in most cases, they take place close to room temperature and humidity conditions.

	Autunite	Torbernite	Meta-autunite	Metatorbernite
Crystal system:	Tetragonal	Tetragonal	Tetragonal	Tetragonal
Crystal class:	4/m2/m2/m	4/m2/m2/m	422	4/m2/m2/m
Space group:	I4/mmm	I4/mmm	P4$_2$22	P4/nmm
Cell dimensions:	a = 7.00	a = 7.06	a = 7.00	a = 6.96
	c = 20.67	c = 20.5	c = 8.44	c = 8.62
Cell content:	Z = 2	Z = 2	Z = 1	Z = 1
Cleavage:	{001} perfect	{001} perfect		
Hardness:	2–2$\frac{1}{2}$	2–2$\frac{1}{2}$	—	2$\frac{1}{2}$
Specific gravity:	3.1–3.2	3.22	—	3.5–3.7
Color:	Lemon to greenish yellow	Emerald to apple-green		
Streak:	Yellowish	Pale green		
Luster:	Vitreous, pearly on {001}	Vitreous, pearly on {001}		
Light transmission:	Translucent	Transparent to translucent		
Optical properties:	uniaxial (−) [commonly appears bi-axial (−)]	uniaxial (−)	uniaxial (−)	uniaxial (+)
refractive indices: ω	1.577–1.578	1.590–1.592	1.595–1.613	1.618–1.649
ε	1.553–1.555 pleochroic in yellows	1.581–1.582 pleochroic in greens and/or blue-greens	1.585–1.600	1.622–1.646

Autunite, $Ca(UO_2)_2(PO_4)_2 \cdot 10-12H_2O$

See also the preceding tabulated data.

Habit: Crystals common; thin, tabular on {001}; common as sub-parallel fanlike groups, as scaly aggregates, and as thick crusts with serrated surfaces, all of which owe their appearance to crystals standing on edge.

Chemistry: Autunite commonly includes small amounts of Mg and/or Ba substituting for Ca. Upon slight heating or, in some cases, merely drying, autunite changes reversibly to meta-autunite; on further heating to $\sim$ 80°C, meta-autunite changes irreversibly to an orthorhombic phase not yet recorded in nature.

Diagnostic features: Strongly fluorescent in yellowish green in ultraviolet light; soluble in acids; thin cleavage flakes may be somewhat flexible. It is, however, difficult to distinguish from several other yellow secondary uranium minerals except by chemical tests or X-ray diffraction.

Occurrence: Autunite is a secondary mineral that typically occurs in the zone of weathering and oxidation of uranium-bearing veins or pegmatites, for example, at Spruce Pine, North Carolina. As noted, most specimens in collections are probably the secondary dehydration product meta-autunite.

Torbernite, $Cu(UO_2)_2(PO_4)_2 \cdot 8-12H_2O$

See also the data given in the introduction to this group of minerals.

Habit: Crystals common; thin to thick, tabular on {001}, typically square in outline, rarely pyramidal; also as subparallel, foliated, micaceous, or scaly aggregates or coatings.

Chemistry: There is no evidence for substitution of Ca, for example, for Cu, but Pb has been recorded. Small amounts of As substitute for P. Loss of water causes apparently irreversible transition to metatorbernite, even in, for example, a dry room.

Diagnostic features: Color and Cu content distinguish torbernite from autunite and other secondary uranium minerals except metatorbernite and *zeunerite* [$Cu(UO_2)_2(AsO_4)_2 \cdot 10-16H_2O$]. In addition, both crystals and cleavage lamellae of torbernite are extremely brittle, and torbernite, unlike autunite, does not fluoresce in ultraviolet radiation.

Occurrence: Torbernite is a secondary mineral often formed as a result of oxidation of uraninite in veins that also contain copper sulfides. Some torbernite is associated with autunite. As noted, most specimens in collections are probably metatorbernite.

Meta-autunite—Metatorbernite,
$Ca(UO_2)_2(PO_4)_2 \cdot 2-6H_2O$—$Cu(UO_2)_2(PO_4)_2 \cdot 8H_2O$

See the data given in the introduction to this group of minerals and the pertinent remarks under the descriptive notes about autunite and torbernite.

Carnotite, $K_2(UO_2)_2(VO_4)_2 \cdot 3H_2O$

Crystal system, class, and space group: Monoclinic; $2/m$; $P2_1/a$.

Cell dimensions and content: $a = 10.47$, $b = 8.41$, $c = 6.91$; $\beta = 103° 40'$; $Z = 2$.

Habit: Crystals rare; occurs chiefly as a powder, or as loosely coherent fine crystalline aggregates; in compact masses or disseminated; rarely as crusts or minute platy crystals flattened on {001}.

Cleavage: {001} perfect.

Hardness: about 2.

Specific gravity: 4–5.

Color and streak: Bright yellow to lemon-yellow.

Luster and light transmission: Dull or earthy. Translucent.

Optical properties:

anisotropic, biaxial (−): $2V = 43°-60°$

refractive indices: α,1.750–1.78; β,1.901–2.06; γ,1.92–2.08

Diagnostic features: Carnotite and *tyuyamunite* [$Ca(UO_2)_2(VO_4)_2 \cdot 5-8H_2O$], both of which are radioactive, are not easily distinguished except by chemical tests or X-ray diffraction.

Occurrence, associations, and uses: Carnotite and tyuyamunite commonly occur together as secondary minerals formed by the action of meteoric waters on uranium and vanadium minerals. They are widespread in sandstones of the Colorado Plateau area in Colorado and adjoining parts of Utah, New Mexico, and Arizona. They occur disseminated, or locally as relatively pure masses, around petrified tree trunks or other vegetal matter. They are mined primarily for their uranium content, but they are also important sources of vanadium.

15 Class VIII: Silicates

The silicates include about a third of all mineral species. Although many of these minerals are quite rare, others make up a large part of the earth's crust. In fact, the crust has been estimated to consist of about 95 percent silicate minerals, of which some 60 percent is feldspar and 12 percent quartz. The predominance of silicates and aluminosilicates reflects the abundance of oxygen, silicon, and aluminum, which are the most common elements in the crust (see page 179).

The number of elements that form silicate minerals is comparatively few. Some additional elements, such as Rb and Sr, do not form specific silicate minerals; instead, they are dispersed in small amounts, replacing more abundant elements in several, for the most part fairly common, silicate minerals. Still other elements that do not form common independent silicates occur associated with other cations (e.g., boron, which forms a number of borosilicates).

The great multiplicity of silicate minerals is due to the variety of silicates that can be formed from the same elements. Many have complex and variable compositions—complexities that hindered the development of a satisfactory classification of silicate minerals until X-ray investigations provided the means for determining their crystal structures.

Prior to the elucidation of their crystal structures, the composition of the silicates was generally interpreted in terms of hypothetical silicic acids, all of which were derived from a theoretical orthosilicic acid, H_4SiO_4. Some of the hypothetical silicic acids were as follows:

Orthosilicic acid	H_4SiO_4
Metasilicic acid	$H_2SiO_3 = (H_4SiO_4 - H_2O)$
Orthodisilicic acid	$H_6Si_2O_7 = (2H_4SiO_4 - H_2O)$
Metadisilicic acid	$H_2Si_2O_5 = (2H_4SiO_4 - 3H_2O)$
Trisilicic acid	$H_4Si_3O_8 = (3H_4SiO_4 - 4H_2O)$

Thence, ratios of silicon to oxygen could be derived to fit any composition. The silicic acid basis had some success in the interpretation of some of the simpler compounds, but it led to manifest absurdities when it was applied to more complex minerals. (For example, the solid-solution end-members albite, $NaAlSi_3O_8$, and anorthite, $CaAl_2Si_2O_8$, fell into different silicate groups.) Consequently, the silicic acid classification of the silicate minerals has been superseded by a classification, outlined below, that is based on crystal structure.

THE STRUCTURE AND CLASSIFICATION OF THE SILICATES

In nearly all silicate structures, the silicon atoms are in fourfold coordination with oxygen. The bonds between silicon and oxygen are so strong that the four oxygens are always at the corners of a tetrahedron of nearly constant dimensions and regular shape, whatever the configuration of the rest of the structure (Figure 15-1). Most of the common silicate minerals can be classified as belonging to one of the different silicate types (see Table 15-1 and Figures 15-2 through 15-5); as shown, the tetrahedra may exist as separate and distinct units or they may be linked by sharing their corners (i.e., their oxygens). Silicate classification for most silicate minerals is based on the types of linkages as follows:

1. *Independent tetrahedral groups:* In this class, the silicon-oxygen tetrahedra are present as separate entities (Figure 15-2a). The resultant composition is SiO_4, which gives such minerals as forsterite, Mg_2SiO_4. This division of the silicates, with the Si:O ratio of 1:4, is known as the *nesosilicates*, from the Greek root meaning island.

2. *Double tetrahedra structures:* Two silicon-oxygen tetrahedra are linked by the sharing of one oxygen between them; the resulting composition is Si_2O_7, with an Si:O ratio of 2:7 (Figure 15-2b); a representative mineral is hemimorphite, $Zn_4Si_2O_7(OH)_2 \cdot H_2O$. Such silicates are known as *sorosilicates*, from the Greek root meaning group.

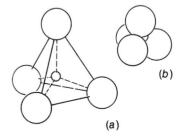

Figure 15-1. The silica tetrahedron. (a) Expanded view showing relatively large oxygen ions at corners and a small silicon ion at the center of the tetrahedron; short dashed lines represent bonds between silicon and surrounding oxygen ions. (b) Tetrahedron with correct spatial arrangement, dashed circle indicates central silicon ion.

Table 15-1.   Structural classification of the silicates

Classification	Structural Arrangement[a]	Formula of Complex Anion	Si:O	Example
Nesosilicates	Independent tetrahedra	$(SiO_4)^{-4}$	1:4	Forsterite, $Mg_2(SiO_4)$
Sorosilicates	Two tetrahedra sharing one oxygen 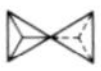	$(Si_2O_7)^{-6}$	2:7	Hemimorphite, $Zn_4(Si_2O_7)(OH)_2 \cdot H_2O$
Cyclosilicates	Closed rings of tetrahedra, each sharing two oxygens			
		$(Si_3O_9)^{-6}$	1:3	Benitoite, $BaTi(Si_3O_9)$
		$(Si_4O_{12})^{-6}$		Axinite, $Ca_3Al_2(BO_3)(Si_4O_{12})(OH)$
	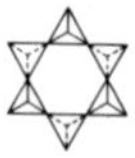	$(Si_6O_{18})^{-6}$		Beryl, $Be_3Al_2(Si_6O_{18})$
Inosilicates	Continuous chains[b] of tetrahedra		1:3 and	
	Single chain, each tetrahedron sharing two oxygens 	$(SiO_3)_m^{-2}$	4:11	Enstatite, $Mg_2(SiO_3)_2$
	Double chain, alternate tetrahedra sharing two and three oxygens 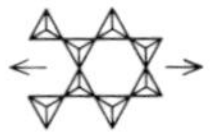	$(Si_4O_{11})_m^{-6}$		Anthophyllite, $Mg_7(Si_4O_{11})_2(OH)_2$
Phyllosilicates	Continuous sheets of tetrahedra, each sharing three oxygens 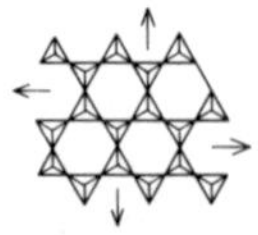	$(Si_2O_5)_m^{-2}$	2:5	Kaolinite, $Al_4(Si_2O_5)_2(OH)_8$
Tektosilicates	Three-dimensional framework of tetrahedra, each sharing all four oxyens (see Figure 15-5)	$(SiO_2)^0$	1:2	Quartz, SiO_2

[a] By sharing an oxygen, silica tetrahedra can polymerize to form larger, complex anions. In the diagrams in this table, oxygens and silicons are not shown as such; an oxygen would be at each apex and a silicon would be at the center of each tetrahedron. The complex anionic groups are drawn as viewed from above; solid lines are in or above, whereas broken lines are below, the plane of the paper.

[b] There is also a triple chain arrangement, giving $(Si_3O_8)_n^{-4}$, that is fundamental to a few macroscopically indistinguishable minerals—e.g., chesterite and jimthompsonite.

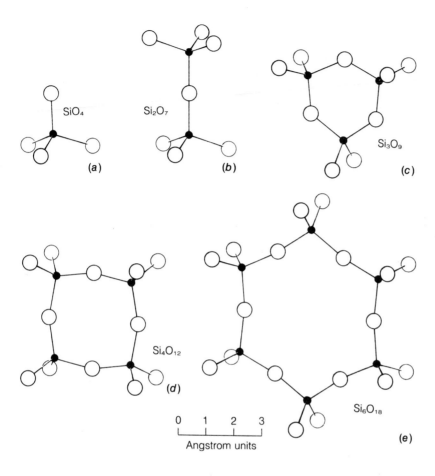

Figure 15-2. Types of linkage of silicon-oxygen tetrahedra. (a) Independent tetrahedra. (b) Double tetrahedra. (c), (d), and (e) Ring structures.

3. *Ring structures:* In this class, two of the oxygens of each tetrahedron are shared with neighboring tetrahedra, and the angular positions of the tetrahedra are such that closed units of a ringlike structure result. Rings of three, four, and six tetrahedra are known (Figures 15-2c, d, and e). Typical examples, each with an Si:O ratio of 1:3, are benitoite, $BaTiSi_3O_9$, with three linked tetrahedra; axinite, $(Ca,Mn,Fe)_3Al_2(BO_3)Si_4O_{12}(OH)$, with four; and beryl, $Be_3Al_2Si_6O_{18}$, with six. This division of the silicates is known as the *cyclosilicates* (Greek *cyclo*, ring).

4. *Chain structures:* In chain structures, tetrahedra are joined together to produce chains of indefinite extent. There are two principal modifications of this structure yielding somewhat different compositions: (a) single chains, in which the Si:O ratio is 1:3 (Figure 15-3a), as characterized by the pyroxenes and pyroxenoids; and (b) double chains, in which alternate tetrahedra in two parallel single chains are cross-linked and the Si:O ratio is 4:11 (Figure 15-3b), as characterized by the amphiboles. The chains may be indefinite in extent, are generally elongated in the c direction of the crystal, and

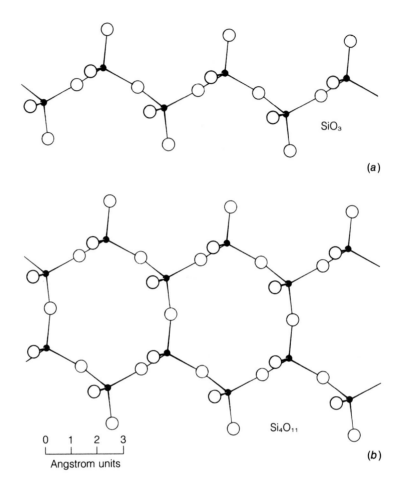

SiO₃

(a)

Si₄O₁₁

(b)

Figure 15-3.   Types of linkage of silicon-oxygen tetrahedra. (a) Single chains. (b) Double chains.

0   1   2   3
Angstrom units

are bonded to each other by the metallic elements. This division of the silicates is known as *inosilicates* (Greek *ino*, thread or chain).

5. *Sheet structures:*   In sheet structures, three oxygens of each tetrahedron are shared with adjacent tetrahedra to form flat sheets of indefinite extent. In essence, this structure is the double-chain inosilicate structure extended indefinitely in two directions instead of only one. The linkage, which gives an Si:O ratio of 2:5 (Figure 15-4), is the fundamental unit in all mica and clay structures. The sheets form a hexagonal planar network that is responsible for the principal characteristics of minerals of this type—their pronounced pseudo-hexagonal habit and their perfect basal cleavage parallel to the planes of the sheets. This division of the silicates is known as *phyllosilicates* (Greek *phyllo*, leaf or sheet).

6. *Three-dimensional networks:*   In three-dimensional networks, every SiO_4 tetrahedron shares all of its corners with other tetrahedra,

giving a three-dimensional network in which the Si:O ratio is 1:2 (Figure 15-5). The forms of silica—quartz, tridymite, cristobalite—have this arrangement. In SiO_2, the positive and negative charges balance. In most silicates of this type, the silicon is partly replaced by aluminum, so the composition is $(Si,Al)O_2$ and the substitution of Al^{+3} for Si^{+4} requires additional positive ions to restore electrical neutrality [e.g., nepheline, $Na(AlSiO_4)$]. The feldspars and the zeolites are also examples of this division of the silicates, which is known as the *tektosilicates* (from the Greek root meaning framework).

Consequently, most of the common silicate minerals can be classified as summarized in Table 15-1. Some minerals, however, contain groupings of more than one type or belong to other types (e.g., the xonotlite [$Ca_6Si_6O_{17}(OH)_2$] type of chain structure and the double ring structure of osumilite [$(K,Na)(Fe,Mg)_2(Al,Fe)_3(Si,Al)_{12}O_{30} \cdot H_2O$]).

The other constituents of a silicate structure—such as additional oxygen atoms, hydroxyl groups, water molecules, and cations—are arranged with the silicate groups in such a way as to produce mechanically stable and electrically neutral structures. Aluminum, after silicon the most abundant cation in the earth's crust, plays an especially noteworthy role. As we mentioned earlier, Al is stable in both fourfold and sixfold coordination. Consequently, it can replace silicon in the SiO_4 groups, and it can also have the same role as the common six-coordination cations Mg^{+2}, Fe^{+2}, Fe^{+3}, etc. In many mineral groups

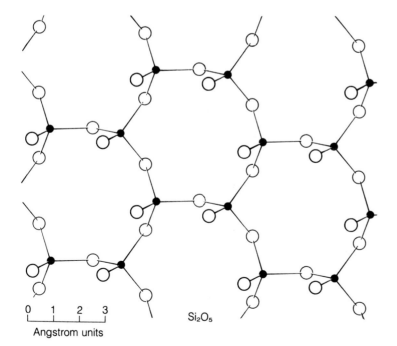

Si_2O_5

0 1 2 3
Angstrom units

Figure 15-4. Types of linkage of silicon-oxygen tetrahedra. Sheet structure.

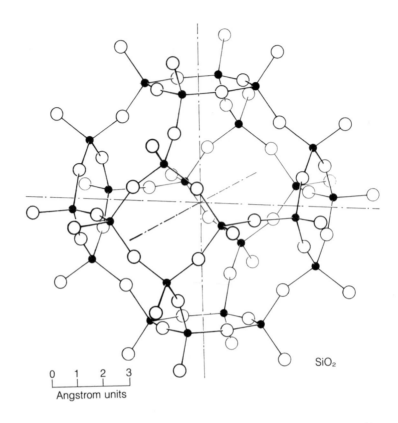

SiO₂

0 1 2 3
Angstrom units

Figure 15-5. Types of linkage of silicon-oxygen tetrahedra. Three-dimensional network.

(e.g., the garnets and the feldspars), aluminum is present entirely in a single coordination; in others (e.g., the amphiboles, the pyroxenes, and the micas), it may be present in both coordinations.

The valence charge on the silicate unit, which determines the number and charge of the other ions present in a structure, can be calculated easily for any unit if you only remember that each silicon has a positive charge of 4 and that each oxygen has a negative charge of 2. Thus, the charge on a single SiO_4 unit is $[1(+4) + 4(-2) =] -4$; on an Si_2O_7 unit, $[2(+4) + 7(-2) =] -6$; on an SiO_3 unit, $[1(+4) + 3(-2) =] -2$; on an Si_4O_{11} unit, $[4(+4) + 11(-2) =] -6$; on an Si_2O_5 unit, $[2(+4) + 5(-2)=] -2$; and on SiO_2, $[1(+4) + 2(-2) =] 0$.

As might be expected, different compositions, habits, and physical properties correspond to different linkages of the tetrahedra in the various types of silicates. Several examples are rather obvious: the bonding within the Si-O framework is much stronger than the bonding between the metal cations and the framework, so the cleavage planes of the silicates are parallel to Si-O chains or sheets. The sheet structure, for instance, produces the platy form of the micas, the chlorites, the kaolins, and several other minerals. The chain structures produce prismatic or fibrous crystals, as exemplified by the pyroxenes and amphiboles. The three-dimensional network structures are commonly manifested by more-or-less equidimensional crystals, and there are characteristic density ranges and refractive index limits that correspond to differences of structure. In general, increased complexity of silicate linkage results in looser packing of the ions, which results in a trend toward lower densities and refractive indices for inosilicates than for, for example, nesosilicates, etc. (For comparable compositions, see Table 15-2.)

Table 15-2. *Relationship between structure type and specific gravity in the magnesium silicate minerals*

Structure Type	Mineral Example	Specific Gravity
Nesosilicate	Forsterite, Mg_2SiO_4	3.22
Inosilicate (single chain)	Enstatite, $Mg_2(SiO_3)_2$	3.18
Inosilicate (double chain)	Anthophyllite, $Mg_7(Si_4O_{11})_2(OH)_2$	2.86
Phyllosilicate	Talc, $Mg_3(Si_2O_5)_2(OH)_2$	2.82

In discussing the composition and structure of the silicate minerals, we often find it convenient to use a general formula that fits all species within a given group. Several of the general formulas in the mineral descriptions that follow this introduction make use of the following symbols:

W = large cations having coordination number greater than six; principally Ca, Na, and K.

X = medium-sized bivalent cations (and lithium) in sixfold coordination; principally Mg and Fe^{+2}.

Y = medium-sized trivalent and quadrivalent cations in sixfold coordination; these include Al, Fe^{+3}, and Ti^{+4}.

Z = small cations in four-coordination; principally Si, commonly replaced by Al (and also by B in a few minerals).

The silicate minerals that are described in detail in this book are:

Subclass Tektosilicates

Silica Group:	
Quartz	SiO_2
Tridymite	SiO_2
Cristobalite	SiO_2
Opal	$SiO_2 \cdot n H_2O$
Feldspar Group:	WZ_4O_8
Sanidine	$KAlSi_3O_8$
Orthoclase	$KAlSi_3O_8$
Microcline	$KAlSi_3O_8$

Subclass Tektosilicates (*Continued*)

Plagioclase Series:	$(Na,Ca)\ Al\ (Al,Si)\ Si_2O_8$
Albite	$Ab_{100}An_0-Ab_{90}An_{10}$
Oligoclase	$Ab_{90}An_{10}-Ab_{70}An_{30}$
Andesine	$Ab_{70}An_{30}-Ab_{50}An_{50}$
Labradorite	$Ab_{50}An_{50}-Ab_{30}An_{70}$
Bytownite	$Ab_{30}An_{70}-Ab_{10}An_{90}$
Anorthite	$Ab_{10}An_{90}-Ab_0An_{100}$
Scapolite Series:	$(Na,Ca)_4\ [(Al,Si)_4O_8]_3\ (Cl,CO_3)$
Feldspathoid Group:	
Leucite	$KAlSi_2O_6$
Nepheline	$(Na,K)\ AlSiO_4$
Sodalite	$Na_8(AlSiO_4)_6Cl_2$
Cancrinite	$Na_6Ca_2\ (AlSiO_4)_6\ (CO_3)_2$
Zeolite Group:	$W_mZ_rO_{2r}\cdot s\,H_2O$
Heulandite	$(Na,Ca)_{2-3}Al_3(Al,Si)_2Si_{13}O_{36}\cdot 12H_2O$
Stilbite	$NaCa_2Al_5Si_{13}O_{36}\cdot 14H_2O$
Laumontite	$CaAl_2Si_4O_{12}\cdot 4H_2O$
Chabazite	$CaAl_2Si_4O_{12}\cdot 6H_2O$
Analcime	$NaAlSi_2O_6\cdot H_2O$
Natrolite	$Na_2Al_2Si_3O_{10}\cdot 2H_2O$

Subclass Phyllosilicates

Clay minerals:	
Kaolinite	$Al_4Si_4O_{10}(OH)_8$
Montmorillonite	$(Na,Ca)_{0.33}(Al,Mg)_2Si_4O_{10}(OH)_2\cdot n\,H_2O$
Illite	$(K,H_3O)(Al,Mg,Fe)_2(Si,Al)_4O_{10}$
	$[(OH)_2,H_2O]$
Vermiculite	$(Mg,Fe,Al)_3(Si,Al)_4O_{10}(OH)_2\cdot 4H_2O$
Serpentine	$(Mg,Fe,Ni)_3Si_2O_5(OH)_4$
Pyrophillite	$Al_2Si_4O_{10}(OH)_2$
Talc	$Mg_3Si_4O_{10}(OH)_2$
Mica Group:	$W(X,Y)_{2-3}(Z_4O_{10})\ (OH)_2$
Muscovite	$KAl_2(AlSi_3O_{10})\ (OH)_2$
Phlogopite	$KMg_3(AlSi_3O_{10})\ (F,OH)_2$
Biotite	$K(Mg,Fe)_3(Al,Fe)Si_3O_{10}(OH,F)_2$
Lepidolite	$K(Li,Al)_3(Si,Al)_4O_{10}(OH,F)_2$
Glauconite	$(K,Na)\ (Fe,Al,Mg)_2(Si,Al)_4O_{10}\ (OH)_2$
Prehnite	$Ca_2Al(AlSi_3O_{10})\ (OH)_2$
Chlorite Series:	$\sim(Mg,Fe,Al)_{5-6}(Si,Al)_4O_{10}(OH)_8$
Stilpnomelane	$\sim K(Fe,Al)_{10}Si_{12}O_{30}(OH)_{12}$
Apophyllite	$KCa_4(Si_4O_{10})_2(OH,F)\cdot 8H_2O$

Subclass Inosilicates (*Continued*)

Amphibole Group:	$W_{0-1}X_2Y_5Z_8O_{22}\ (OH,F,Cl)_2$
Anthophyllite Series:	$(Mg,Fe)_7Si_8O_{22}(OH)_2$

Subclass Inosilicates (*Continued*)

Cummingtonite Series:	$(Mg,Fe)_7Si_8O_{22}(OH)_2$
Tremolite-Actinolite Series:	$Ca_2(Mg,Fe)_5Si_8O_{22}(OH)_2$
"Hornblende Series":	$Ca_2(Mg,Fe)_4Al(Si_7Al)O_{22}(OH,F)_2$
Alkali–Amphibole Series:	$\sim Na_2(Mg,Fe,Al)_5Si_8O_{22}(OH)_2$
Pyroxene Group:	$(W,X,Y)_2Z_2O_6$
Enstatite-Hypersthene Series:	$(Mg,Fe)_2Si_2O_6$
Pigeonite	$\sim(Mg,Fe,Ca)\ (Mg,Fe)Si_2O_6$
Diopside–Hedenbergite Series:	$Ca(Mg,Fe)\ Si_2O_6$
Augite	$(Ca,Na)(Mg,Fe,Al,Ti)(Si,Al)_2O_6$
Acmite	$NaFeSi_2O_6$
Jadeite	$Na(Al,Fe)Si_2O_6$
Spodumene	$LiAlSi_2O_6$
Pyroxenoid Group:	
Wollastonite	$CaSiO_3$
Pectolite	$NaCa_2Si_3O_8(OH)$
Rhodonite	$(Mn,Fe,Mg,Ca)SiO_3$

Subclass Cyclosilicates

Axinite	$(Ca,Mn,Fe)_3Al_2(BO_3)Si_4O_{12}(OH)$
Beryl	$Be_3Al_2Si_6O_{18}$
Cordierite	$Al_3(Mg,Fe)_2(AlSi_5O_{18})$
Tourmaline	$WX_3Y_6(BO_3)_3(Si_6O_{18})(OH,F)_4$

Subclass Sorosilicates

Lawsonite	$CaAl_2Si_2O_7(OH)_2 \cdot H_2O$
Pumpellyite	$Ca_2MgAl_2(SiO_4)(Si_2O_7)(OH)_2 \cdot H_2O$
Melilite Group:	
Gehlenite	$Ca_2Al(AlSiO_7)$
Akermanite	$Ca_2Mg(Si_2O_7)$
Hemimorphite	$Zn_4Si_2O_7(OH)_2 \cdot H_2O$
Vesuvianite	$Ca_{10}Mg_2Al_4(SiO_4)_5(Si_2O_7)_2(OH)_4$
Epidote Group:	$W_2(X,Y)_3Z_3O_{12}(OH)$
Zoisite	$Ca_2Al_3Si_3O_{12}(OH)$
Clinozoisite	$Ca_2Al_3Si_3O_{12}(OH)$
Epidote	$Ca_2(Al,Fe)_3Si_3O_{12}(OH)$
Allanite	$(Ce,Ca,Y)_2(Al,Fe,Mg)_3Si_3O_{12}(OH)$

Subclass Nesosilicates

Olivine Series:	$(Mg,Fe)_2SiO_4$
Willemite	Zn_2SiO_4
Zircon	$ZrSiO_4$
Thorite	$ThSiO_4$
Garnet Group:	$X_3Y_2(SiO_4)_3$
Almandine	$Fe_3Al_2(SiO_4)_3$
Pyrope	$Mg_3Al_2(SiO_4)_3$

Subclass Nesosilicates (*Continued*)

Spessartine	$Mn_3Al_2(SiO_4)_3$
Grossular	$Ca_3Al_2(SiO_4)_3$
Andradite	$Ca_3Fe_2(SiO_4)_3$
Uvarovite	$Ca_3Cr_2(SiO_4)_3$
Titanite	$CaTiSiO_5$
Aluminum Silicate Group:	
Andalusite	Al_2SiO_5
Sillimanite	Al_2SiO_5
Kyanite	Al_2SiO_5
Staurolite	$(Fe,Mg,Zn)_2Al_9(Si,Al)_4O_{22}(OH)_2$
Topaz	$Al_2SiO_4(F,OH)_2$
Humite Group:	
Norbergite	$Mg_3(SiO_4)(F,OH)_2$
Chondrodite	$(Mg,Fe)_5(SiO_4)_2(F,OH)_2$
Humite	$(Mg,Fe)_7(SiO_4)_3(F,OH)_2$
Clinohumite	$(Mg,Fe)_9(SiO_4)_4(F,OH)_2$
Datolite	$CaBSiO_4(OH)$
Chloritoid	$(Fe,Mg,Mn)_2(Al,Fe)Al_3O_2(SiO_4)_2(OH)_4$

Silicate of Unknown Structure

Chrysocolla	$(Cu,Al)_2H_2Si_2O_5(OH)_4 \cdot nH_2O$

SUBCLASS TEKTOSILICATES

The tektosilicates include some of the most important rock-forming minerals. The simplest is quartz, SiO_2, in composition an oxide, but included here because its structure consists of a framework of SiO_4 tetrahedra with each oxygen linked to a silicon atom in a neighboring tetrahedron. Except for quartz and its polymorphs, all the minerals in this subclass are aluminosilicates since, in order to produce a net negative charge on the framework, some of the Si^{+4} is replaced by Al^{+3}. Up to half the Si^{+4} may be replaced by Al^{+3}; the most common Si:Al ratios are 1:1 and 3:1, which correspond to formula units of the type $AlSiO_4^{-1}$ and $AlSi_3O_8^{-1}$, respectively.

The cations that balance the negative charge on these formula units are large ions (radius approximately 1 Å or greater) with coordination numbers of eight or greater. Smaller six-coordination cations, such as magnesium and iron, do not form tektosilicates, which is understandable when we consider the net negative charge on each oxygen in the tektosilicate structure; for $AlSiO_4^{-1}$, it is one-fourth, for $AlSi_3O_8^{-1}$, it is one-eighth. A bivalent ion in six-coordination, such as Mg^{+2}, requires a contribution of one-third of a negative charge from each of the surrounding oxygens, which is not available in a tektosilicate structure. Only univalent or bivalent cations with a coordination of eight or more can compensate the negative charge on the aluminosilicate framework.

Hence, the tektosilicates are aluminosilicates of sodium, potassium, calcium, and barium.

All the tektosilicates are colorless, white, or pale gray when free from inclusions. They also have rather low densities, as a result of the comparatively open nature of the tektosilicate structure. Their hardnesses are rather uniform, about 4–6. On the whole the tektosilicates form a remarkably homogeneous subclass, with many similarities in composition and properties common to all minerals within it.

SILICA GROUP

Silica, SiO_2, occurs in nature as *quartz, tridymite, cristobalite, coesite, stishovite,* and *lechatelierite.* Of these, quartz is very common; tridymite and cristobalite are widely distributed, particularly in volcanic rocks, and can hardly be termed rare; coesite and stishovite occur in only a few loci, such as in meteorite impact craters, and are rare; and lechatelierite, not a mineral but glass, which occurs, for example, in fulgurites, is also rare. Of these, only quartz, tridymite, and cristobalite are described further in this book.

In addition, there is a natural hydrated silica ($SiO_2 \cdot n\,H_2O$), *opal.* Opal, which is relatively common and widespread, is also described in this book.

Each of the crystalline phases has its own stability field (Figure 15-6). At atmospheric pressure: quartz is the stable form up to 867°C; tridymite is stable between 867°C and 1470°C; cristobalite is stable from 1470°C

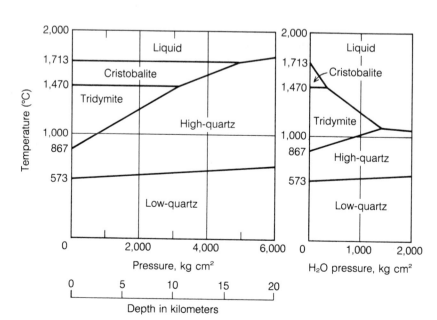

Figure 15-6. Stability relations of the different forms of SiO_2; dry system on the left, system under water vapor pressure on the right.

up to the melting point at 1713°C; and, from 1713°C to the boiling point, liquid silica is the stable phase. The relatively great difference in density between quartz and the other polymorphs means that pressure has a marked effect on inversion temperatures. It also means that tridymite is not stable above about 3000 kg cm^{-2}, nor cristobalite above about 5000 kg cm^{-2}. The absence of tridymite and cristobalite from plutonic rocks, even those that may have crystallized at high temperatures, is thus readily explicable, since pressures of the above order are attained at comparatively shallow depths in the crust.

Each of these three polymorphs of silica is built of SiO_4 tetrahedra that are linked together to form a three-dimensional network. The pattern of linkage, however, is different for each; hence, the difference in their crystal structures and their properties. Note, for example, that the just-mentioned differences in specific gravity values may be attributed to the fact that cristobalite and tridymite have comparatively open structures, whereas the atoms in quartz are more closely packed.

Each of these three polymorphs has a high- and a low-temperature modification. In quartz, for example, the change from the one to the other takes place at 573°C at atmospheric pressure. Similarly, although not shown on the diagram, there are a high-tridymite that changes into low-tridymite at 117°C and a high-cristobalite that changes into low-cristobalite between 200°C and 268°C. The inversion from the high- to the low-temperature forms of the individual species is of quite another order than inversions between the species; this is true because the three minerals have their SiO_4 tetrahedra linked together according to different schemes, and the linkage of each has to be broken down completely and rearranged for transformation from one to another. On the other hand, the change from a high-temperature to a low-temperature form of the same species does not alter the way in which the tetrahedra are linked; they merely undergo a displacement and rotation that alters the symmetry of the structure without breaking any of the links. All of the high-temperature modifications are more symmetric than their low-temperature modifications.

The high-low transformation of each mineral takes place rapidly at the transition temperature and is reversible; actually the high- to low-cristobalite inversion is a bit slow; although it takes place at 268°C on heating, it appears to take place over a rather broad temperature range on cooling. Contrariwise, the changes from one polymorphic form to another are extremely sluggish, and the existence of tridymite and cristobalite as minerals show that they can remain unchanged almost indefinitely at ordinary temperatures. As we pointed out in Chapter 3, the presence of foreign elements in the structure also may have a stabilizing effect on tridymite and cristobalite. The few comprehensive analyses of these minerals show the presence of Na and Al, suggesting a substitution of NaAl for Si in these open structures. Quartz, on the other hand, is generally very pure SiO_2.

Two other phenomena of great significance also merit mention here:

1. Even at temperatures below 867°C, especially when crystallization takes place rapidly (e.g., in the presence of mineralizers such as hot gases), cristobalite and/or tridymite may crystallize, although quartz is the stable phase.

2. High-quartz and low-quartz are formed only within their stability fields, never at higher temperatures.

From these facts, we can draw the following conclusions: quartz in an igneous rock signifies that its crystallization from magma took place below 867°C (with due regard for the effect of pressure); on the other hand, the presence of cristobalite or tridymite proves nothing as to the temperature of crystallization.

As we already pointed out, at ordinary temperatures, quartz is always present as low-quartz. But, by the crystal form (Figure 15-7), the nature of the twinning, and other properties, we can often determine the identity of the original form. For example, it can be shown that in the quartz-bearing igneous rocks, quartz crystallized originally as high-quartz (i.e., above 573°C), whereas in many quartz veins and some pegmatites, it crystallized originally as low-quartz (i.e., below 573°C).

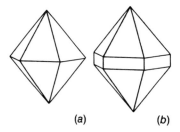

(a)   (b)

Figure 15-7. Quartz, common high-temperature (high-quartz) crystal habits. (a) $\{10\bar{1}1\}$ hexagonal dipyramid; (b) $\{10\bar{1}1\}$ and a short $\{10\bar{1}0\}$ hexagonal prism.

Quartz, SiO_2

Crystal system, class, and space group (low-quartz): Trigonal; 32; $P3_121$-right hand, $P3_221$-left hand.

Cell dimensions and content: $a = 4.9124$, $c = 5.4052$; $Z = 3$.

Habit: Crystals common; typically prismatic, terminated by two different unit rhombohedra. When these two are equally developed, the appearance is that of a hexagonal dipyramid; in most cases, however, inequality in the rate of growth has caused one rhombohedron form to be better developed than the other and, consequently, alternate rhombohedron faces are similar, whereas adjacent ones are different (Figure 15-8c). Prism faces are commonly striated horizontally, the striations being due to the incipient development of several rhombohedron faces. The wide diversity of shape of quartz crystals is the result of unequal development of the prisms and rhombohedra, and unequal development of like faces through differences in rate of growth. The trigonal pyramids appear as small faces truncating the corners between rhombohedron and prism faces; the trapezohedron faces generally appear in combination with the trigonal pyramids, beveling the edges between the pyramids and the prism faces (Figures 15-8c and d).

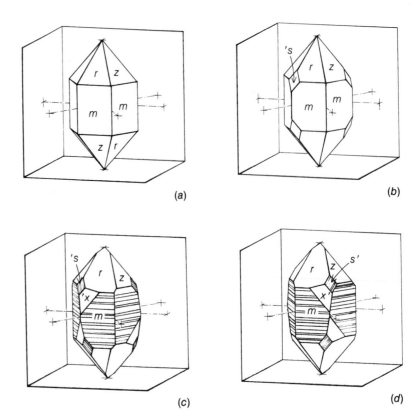

Figure 15-8.   Common forms in quartz crystals. (a) Prism and two rhombohedra. (b) Prism, two rhombohedra, and left trigonal pyramids. (c) Prism, two rhombohedra, left trigonal pyramid, and left trigonal trapezohedron. (d) Prism, two rhombohedra, right trigonal pyramid, and right trigonal trapezohedron.

$m = \{10\bar{1}0\}$ $'s = \{2\bar{1}\bar{1}1\}$
$r = \{10\bar{1}1\}$ $'x = \{6\bar{1}\bar{5}1\}$
$z = \{01\bar{1}1\}$ $s' = \{11\bar{2}1\}$
$\qquad\quad x' = \{51\bar{6}1\}$

Twinning: General, but seldom observable in the external form of the crystals. Common twins are the Dauphiné type, a penetration twin with the c axis as twin axis (Figures 2-78a and c and Figure 15-9), and the Brazil type, a penetration twin with $\{11\bar{2}0\}$ as the twin plane (Figure 2-78b). Contact twins with $\{11\bar{2}2\}$ as twin plane (Japanese Law) (Figure 15-10) are of less common occurrence. The Brazil type necessarily involves right- and left-handed quartz, whereas the other types do not. Twinning is the cause of a large amount of rejected and wasted material in the manufacture of piezoelectric plates since these plates must be free of twinning.

Common forms and angles: (Figures 15-6, 15-7, and 15-8):
$(10\bar{1}1) \wedge (\bar{1}101) = 85°46'$ $(10\bar{1}0) \wedge (01\bar{1}1) = 66°52'$
$(10\bar{1}1) \wedge (01\bar{1}1) = 46°16'$ $(10\bar{1}0) \wedge (11\bar{2}1) = 37°58'$
$(10\bar{1}0) \wedge (10\bar{1}1) = 38°13'$ $(10\bar{1}0) \wedge (51\bar{6}1) = 12°1'$

Cleavage, parting, and fracture: Generally none. Some quartz exhibits an indistinct rhombohedral parting. Conchoidal.

Hardness: 7.

Specific gravity: 2.65 in macrocrystalline varieties; lower (~2.60) in cryptocrystalline varieties.

Color and streak: Typically colorless or white, but can be practically any color. White.

Luster and light transmission: Vitreous in macrocrystalline varieties, commonly waxy or dull in cryptocrystalline varieties. Transparent to subtranslucent.

Optical properties:

anisotropic, uniaxial (+)

refractive indices: ω, 1.544; ϵ, 1.553

Chemistry: The composition of all pure quartz is close to 100 percent SiO_2. The structure is such that there is very limited space for accommodating extra atoms, and silicon cannot be replaced readily by any other common quadrivalent cation.

Varieties: Quartz occurs in two distinct types, the macrocrystalline and the cryptocrystalline varieties (Figure 15-11), the latter often being classed as the subspecies chalcedony. Chalcedony consists of quartz crystallites, commonly fibrous in form and with submicroscopic pores that cause it to have a somewhat lower density than macrocrystalline quartz.

Many names are currently applied to different varieties of quartz; some of the more common names and characteristics of the material to which they are applied are as follows:

Macrocrystalline Varieties

Rock Crystal: transparent colorless material (at some localities, this variety has been given such names as *Herkimer diamonds*).

Milky: milk-white and almost opaque, luster typically somewhat greasy; occurs in large masses in veins and pegmatites.

Amethyst: transparent purple material, much used as an ornamental and semiprecious stone; most occurs lining cavities in volcanic rocks.

Rose: rose-red or pink, rarely as crystals.

Yellow (citrine): yellow and transparent; cut stones are sometimes sold as topaz. Amethyst on heating often turns yellow and much citrine is actually heat-treated amethyst.

Smoky: transparent and semitransparent brown, gray to nearly black varieties; sometimes called *cairngorm* because material from Cairngorm in Scotland has been widely used as an ornamental stone. Radiation from radioactive material often causes a smoky appearance in colorless quartz, and this process may be responsible for much of the smoky quartz in nature.

Cryptocrystalline Varieties (Chalcedony)

Carnelian and Sard: red and reddish brown chalcedony.

Heliotrope and Bloodstone: green chalcedony with red spots.

Figure 15-9. Quartz. Smoky quartz with hexagonal prism $\{10\bar{1}0\}$ positive and negative rhombohedra $\{10\bar{1}1\}$ and $\{01\bar{1}1\}$ and left trigonal trapezohedron $\{6\bar{1}5\bar{1}\}$, twinned according to the Dauphiné Law with c as twin axis giving repetition of the left trigonal trapezohedron faces with a rotation of 60° about c; Switzerland, 8 cm wide. (Courtesy of Royal Ontario Museum.)

Figure 15-10. Quartz. Twinned crystal according to Japanese Law with twin plane $(11\bar{2}2)$; forms are hexagonal prism $\{10\bar{1}0\}$, striated parallel to a, and positive and negative rhombohedra $\{10\bar{1}1\}$ and $\{01\bar{1}1\}$; Kai Province, Japan; 6 cm wide. (Courtesy of Royal Ontario Museum.)

398

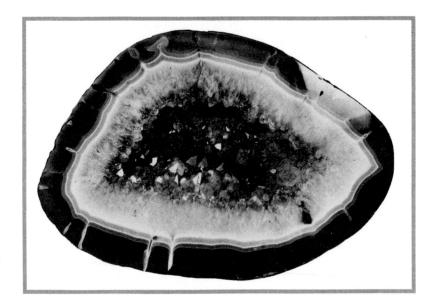

Figure 15-11. Quartz geode. A cavity in basalt was first lined with cryptocrystalline chalcedony, then with banded agate, and finally with coarsely crystalline quartz; Uruguay, South America. Greatest dimension is 18 cm. (Courtesy of the American Museum of Natural History.)

Agate: banded forms of chalcedony, the banding being due to interrupted deposition in cavities, the bands typically being parallel to the walls of the cavity (Figure 15-11). Some natural agate shows attractive color banding, but much of the colored agate used for ornaments has been artificially dyed, the slightly porous nature of the chalcedony allowing it to absorb material from solution. *Onyx* is the name generally given to plane-banded, alternately light and dark agate; it is used widely for carved cameos.

Moss Agate: white or cream-colored chalcedony enclosing brown or black dendritic mosslike aggregates of manganese oxides.

Jasper: opaque chalcedony, generally red, yellow, or brown, the color being due to included colloidal particles of iron oxides.

Chert and Flint: massive opaque chalcedony, typically white, pale yellow, brownish, gray, or black, and occurring as nodules or extensive beds in sedimentary rocks. The wide occurrence, the hardness and toughness, the conchoidal fracture, and the sharp cutting edge led to the widespread use of flint and chert by primitive peoples for tools and weapons.

Silicified Wood: generally consists of reddish or brown chalcedony.

Tiger's eye: quartz pseudomorphous after asbestos (e.g., riebeckite, var. crocidolite) and retaining the fibrous structure of that mineral. Most of it is yellow and brown or yellow and blue. It is extensively used as an ornamental stone, especially in jewelry.

Diagnostic features: Macrocrystalline quartz is generally rather easily recognized by its crystal form, hardness, and lack of cleavage; chalcedonic varieties are identified by their hardness and typical dense structure. Quartz is piezoelectric and pyroelectric. It is insoluble in all acids except hydrofluoric acid.

Occurrence and uses: Quartz is stable over practically the whole range of geological conditions, and because SiO_2 is the most abundant oxide in the earth's crust, it is a very common mineral. It is present in silica-rich igneous rocks, both volcanic and plutonic, and it makes up a large part of hydrothermal veins and granite pegmatites. Since it is hard and extremely resistant to chemical weathering, it is the most abundant detrital mineral and the basic constituent of most sandstones. It is an important constituent in most metamorphic rocks. The different varieties of chalcedony occur in sedimentary rocks and in veins and cavities.

The uses of quartz are many and diverse. Enormous tonnages are used in the construction industry: quartzite and sandstone are used as building stone, as aggregate in concrete, and as sand in mortar and cement. Large amounts are used as a flux in metallurgy; in the manufacture of glass, ceramics, and refractories; as an abrasive; and as a filler. Many of the colored varieties are cut and polished as ornamental and semiprecious stones. Fused silica is a useful material because of its chemical inertness and its low coefficient of expansion—dishes and crucibles of fused silica can be heated and then chilled rapidly without danger of breakage. One of the most interesting applications of quartz crystals makes use of their piezoelectric properties to measure pressures and to control the frequency of electrical impulses; although the total amount used annually for these purposes is comparatively small, the supply of satisfactory natural crystals (mainly from Brazil) is not always adequate, and so laboratory methods of growing suitable crystals have been developed.

Tridymite, SiO_2

Crystal system, class, and space group: Orthorhombic; *222; C222₁* (low-tridymite). [Natural high-tridymite is monoclinic (pseudohexagonal).]

Cell dimensions and content: $a = 9.91$, $b = 17.18$, $c = 40.78$; $Z = 160$.

Habit: Small (typically 1 mm or less) pseudohexagonal platy crystals and aggregates of wedge-shaped twins (inversion pseudomorphs after hexagonal high-tridymite).

Cleavage and fracture: None. Conchoidal.

Hardness: $6\frac{1}{2}$–7.

Tenacity: Extremely brittle.

Specific gravity: 2.26–2.27.

Color and streak: Colorless or white. White.

Luster and light transmission: Vitreous; pearly on {0001} of pseudomorph. Transparent.

Optical properties:

anisotropic, biaxial (+): $2V = 35°–90°$

refractive indices: α, 1.469–1.479, β, 1.469–1.480; γ, 1.473–1.483

Chemistry and alteration: A small amount (2–3 percent) of Si^{+4} may be replaced by Al^{+3}, and electrical neutrality in the structure is maintained by the introduction of cations such as Na^+ and Ca^{+2}; most analyses of natural tridymite show about 95 percent SiO_2. Although its stability field is above 867°C, most tridymite in rocks is probably formed at lower temperatures, at least in part, because of the presence of foreign ions in the structure that may extend its stability field to lower temperatures.

Tridymite is commonly altered to quartz.

Diagnostic features: Shape and size of crystals and mode of occurrence.

Occurrence and associations: Tridymite occurs chiefly as phenocrysts in and as a groundmass component of silicic volcanic rocks (including pyroclastics) and filling vesicles, fractures, and other cavities in these rocks—especially in rhyolites, dacites, and trachytes. In many instances, it can be shown to have formed originally as high- (β-) tridymite. It is commonly associated with such diverse minerals as sanidine, cristobalite, fayalite, cassiterite, kaolinite, and alunite. It also has been recorded as a constituent of silicified wood, apparently formed by crystallization of opaline silica at ordinary temperatures.

Cristobalite, SiO_2

Crystal system, class, and space group: Tetragonal; 422; $P4_32_12$ [low-(α-) cristobalite].

Cell dimensions and content: $a = 4.97$, $c = 6.91–6.93$; $Z = 4$.

Habit: Crystals uncommon; small octahedral crystals (usually 1 mm or less), many of which are twinned on {111} and aggregated in rounded forms. [The crystals are inversion pseudomorphs after isometric high-cristobalite.]

Cleavage: None.

Hardness: $6\frac{1}{2}$.

Specific gravity: 2.32–2.34.

Color and streak: White.

Luster and light transmission: Vitreous. Translucent to subtranslucent.

Optical properties:

anisotropic, uniaxial (−)

refractive indices: ω, 1.487; ϵ, 1.484

Chemistry: Ranges in composition as does tridymite and for the same reasons.

Diagnostic features: Shape and size of crystals and other habits; mode of occurrence.

Occurrence and associations: Typically associated with tridymite, sanidine, augite, and fayalite in rhyolites, trachytes, and dacites—as fibrous spherulites and/or filling vesicles and other cavities; also as a component of opal.

Opal, $SiO_2 \cdot nH_2O$

Crystal system: None, amorphous.

Crystal structure: Essentially amorphous to X-rays, but some patterns show weak and diffuse reflections that indicate the presence of cristobalite groups.

Habit: Massive, commonly in rounded and botryoidal forms; stalactitic, pisolitic, or as crusts.

Hardness: $5\frac{1}{2}$.

Specific gravity: 2.0–2.25.

Color and streak: Colorless (var. *hyalite*) or white; also gray, brown, or red, the color being due to fine-grained impurities; or with a rich iridescence and play of colors (var. *precious opal*). White.

Luster and light transmission: Vitreous or waxy, colored varieties commonly somewhat resinous. Translucent.

Opalescence: Investigations of opal using electron microscopy have shown that it consists of a close-packed aggregate of amorphous hydrated silica spheres, each a few tenths of a micron in diameter. A play of colors results when the spheres are uniform in size and regularly packed (Figure 15-12). Sphere diameter and refractive index determine the range of colors displayed. (The play of color is absent in nonprecious opal because of irregular size and/or packing of the constituent spheres.)

Optical properties:
isotropic
refractive index: 1.435–1.460

Chemistry: The water content differs from specimen to specimen, but it is generally between 3 and 10 percent.

Diagnostic features: Distinguished from chalcedony, which it may resemble and into which it has commonly been transformed by crystallization, by mode of occurrence, form, and low density.

Occurrence and uses: Opal is deposited at relatively low temperatures from silica-bearing waters; much of it appears to represent solidified colloidal silica. It occurs in fissures and cavities in many different rocks. It is commonly associated with igneous rocks, having been deposited by thermal waters; extensive deposits (siliceous sinter or

Figure 15-12. Opal. Packing of spheres in an Australian opal as exhibited in a scanning electron micrograph. Spheres are approximately 3000 Å in diameter. (Electron micrograph courtesy of Gemological Institute of America.)

"geyserite") are formed around hot springs, as at Yellowstone National Park, Wyoming. It is commonly concretionary or constituting replacement fossils or pseudomorphs. Opal is also the form of silica secreted by sponges, radiolaria, and diatoms; diatomite occurs in extensive beds in many parts of the world.

Diatomite is used extensively as a filtering medium, as a mild abrasive, and as an insulator against heat, cold, and sound. In the United States, large amounts are mined at Lompoc, California. Precious opal is prized as a gemstone; the finest opals come from deposits in sandstone in Australia.

FELDSPAR GROUP

The feldspars are the most abundant of all minerals. Although they are all closely related in form and physical properties, they fall into two subgroups: (1) the potassium and barium feldspars, which are monoclinic or very nearly monoclinic in symmetry, and (2) the sodium and calcium feldspars (the plagioclases), which are definitely triclinic.

The general formula for the feldspars can be written WZ_4O_8, in which W may be Na, K, Ca, and/or Ba, and Z is Si and/or Al, the Si:Al ratio ranging from 3:1 to 1:1. Since all feldspars contain a certain minimum amount of Al, the general formula may be somewhat more specifically stated as $WAl(Al,Si)Si_2O_8$, the variable (Al,Si) being balanced by variation in the proportions of univalent and bivalent cations.

The structure of the feldspars is a continuous three-dimensional network of SiO_4 and AlO_4 tetrahedra, with the positively charged sodium, potassium, calcium, and barium situated in the interstices of the negatively charged network. The network of SiO_4 and AlO_4 tetrahedra is to some degree elastic, and thus it can adjust itself to the sizes of the W cations: with the relatively large cations (K,Ba), the symmetry is monoclinic or triclinic only slightly off monoclinic; with the smaller cations (Na,Ca), the structure is slightly distorted and the symmetry is triclinic.

The barium feldspars are rare. Thus, even though minor amounts of barium may be present in potassium feldspars, which are common, we omit the barium feldspars from further consideration. We shall, therefore, discuss the feldspars as a three-component system, the components being $KAlSi_3O_8$ (Or), $NaAlSi_3O_8$ (Ab), and $CaAl_2Si_2O_8$ (An). Complexities are introduced by both solid-solution relations that exist among these three components and the existence of polymorphic forms. The following species are recognized:

Sanidine	$KAlSi_3O_8$	Monoclinic
Orthoclase	$KAlSi_3O_8$	Monoclinic
Microcline	$KAlSi_3O_8$	Triclinic

Plagioclase Series End-members:	$(Na,Ca)Al(Al,Si)Si_2O_8$	Triclinic
Albite	$NaAlSi_3O_8$	Triclinic
Anorthite	$CaAl_2Si_2O_8$	Triclinic

The potassium feldspar minerals occur in several distinct forms having different but intergradational optical and physical properties. Sanidine, the monoclinic high-temperature polymorph, occurs chiefly in volcanic rocks. Orthoclase, another monoclinic variety, and microcline (triclinic) occur in a wide variety of igneous and metamorphic rocks that have crystallized at intermediate to low temperatures. *Adularia*, which may be either monoclinic or triclinic, is the name given to a form with a distinctive habit, one simulating rhombohedra, that occurs chiefly in low-temperature hydrothermal veins.

Research has clarified the relationships among these forms. Microcline and sanidine are polymorphs with an order-disorder relationship; the Si and Al atoms are randomly distributed over their positions in sanidine, but they are ordered in microcline. The disordered form is the more stable polymorph above about 700°C, and microcline has been transformed into sanidine by hydrothermal treatment at this temperature; the reverse transformation, however, has not been achieved in the laboratory, evidently because of the high activation energy required for ordering of the Si and Al atoms. Orthoclase and adularia are structurally intermediate between sanidine and microcline. Much orthoclase probably crystallized originally as sanidine, which subsequently underwent an ordering of some of its Al and Si atoms, but adularia is evidently a metastable form that has developed under conditions of rapid crystallization within the stability field of microcline. Such rapid crystallization would presumably prevent the attainment of an ordered arrangement of Si and Al.

At temperatures above about 660°C, complete solid solution exists between $KAlSi_3O_8$ and $NaAlSi_3O_8$ (Figure 5-8). At lower temperatures, solid solutions intermediate between potassium-feldspar and albite are unstable and, under conditions of slow cooling, they exsolve into oriented intergrowths, commonly subparallel lamellae, that are alternately sodium-rich and potassium-rich in composition. Such intergrowths are called *perthites* (Figure 15-13) or *antiperthites*. In some cases, especially in those alkali feldspars in the range $Or_{25}Ab_{75}$-$Or_{60}Ab_{40}$, exsolution commonly yields extremely fine submicroscopic lamellae; these are frequently called *anorthoclase cryptoperthites*. In perthites, the plagioclase occurs as uniformly oriented films, veins, or patches within the orthoclase (or microcline); in the antiperthites, the relationship is the reverse. Perthite, when heated for a long time at, for example, 1000°C, becomes homogeneous. Not all perthites, however, have been formed by exsolution; some are the product of partial metasomatic replacement of originally homogeneous potassium-feldspars by sodium-bearing solutions.

Figure 15-13. Perthite. Intergrowth of pink microcline (dark) with white/colorless plagioclase (light); large surface is {001} cleavage, smaller is {010} cleavage; a axis horizontal; North Burgess Township, Lanark County, Ontario. (Courtesy of Queen's University.)

X-ray examination of potassium feldspars and of albite provides the following explanation for lamellar perthitic intergrowths. The framework of linked SiO_4 and AlO_4 tetrahedra, being similar for the monoclinic and triclinic forms, is continuous throughout the structure. At high temperatures, the K and Na ions are distributed randomly in the framework, thus producing an essentially homogeneous crystal. At lower temperatures, ordering may occur with the formation of potassium-rich and sodium-rich lamellae, producing alternate sheets with monoclinic or triclinic (but nearly monoclinic) and triclinic symmetry, respectively. The a unit length of potassium feldspar and of albite is markedly different (8.57 Å and 8.14 Å), whereas the b and c unit lengths are almost the same (12.98 Å and 7.20 Å versus 12.79 Å and 7.16 Å). These relations account for the lamellae being approximately parallel to {100}; that is, the similar b and c lengths coincide in the {100} plane, whereas the a lengths run through the lamellae and shorten in the albite regions and lengthen in the potassium feldspar regions.

The system Ab-An exhibits complete solid solution at high temperatures, but there are complications at low temperatures. Very briefly, the plagioclase feldspars exist in both high- and low-temperature forms, and there is only incomplete solid solution between the low-temperature forms.

There is essentially no solid solution between Or and An; hence, feldspars intermediate between these two components do not occur.

Phase relations in the three-component system Or-Ab-An are illustrated in Figure 15-14. The composition of any feldspar is conveniently expressed by the use of the symbols Or, Ab, and An for the components; thus, for example, $Or_{26}Ab_{66}An_8$ is the composition of a possible anorthoclase.

Sanidine, Orthoclase, $KAlSi_3O_8$

Crystal system, class, and space group: Monoclinic; $2/m$; $C2/m$.

Cell dimensions and content: $a \cong 8.56$, $b \cong 13.00$, $c = 7.19$; $\beta = 116° 01'$; $Z = 4$.

Habit: Crystals uncommon; typically short prismatic, somewhat flattened parallel to {010}, or elongated parallel to a with {010} and {001} prominent (Figures 2-42e, 15-15a, and 15-15b).

Twinning: Common on the Carlsbad Law (composition plane {010}, Figure 15-15c), less common on the Baveno Law (composition plane {021}, Figure 15-15d) and Manebach Law (composition plane {001}, Figure 15-15e).

Common forms and angles:

$(110) \wedge (1\bar{1}0) = 61°16'$		$(001) \wedge (\bar{2}01) = 80°07'$
$(130) \wedge (\bar{1}30) = 58°46'$		$(001) \wedge (021) = 44°51'$
$(001) \wedge (\bar{1}01) = 50°06'$		$(021) \wedge (0\bar{2}1) = 89°42'$

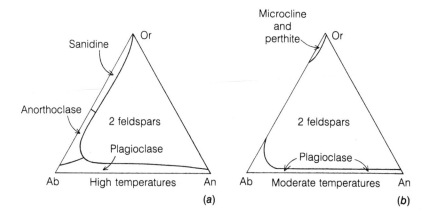

Figure 15-14. Phase relations in the Or-Ab-An system, (a) at high temperatures, (b) at moderate temperatures.

Cleavage: {001} perfect, {010} good.
Hardness: 6.
Specific gravity: 2.56.
Color and streak: Sanidine commonly colorless; orthoclase white, cream, or pink. White.
Luster and light transmission: Subvitreous, pearly on cleavage surfaces. Transparent to subtranslucent.

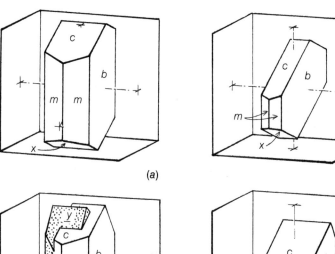

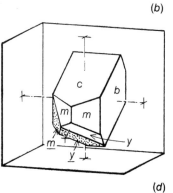

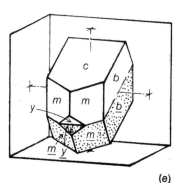

Figure 15-15. Forms in orthoclase (monoclinic). (a), (b) Simple crystals. (c) Carlsbad twin. (d) Baveno twin. (e) Manebach twin. Forms lettered as follows: $b\{010\}$, $c\{001\}$, $m\{110\}$, $x\{\bar{1}01\}$, $y\{\bar{2}01\}$.

Optical properties:
  Sanidine: biaxial (−): 2V = 18°–54°
    refractive indices: α, 1.518–1.524; β, 1.522–1.529; γ, 1.522–
      1.530
  Orthoclase: biaxial (−): 2V—typically ∼ 60°
    refractive indices: α, 1.518–1.520; β, 1.522–1.524; γ, 1.522–
      1.524

Diagnostic features: Has a glassy appearance, is commonly tabular parallel to {010}, and occurs in volcanics and near-surface dike rocks. Sanidine and orthoclase can be distinguished macroscopically from the plagioclase feldspars by the absence of twinning striations. Nonetheless, staining procedures are also rather frequently employed. The distinction between orthoclase and microcline is not so easily made; occurrence, however, is a useful guide since much (but not all!) of the potassium feldspar of igneous rocks is orthoclase, whereas practically all that of pegmatites and hydothermal veins is microcline.

Occurrence: Sanidine occurs in potassium-rich volcanic rocks such as rhyolite and trachyte. Orthoclase is the characteristic potassium feldspar of igneous rocks, occuring both alone and in perthitic intergrowth with albite; it also occurs in some metamorphic rocks. The variety *adularia* is formed at comparatively low temperatures, typically in veins; it is characterized by simple crystals (combinations of {110}, {001}, and {$\bar{1}$01}), many of which have the appearance of rhombohedra.

Microcline, $KAlSi_3O_8$

Crystal system, class, and space group: Triclinic; $\bar{1}$; C$\bar{1}$.
Cell dimensions and content: a = 8.58, b = 12.97, c = 7.22; α = 90°39′, β = 115°56′, γ = 87°47′; Z = 4.
Common forms and angles: Similar to those of orthoclase.
Habit: Crystals common; otherwise similar to orthoclase.
Twinning: Commonly polysynthetically twinned on both the albite and pericline laws, giving a characteristic grating or gridiron structure on all sections except those in the prism zone [001]; this twinning, often called *grid twinning,* is best seen by microscopic examination using doubly polarized light (Figure 15-16). Microcline may also be twinned according to the Carlsbad, Baveno, and Manebach Laws, which are described under Sanidine-Orthoclase. Carlsbad twinning is common, whereas Baveno and Manebach twinning are rare.
Cleavage: {110} perfect, {010} good.
Hardness: 6.
Specific gravity: 2.56.

Figure 15-16.   Microcline "grid twinning" (∼50x). (From *Optical Mineralogy* by W. R. Phillips and D. T. Griffen. W. H. Freeman and Company, 1981.)

Color and streak: White, cream, or pink; also light green (var. *ama-zonite*). White.

Luster and light transmission: Vitreous, pearly on cleavage surfaces. Translucent.

Optical properties:

anisotropic, biaxial (−): 2V—typically ~ 80°

refractive indices: α, 1.514–1.516; β, 1.518–1.519; γ, 1.521–1.522

grid twinning is distinctive

Diagnostic features: Although a light green color is diagnostic, most microcline is not green. It can, however, be distinguished from orthoclase by its optical properties. Occurrence also affords a useful, albeit hazardous, guide (see Orthoclase).

Occurrence and uses: Microcline, being an ordered polymorph, is usually formed at lower temperatures than is orthoclase, the disordered polymorph. Thus, microcline is the common potassium feldspar of pegmatites and hydrothermal veins. Crystals several feet long are not uncommon in pegmatites. Microcline also occurs in metamorphic rocks.

Large quantities of microcline and microcline perthite are mined from granite pegmatites and used in the manufacture of glass, porcelain, and enamel (e.g., from pegmatites in North Carolina).

Amazonite, also called *amazonstone*, is sometimes cut and polished for use as an ornamental stone.

Plagioclase Series, $(Na,Ca)Al(Al,Si)Si_2O_8$

Crystal system, class, and space group: Triclinic; $\bar{1}$; $C\bar{1}$ (albite), $P\bar{1}$ (anorthite).

Cell dimensions and content: a $\cong$ 8.144, b $\cong$ 12.787, c $\cong$ 7.160; α = 94°16′, β = 116°35′, γ = 87°40′; Z = 4 (albite). a $\cong$ 8.177, b $\cong$ 12.877, c $\cong$ 14.169; α = 93°10′, β = 115°51′, γ = 91°13′; Z = 8 (anorthite).

Habit: Crystals common for albite and oligoclase; uncommon for andesine, labradorite, and bytownite; rare for anorthite; typically tabular parallel to {010} (Figure 15-17a); commonly as irregular grains and cleavable masses.

Twinning: Nearly universally polysynthetically twinned according to one or both of the albite (Figures 15-17b, c, d, and e) and pericline laws; Carlsbad, Baveno, and Manebach twins also common.

Cleavage: {001} perfect, {010} good.

Hardness: 6.

Specific gravity: 2.62–2.76 (Figure 15-18).

Color and streak: White or gray; rarely reddish or reddish-brown; also play of colors in some plagioclases (see *Diagnostic features*). White.

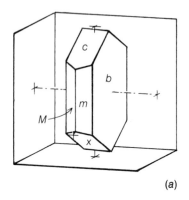

(a)

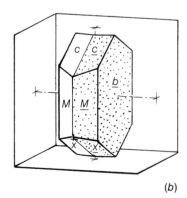

(b)

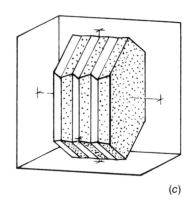

(c)

Figure 15-17.   Forms and twinning in plagioclase. (a) Single crystal; (b) albite twin; (c) polysynthetic twinning; (d) albite twinning on cleavage surface of plagioclase from Tvedestrand, Norway; 5 cm across; (e) twinning lamellae as seen in thin-section using analyzed polarized light (~ 300x). Forms lettered as follows: $b\{010\}$, $c\{001\}$, $m\{110\}$, $M\{1\bar{1}0\}$, $x\{\bar{1}01\}$. (Photograph for part e courtesy of B. J. Skinner.)

(d)

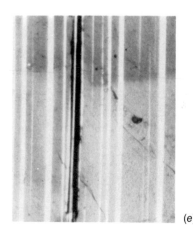

(e)

Luster and light transmission: Vitreous, pearly on cleavage surfaces. Translucent.

Optical properties:
    anisotropic, biaxial ($\pm$): $2V = 77$–$90°$ (see Figure 15-18)
    refractive indices: see Figure 15-18
    polysynthetic twinning is common

Chemistry: There is a continuous series from pure albite, $NaAlSi_3O_8$ (Ab), to pure anorthite, $CaAl_2Si_2O_8$ (An). The series is arbitrarily divided into six species or subspecies as follows:

Albite	An_0-An_{10}
Oligoclase	An_{10}-An_{30}
Andesine	An_{30}-An_{50}
Labradorite	An_{50}-An_{70}
Bytownite	An_{70}-An_{90}
Anorthite	An_{90}-An_{100}

Diagnostic features: Distinguished from the potassium feldspars by the twinning striations on basal cleavage surfaces. Differentiation between the individual species or subspecies within the plagioclase series is best done optically or by X-ray diffraction, but careful density determinations give a good indication of composition. Rock type is also a useful guide (see *Occurrence and uses*).

Some albite, oligoclase, and labradorite in coarse cleavages commonly exhibit a play of colors in shades of blue or blue-green, yellow, and brown. Those with bulk compositions in the albite-oligoclase range, which are typically light colored, are called *peristerite*.

Occurrence and uses: Most anorthite occurs in contact metamorphosed limestones. Bytownite and labradorite are characteristic of igneous rocks of gabbroic composition and of the anorthosites; andesine, of andesites and diorites; oligoclase, of monzonites and granodiorites; albite, of granites and granitic pegmatites. The albite of pegmatites is of two distinct types—massive and lamellar, the latter widely referred to as *cleavelandite*. The plagioclases are also common in metamorphic rocks; in low-grade schists and gneisses,

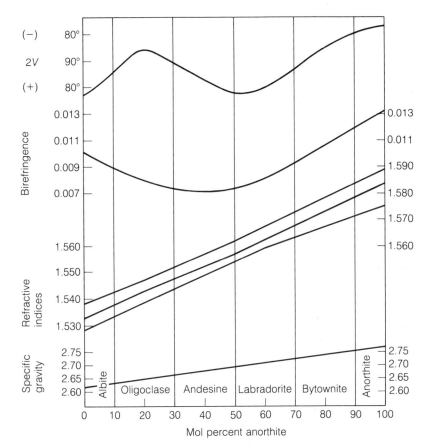

Figure 15-18. The specific gravity and optical properties of the plagioclase series (low temperature).

the plagioclase is typically albite; in medium-grade rocks, it is typically oligoclase or andesine. Pure or nearly pure albite occurs as veins in a few schists.

Albite and oligoclase are mined from some pegmatites and used in the manufacture of ceramics.

Scapolite Group, $Na_4Al_3Si_9O_{24}Cl$—$Ca_4Al_6Si_6O_{24}(CO_3,SO_4)$

Crystal system, class and space group: Tetragonal; 4/m; I4/m.
Cell dimensions and content: $a = 12.11$–12.23, $c = 7.55$–7.68; $Z = 2$.
Habit: Crystals uncommon; most, which tend to be stubby, are combinations of prisms and pyramids, typically rather large, with uneven faces; commonly massive, granular.
Cleavage: {100}, {110} distinct, commonly giving a splintery woodlike appearance to massive specimens.
Hardness: 6.
Specific gravity: 2.56–2.77, higher with greater Ca content.
Color and streak: Generally white, gray, or greenish; also pink, yellow, or blue. White.
Luster and light transmission: Vitreous. Transparent to translucent.
Optical properties:
anisotropic, uniaxial (−)
refractive indices: ω, 1.540–1.600; ϵ, 1.535–1.565
optical properties may be used to identify members of the series
Chemistry: The scapolite series ranges in composition from sodium-rich *marialite* to calcium-rich *meionite*. *Dipyre* and *mizzonite* are names sometimes applied to intermediate members. In composition, scapolite is closely related to plagioclase, the only difference being the presence of small amounts of Cl^{-1} and CO_3^{-2} (and in some cases also SO_4^{-2}) and the additional sodium and calcium required to neutralize the additional negative charge.
Diagnostic features: Crystals, where present, are characteristic; woodlike cleavage. Some scapolite fluoresces orange or yellow in ultraviolet light.
Occurrence: Scapolite typically occurs in metamorphosed limestones, locally in considerable quantities, especially around intrusive igneous rocks. It also occurs in schists and gneisses, commonly replacing plagioclase.
Name: The name *wernerite* was proposed for this mineral at about the same time that *scapolite* was. Although both names are still used, wernerite is now generally restricted in its application to scapolites of intermediate composition.

FELDSPATHOID GROUP

The feldspathoids are a group of sodium and potassium aluminosilicates that appear in place of the feldspars when an alkali-rich magma is deficient in silica. They never occur together in equilibrium with primary quartz because they will react with free silica to give feldspar. Although the feldspathoids are all tektosilicates, the structures of the different species within the group are less closely interrelated than the structures of the feldspars are. In fact, they crystallize in several systems, although the common species are either isometric or hexagonal. The most common species of the feldspathoid group are:

Leucite	$KAlSi_2O_6$
Nepheline	$(Na,K)AlSiO_4$
Cancrinite	$Na_6Ca_2(AlSiO_4)_6(CO_3)_2$
Sodalite	$Na_8(AlSiO_4)_6Cl_2$

Analcime, $NaAlSi_2O_6 \cdot H_2O$, is sometimes included with the feldspathoids, chiefly because it also occurs as a primary mineral in silica-deficient igneous rocks. Herein, it is classed, along with other hydrated tektosilicates, in the zeolite group.

The feldspathoids are readily attacked by acids. This characteristic is evidently dependent on the comparatively high Al:Si ratio; the Al can be removed in solution and then the lattice collapses, generally with the formation of gelatinous silica. This formation of gelatinous silica is the basis of a useful test for the presence of feldspathoids in an igneous rock. This test may be made by spreading a film of syrupy (85 percent) phosphoric acid on a smooth surface of the rock, allowing it to stand for three minutes, dipping it in water to remove the acid, and then immersing it in a 0.25 percent solution of methylene blue for one minute. This treatment stains nepheline, sodalite, cancrinite, and analcime deep blue; leucite, however, is not affected.

Leucite, $KAlSi_2O_6$

Crystal system, class, and space group: $> \sim 625°C$, isometric; $4/m\bar{3}2/m$; Ia3d. $< \sim 625°C$, tetragonal; $4/m$; $I4_1/a$.
Cell dimensions and content: Isometric; $a = 13.4$; $Z = 16$. Tetragonal; $a \cong 13.1$, $c \cong 13.8$; $Z = 16$.
Habit: Crystals common; typically as trapezohedra (Figure 15-19).
Twinning: {110} repeated.
Cleavage: {110} indistinct.
Hardness: 6.
Specific gravity: 2.47.
Color and streak: Colorless, white, or gray. White.

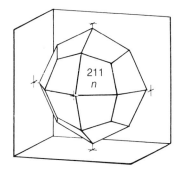

Figure 15-19. Leucite. Typical crystal form, the trapezohedron.

Luster and light transmission: Vitreous to dull. Translucent.

Optical properties:

anisotropic, uniaxial (+)

refractive indices: ω, 1.508–1.511; ϵ, 1.509–1.511

Chemistry: A small amount of the potassium may be replaced by sodium.

Alteration: In many occurrences leucite has altered to *pseudoleucite*, a pseudomorph consisting chiefly of a mixture of orthoclase and nepheline plus or minus analcime.

Diagnostic features: Readily recognized by its crystal form and its occurrence as phenocrysts in volcanic rocks. Analcime resembles leucite in crystal form, but well-crystallized analcime occurs in cavities and vugs, not as phenocrysts.

Occurrence: Leucite is abundant in the volcanic rocks of a few regions. The best specimens are the fresh crystals in the lavas of Mount Vesuvius. Leucite does not occur in plutonic rocks, and laboratory investigations have indicated that leucite is not a stable phase in the system K_2O-Al_2O_3-SiO_2 at pressures above 2500 kg cm^{-2} (see also Figure 5-4).

Nepheline, (Na,K)AlSiO$_4$

Crystal system, class, and space group: Hexagonal; 6; $P6_3$.

Cell dimensions and content: a = 10.01, c = 8.405; Z = 8.

Crystal structure: The structure of nepheline is similar to that of tridymite, with half of the silicon being replaced by aluminum and electroneutrality being preserved by the addition of an equivalent number of sodium ions to voids in the structure. The replacement of Si by Al gives the c axis a polar character not present in the tridymite structure.

Habit: Crystals uncommon; typically simple hexagonal prisms (showing rectangular and hexagonal cross sections in rocks); occurs more commonly as shapeless grains and irregular masses.

Cleavage: {10$\bar{1}$0} indistinct.

Hardness: 6.

Specific gravity: 2.55–2.65.

Color and streak: White, gray, or brown. White.

Luster and light transmission: Vitreous or greasy. Transparent to translucent.

Optical properties:

anisotropic, uniaxial (−)

refractive indices: ω, 1.529–1.549; ϵ, 1.526–1.544

Chemistry: The pure compound NaAlSiO$_4$ can be made artificially. Natural nephelines contain potassium, several of them approaching the formula KNa$_3$(AlSiO$_4$)$_4$, apparently as a reflection of the space requirements of the nepheline structure, in which one alkali position in four is larger than the other three and preferentially accommodates potassium. A little calcium is also generally present in natural nephelines.

Diagnostic features: Coarsely crystallized with a greasy luster. Nepheline is readily decomposed by hydrochloric acid, with the separation of gelatinous silica. In syenites, it tends to weather out leaving feldspar in relief.

Occurrence and use: Nepheline occurs in both plutonic and volcanic rocks and in pegmatites associated with nepheline syenites.

Pure nepheline and nepheline-feldspar concentrations are useful raw materials for the manufacture of glass and ceramics. Considerable amounts are mined for this purpose in the Peterborough District, Ontario.

Sodalite, $Na_8(AlSiO_4)_6Cl_2$

Crystal system, class, and space group: Isometric; $\bar{4}3m$; $P\bar{4}3m$.

Cell dimensions and content: $a = 8.83–8.91$; $Z = 1$.

Habit: Crystals rare, as small dodecahedra; more commonly massive and granular.

Cleavage: {110} indistinct.

Hardness: 6.

Specific gravity: 2.3.

Color and streak: Commonly azure blue, but may be white, gray, or pink. White.

Luster and light transmission: Vitreous. Translucent.

Optical properties:
isotropic
refractive index: 1.483–1.487

Diagnostic features: Blue color and translucency generally characteristic. (It roughly resembles azurite, but azurite is readily distinguished by its lower hardness and effervescence in acid.)

Occurrence and uses: Sodalite occurs in both volcanic and plutonic rocks of the nepheline syenite family, but more commonly in the plutonic types, generally in association with nepheline. Blue sodalite occurs in considerable quantity in pegmatitic nepheline syenite near Bancroft, Ontario and at Ice River, British Columbia. It is sometimes cut and polished as an ornamental stone.

Related species: Lazurite $(Na,Ca)_{7–8}(Al,Si)_{12}(O,S)_{24}[SO_4,Cl_2(OH)_2]$ also belongs to the sodalite group of minerals; it has a brilliant blue color and is the chief component of the decorative and jewelry stone known as *lapis lazuli*. The pigment *ultramarine,* formerly made by crushing lazurite, is now manufactured synthetically.

Cancrinite, $Na_6Ca_2(AlSiO_4)_6(CO_3)_2$

Crystal system, class, and space group: Hexagonal; 6; $P6_3$.
Cell dimensions and content: a = 12.63–12.78, c = 5.11–5.19;
   Z = 1.
Habit: Crystals rare; prismatic; typically massive.
Cleavage: $\{10\bar{1}0\}$ perfect.
Hardness: 6.
Specific gravity: 2.4–2.5.
Color and streak: White, pink and yellow. White.
Luster and light transmission: Vitreous. Translucent.
Optical properties:
   anisotropic, uniaxial (−)
   refractive indices: ω, 1.490–1.530; ϵ, 1.488–1.503
Chemistry: The composition ranges somewhat. In addition to the ele-
   ments indicated in the formula, most cancrinite contains small
   amounts of sulfate and chloride.
Diagnostic features: Color, where yellow. Cancrinite can be distinguished from other feldspathoids by its effervescence in warm
   dilute HCl.
Occurrence and associations: Cancrinite occurs solely in plutonic
   rocks of the nepheline syenite family, evidently because carbonate
   ions may be incorporated into the structure only under considerable
   pressure of carbon dioxide. It is generally associated with other
   feldspathoids, especially nepheline and sodalite.

ZEOLITE GROUP

The zeolites comprise a group of tektosilicate minerals that are of special
interest because they can lose a part or all of their water without a
change of crystal structure. They can absorb other compounds in place
of the removed water, and they are also capable of undergoing cation
exchange. They form a well-defined group of species that are closely
related to one another in composition, in conditions of formation, and
hence in mode of occurrence. They are not, however, a group of species
that are related in crystal structure as, for example, the feldspars are.

The general formula, is $W_mZ_rO_{2r} \cdot sH_2O$, in which W is chiefly Na
and Ca (K, Ba, and Sr to a lesser extent), Z is Si + Al where Si:Al is 1 or
greater, and s is variable; also, the ratio Al_2O_3:(CaO + Na_2O) is always
1:1, and the (Al + Si):O ratio is always 1:2. The compositional ranges
of individual zeolite species, however, are narrow, which is indicative
of a narrow range of stability for each of the species and also may serve
as an explanation for the existence of a large number of species within
the group.

The atomic substitutions in the zeolite group are of two types: the
first is similar to that in the feldspars—KSi for BaAl and NaSi for CaAl;

the second alters the number of cations—Ba for K_2 and Ca for Na_2. For instance, in thomsonite, the usual number of cations is represented by the formula $NaCa_2Al_5Si_5O_{20} \cdot 6H_2O$, but it is possible to replace half the Ca by Na_2, thus raising the total number of positive ions to four and changing the formula to $Na_3CaAl_5Si_5O_{20} \cdot 6H_2O$. This change is possible because the structure contains a sufficient number of suitable spaces to accomodate the additional ions. Both types of replacement, however, must be borne in mind in considering the composition of the individual zeolites.

All the zeolites are tektosilicates with particularly open, wide-meshed $(Si,Al)O_2$ frameworks. As in the feldspars, the cavities in this framework contain the cations that balance the negative charge of the framework. A characteristic feature of the zeolites, however, is the ease with which exchange of the cations can take place. The cations of many natural zeolites can be replaced artificially by thallium, potassium, silver, or sodium by simply placing a crystal in a solution that contains these ions. Nonetheless, the tektosilicate framework is held together by strong bonds and is so rigid that the individual crystals retain their shape during the change; the channels through the framework are large enough to enable the ions to pass freely. Also, as we have already noted, the water molecules can be removed without breaking down the structure; this phenomenon is in striking contrast with the behavior of the majority of hydrated inorganic salts, for which removal of even a part of the water results in the complete breakdown of the structure. On heating, the water is given off continuously rather than in certain amounts at definite temperatures, as is the case for most hydrated compounds. The reason for this difference is easy to see; in zeolites, the strongly bonded framework is responsible for a high stability, whereas, in the majority of inorganic salts, the structure is a simple regular packing of relatively small ions. Nonetheless, the removal of water molecules is not achieved without some effect on the zeolite structure; as part of the water is removed, there is generally a redistribution of energy within the structure that causes the remaining water to be held more firmly and/or the disturbed energy equilibrium may be restored by movements of the atoms of the framework, which has some flexibility. The movements generally result in a decrease in length of one or more sides of the unit cell.

Zeolites may be deposited from late magmatic solutions, by hydrothermal solutions, or they may be formed during diagenesis. Many are products of alteration of feldspars, feldspathoids, or volcanic glass. They are especially common as amygdules, fracture fillings, and fillings in other cavities in many kinds of rocks, but especially in basalts. They are characteristic of a low-temperature low-pressure metamorphic facies known as the zeolite facies. They also occur in some metalliferous veins, especially where the gangue is calcite, and in the deposits of hot springs, where they have in some cases been formed within historic time and are still being formed. The zeolites are commonly associated with calcite, prehnite, pectolite, apophyllite, and datolite. Zeolites are also abundant constituents of some, albeit rather uncommon, sedimentary and/or pyroclastic rocks, apparently having been formed from volcanic ash during diagenesis or low-grade metamorphism. Although about 20 different zeolite minerals have been recorded as occurring in sedimentary rocks, only a few make up the major part of such rocks; these include analcime, chabazite, heulandite, clinoptilolite, and laumontite.

The unique physical and chemical properties of zeolites—their capability of undergoing cation exchange and reversible dehydration, and their being open framework structures liable to act as molecular sieves—render the extensive zeolite-bearing sedimentary deposits of potential commercial value.

Among other things, zeolites are used as soil conditioners, as ion exchangers in pollution abatement, and as acid-resistant adsorbants in gas drying. The name *zeolite* is derived from a Greek word meaning to boil, which refers to the swelling up and apparent boiling that occurs when these minerals are heated and water is driven off.

Heulandite, $(Na,Ca)_{2-3}Al_3(Al,Si)_2Si_{13}O_{36} \cdot 12H_2O$

Crystal system, class, and space group: Monoclinic; $2/m$; $C2/m$.
Cell dimensions and content: $a = 7.46$, $b = 17.84$, $c = 15.88$; $\beta = 91°26'$; $Z = 4$.
Habit: Crystals common; trapezoidal crystals tabular parallel to $\{010\}$ (Figure 15-20).
Cleavage: $\{010\}$ perfect.
Hardness: 4 on most surfaces; 3 on cleavage surfaces.
Specific gravity: 2.2.
Color and streak: White, commonly stained red or brown by iron oxides. White.
Luster and light transmission: Vitreous, pearly on cleavage surfaces. Translucent.
Optical properties:
  anisotropic, biaxial (+): $2V = 30°$
  refractive indices: α, 1.487–1.505; β, 1.487–1.505; γ, 1.488–1.512
Chemistry: Heulandite ranges in composition toward the isotructural mineral *clinoptilolite* by the substitution of K for Na. Some heulandite shows considerable substitution of Sr and Ba for Ca.
Diagnostic features: Form of crystals is characteristic and distinctive. Heulandite gives off water rather violently when heated in a flame. It gelatinizes in HCl.
Occurrence: Heulandite is commonly associated with other zeolites, especially stilbite, in cavities in basaltic rocks, for example, in the basaltic sills and lava flows of northern New Jersey. It is also a widespread alteration product of intermediate-to-acid volcanic

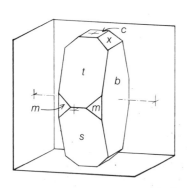

Figure 15-20. Forms in typical heulandite crystals: $c\{001\}$, $m\{110\}$, $x\{011\}$, $s\{\bar{1}01\}$, $t\{101\}$.

glass, and, in addition, it occurs as an authigenic mineral in sedimentary rocks, especially in tuffaceous sandstones; much of this material, however, is in the composition range of clinoptilolite.

Stilbite, $NaCa_2Al_5Si_{13}O_{36} \cdot 14H_2O$

Crystal system, class, and space group: Monoclinic; $2/m$; $C2/m$.
Cell dimensions and content: $a = 13.63$, $b = 18.17$, $c = 11.31$; $\beta = 129°10'$; $Z = 4$.
Habit: Crystals common; typically rough sheaflike crystals that are aggregates of cruciform penetration twins (Figure 15-21).
Cleavage: {010} perfect.
Hardness: 4.
Specific gravity: 2.1–2.2.
Color and streak: White, cream, yellow-orange, or pink. White.
Luster and light transmission: Vitreous, pearly on cleavage surfaces. Transparent to translucent.
Optical properties:
anisotropic, biaxial (−): $2V = 30°–49°$
refractive indices: α, 1.482–1.500; β, 1.491–1.507; γ; 1.493–1.513
Chemistry: Most stilbite contains some sodium replacing calcium.
Diagnostic features: Sheaflike form of crystals.
Occurrence: Stilbite occurs in cavities in basalts, commonly in association with heulandite.

Figure 15-21. Stilbite. Sheaflike cluster of crystals from Poona, India; height ~ 12 cm. (Photograph by S. C. Chamberlain. By permission, from Chamberlain, S. C. and King, V. T. 1981. "The William W. Pinch Collection." *Rocks and Minerals*, **56,** 49–66.)

Laumontite, $CaAl_2Si_4O_{12} \cdot 4H_2O$

Crystal system, class, and space group: Monoclinic; 2 or m; $C2$ or Cm, respectively.
Cell dimensions and content: $a = 14.90$, $b = 13.17$; $c = 7.55$; $\beta = 111° 30'$; $Z = 4$.
Habit: Crystals common; typically small prismatic crystals with oblique terminations; also massive.
Twinning: Common on {100}.
Cleavage: {010}, {110} perfect.
Hardness: 4.
Specific gravity: 2.25–2.30.
Color and streak: White.
Luster and light transmission: Vitreous or chalky. Translucent.
Optical properties:
anisotropic, biaxial (−): $2V = 25°–47°$
refractive indices: α, 1.502–1.514; β, 1.512–1.522; γ, 1.514–1.525

Other properties: Laumontite is strongly pyroelectric (its crystals have no center of symmetry).

Alteration: Laumontite readily loses about one-eighth of its water, becoming chalky and friable (var. *leonhardite*). (Most specimens in collections are actually leonhardite.)

Diagnostic features: Crystal form and characteristic alteration. Laumontite gelatinizes with HCl.

Occurrence: Laumontite typically occurs as small, stout prismatic crystals in veins and cavities of igneous rocks; it has also been found in large amounts in some tuffaceous sediments as an alteration product of plagioclase and glass, in a few places constituting extensive beds of impure laumontite rock.

Chabazite, $CaAl_2Si_4O_{12} \cdot 6H_2O$

Crystal system, class, and space group: Trigonal: $\bar{3}2/m$; $R\bar{3}m$.

Cell dimensions and content: $a = 13.78$, $c = 14.97$; $Z = 6$.

Habit: Crystals common; typically in simple rhombohedra that look like cubes, the angles between the rhombohedral faces being close to 90°.

Twinning: Penetration twins common, twin axis c.

Cleavage: $\{10\bar{1}1\}$ indistinct.

Hardness: 4.

Specific gravity: 2.05–2.15.

Color and streak: Colorless or white; less commonly tinted yellow or pink. White.

Luster and light transmission: Vitreous. Transparent to translucent.

Optical properties:
anisotropic, uniaxial $(-)$
refractive indices: ω, 1.472–1.494; ϵ, 1.470–1.485

Chemistry: Most chabazite contains sodium; a series exists between chabazite and its Na analog, *herschelite*. Also, K may be present up to 5.8 percent and Sr up to 5.7 percent.

Diagnostic features: Distinguished from the other common zeolites by crystal form. It is harder than calcite and does not effervesce, although it does decompose, in HCl.

Occurrence: Chabazite typically occurs lining cavities in basalts and andesites, but it has also been reported as a major constituent of tuffaceous rocks, where it was evidently formed by the alteration of calcic plagioclase.

Analcime, $NaAlSi_2O_6 \cdot H_2O$

Crystal system, class, and space group: Isometric; $4/m\bar{3}2/m$; $I4_1/acd$.

Cell dimensions and content: $a = 13.67–13.75$; $Z = 16$.

Habit: Typically in trapezohedral crystals; also granular and massive as a rock-forming mineral.

Cleavage and fracture: None. Uneven.

Hardness: 5.

Specific gravity: 2.26.

Color and streak: Colorless or white; uncommonly orange. White.

Luster and light transmission: Vitreous. Transparent to translucent.

Optical properties:

    isotropic

    refractive index: 1.479–1.493

Chemistry: Analcime forms a series with *pollucite* [$(Cs,Na)_2Al_2Si_4O_{12} \cdot H_2O$]. In analcime that has been formed at high temperatures, up to about one-tenth of the sodium may be replaced by potassium; at low temperatures, there is very little substitution. Laboratory studies have shown that synthetic analcime is stable only up to about 525°C; the presence of K-bearing analcime as a primary mineral in some igneous rocks suggests that potassium in solid solution may stabilize it to higher temperatures. The calcium analog, *wairakite* ($CaAl_2Si_4O_{12} \cdot 2H_2O$), is monoclinic and pseudocubic.

Diagnostic features: Distinguished from other zeolites by crystal form. It resembles leucite, but it may be distinguished by its mode of occurrence (see Leucite). Gelatinizes in HCl.

Occurrence: Analcime occurs as well-formed crystals in veins and cavities of igneous rocks; it is rather widespread as a primary mineral in the groundmass of alkali-rich basic igneous rocks; it is widespread and abundant in some sedimentary and pyroclastic formations, where it has been formed by the action of alkaline waters rich in sodium on volcanic ash and clay minerals (e. g., in the Green River Formation of Utah, Colorado, and Wyoming).

Name and classification: The mineral is also known as *analcite:* however, *analcime* is the name accepted by IMA (being the original form proposed by Haüy). Analcime is not considered to be a zeolite by some mineralogists. We consider it to be a member of the group, even though its structure and chemistry and some of its origins are perhaps more like those of the feldspathoids.

Natrolite, $Na_2Al_2Si_3O_{10} \cdot 2H_2O$

Crystal system, class, and space group: Orthorhombic; *mm2; Fdd2.*

Cell dimensions and content: a=18.38, b=18.71, c=6.63; Z=8.

Habit: Crystals common; prismatic, generally elongated (to needle-like) and terminated by pyramid faces; as divergent clusters of crystals; also in radiating nodular forms and as compact masses.

Cleavage: {110} perfect.

Hardness: 5.

Specific gravity: 2.25.

Color and streak: Colorless or white. White.

Luster and light transmission: Vitreous. Transparent to translucent.

Optical properties:

anisotropic, biaxial (+): $2V = 60°$

refractive indices: α, 1.473–1.483; β, 1.476–1.486; γ, 1.485–1.496

Diagnostic features: Characteristic crystal form distinguishes it from other relatively commonly occurring zeolites, but several other less common zeolites are essentially indistinguishable by macroscopic means. Natrolite gelatinizes in HCl.

Occurrence: Natrolite generally occurs in association with other zeolites in cavities in basalts and as a very late crystallization constituent or alteration product of nepheline in igneous rocks.

SUBCLASS PHYLLOSILICATES

The basic structural feature of all minerals in this subclass is the presence of SiO_4 tetrahedra linked by sharing three of the four oxygens, and thereby forming sheets with a pseudo-hexagonal network. This results in a composition for the sheet of $(Si_4O_{10})^{-4}$; Al may replace up to half the Si, giving sheets such as $(AlSi_3O_{10})^{-5}$ and $(Al_2Si_2O_{10})^{-6}$. These sheets are variously referred to as "the silica layer," "the silica sheet," or "the tetrahedral layer."

In all of the phyllosilicates except apophyllite and a few other relatively rare minerals, this tetrahedral layer is combined with another sheetlike grouping of cations (generally aluminum, magnesium, or iron) in six-coordination with oxygen and hydroxyl anions. Six-coordination means that the anions are arranged around the cations in an octahedral pattern, one anion at each solid corner of an octahedron and a cation at the center. By the sharing of anions between adjacent octahedra, a planar network results and is often referred to as "the octahedral layer" (remember that octahedral refers to the arrangement of the anions, not their number). The minerals $Al(OH)_3$ (gibbsite) and $Mg(OH)_2$ (brucite) (Figure 10-18) have this type of structure, and the Al-OH layers and Mg-OH layers in the phyllosilicates are often known as gibbsite layers and brucite layers, respectively. The gibbsite layer has a dioctahedral arrangement; that is, there are two cations for each six OH anions, whereas a brucite layer is trioctahedral, there being three cations for each six OH anions.

The dimensions of the tetrahedral and the octahedral layers are closely similar and, consequently, composite tetrahedral-octahedral layers are readily formed, either one of each layer (a two-layer structure) or an octahedral layer sandwiched between two tetrahedral layers (a three-layer structure). In each tetrahedral layer, the free oxygen ion at the apex of each SiO_4 tetrahedron is located above and at the center of a triangle formed by the other three oxygens. Thus, the free oxygens fall in hexagonal rings with the same spacings as the silicons. This pattern corresponds to that of the hydroxyl ions on the surface of an octahedral layer. It is thus possible for octahedral and tetrahedral layers to be linked by sharing in common the oxygen and hydroxyl ions of the individual sheets (Figures 15-22, 15-25, 15-26, and 15-28). If only one surface of an octahedral layer is shared with a tetrahedral layer, a two-layer mineral results (e.g., kaolinite); if both surfaces are shared, a three-layer mineral results (e.g., talc, pyrophyllite, muscovite, chlorite).

The composite octahedral-tetrahedral layers are always stacked in the direction of the c axis in the crystal; in the ab plane the crystals are pseudo-hexagonal (most phyllosilicates are monoclinic or triclinic), reflecting the hexagonal nature of the layers. All of the phyllosilicates have perfect basal cleavage, which takes place between the composite layers.

Although some of the phyllosilicates are stable to rather high temperatures, many of them are formed at ordinary temperatures as the result of sedimentary, diagenetic, or weathering processes. It seems that sheet structures are more readily formed under such conditions than are the other silicate types.

THE CLAY MINERALS

The term *clay* implies an earthy, fine-grained material that develops plasticity when mixed with a limited amount of water. Chemical analyses of clays show that they are made up of hydrous aluminosilicates, commonly with appreciable amounts of iron, magnesium, calcium, sodium, and potassium. All are very fine grained, frequently forming colloidal solutions; the upper limit of clay size is generally placed at a particle diameter of 0.004 mm. Genetically, most clays are the product of sedimentation, diagenesis, or weathering, but they can also be formed by hydrothermal activity.

The characteristic minerals of most clays are phyllosilicates that belong to one of the following five groups: (1) the kaolin (or *kandite*) group, (2) the montmorillonite (or *smectite*) group, (3) the clay mica (or *illite*) group, (4) vermiculite, and (5) the chlorites. (The chlorites are not, however, considered to be clay minerals.) Minerals of the kaolin group have compositions corresponding to the formula $Al_4Si_4O_{10}(OH)_8$, and structures of Si_4O_{10} sheets alternating with gibbsite-type sheets. Minerals of the montmorillonite group have structures similar to pyrophyllite, but with exchangeable cations and a variable number of water molecules between the layers, which results in swelling when these minerals are immersed in water. Common clay mica, illite, is essentially fine-grained muscovite; it is commonly mixed or interlayered with montmorillonite. Vermiculite is essentially a trioctahedral analog of montmorillonite, with a talc structure. Chlorite in clays is typically mixed with other clay minerals, and it is then often difficult to detect.

The clay minerals have many physical features in common. They do not occur as macroscopic or, except for the kaolin minerals, as even microscopic crystals; instead, they occur as earthy masses, which characteristically adhere to one's tongue. As a result, hardness is not a diagnostic property, since all masses easily disaggregate when scratched and thus appear very soft. (The true hardness of these minerals is about 2–3.) The water content of the clay minerals varies with the humidity of the atmosphere, and their apparent density varies accordingly, decreasing with the water content. Consequently, it is extremely difficult to distinguish between the clay minerals on the basis of their physical properties, and, indeed, the positive identification of the minerals in the clay fraction of sediments and soils is one of the most exacting problems for a mineralogist; in fact, it generally requires a combination of optical, X-ray, and, in some cases, differential thermal analysis and chemical tests.

Kaolinite, $Al_4Si_4O_{10}(OH)_8$

Crystal system, class, and space group: Triclinic; *1; P1.*
Cell dimensions and content: $a = 5.15$, $b = 8.95$, $c = 7.37$; $\alpha = 91°48'$, $\beta = 104°30'$, $\gamma = 90°$; $Z = 1$ (Figure 15-22).
Habit: Typically in earthy aggregates; pseudo-hexagonal platy crystals are generally distinguishable under the electron microscope (Figure 15-23).
Cleavage: {001} perfect, but not observable to the unaided eye because of small grain size.
Hardness: 2.
Specific gravity: 2.6.

Figure 15-22.   Structure of kaolinite, $Al_4Si_4O_{10}(OH)_8$. A tetrahedral sheet Si_4O_{10} linked to octahedral $Al_4O_4(OH)_8$, sometimes called the "gibbsite-type layer."

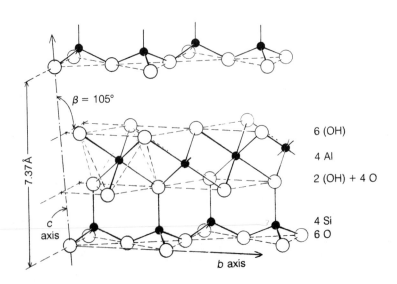

$\beta = 105°$

7.37 Å

6 (OH)

4 Al

2 (OH) + 4 O

4 Si
6 O

c axis

b axis

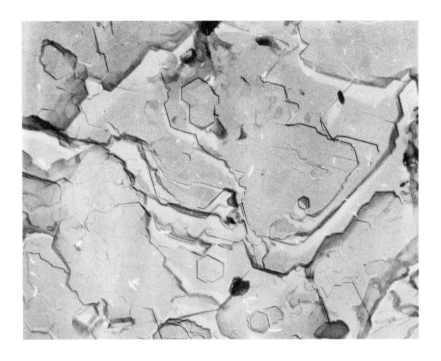

Figure 15-23. Kaolinite. Electron micrograph of kaolinite from Cold Spring clay deposit near Vesuvius, Virginia. Preshadowed carbon replica of fracture surface; magnification ~ 48,000x. (Courtesy of Mineral Constitution Laboratories, Pennsylvania State University.)

Color and streak: White, often stained reddish, brown, or gray by impurities. White.

Luster and light transmission: Pearly if coarsely crystalline, but usually dull and earthy. Subtranslucent.

Optical properties:
anisotropic, biaxial (−): $2V = 23°–60°$
refractive indices: α, 1.553–1.565; β, 1.559–1.569; γ, 1.560–1.570
generally nearly opaque

Chemistry: The composition of kaolinite corresponds closely to the formula, there being little or no atomic substitution. Kaolinite is the commonest of four polytypes, the others being *dickite, nacrite,* and *metahalloysite.* Dickite and nacrite, both monoclinic, are rare; metahalloysite is moderately common. The polytypes differ in the stacking of the basic structural unit (the kaolin layer), which consists of a tetrahedral layer linked to a gibbsite-type layer. Regular sequences of one, two, and six kaolin layers are found in kaolinite, dickite, and nacrite, respectively. Metahalloysite is derived from halloysite $[Al_4Si_4(OH)_8O_{10} \cdot 8H_2O]$, which is made up of an irregular sequence of kaolin layers plus interlayer water. In essence, *metahalloysite* is halloysite from which most of the interlayer water has been removed; the irregular stacking, which persists, is characteristic.

Occurrence and uses: Kaolinite is formed by the decomposition of other aluminosilicates, especially the feldspars, either by weathering or by hydrothermal activity. Large deposits have been formed by the hydrothermal alteration of feldspar in granites and granite pegmatites, as in England (Cornwall), Czechoslovakia, and China (whence the name *kaolin* came). Some deposits have been formed by erosion of kaolinized granite and redeposition of the kaolinite, as in Georgia and North Carolina.

Kaolinite is an important industrial mineral, being used as a filler in paper and as an essential raw material in the manufacture of ceramics.

Montmorillonite, $(Na,Ca)_{0.33}(Al,Mg)_2Si_4O_{10}(OH)_2 \cdot nH_2O$

Crystal system, class, and space group   Monoclinic; $2/m$; $C2/m$(?).
Cell dimensions: $a \cong 5.17$, $b \cong 8.94$, $c \cong 9.95$; $\beta \cong 99° 50'$.
Habit: In earthy masses.
Cleavage: {001} perfect but not macroscopically visible.
Hardness: $2-2\frac{1}{2}$.
Specific gravity: 2.0–2.7, less with higher water contents.
Color and streak: Typically gray, greenish gray, or buff; may be white, yellow, yellow-green, pink, or brown. White.
Luster and light transmission: Greasy or dull. Subtranslucent.
Optical properties:
   anisotropic, biaxial (−): $2V = 0°-30°$
   refractive indices: α, 1.48–1.57; β, 1.50–1.59; γ, 1.50–1.60
Chemistry: The name *montmorillonite* is applied to the mineral approximating the composition given here and to a group of minerals of similar structure and properties that may be derived from it by atomic substitution. The composition of all montmorillonite group minerals may be expressed by the formula $X_{0.33}Y_{2-3}Z_4O_{10}(OH)_2 \cdot nH_2O$ in which X, the exchangeable ions, may be Ca/2, Li, or Na; Y may be Al, Cr, Fe, Li, Mg, Ni, or Zn; and Z may be Al or Si. The fact that some of these lead to a net negative charge on the layers is generally compensated by cations such as Ca^{+2}, Na^+, and H_3O^+ (i.e., $H^+ + H_2O$) that are adsorbed between the layers, which explains the swelling of montmorillonite when it is immersed in water, and also its property of *cation exchange*, whereby cations in solution can be exchanged for cations adsorbed in the mineral.

In general usage, montmorillonite as an individual species is the aluminum-rich form approximating the given formula; *nontronite* is an iron-rich species, greenish yellow when pure, in which the nontetrahedral Al is largely replaced by Fe^{+3}; *saponite* is the group member in which the nontetrahedral Al is largely replaced by Mg; *sauconite* is like saponite but the Al is largely replaced by Zn plus or minus Mg and Fe^{+3}.

Diagnostic features: Its claylike character, its soapy feel; the property of swelling and forming a gel-like mass in water.

Occurrence and uses: Montmorillonite is most commonly formed by the alteration of beds of volcanic ash. *Bentonite* is a rock consisting largely of montmorillonite. Montmorillonite is also an important constituent of some soils.

The physical properties of montmorillonite make it an important industrial mineral. It is much used in drilling muds because of the gel-like suspension it forms in water. It is used as a binder for pelletizing iron ore fines and as a plasticiser in molding sands for foundries because a small amount will give plasticity to a large quantity of inert material. Its swelling properties, so useful in industry, often cause serious complications in civil engineering operations, where montmorillonite-bearing material may have to be eliminated during excavation.

Most of the world's supply of bentonite is mined from deposits in Wyoming, Texas, and Mississippi.

Illite, $(K,H_3O)(Al,Mg,Fe)_2(Si, Al)_4O_{10}[(OH)_2,H_2O]$

Crystal system, class, and space group: Monoclinic; $2/m$; $C2/m(?)$.

Cell dimensions and content: $a = 5.2$, $b = 9.0$, $c = 10.0$; β, variable (?); $Z = 1$.

Habit: Typically microcrystalline.

Cleavage: {001} perfect.

Hardness: 1–2.

Specific gravity: 2.6–2.9.

Color and streak: White, light gray, or buff, commonly stained to other hues. White.

Luster and light transmission: Silky to dull. Generally opaque in mass.

Optical properties:

anisotropic, biaxial (−): $2V < 10°$

refractive indices: α, 1.54–1.59; β, 1.57–1.63; γ, 1.57–1.63

Chemistry: Illite is the name generally applied to a group of clay mica minerals. They are closely similar to muscovite in composition but typically show less K and more Si. Other minerals thought to belong to the clay-mica group have been given such names as *hydromica, hydromuscovite, hydrobiotite,* and *brammalite.*

Occurrence: Illite is the predominant constituent of most argillaceous sediments and sedimentary rocks (e.g., shale) and also is present in many soils. The illite of sedimentary rocks may be detrital and/or diagenetic.

Vermiculite, $(Mg,Fe,Al)_3(Al,Si)_4O_{10}(OH)_2 \cdot 4H_2O$

Crystal system, class, and space group: Monoclinic; m; Cc.

Cell dimensions and content: $a \cong 5.3$, $b \cong 9.2$, $c \cong 29$; $\beta = 97°$; $Z = 2$.

Habit: Typically as pseudomorphs after phlogopite or biotite.

Cleavage: {001} perfect.

Hardness: $1\frac{1}{2}$.

Specific gravity: 2.4.

Color and streak: Yellow to brown. White.

Luster and light transmission: Pearly or submetallic. Translucent.

Optical properties:
anisotropic, biaxial (−): $2V = 0°–8°$
refractive indices: α, 1.525–1.564; β, 1.545–1.583; γ, 1.545–1.583
slightly pleochroic

Chemistry: Chemically and structurally, vermiculite is related to talc in much the way that montmorillonite is related to pyrophyllite. In fact, vermiculite has many of the same properties as montmorillonite (e.g., cation exchange capacity, a (001) spacing that varies with the water content, and the identity of exchangeable cations present).

Diagnostic features: The rapid and large (up to 30 times original volume) expansion that vermiculite will undergo, in a direction parallel to the c axis, when it is heated quickly to 250–300°C.

Occurrence and uses: Vermiculite occurs rather commonly as the product of weathering and of hydrothermal alteration of phlogopite or biotite. It is also present in the clay fraction of some soils. Some of the soil vermiculites have been found to be dioctahedral.

Large quantities of vermiculite are mined and are expanded by heating for use as an insulator, both acoustical and thermal, in building construction. It is also mixed with cement and plaster to make lightweight concrete and plaster. In its expanded form, it is mixed with soil for germinating mixes.

Serpentine, $(Mg,Fe, Ni)_3Si_2O_5(OH)_4$

Serpentine as a mineral name applies to a group of minerals including chrysotile, antigorite, and lizardite. The structures of the diverse polytypes are similar to those of minerals of the kaolinite group. In at least some serpentines, however, the layers are curved to form tubular structures (Figure 15-24). It is now thought by many mineralogists that these shapes may reflect growth involving either radial or coupled dislocations. The physical properties of chrysotile and antigorite are extremely similar, but antigorite generally has a lamellar or platy structure whereas most chrysotile is fibrous (serpentine asbestos is chrysotile). *Lizardite* is the name usually applied to massive serpentine that constitutes much of the host rock of chrysotile veins.

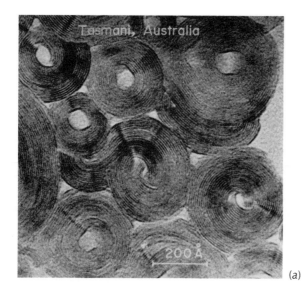

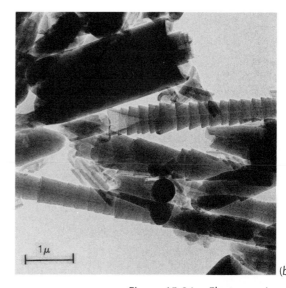

(a)

(b)

Figure 15-24. Electron micrographs of serpentine. (a) Perpendicular to the fiber axis; note that the characteristic hollow cyclinder form is evident in all but the fiber near the center, which, instead, exhibits an unusual growth pattern. (b) Parallel to the fiber axis; these conical ("cone-in-cone") and cylindrical fibrils are in synthetic chrysotile in this micrograph. (Courtesy of Keiji Yada.)

Crystal system, class, and space group: Generally reported as monoclinic, but some stacking sequences yield orthorhombic cells; the crystal class and space group most frequently cited are *m* and *Cm*, respectively.

Habit: Macroscopic crystals are unknown; as noted, the serpentine minerals occur in structureless masses, except where asbestiform.

Cleavage: None observable.

Hardness: Variable, $2\frac{1}{2}$–5.

Specific gravity: 2.5–2.6.

Color and streak: Typically green, also yellow, brown, reddish brown, and gray. White.

Luster and light transmission: Waxy or greasy in massive varieties, silky in fibrous material. Transparent to subtranslucent.

Optical properties:
Antigorite: biaxial (−): $2V=30°$–$60°$
indices: α, 1.555–1.567; β, 1.560–1.573, γ, 1.560–1.574
Chrysotile: biaxial (−): $2V=30°$–$50°$
indices: α, 1.529–1.559; β, 1.530(?)–1.564; γ, 1.537–1.567

Chemistry and alteration: The composition of most serpentine corresponds closely to the preceding formula; in addition, Al is common, probably through substitution of Al-Al for Mg-Si. Serpentine decomposes at about 500°C to forsterite, talc, and water; its presence in a rock may, therefore, indicate an origin below this temperature.

Diagnostic features: The physical properties, especially the color and the luster. Serpentine asbestos can be distinguished from amphibole asbestos by the large amount of water that the former gives off at comparatively low temperatures (~500°C). Also, amphibole asbestos tends to be more brittle and readily powders, whereas serpentine asbestos tends to mat.

Occurrence and uses: Serpentine may be formed by the alteration of olivine and enstatite under conditions of low- and medium-grade metamorphism. Some serpentine occurs as large rock masses generally referred to as serpentinites.

Chrysotile provides the greater part of the asbestos of commerce. The combination of fibrous structure, low heat conductivity, high electrical resistance, and chemical inertness gives asbestos wide industrial application. The greatest consumption is in asbestos-cement products, such as roofing shingles and sheets. The mines of the Thetford Mines, Black Lake, and Asbestos districts of southeastern Quebec are the largest producers of chrysotile in the world.

Massive serpentine is sometimes cut and polished as an ornamental stone.

Varieties and related species: The nickel-bearing variety of serpentine, with an apple-green color, is sometimes referred to by the name *garnierite*, actually a general term widely applied to a number of hydrous nickel silicates. Garnierite, an alteration product of nickel-bearing peridotites, is mined as an ore of nickel in New Caledonia. *Greenalite*, $(Fe^{+2},Fe^{+3})_{2-3}Si_2O_5(OH)_4$, is a member of the series that is an important constituent of the iron ores of the Mesabi District, Minnesota.

Pyrophyllite, $Al_2Si_4O_{10}(OH)_2$

Crystal system, class, and space group: Monoclinic; 2/m; C2/c.
Cell dimensions and content: a = 5.15, b = 8.92, c = 18.59; β = 99° 55'; Z = 2 (see Figure 15-25).
Habit: Crystals uncommon; typically in radiating spherulitic aggregates of small platy crystals or in compact fine-grained masses.
Cleavage: {001} perfect; cleavage lamellae flexible but not elastic.
Hardness: $1-1\frac{1}{2}$.
Specific gravity: 2.84.
Color and streak: White or pale yellow, commonly stained reddish, bluish, or brownish by iron oxides. White.
Luster and light transmission: Pearly on cleavage surfaces, otherwise greasy or dull. Transparent to translucent.
Optical properties:
   anisotropic, biaxial (−): 2V = 53°–62°
   refractive indices: α, 1.534–1.556; β, 1.586–1.589; γ, 1.596–1.601
Diagnostic features: Macroscopically discernible properties of pyro-

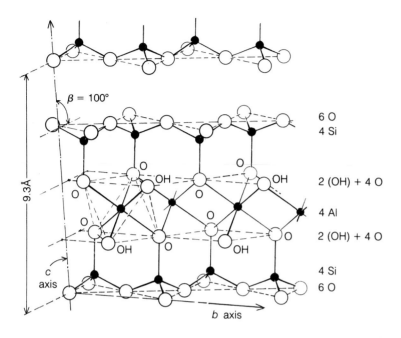

$\beta = 100°$

6 O
4 Si

2 (OH) + 4 O

4 Al

2 (OH) + 4 O

4 Si
6 O

9.3Å

c axis

b axis

Figure 15-25. Structure of pyrophyllite, Al$_4$(Si$_4$O$_{10}$)$_2$(OH)$_4$. Octahedral "gibbsite-type layer" linked to two tetrahedral Si$_4$O$_{10}$ sheets.

phyllite are practically identical to those of talc. It is said, however, that powdered pyrophyllite in water gives a pH of 6, whereas powdered talc in water gives a pH of 9.

Occurrence and uses: Pyrophyllite occurs in quartz-bearing hydrothermal veins and in low- and medium-grade metamorphic rocks that are rich in aluminum. The industrial uses of pyrophyllite are similar to those of talc and, in mineral trade statistics, the two minerals are generally reported together. Pyrophyllite, however, has the distinction of being the best material as a carrier of insecticides, such as DDT, and most insecticidal dusting powders contain it. There has been a considerable production of pyrophyllite in the Central Piedmont area of North Carolina. Also, most of the material called *agalmatolite*, used extensively in Chinese carvings, is fine-grained pyrophyllite.

Talc, Mg$_3$Si$_4$O$_{10}$(OH)$_2$

Crystal system, class, and space group: Monoclinic; 2/m; C2/c. (Some talc is triclinic.)

Cell dimensions and content: a = 5.27, b = 9.12, c = 18.85; β = 100° 00'; Z = 4.

Habit: Foliated masses and compact fine-grained aggregates.

Cleavage: {001} perfect; cleavage lamellae flexible but not elastic.

Hardness: 1. (The extreme softness and cleavability of talc result from the absence of bonding except by van der Waals forces between the uncharged layers in the structure.)

Specific gravity: 2.82.

Color and streak: Characteristically pale green; also white or gray; in some cases stained reddish or brown by iron oxides. White.

Luster and light transmission: Pearly on cleavage surfaces, otherwise greasy or dull. Translucent.

Optical properties:
anisotropic, biaxial (−): $2V = 0°–30°$
refractive indices: α, 1.538–1.550; β, 1.575–1.594; γ, 1.565–1.600

Diagnostic features: Distinguished macroscopically from other minerals, except pyrophyllite, by its extreme softness, soapy feel, and color.

Occurrence and uses: Much talc is a mineral of low- and medium-grade metamorphic rocks rich in magnesium. It may be a major rock-forming mineral, especially hydrothermally altered ultramafic rocks, such rocks being known as *steatites*. Soapstones and talc schists are other talc-rich rocks.

Talc has many commercial uses; the most familiar, although one that consumes comparatively little of the mineral, is as talcum powder. Talc has low conductivity for heat and electricity, is fire resistant, hardens when heated to a high temperature, and is not attacked by acids, properties that are valuable in many industrial applications. It is used as a lubricant, in ceramics (especially for electrical porcelain and refractories), and as a filler in paint, paper, and rubber. Slabs of soapstone are used for electrical switchboards and for acid-proof table tops and sinks.

MICA GROUP

The micas constitute an isomorphous group within the phyllosilicates. Their structures (see Figure 15-26) are closely related to those of pyrophyllite and talc. Two sheets of linked SiO_4 tetrahedra are juxtaposed with the vertices of the tetrahedra pointing inward; these vertices are cross-linked either with Al, as in muscovite, or with Mg, as in phlogopite; hydroxyl groups, which are also present, complete the six-coordination of the Al or Mg. A firmly bound layer is thus produced with the bases of the silica tetrahedra on both of its outer sides. This layer has a negative charge because Al is substituting for Si in some of the tetrahedra (this difference is the main one between, for example, muscovite and pyrophyllite). This negative charge is compensated by potassium, or less commonly by sodium or calcium, ions that are located between the layers and serve to link the layers together. The resulting ionic linkage between the layers accounts for the greater hardness of the

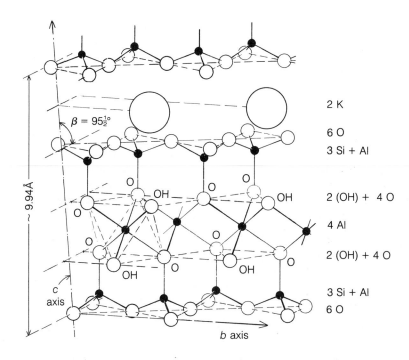

β = 95½°

~ 9.94Å

c axis

b axis

2 K

6 O

3 Si + Al

2 (OH) + 4 O

4 Al

2 (OH) + 4 O

3 Si + Al

6 O

O — OH

O

O

O

OH

O

O

OH

O

O

O

OH

O

Figure 15-26. Structure of muscovite, $K_2Al_4(Si_6Al_2)O_{20}(OH)_4$. Pyrophyllite layers with one aluminum substituted for one out of four silicons in each tetrahedral layer, linked together by potassium atoms in twelvefold coordination with oxygen.

micas as contrasted to that of talc and pyrophyllite, in which the individual layers are uncharged and hence have practically no intercohesion.

The individual species in the mica group include a number of polytypes that differ in the number and orientation of the layers in their unit cells. The geometry of the structure is such that six distinct polytypes are possible (see Table 15-3). These polytypes can only be distinguished by X-ray diffraction techniques or, in a few cases, by optical properties.

According to Deer et al. (vol. 3, p. 6; see References for Part II), the

Table 15-3. Mica polytypes

Designation	Crystal System	Crystal Class	Space Group	~ Cell Dimensions				Number of Layers
				a(Å)	b(Å)	c(Å)	$β$	
1M	Monoclinic	2/m	C2/m	5.3	9.2	10	100°	1
$2M_1$	Monoclinic	2/m	C2/c	5.3	9.2	20	95°	2
$2M_2$	Monoclinic	2/m	C2/c	9.2	5.3	20	98°	2
2O	Orthorhombic	mm2	$Ccm2_1$	5.3	9.2	20	90°	2
3T	Hexagonal (Trigonal)	32	$P3_112$ or $P3_212$	5.3	—	30	—	3
6H	Hexagonal	622	$P6_122$ or $P6_522$	5.3	—	60	—	6

Data from J. V. Smith and H. S. Yoder, Jr., 1956, "Experimental and theoretical studies of the mica polymorphs." *Mineralogical Magazine*, **31**, 209–235.

most common polytypes are as follows: muscovite, $2M_1$; phlogopite and biotite, $1M$ (also $2M_1$ and $3T$); lepidolite, $1M$ and $2M_1$ (also $3T$); and glauconite, $1M$, $2M$, and $3T$. In our descriptions of muscovite, phlogopite, and biotite, the structural data relate to the first listed polytype.

The general formula for the group is $W(X,Y)_{2-3}Z_4O_{10}(OH,F)_2$, in which W is chiefly K (Na in paragonite) but may also include Ba, Ca, Rb, etc.; X and Y represent Al, Mg, Fe^{+2}, Fe^{+3}, and Li, but also commonly include minor Mn, Cr, Ti, etc.; and Z is almost always Si and Al with the Si:Al ratio typically being about 3:1, but may also include Be and Fe^{+3}. (Thus, Al is both four- and six-coordinated in these minerals.) In the following list, the formulas have been simplified to an ideal type for each described species (i.e., the formulas neglect most of the variations due to atomic substitution):

Muscovite	$KAl_2(AlSi_3O_{10})(OH)_2$
Phlogopite	$KMg_3(AlSi_3O_{10})(F,OH)_2$
Biotite	$K(Mg,Fe)_3(Al,Fe)Si_3O_{10}(OH,F)_2$
Lepidolite	$K(Li,Al)_3(Si,Al)_4O_{10}(OH,F)_2$
Glauconite	$(K,Na)(Fe,Mg,Al)_2(Si,Al)_4O_{10}(OH)_2$

As a group, the micas are characterized by their perfect basal cleavage that gives thin, flexible, and elastic sheets. Crystals, typically tabular with prominent basal planes, are hexagonal in outline with the angles (110) $\wedge$ ($1\bar{1}0$) and (110) $\wedge$ (010) close to 60°. A blow with a dull-pointed instrument on a cleavage plate of any of the micas develops a six-rayed percussion figure with the line parallel to {010} more distinct than the other two lines, which are parallel to the prism faces.

Muscovite, $KAl_2(AlSi_3O_{10})(OH)_2$

Crystal system, class, and space group: Monoclinic; 2/m; C2/c.
Cell dimensions and content: a $\cong$ 5.3, b $\cong$ 9.2, c $\cong$ 20.0, $\beta \cong$ 95°; Z = 2. (See Figure 15-26.)
Habit: Crystals common, but muscovite occurs typically in lamellar masses or small flakes; crystals tabular with a hexagonal outline (Figure 15-27).
Cleavage: {001} perfect, cleavage lamellae flexible and elastic; may also show parting parallel to {010} and {110}.
Hardness: $2\frac{1}{2}$ (on cleavage), 4 (across cleavage).
Specific gravity: 2.8–2.9.
Color and streak: Colorless or pale shades of green, gray or brown (in thin sheets). White.
Luster and light transmission: Vitreous to pearly. Transparent.
Optical properties:
anisotropic, biaxial (−): 2V = 30°–47°
refractive indices: α, 1.552–1.574; β, 1.582–1.610; γ, 1.587–1.610

Figure 15-27. Muscovite "books" atop albite from a pegmatite of the Amelia District of Virginia (~0.5x). (Courtesy of Smithsonian Institution, Museum of Natural History.)

Chemistry: Muscovite may range considerably in composition as a result of atomic substitutions. Some Na is generally present replacing K; replacement is greater the higher the temperature of formation, and the sodium content of muscovite may thus be a useful geological thermometer. (When Na content exceeds K content, the mineral is called *paragonite*.) Aluminum in six-coordination may be partly replaced by Mg and Fe^{+2}, or less commonly by Cr^{+3} (giving a bright green color, var. *fuchsite*) or V^{+3} (*roscoelite*). The ratio Si:Al in four-coordination ranges from 3:1 to about 7:1; high-silica-content muscovite is generally referred to by the variety name *phengite*. Some F may be present replacing OH.

Diagnostic features: The perfect cleavage yielding flexible and elastic foliae identifies muscovite as a mica. Brown varieties closely resemble phlogopite, and chemical tests may be necessary to distinguish them; in the field, however, occurrences and associations of these two micas are generally quite different.

Occurrence and use: Muscovite occurs in a variety of associations. In igneous rocks it is confined to a few granites, but it is also a common mineral in granite pegmatites (which provide the natural material used in industry). It is a common and abundant mineral in phyllites, schists, and gneisses of low- and medium-grade metamorphism. Extremely fine-grained muscovite, generally called *sericite*, is common as an alteration product of other minerals, for example, feldspar. Muscovite itself is fairly resistant to alteration and weathering, and thus it occurs as a detrital mineral in some clastic sediments. As we have mentioned previously, clay-mica (*illite*) evidently forms in the sedimentary environment, in which it makes up a large part of several argillaceous sediments and sedimentary rocks.

   The combination of perfect cleavage, flexibility, elasticity, low thermal conductivity, infusibility, and high dielectric strength makes muscovite a valuable mineral that is widely used in industry. About 90 percent of the production of sheet muscovite is used in the electrical industry for condensers, as insulating material between commutator segments, and in heating elements. Ground mica is used as a filler and as a dusting medium to prevent such materials as rubber goods, asphalt roofing, and asphalt tiles from sticking together. The chief producer of sheet muscovite is India, but considerable amounts have also been mined in New Hampshire and North Carolina.

Phlogopite, $KMg_3(AlSi_3O_{10})(F,OH)_2$
Biotite, $K(Mg,Fe)_3(Al,Fe)Si_3O_{10}(OH,F)_2$

Crystal system, class, and space group: Monoclinic; $2/m$; $C2/m$.
Cell dimensions: $a = 5.3$, $b = 9.2$, $c \cong 10.0$; $\beta \cong 99°18'$; $Z = 1$.

Habit: Crystals common; typically as pseudo-hexagonal prisms; more commonly as lamellar plates without crystal outline.

Cleavage: {001} perfect, giving flexible and elastic lamellae.

Hardness: $2\frac{1}{2}$ (on cleavage), 4 (across cleavage).

Specific gravity: 2.8–3.4, higher with greater iron contents.

Color and streak: Pale yellow to brown, in some cases greenish or reddish for phlogopite; dark green, brown, or black for biotite. White to gray.

Luster and light transmission: Vitreous to pearly, sometimes submetallic. Transparent to translucent.

Optical properties: (biotite)
anisotropic, biaxial (−): $2V = 0°–25°$
refractive indices: α, 1.565–1.625; β, 1.605–1.696; γ, 1.605–1.696
pleochroic, darker parallel to cleavage

Other properties: Some mica exhibits asterism; that is, a point of light viewed through a cleavage sheet shows a six-rayed star caused by regularly arranged microscopic inclusions.

Chemistry: Phlogopite and biotite are arbitrary divisions of a single phase of variable composition, phlogopite being the pale-colored, low-iron part of the series. The composition of the series is exceedingly variable: K can be replaced in part by Na, Ca, Ba, Rb, or Cs; Mg can be completely replaced by Fe^{+2} and Fe^{+3} and in part by Ti and Mn; the Si:Al ratio is somewhat variable; and some of the OH is replaceable by F. A marked correlation exists between composition and geological environment; for example, in igneous rocks, the iron content of biotite tends to increase with silica content of the rock (ultrabasic rocks containing phlogopite and granites and granite pegmatites containing iron-rich biotites).

Diagnostic features: The perfect cleavage, yielding flexible and elastic sheets, identify these minerals as micas. The colors serve to identify them, at least tentatively, more explicitly.

Occurrence and uses: Phlogopite occurs in ultrabasic igneous rocks (e.g., periodotites), in some marbles, and in magnesium-rich pegmatites. Biotite occurs in many igneous rocks but most commonly in the intermediate and acidic families; it is also an important constituent of metamorphic rocks, especially schists, gneisses, and hornfelses.

Phlogopite is used industrially in the same way as is muscovite and, for some purposes, it is superior to muscovite. It is preferred for electrical commutators since it wears at about the same rate as the copper segments. It also has greater heat resistance and can withstand temperatures up to 1000°C. Phlogopite in which all the OH is replaced by F has been made in the laboratory in sufficiently large crystals for industrial use, but the natural mineral is less expensive. Canada and Madagascar are the principal producers of natural phlogopite for industrial use.

Lepidolite, $K(Li,Al)_3(Si,Al)_4O_{10}(F,OH)_2$

Crystal system, class, and space group: Monoclinic 2/m; C2/m.
Cell dimensions and content: a = 5.2, b = 9.0, c = 10.1; β = 100° 50'; Z = 1.
Habit: Crystals uncommon; typically in medium- to fine-grained crystalline aggregates (good crystals and large plates are much less common for lepidolite than for the other micas).
Cleavage: {001} perfect, giving flexible and elastic flakes.
Hardness: $2\frac{1}{2}$ (on cleavage), 4 (across cleavage).
Specific gravity: 2.8–2.9.
Color and streak: Commonly pale lilac; less commonly colorless, pale yellow, or pale gray. White.
Luster and light transmission: Vitreous to pearly. Transparent.
Optical properties:
anisotropic, biaxial (−): 2V = 0°–58°
refractive indices: α, 1.525–1.548; β, 1.551–1.585; γ, 1.554–1.587
Chemistry: Lepidolite ranges considerably in composition; K is commonly replaced in part by Na, Rb, and Cs; the relative amounts of Li and Al in six-coordination may range widely; some Si may be replaced by Al; and some OH may be replaced by F.
Diagnostic features: Distinguished from the other micas by its occurrence in complex granite pegmatites and by its pale lilac color.
Occurrence and uses: Lepidolite occurs in a few granites and in complex granite pegmatites, typically those that also contain other lithium minerals.

Lepidolite has been mined from complex granite pegmatites as an ore of lithium.

Glauconite, $(K,Na)(Fe,Al,Mg)_2(Si,Al)_4O_{10}(OH)_2$

Crystal system, class, and space group: Monoclinic; 2/m; C2/m.
Cell dimensions and content: a = 5.29, b = 9.19, c = 10.03; $\beta \cong 100°$; Z = 1.
Habit: Crystals very rare; typically as small granules.
Cleavage: {001} perfect, but not macroscopically visible.
Hardness: 2.
Specific gravity: 2.5–2.8.
Color and streak: Green to black. Green.
Luster and light transmission: Earthy and dull. Subtranslucent.
Optical properties:
anisotropic, biaxial (−): 2V = 0°–20°
refractive indices: α, 1.56–1.61; β, 1.61–1.65; γ, 1.61–1.65; pleochroic in greens
Chemistry: Structurally, glauconite is a dioctahedral mica and can be

considered as muscovite, in which part of the Al in six-coordination is replaced by Mg and Fe^{+2}, thereby reducing the positive charge on the octahedral layer; this charge reduction is compensated for by the replacement of Al in four-coordination by Si. Much of the remaining Al in six-coordination may be replaced by Fe^{+3}. Most analyses show excess water, at least much of which is probably adsorbed water not driven off by common analytical procedures.

Diagnostic features: Color and habit; readily attacked by HCl.

Occurrence: Glauconite is an authigenic (perhaps in part biochemical) mineral of marine sedimentary rocks; those in which it is the predominent mineral are widely referred to as greensands. At the present time, it is forming on the sea floor in some areas where clastic sedimentation is slight or not taking place. Its occurrence in the geological column is commonly linked with disconformities. Although glauconite has been recorded as forming from detrital biotite, most of it has probably crystallized directly from an aluminosilicate gel.

Greensands of the coastal plain of, for example, New Jersey have been used locally as fertilizer. They are also used as water softeners because they have high base exchange capacities and generally regenerate rapidly.

Related species: The mineral *celadonite* is apparently identical in structure and composition to glauconite, but its mode of occurrence is quite different; it occurs as blue-green earthy material in vesicular cavities in basalt, apparently having been formed by hot solutions during the late cooling stages of the rock.

Prehnite, $Ca_2Al(AlSi_3O_{10})(OH)_2$

Crystal system, class, and space group: Orthorhombic; *mmm; Pncm.*

Cell dimensions and content: $a = 4.61$, $b = 5.47$, $c = 18.48$; $Z = 2$.

Habit: Crystals uncommon; generally pseudocubic or tabular parallel to {001}; typically massive, globular, or stalactitic with a crystalline surface.

Cleavage: {001} good.

Hardness: $6\frac{1}{2}$.

Specific gravity: 2.9.

Color and streak: Characteristically pale green, but for some specimens white or gray. White.

Luster and light transmission: Vitreous. Transparent to translucent.

Optical properties:
anisotropic, biaxial (+): $2V = 60°–70°$
refractive indices: α, 1.610–1.637; β, 1.615–1.647; γ, 1.632–1.673

Chemistry: Fe^{+3} can replace Al^{+3} up to 8 percent Fe_2O_3.

Diagnostic features: Color and habit. It gelatinizes with HCl.

Occurrence and associations: Prehnite occurs chiefly in cavities in basic igneous rocks, generally associated with zeolites; also in low-grade metamorphic rocks and as a secondary mineral after calcic plagioclase in altered igneous rocks. Magnificent specimens of prehnite have been found in the trap-rock quarries of northern New Jersey and northern Virginia.

Chlorite Group, $A_{5-6}Z_4O_{10}(OH)_8$, in which A = Al, Fe, Li, Mg, Mn and/or Ni; and Z = Al, B, Si, and/or Fe.

Crystal system, class, and space group: Monoclinic; $2/m$ [triclinic polytypes are also known; see Table 15-4]; $C2/m$.

Cell dimensions and content (clinochlore): a = 5.3, b = 9.2, c = 14.3; β = 97° 06'; Z = 1 (see Figure 15-28).

Habit: Rare pseudo-hexagonal crystals; tabular parallel to {001}; typically as scaly aggregates and as fine-grained and earthy masses.

Cleavage: {001} perfect, cleavage flakes flexible but not elastic.

Hardness: $2\frac{1}{2}$ (on cleavage).

Specific gravity: 2.6–3.3, higher with greater iron content (Figure 15-29); 2.7–2.9 in common varieties.

Color and streak: Characteristically green (hence the name), manganese-bearing varieties orange to brown, chromium-bearing varieties violet. White, pale green.

Luster and light transmission: Vitreous to earthy. Transparent to translucent.

Optical properties (clinochlore):
   anisotropic, biaxial (+): $2V$ = 0°–40°
   refractive indices: α, 1.57–1.59; β, 1.57–1.60; γ, 1.58–1.61
   slightly pleochroic in greens and colorless

Chemistry: As might be suspected from the formula, the chlorites form an extensive group with a high degree of atomic substitution. In the

Table 15-4. *Chlorite polytypes*

Crystal System	Crystal Class	Space Group	~Cell Dimensions			β	Number of Layers
			$a(\text{Å})$	$b(\text{Å})$	$c(\text{Å})$		
Monoclinic	$2/m$	$C2/m$	5.3	9.2	14.3	97°06'	of Layers
Triclinic	1	$C1$	5.3	9.2	14.3	97°06'	1
Triclinic	1	$C1$	5.3	9.2	28.6	97°06'	2
Triclinic	1	$C1$	9.2	5.3	42.9	86°[a]	3

Data from G. W. Brindley, B. M. Oughton, and K. Robinson, "Polymorphism of the chlorites, I. Ordered structures." *Acta Crystallographica*, **3**, 408–416.
[a] Recorded as angle α.

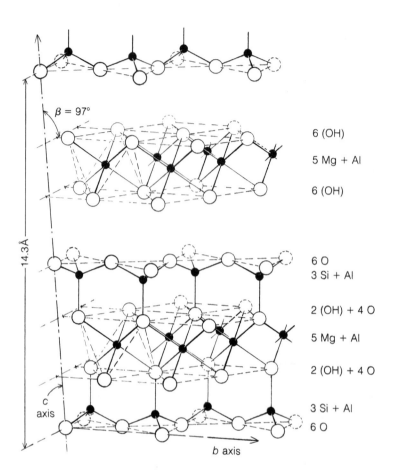

Figure 15-28. Structure of chlorite, $(Mg,Fe,Al)_6(Al,Si)_4O_{10}(OH)_8$ [shown here as $Mg_{10}Al_2(Al_2Si_6)$-$O_{20}(OH)_{16}$]. A structure similar to pyrophyllite but with five Mg and one Al in the octahedral positions and an additional octahedral layer of $Mg_5Al(OH)_{12}$ between the pyrophyllitelike layers. This structure may be viewed alternatively as consisting of interlayered layers of the brucite-type $[Mg_5Al(OH)_{12}]$ and talclike layers $[Mg_5Al(Si,Al)_8$-$O_{20}(OH)_4]$. It is also noteworthy that the fundamental serpentine unit, $Mg_6Si_4O_{10}(OH)_4$, is essentially the talc layer minus the bottom (in the diagram) $AlSi_3O_8$ sheet.

general formula, Mg and Fe are mutually replaceable; Al in six-coordination ranges from 0 to 2; Al in four-coordination ranges from 0.5 to 2, Si ranges from 2 to 3.5, and Fe^{+3} is also commonly present. Some varieties of chlorite contain appreciable amounts of chromium, as well as the indicated nickel and manganese. Thus, the chemical composition and physical properties of minerals of the chlorite series differ widely, and a large number of names are currently accepted. Clinochlore $[(Mg,Fe)_5Al(Si_3Al)O_{10}(OH)_8]$, which forms a series with chamosite $[(Fe,Mg)_5Al(Si_3Al)_{10}(OH,O)_8]$, is probably the most common.

The structure of chlorite, illustrated in Figures 15-28 and 15-29, consists of interlayered talclike and brucitelike layers. Alternatively, it may be viewed as analogous to the mica structure, with the brucite layer in the chlorite thus corresponding to, for example, the potassium ions in muscovite.

Diagnostic features: The green color, the micaceous cleavage, and the nonelastic nature of cleavage flakes.

Occurrence: Chlorite is an abundant mineral in low-grade schists and is common in igneous rocks as an alteration product of biotite and other ferromagnesian minerals. Hydrothermal alteration of pre-existing rocks has often resulted in the formation of chlorite in large amounts. Chlorite is also present in the clay mineral fraction of many sediments. *Chamosite,* an iron-rich chlorite, is a common constituent of some sedimentary iron ores.

Stilpnomelane, $\sim K(Fe,Al)_{10}Si_{12}O_{30}(OH)_{12}$

Crystal system, class, and space group: Triclinic; *1; P1.*
Cell dimensions and content: a = 21.724, b = 21.724, c = 17.740; $\alpha = 124° 08'$, $\beta = 95° 52'$, $\gamma = 120° 00'$; Z = 16.
Habit: Crystals rare; typically as thin plates, commonly in radiating or sheaflike aggregates; also as encrustations.
Cleavage: {001} perfect to brittle flakes; {010} indistinct.
Hardness: 3–4.
Specific gravity: 2.59–2.96.
Color and streak: Red-brown, golden brown, dark green, bronzy, black; in some cases color-zoned. White to tan.
Luster and light transmission: Pearly to submetallic. Translucent to subtranslucent.
Optical properties:
 anisotropic, biaxial [uniaxial] (−): $2V \cong 0°–40°$

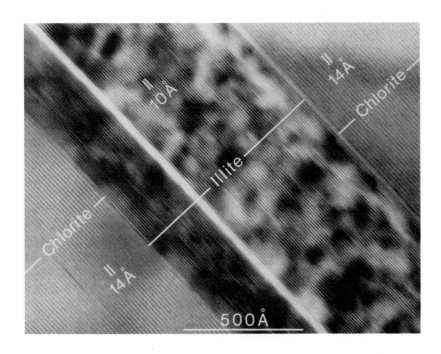

Figure 15-29. Transmission electron micrograph of interleaved layer silicates with layers oriented perpendicular to the micrograph. Layers of chlorite alternate with illite (mottled appearance). The 14 Å layers of chlorite can be seen to consist of separate talclike and brucitelike sublayers. (Courtesy of Jung Lee.)

refractive indices: α, 1.543–1.634; β, 1.576–1.745; γ, 1.576–1.745

pleochroic

Chemistry: Mn commonly substitutes for Fe, and some mineralogists have described different specimens as Mn- (versus Fe-) stilpnomelane.

Diagnostic features: The {001} cleavage is less perfect than that of biotite and gives brittle flakes; also, biotite does not have a second cleavage.

Occurrence and associations: With chlorite, epidote, almandine, actinolite, etc. in regional metamorphic rocks derived from graywackes; with lawsonite and glaucophane in relatively low-temperature, high-pressure metamorphic rocks; and with iron-rich silicates and hematite in metamorphosed iron formations of, for example, the Lake Superior Region.

Apophyllite, $KCa_4(Si_4O_{10})_2(OH,F) \cdot 8H_2O$

Crystal system, class, and space group: Tetragonal; $4/m2/m2/m$; $P4/mnc$.

Cell dimensions and content: $a = 8.960$, $c = 15.78$; $Z = 2$.

Crystal structure: Instead of having the regular planes of linked six-membered rings present in the previously described phyllosilicates, apophyllite has regular four-membered rings linked together by irregular eight-membered rings, with the whole forming a sheet with the usual phyllosilicate formula $(Si_4O_{10})^{-4}$.

Habit: Crystals common; combinations of {110}, {101}, and {001}; crystals appear to be isometric combinations of the cube and octahedron, but the true symmetry is revealed by the unidirectional cleavage and the difference in luster between basal planes and other faces; also tabular, flattened parallel to {001}. Prisms are typically striated parallel to c.

Common forms and angles: (001) $\wedge$ (101) = 60° 25′

Cleavage: {001} perfect, {110} indistinct.

Hardness: 5.

Specific gravity: 2.35.

Color and streak: Colorless or white; in some cases, tinted pink, green, or yellow. White.

Luster and light transmission: Pearly on {001}, otherwise vitreous. Transparent to translucent.

Optical properties:
anisotropic, uniaxial (+)
refractive indices: ω, 1.531–1.536; ϵ, 1.533–1.538
low birefringence.

Chemistry: Most analyses show a little sodium replacing potassium.

Predominance of F versus OH leads to *fluorapophyllite* and *hydroxyapophyllite*, respectively.

Diagnostic features: Crystal form, cleavage, and associations.

Occurrence and associations: Apophyllite occurs in hydrothermal veins and in cavities in basalts; in the latter, it is commonly associated with zeolites, as in northern New Jersey, northern Virginia, and Nova Scotia.

SUBCLASS INOSILICATES

The inosilicates are a subclass in which the SiO_4 tetrahedra are linked to form chains of indefinite extent. There are two main types of chain structure: single chains, in which each tetrahedron is linked to the next by a common oxygen to produce an overall composition $(SiO_3)^{-2}$; and double chains, in which two single chains that are side by side are cross-linked through alternate tetrahedra to produce an overall composition $(Si_4O_{11})^{-6}$. Each type is represented by an important group of rock-forming minerals: the pyroxenes are single-chain structures; the amphiboles are double-chain structures.

Minerals of the two groups have marked similarites in structure and in physical and chemical properties. In each, the chains are aligned with their elongation in the c direction of the containing crystals, and adjacent chains are bonded by ions (mainly Ca, Mg, and/or Fe) that lie between the chains. Furthermore, the unit cells of analogous minerals of the two groups are nearly identical except for a doubling of the period in the b direction in the amphiboles, for example:

	a	b	c	β
Tremolite (amphibole)	9.85 Å	18.1 Å	5.3 Å	104° 40'
Diopside (pyroxene)	9.75 Å	8.93 Å	5.25 Å	105° 50'

[You should note, however, that the orthorhombic members of both groups have about double the spacing of {100} as that of the monoclinic members.]

Bonding between the chains is considerably weaker than that within the chains; consequently, both the pyroxenes and the amphiboles have good prismatic cleavage, and they commonly occur in crystals in which {110} is a prominent form ({210} in orthorhombic species). The crystal form and the cleavage are important distinguishing features between the two groups; in fact, in many cases, they are the only features that can be readily used to tell amphibole from pyroxene macroscopically: in amphiboles, (110) $\wedge$ ($1\bar{1}0$) is about 56°; in pyroxenes, it is about 93°. Therefore, the cross-sections of pyroxene crystals outline nearly square cleavage prisms whereas those of amphibole crystals exhibit diamond-shaped cleavage prisms, (Figure 15-30). The basis for the difference can

442

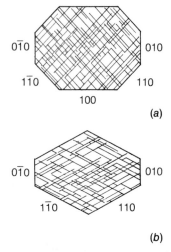

Figure 15-30.   Cross sections of (a) pyroxene, and (b) amphibole crystals, showing characteristic crystal forms and traces of the prismatic cleavages {110}.

be understood by studying the diagrams (Figures 15-31 a and b). The chains have trapezium shaped cross-sections, the length of which in the b direction is one half as great for the pyroxenes as for the amphiboles. As shown by the broken lines in Figure 15-31, cleavage takes place in a diagonal manner, with no transection of the chains.

As we have already mentioned, in addition to the structural similarities between the pyroxenes and the amphiboles, there also are many similarities in their chemical compositions and physical properties. Several analogous species in the two groups have similar compositions. Both contain the same elements, so qualitative chemical tests [except for the presence of (OH) groups, present only in the amphiboles] fail to distinguish them. In fact, it was early argued that such pairs of species were actually polymorphs. This argument, however, was settled in 1930 when the structures were elucidated, and the presence of (SiO_3) chains within the pyroxenes and (Si_4O_{11}) chains within the amphiboles were established. Physical properties—such as color, luster, hardness, and density—are also similar. For macroscopic identification, the most useful property is the difference in cleavage angle. Recognition of this difference is enhanced by the fact that the perfect cleavage of amphiboles typically yields bright, glittering surfaces, whereas the only fairly good cleavage of pyroxenes is seldom so expressed. In some cases, however, the form of the crystals is also indicative; pyroxene crystals are generally short prisms, whereas amphibole crystals tend to be long prisms, commonly acicular or fibrous. Furthermore, their modes of occurrence differ somewhat; pyroxenes are characteristic of basic and ultrabasic igneous rocks, whereas the amphibole hornblende is more common in intermediate igneous rocks. Also, amphiboles are common in metamorphic rocks, especially those of medium grade (in fact, many amphiboles occur only in these rocks), whereas pyroxenes are relatively uncommon in metamorphic rocks, being largely confined to those of high grade origin.

AMPHIBOLE GROUP

The amphibole group comprises a complex group of silicates (Fleischer, 1980, lists 57 species) that, although falling in both the orthorhombic and monoclinic systems, are closely related in crystallography and other physical properties as well as in chemical composition. A general formula of members of the amphibole group is $W_{0-1}X_2Y_5Z_8O_{22}(OH,F,Cl)_2$, in which

W = Ca, Na, K
X = Ca, Fe^{+2}, Li, Mg, Mn, Na
Y = Al, Cr, $Fe^{+2,}$ Fe^{+3}, Mg, Mn, Ti
Z = Al, Si, Ti

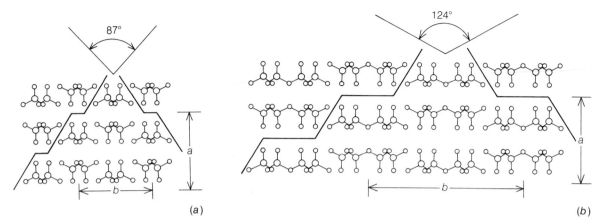

(a)

(b)

Figure 15-31. The relationship between atomic structure and cleavage in the pyroxenes (a) and the amphiboles (b). In these projections, along c, only the silica tetrahedra chains are shown. The planes of weakness are indicated by the solid lines; the resulting cleavage directions are indicated above the main diagrams. (Modified after W. F. de Jong, *General Crystallography (A Brief Compendium)*. Copyright © 1959. W. H. Freeman and Company.)

There is a discussion of the nomenclature of the group, which has been approved by the IMA Commission on New Minerals and Mineral Names, by Leake (*American Mineralogist,* **63**, 1023–1052).

Briefly, the amphiboles can be categorized in four groups: (1) the iron-magnesium-manganese group, which includes orthohombic anthophyllite, gedrite, and holmquistite and the monoclinic cummingtonite series and clinoholmquistite; (2) the calcic amphibole group, which includes, among others, the tremolite-actinolite series, magnesio- and ferro-hornblende, pargasite, hastingsite, and edenite; (3) the sodic-calcic group, which includes the taramites, katophorites, barroisites, winchites, and richterites; and (4) the alkali-amphibole group, which includes the glaucophanes, riebeckites, eckermannites, arfvedsonites, crossite, and kozulite.

As just noted, in the general formula W may be considered to stand for the large cations (K, however, may be present in only small amounts); X in most amphiboles is predominantly the smaller cation Fe^{+2} and Mg and, in some cases, Mn^{+2}, with the Fe and Mg completely interchangeable except in anthophyllite and cummingtonite; the Y is most commonly Fe^{+3}, Ti, and Al in six-coordination; and Z is largely Si and Al in four-coordination, with Al never exceeding the ratio indicated by $AlSi_3O_{11}$. Structurally, there are four (and in a few amphiboles, five) cation sites, which are generally referred to as $M1$, $M2$, $M3$, and $M4$ $(+A)$. Thus, the composition of an amphibole may be extremely complex. Nevertheless, on the basis of composition, the most frequently encountered amphiboles may be conveniently grouped as follows:

Orthorhombic
    Anthophyllite Series $(Mg,Fe)_7Si_8O_{22}(OH)_2$
Monoclinic
    Cummingtonite Series $(Mg,Fe)_7Si_8O_{22}(OH)_2$
    Tremolite-Actinolite Series $Ca_2(Mg,Fe)_5Si_8O_{22}(OH)_2$

"Hornblende Series"	$Ca_2(Mg,Fe)_4Al(Si_7Al)O_{22}(OH,F)_2$
Alkali Amphibole Group	
Glaucophane Series	$Na_2(Mg,Fe)_3Al_2Si_8O_{22}(OH)_2$
Riebeckite Series	$Na_2(Fe,Mg)_3Fe_2Si_8O_{22}(OH)_2$
Arfvedsonite Series	$Na_3(Fe,Mg)_4FeSi_8O_{22}(OH)_2$

In the anthophyllite series, minerals with the $Mg:(Mg + Fe^{+2})$ ratio from 0.1 to 0.89 are properly termed anthophyllite; those with the ratio from 0.9 to 1.0 are *magnesioanthophyllite*; and those with the ratio $Fe^{+2}:(Fe^{+2} + Mg)$ from 0.9 to 1.0 are *ferroanthophyllite*. Al is commonly present and can replace Si up to the aforementioned $AlSi_3O_{11}$ general maximum. Other substitutions are minor. Members of the anthophyllite series occur largely, if not wholly, in metamorphic rocks.

In the cummingtonite series, minerals with the $Mg:(Mg + Fe^{+2})$ ratio from 0.3 to 0.69 are cummingtonite proper; those with the ratio from 0.7 to 1.0 are called *magnesiocummingtonite*; and those with the ratio from 0 to 0.29 are *grunerite*. Although, or perhaps because, the cummingtonite and anthophyllite series overlap in chemical composition, the minerals occur side by side in many rocks. Members of the cummingtonite series are also more-or-less restricted in occurrence to metamorphic rocks.

The tremolite-actinolite series includes tremolite, $Mg:(Mg + Fe^{+2})$ $= 1.0-0.9$; actinolite, $Mg:(Mg + Fe^{+2}) = 0.89-0.5$; and ferro-actinolite, $Mg:(Mg + Fe^{+2}) = 0.5-0$. In addition, Al and Fe^{+3} may substitute for Y, Mg and Si may be replaced in part by Al and/or Ti, F may be present in lieu of some OH, and Na may occur. Members of this series are also most common in metamorphic rocks.

The series that we refer to as the "hornblende series" is more correctly called the *magnesiohornblende-ferrohornblende series*. In magnesiohornblende, the $Mg:(Mg + Fe^{+2})$ ratio ranges from 0.5–1.0; in ferrohornblende, the ratio ranges up to 0.49. Hornblende is the name applied to the dark gray or essentially black-to-greenish black amphiboles that occur in many igneous rocks.

The alkali-amphibole group includes three series whose individual members are relatively common in either metamorphic or alkalic igneous rocks. In the glaucophane series, those members with $Mg:(Mg + Fe^{+2}) = 0.5-1.0$ are called *glaucophane*, and those with $Mg:(Mg + Fe^{+2}) = 0.0-0.5$ are called *ferroglaucophane*. In the riebeckite series, the same ratios give magnesioriebeckite and riebeckite, respectively. In the arfvedsonite series, arfvedsonite has ratio values of $Mg:(Mg + Fe^{+2}) = 0.0-0.5$ and $Fe^{+3}:(Fe^{+3} + Al) = 0.5-1.0$, whereas magnesio-arfvedsonite has $Mg:(Mg + Fe^{+2}) = 0.05-1.0$ and the same $Fe^{+3}:(Fe^{+3} + Al)$ ratio. [Lower $Fe^{+3}:(Fe^{+3} + Al)$ ratios obtain for eckermannite and ferro-eckermannite.]

In the following descriptions, most of the data pertain to commonly encountered members of the just noted series (and group), *not* to end-members.

Anthophyllite, $(Mg,Fe)_7Si_8O_{22}(OH)_2$

Crystal system, class, and space group: Orthorhombic; $2/m2/m2/m$; Pnma.

Cell dimensions and content: $a \cong 18.56$, $b \cong 18.08$, $c \cong 5.28$; $Z = 4$.

Habit: Single crystals uncommon; typically in aggregates of prismatic crystals; also fibrous and asbestiform.

Cleavage: {210} perfect, $(210) \wedge (2\overline{1}0) = 55°$.

Hardness: 6.

Specific gravity: 2.86–3.28, higher with greater iron content.

Color and streak: White, gray, or brown. White.

Luster and light transmission: Vitreous, somewhat silky in fibrous varieties. Transparent to translucent.

Optical properties:
anisotropic, biaxial ($\pm$): $2V = 70°–90°$
refractive indices: α, 1.60–1.69, β, 1.61–1.71; γ, 1.62–1.72

Chemistry: See the introductory remarks for the amphibole group.

Diagnostic features: Rather distinctive color and appearance when compared to other amphiboles, except cummingtonite; it can be readily distinguished from cummingtonite only by careful density determination or by optical or X-ray means.

Occurrence, associations, and uses: Anthophyllite occurs in magnesium-rich metamorphic rocks of medium grade, commonly associated with talc or cordierite.

Asbestiform anthophyllite is used in industry, but the fibers are generally brittle and of relatively low tensile strength; consequently, it is unsuitable for spinning and is used mainly in asbestos cement and as an insulating material.

Related species: The name *gedrite* has been applied to aluminian anthophyllites.

Cummingtonite, $(Mg,Fe)_7Si_8O_{22}(OH)_2$

Crystal system, class, and space group: Monoclinic; $2/m$; C2/m.

Cell dimensions and content: $a = 9.6$, $b = 18.3$, $c = 5.3$; $\beta = 102°$ 20'; $Z = 2$.

Habit: Single crystals uncommon; typically in aggregates of fibrous crystals, commonly radiating.

Cleavage: {110} perfect; $(110) \wedge (1\overline{1}0) \cong 55° 50'$.

Hardness: 6.

Specific gravity: 3.2–3.6, higher with greater iron content.

Color and streak: Pale greenish to dark brown. White.

Luster and light transmission: Vitreous to silky. Translucent.

Optical properties:
    anisotropic, biaxial (+): $2V = 60°–90°$
    refractive indices: α, 1.63–1.66; β, 1.64–1.68; γ, 1.66–1.70
Chemistry: See the introductory remarks for the amphibole group.
Diagnostic features: Rather characteristic color serves to distinguish it from the other amphiboles, except some anthophyllite. Its density is generally greater than that of anthophyllite.
Occurrence: Cummingtonite occurs in calcium-poor iron-rich metamorphic rocks of medium grade, commonly in association with ore deposits. A few occurrences in igneous rocks have been recorded.

Tremolite-Actinolite, $Ca_2(Mg,Fe)_5Si_8O_{22}(OH)_2$

Crystal system, class, and space group: Monoclinic; $2/m$; $C2/m$.
Cell dimensions and content: $a \cong 9.85$, $b \cong 18.1$, $c \cong 5.3$; $\beta \cong 104° 40'$ (tremolite); $Z = 2$.
Habit: Crystals common, but typically in aggregates of long prismatic crystals; also fibrous and asbestiform; rarely dense and finegrained (var. *nephrite*).
Cleavage: {110} perfect; (110) $\wedge$ ($1\overline{1}0$) $\cong 56°$.
Specific gravity: 2.98–3.46, higher with greater iron content (Figure 15-32).
Color and streak: White in tremolite; green with greater iron content (actinolite); manganiferous varieties of tremolite, pink or pale violet. White.
Luster and light transmission: Vitreous. Transparent to translucent.
Optical properties:
    anisotropic, biaxial (−): $2V = 84°–88°$ (tremolite); 75°–84° (actinolite)
    refractive indices: see Figure 15-32
    actinolite is pleochroic in greens and colorless
Chemistry: See the statement in the introductory remarks about the amphiboles.
Diagnostic features: Distinctive color and habit. Tremolite resembles wollastonite and sillimanite, but wollastonite is decomposed by HCl, and sillimanite has only one direction of cleavage.
Occurrence and uses: Tremolite and actinolite are common minerals in low- and medium-grade metamorphic rocks, tremolite being characteristic of metamorphosed dolomitic limestones and actinolite occurring in rocks richer in iron.
        Asbestiform varieties of tremolite are mined to some extent, but they are less valuable industrially than serpentine asbestos. Nephrite, the amphibole form of jade (the other being the pyroxene jadeite), is a variety of actinolite; it is carved into ornaments and,

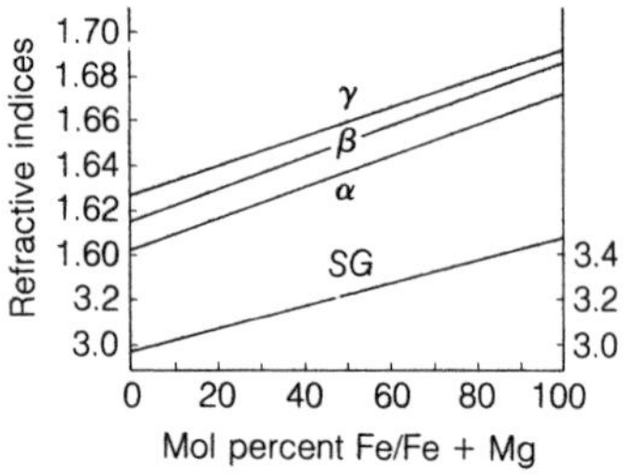

Figure 15-32.   Relationship between specific gravity, refractive indices, and composition in the tremolite-actinolite series.

before the discovery of metals, it was used extensively for tools and weapons.

Hornblende, $Ca(Mg,Fe)_4Al(Si_7Al)O_{22}(OH,F)_2$

Crystal system, class, and space group: Monoclinic; $2/m$; $C2/m$.

Cell dimensions and content: $a \cong 19.8$, $b \cong 18.0$, $c \cong 5.3$; $\beta = 105° 30'$; $Z = 2$.

Habit: Crystals common; prismatic, typically with hexagonal cross sections (combination of {110} and {010}—see Figure 15-33); also as irregular grains.

Twinning: Common, twin plane {100}.

Common forms and angles (Figure 15-33):

$(110) \wedge (1\bar{1}0) = 55° 35'$ $(001) \wedge (\bar{1}01) = 31° 37'$

$(011) \wedge (0\bar{1}1) = 31° 44'$ $(031) \wedge (0\bar{3}1) = 80° 56'$

$(001) \wedge (100) = 74° 29'$ $(\bar{1}01) \wedge (011) = 34° 41'$

Cleavage: {110} perfect; $(110) \wedge (1\bar{1}0) = 55° 35'$.

Hardness: 6.

Specific gravity: 3.0–3.4, higher with greater iron content.

Color and streak: Dark green, dark brown, black. Gray or pale green.

Luster and light transmission: Vitreous. Translucent to subtranslucent.

Optical properties:

anisotropic, biaxial (−): $2V = 35°$–$90°$

refractive indices: α, 1.61–1.70; β, 1.62–1.70; γ, 1.63–1.71

pleochroic in greens and/or browns

Chemistry: Hornblende ranges greatly in its composition. As a consequence, the IMA subcommittee recognized several hornblendes in addition to those noted in the introductory statement in this book and also in addition to those included within the confines of the given formula (e.g., actinolitic hornblende, tschermakitic hornblende, paragasitic hornblende, and hastingsitic hornblende).

Diagnostic features: Distinguished from pyroxene by its cleavage angles of approximately 56° and 124° and the six-sided cross section of its crystals. Color is generally darker than that of other species of amphibole.

Occurrence: Hornblende is an important and widespread rock-forming mineral. It is especially common in medium-grade metamorphic rocks, being an essential contituent of hornblende schists and amphibolites. In igneous rocks, it is more common in plutonic rocks than in volcanic rocks, evidently because incorporation of OH in the structure is favored by crystallization under pressure; it occurs especially in diorites and syenites and in pegmatites associated with these rocks.

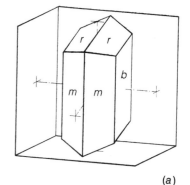

(a)

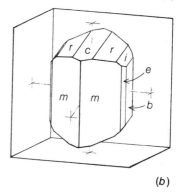

(b)

Figure 15-33. Typical crystal forms of hornblende: $b\{010\}$, $c\{001\}$, $r\{011\}$, $i\{031\}$, $m\{110\}$, $e\{130\}$.

Alkali Amphibole Group
  Glaucophane Series, $Na_2(Mg,Fe)_3Al_2Si_8O_{22}(OH)_2$
  Riebeckite Series, $Na_2(Fe,Mg)_3Fe_2Si_8O_{22}(OH)_2$
  Arfvedsonite Series, $Na_3(Fe,Mg)_4FeSi_8O_{22}(OH)_2$

Crystal system, class, and space group: Monoclinic; 2/m; C2/m.
Cell dimensions and content: $a \cong 9.6$, $b \cong 17.7$, $c \cong 5.3$; $\beta \cong 103°$
  50'; $Z = 2$ (glaucophane). $a \cong 9.8$, $b \cong 18.0$, $c \cong 5.3$; $\beta \cong 103°$;
  $Z = 2$ (riebeckite). $a \cong 9.9$, $b \cong 18.0$, $c \cong 5.3$; $\beta \cong 104°$; $Z \cong 2$
  (arfvedsonite).
Habit: Crystals uncommon; prismatic to acicular; also fibrous or asbestiform. In igneous rocks, riebeckite commonly occurs as mosslike aggregates of tiny grains.
Cleavage: {110} perfect, (110) $\wedge$ (1$\bar{1}$0) $\cong$ 56°.
Hardness: 6.
Specific gravity: 3.0–3.4, higher with greater iron content.
Color and streak: Pale blue, lavender blue, greenish blue, dark blue to
  black, deepening with iron content. White to blue-gray.
Luster and light transmission: Vitreous; silky in fibrous varieties. Translucent.
Optical properties:
  anisotropic, biaxial (−): $2V = 0°$–$50°$ (glaucophane); biaxial (±):
    $2V = 50°$–$90°$ (riebeckite); biaxial (−): $2V = 0$–$70°$ (arfved-
    sonite).
  refractive indices: α, 1.59–1.64; β, 1.60–1.65; γ, 1.62–1.66 (glau-
    cophane). α, 1.69–1.71; β, 1.69–1.70; γ, 1.70–1.72 (rie-
    beckite). α, 1.67–1.70; β, 1.68–1.71; γ, 1.68–1.72 (arf-
    vedsonite).
  pleochroic in nearly colorless, violet, and blue (glaucophane)
  pleochroic in deep blues and yellowish greens (riebeckite)
  pleochroic in yellows and green to indigo (arfvedsonite)
Chemistry: In addition to the statement in the introductory remarks
  about the amphibole group, it is noteworthy that the IMA subcommittee recommendation places crossite between the glaucophane
  and riebeckite series. On the basis of the $Fe^{+3}:(Fe^{+3}+Al)$ ratio, the
  name *crossite* is applied to these amphiboles with ratios between
  0.30 and 0.70.
Diagnostic features: Characterized among the amphiboles by their
  blue color; in dark-colored varieties, it may be necessary to crush
  the mineral to observe the blue tint.
Occurrence and uses: Glaucophane is confined to low- and medium-
  grade metamorphic rocks, typically those thought to have been
  formed under moderately high-pressure conditions; it is an essential
  mineral in the glaucophane schists, which are widely distributed in

California. Riebeckite occurs in both metamorphic and igenous rocks; in the latter, it occurs especially in sodium-rich granites, rhyolites, and associated pegmatites. Arfvedsonite occurs in alkalic igneous rocks.

Riebeckite occurs in an asbestiform variety known as *crocidolite* or blue asbestos, which is mined in South Africa and Western Australia. Some of the South African crocidolite that has been partly or completely replaced by quartz is used as an ornamental stone under the name *tiger's eye*; some tiger's eye retains the blue color of the original mineral, but more commonly the iron has oxidized to a golden brown color.

PYROXENE GROUP

The pyroxenes, according to Deer et al. (vol. 2A, p. 3), "are the most important group of rock-forming ferromagnesian silicates." They are a group of minerals that are closely related structurally, in physical properties, and in chemical composition, even though they crystallize in two different systems—orthorhombic and monoclinic. In all species of the group, the fundamental and common form is the prism—{110} in monoclinic members, {210} in orthorhombic—with interfacial angles of about 87° and 93°. Also, there are good cleavages parallel to these prism faces.

The chemical composition of the pyroxenes can be expressed by the general formula $(W,X,Y)_2Z_2O_6$, or more precisely as $(W)_{1-p}(X,Y)_{1+p}Z_2O_6$, in which W, X, Y, and Z indicate elements having similar ionic radii and capable of replacing each other within the structure. In the pyroxenes, these elements may be:

W = Ca, Na
X = Mg, Fe^{+2}, Mn^{+2}, Ni, Li
Y = Al, Fe^{+3}, Cr, Ti
Z = Si, Al

The proportion of W atoms is generally close to 1 or 0. Of the X group, manganese is generally present in minor amounts except in the rare pyroxene johannsenite ($CaMnSi_2O_6$), and Li occurs as a major constituent only in spodumene ($LiAlSi_2O_6$). Of the Y group, Ti is present only in minor amounts, replacing Al and Fe^{+3}. Z is generally Si; in natural pyroxenes, the Si has been found to be replaced by Al up to an Si:Al ratio of 3:1.

From the structural standpoint, the pyroxenes consist of continuous chains of silica (i.e., Z) tetrahedra that are linked laterally by W, X, and Y cations, which occupy two kinds of sites, $M1$ and $M2$ (Figure 15-34). Consequently, pyroxenes can be expressed by the structural formula

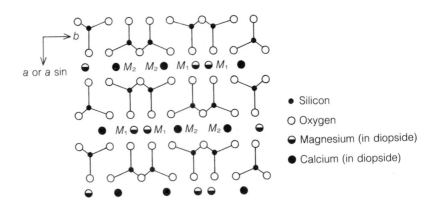

Figure 15-34. Schematic structural diagram of pyroxene, projected along c, showing silica chains and the M_1 and M_2 cation sites.

$(M2)$ $(M1)(Si,Al)_2O_6$. The $M2$ site, which is typically occupied by large cations such as Ca (in, for example, diopside) and Na (in, for example, jadeite and acmite), is located more or less between bases of tetrahedra; the $M1$ site, which is occupied by the smaller cations, lies between apices of tetrahedra. Consequently, the coordination around $M2$ cations is irregular and differs according to the cation present, whereas the coordination around $M1$ cations is essentially octahedral. Deer et al. (op. cit.) give a well-illustrated summary of diverse ways of viewing the general structure and several variants of it.

The following table gives the names that have been applied to the common members of the group:

Orthorhombic	
Enstatite	$Mg_2Si_2O_6$
Bronzite	$(Mg,Fe)_2Si_2O_6$
Hypersthene	$(Mg, Fe)_2Si_2O_6$
Monoclinic	
Clinoenstatite	$Mg_2Si_2O_6$
Clinohypersthene	$(Mg,Fe)_2Si_2O_6$
Pigeonite	similar to clinohypersthene, with some Ca
	$[\sim(Mg,Fe^{+2}, Ca)(Mg, Fe^{+2})Si_2O_6]$
Diopside	$CaMgSi_2O_6$
Hedenbergite	$CaFeSi_2O_6$
Augite	intermediate between diopside and hedenbergite, typically with some Al
	$[\sim(Ca,Na)(Mg,Fe^{+2},Fe^{+3},Ti,Al)(Si,Al)_2O_6]$
Acmite (Aegirine)	$NaFeSi_2O_6$
Jadeite	$Na(Al,Fe)Si_2O_6$
Spodumene	$LiAlSi_2O_6$
Johannsenite	$CaMnSi_2O_6$
Omphacite	$\sim(Ca,Na)(Mg,Fe^{+3},Fe^{+3},Al)Si_2O_6$
Aegirine-Augite	$\sim(Na,Ca)(Fe^{+3},Fe^{+2},Mg,Al)Si_2O_6$

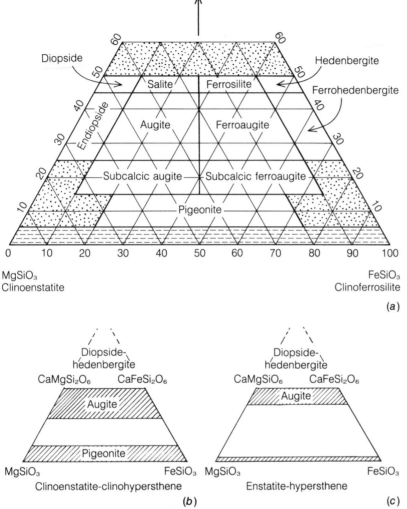

Figure 15-35. (a) Composition fields of the Ca-, Mg-, Fe-clinopyroxenes: very few known natural pyroxenes have compositions that fall within the stippled areas; natural orthopyroxenes have compositions that fall within the dashed-line area. (b) and (c) Composition and phase and phase relations in the common pyroxenes at high and medium temperatures, respectively; the unshaded areas represent the miscibility gap between high-calcic and low-calcic pyroxenes, which increases with decreasing temperature. (Part a is modified after A. Poldervaart and H. H. Hess, 1951, "Pyroxenes in the crystallization of basaltic magmas." *Journal of Geology,* **59,** 472–489.)

The common rock-forming pyroxenes may be considered as phases in the ternary system $MgSiO_3$-$FeSiO_3$-$CaSiO_3$, with $CaSiO_3$ 50 percent or less (Figure 15-35).

The Orthopyroxenes
   Enstatite, $Mg_2Si_2O_6$
   Bronzite, $(Mg,Fe)_2Si_2O_6$
   Hypersthene, $(Mg,Fe)_2Si_2O_6$
   Orthoferrosilite, $Fe_2Si_2O_6$

Crystal system, class, and space group: Orthorhombic; *2/m2/m2/m: Pbca.*

Cell dimensions and content: $a \cong 18.2$, $b \cong 8.8$, $c \cong 5.2$; $Z = 16$.

Habit: Well-formed crystals uncommon; typically as irregular grains or coarse cleavable masses; also as coronas around olivines in some gabbros and ultramafic igneous rocks.

Cleavage: $\{210\}$ good, $(210) \wedge (2\bar{1}0) = 91° 48'$.

Hardness: 6.

Specific gravity: 3.18–3.9, higher with greater Fe content.

Color and streak: White or pale green in low-iron varieties (enstatite); brownish green, light-to-dark brown, to black with greater Fe content (hypersthene); may be bronzy, especially on slightly altered surfaces (*bronzite*). White to gray.

Luster and light transmission: Vitreous. Translucent.

Chemistry and alteration: Enstatite, bronzite, and hypersthene are the most frequently applied names for members of a solid-solution series that extends from enstatite, $Mg_2Si_2O_6$, to orthoferrosilite, $Fe_2Si_2O_6$. The series is usually divided as follows: Enstatite-Fs_{0-12}, Bronzite-Fs_{12-30}, Hypersthene-Fs_{30-50}, Ferrohypersthene-Fs_{50-70}, Eulite-Fs_{70-88}, and Ferrosilite-Fs_{88-100}. (The divisions at Fs_{12} and Fs_{88}, rather than at Fs_{10} and Fs_{90}, are based on changes of optical sign.)

Orthoferrosilite is extremely rare in nature, apparently because of the fact that at high temperatures, fayalite (Fe_2SiO_4) and quartz (SiO_2) will crystallize in its stead. The other members of the series are relatively common, having the occurrences noted later.

When heated above about 1000°C, enstatite changes to proto-enstatite, an orthorhombic phase with a different structure *Pbcn*. On rapid cooling, protoenstatite generally changes to clinoenstatite, a monoclinic phase with a structure ($P2_1/c$) similar to that of pigeonite. Actually, these compositions have four polymorphs—enstatite and protoenstatite (both orthorhombic) and high- and low-clinoenstatite (both monoclinic). In nature, the orthorhombic forms (i.e., enstatite, etc.) are by far the most common. To date, protoenstatite has not been proved to exist naturally.

Essentially all of the orthopyroxenes contain one or more of Ca, Mn, Fe^{+3}, Ti, Al, Cr, and Ni, in most cases at a level below 15 mol percent of their oxides. In some cases, the identity of these contained ions can be correlated with geological occurrence (e.g., high Al in metamorphosed shales); in other cases, they have been roughly correlated with temperature of formation (e.g., higher Ca contents apparently reflect higher temperatures of formation).

Enstatite and bronzite, in particular, readily alter to serpentine.

Optical properties:

  Enstatite: biaxial (+): $2V = 55°–90°$
    α, 1.654; β, 1.655; γ, 1.665 (Fs_0)
  Bronzite: biaxial (−): $2V = 63°–90°$
    α, 1.677; β, 1.685; γ, 1.690 (Fs_{20})
  Hypersthene: biaxial (−): $2V = 54°–63°$
    α, 1.699; β, 1.711; γ, 1.714 (Fs_{40})

Enstatite is colorless, bronzite and hypersthene are pleochroic in pale red-brown and light green.

Diagnostic features: Dark-colored varieties not easily distinguished from augite. The pyroxene cleavage and the color serve to characterize enstatite and bronzite.

Occurrence: The orthorhombic pyroxenes typically occur in basic and ultrabasic rocks low in calcium, such as pyroxenites, peridotites, norites, and some basalts and andesites. Some of these rocks are nearly 100 percent orthopyroxene and are appropriately called bronzitites or hypersthenites. A variety of a granitic rock that contains hypersthene or ferrohypersthene is called charnockite. Most orthorhombic pyroxenes of igneous rocks contain low-to-moderate amounts of iron. Orthopyroxenes, especially ferrohypersthene and eulite, also occur in some high-grade metamorphic rocks. Compositions closest to $MgSiO_3$ are recorded for enstatite in meteorites.

Pigeonite, $\sim (Mg,Fe^{+2},Ca)(Mg,Fe^{+2})Si_2O_6$

Crystal system, class, and space group: Monoclinic; $2/m$; $P2_1/c$.

Cell dimensions and content: $a \cong 9.70$, $b \cong 8.95$, $c \cong 5.24$; $\beta \cong 108°40'$; $Z = 4$.

Habit: Crystals rare; short prismatic; most commonly disseminated as small grains or phenocrysts in basaltic rocks.

Twinning: Common on {100} and {001}, simple or lamellar.

Cleavage and parting: {110} good, $(110) \wedge (1\bar{1}0) \cong 87°$; {100}, {010} and {001}.

Hardness: 6.

Specific gravity: 3.3–3.75.

Color and streak: Brown, purplish, greenish, black. White to black.

Luster and light transmission: Dull to vitreous. Translucent to nearly opaque.

Optical properties:
anisotropic, biaxial (+): $2V = 0°–30°$
refractive indices: α, 1.682–1.732; β, 1.684–1.732, γ, 1.705–1.751
most are slightly pleochroic

Chemistry: Pigeonite has been described in a variety of ways; for example, as a Ca-poor, nonaluminous augite and as a Ca-rich clinohypersthene. $CaSiO_3$ is in the 5–15 percent range. The most common differences reflect Fe:Mg ratios, which range from about 70:30 to 30:70. Al, Cr, Na, Mn, and Ti are commonly present in minor amounts.

Diagnostic features: Not distinguishable from augite by macroscopic means; occurrence may be suggestive. Optical data are definitive.

Occurrence: Pigeonite occurs in volcanic and near-surface intrusive igneous rocks, particularly dolerites, dacites, and andesites. It also has been reported from some meteorites.

Diopside, $CaMgSi_2O_6$
Hedenbergite, $CaFeSi_2O_6$
Augite, $(Ca,Na)(Mg,Fe,Al,Ti)(Si,Al)_2O_6$

Crystal system, class, and space group: Monoclinic; $2/m$; $C2/c$.

Cell dimensions and content: $a = 9.75$, $b = 8.93$, $c = 5.25$; $\beta = 105°50'$ (diopside). $a = 9.84$, $b = 9.03$, $c = 5.25$; $\beta = 104°48'$ (hedenbergite). $a \cong 9.8$, $b \cong 9.0$, $c \cong 5.25$; $\beta \cong 105°$ (augite). $Z = 4$.

Habit: Crystals of diopside and augite common; of hedenbergite, uncommon; typically short prismatic, most of which are nearly square in cross section as a result of the predominance of $\{110\}$; also granular.

Twinning: Common, twin plane $\{100\}$ (Figure 2-75a).

Figure 15-36. Typical crystal forms of augite: $a\{100\}$, $b\{010\}$, $c\{001\}$, $m\{110\}$, $p\{\bar{1}01\}$, $u\{111\}$, $s\{\bar{1}11\}$, $o\{\bar{2}21\}$. (*d*) Typical crystal showing the basal parting.

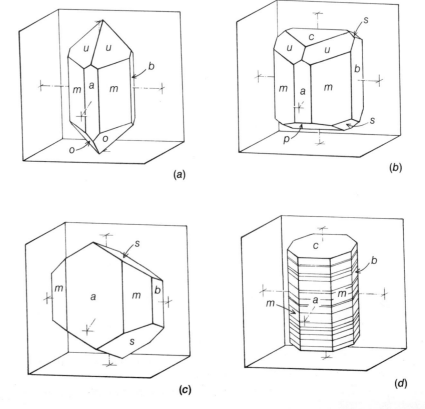

(a)

(b)

(c)

(d)

Common forms and angles (diopside—see Figures 2-42*d* and 15-36):

$(110) \wedge (1\bar{1}0) = 92°50'$ $(021) \wedge (0\bar{2}1) = 97°10'$

$(001) \wedge (100) = 74°10'$ $(001) \wedge (111) = 33°49'$

$(001) \wedge (\bar{1}01) = 31°19'$ $(011) \wedge (0\bar{1}1) = 59°04'$

Cleavage and parting: {110} good, $(110) \wedge (1\bar{1}0) \cong 93°$. Well-developed parting parallel to {100} (var. *diallage*) or parallel to {001}.

Hardness: 6.

Specific gravity: 3.25–3.55, increasing with Fe content.

Color and streak: Colorless or white in rare iron-free varieties, usually dark green to black. White to gray.

Luster and light transmission: Vitreous. Transparent to translucent.

Optical properties:

Diopside: biaxial (+): $2V = 58°$

refractive indices: α, 1.664; β, 1.672; γ, 1.694 (Di_{100})

Hedenbergite: biaxial (+): $2V = 63°$

refractive indices: α, 1.732; β, 1.739, γ, 1.757 (Di_0)

Augite: biaxial (+): $2V = 25°–60°$

refractive indices: α, 1.671–1.735; β, 1.672–1.741; γ, 1.703–1.761

weakly to strongly pleochroic (diopside and hedenbergite)

weakly pleochroic (augite)

Chemistry and alteration: Diopside and hedenbergite are end-members of a solid-solution series. The other minerals of this complete series are salite ($Hd_{10 \text{ or } 20–50}$) and ferrosalite ($Hd_{50–80 \text{ or } 90}$). *Endiopside* is the name given to magnesium-rich (i.e., calcium-poor) diopsides that range toward clinoenstatite; *ferrohedenbergite* is the name of analogous hedenbergites that range toward clino-ferrosilite. Hedenbergite also constitutes a continuous solid-solution series with *johannsenite* ($CaMnSi_2O_6$). In addition, in diopsides, Al may replace Si up to about 10 percent, Ti may be present up to about 2 percent, Cr is commonly present, and several other ions have been recorded in minor quantities.

As you can see in the diagram (Figure 15-34), augite is related to salite in much the same way that endiopside is related to diopside; that is, augite may be looked on as a calcium-poor (and aluminum-bearing) member of the diopside-hedenbergite series. In any case, there is a chemical continuum among these minerals and ferroaugite. Like the amphibole hornblende, augite may contain many other ions that replace those given in the formula. Some of the substitutions have led to the naming of such varieties as *ferroaugite, ferrian augite, titanaugite, sodian augite,* and *subcalcic augite.* Most common augites can contain between about 3.0 and 7.5 percent Al. (Ferroaugites tend to contain less and titanaugites commonly contain up to more than twice that amount.) In many cases, such substitutions can be correlated with genesis, even with, for example, states of magmatic differentiation.

These pyroxenes may exhibit alteration products—commonly

chlorite, less commonly epidote with or without carbonate mineral(s). Augite has often altered to an amphibole called *uralite*. [According to Fleischer (op. cit., p. 163), the name *uralite* may be applied to any amphibole that is pseudomorphous after any pyroxene.]

Diagnostic features: Characterized by crystal form, cleavage, and color. Diopside is white to pale green, in contrast to the darker color of augite and hedenbergite.

Occurrence: Augite is perhaps the most important ferromagnesian mineral of igneous rocks; it is especially abundant in basic and ultrabasic rocks, being characteristic of gabbros, and basalts, and some pyroxenites, and it also occurs in many andesites.

Diopside and hedenbergite occur chiefly in medium- and high-grade metamorphic rocks, especially those rich in calcium. White and light green diopside characteristically occur in metamorphosed dolomitic limestones; hedenbergite is commonly associated with skarn ore deposits formed at high temperatures. Pyroxenes from this series also occur in igneous rocks, chiefly those that are ultramafic.

Acmite (Aegirine), $NaFeSi_2O_6$

Crystal system, class, and space group: Monoclinic; $2/m$; $C2/c$.

Cell dimensions and content: $a = 9.658$, $b = 8.795$, $c = 5.294$; $\beta = 107° 25'$; $Z = 4$.

Habit: Crystals uncommon; typically prisms, some elongate and terminated with blunt-to-acute pyramids, commonly vertically striated; also as capillary or acicular fibers or irregular grains.

Twinning: Commonly twinned, twin plane {100}.

Cleavage: {110} good.

Hardness: 6.

Specific gravity: 3.5–3.6.

Color and streak: Dark green, green-brown, or nearly black. Gray.

Luster and light transmission: Vitreous. Translucent.

Optical properties:
anisotropic biaxial $(-)$: $2V = 58°–90°$
refractive indices: α, 1.72–1.78; β, 1.74–1.82; γ, 1.76–1.84
pleochroic in green and yellow-greens

Chemistry: Relations among the Na-rich and Ca-rich amphiboles are shown in Figure 15-37. As you can see, extensive solid solution with diopside-hedenbergite gives aegirine-augite.

Diagnostic features: Occurrence; resembles augite, but is more apt to occur in long prismatic crystals.

Occurrence: Acmite, called *aegirine* by many mineralogists, and aegirine-augite occur in sodium-rich igneous rocks, especially those of the nepheline syenite family, but they also occur in a few granites and other syenites and in pegmatites associated with those rocks.

Name: The name *aegirine* is often used as a synonym of acmite. Some mineralogists distinguish between acmite and aegirine on the basis of color differences, calling the brown variety acmite and the green-to-black material aegirine. Also, the brown crystals tend to be pointed {221} and {661}, whereas the greenish crystals are typically blunt ended {111}.

Jadeite, Na(Al,Fe)Si$_2$O$_6$

Crystal system, class, and space group: Monoclinic; 2/m; C2/c.
Cell dimensions and content: a = 9.42, b = 8.56, c = 5.22; β = 107° 35'; Z = 4.
Habit: Crystals rare; typically fine-grained granular or dense masses.
Twinning: {100} and {001}, simple or lamellar.
Cleavage: {110} good, but seldom visible in hand specimens; fine-grained material exceedingly tough.
Hardness: 6–6$\frac{1}{2}$.
Specific gravity: 3.25–3.35.
Color and streak: Typically green; also white, violet, blue, or brown. White.
Luster and light transmission: Vitreous, somewhat waxy or greasy in polished specimens. Translucent.
Optical properties:
   anisotropic, biaxial (+): 2V = 65°–75°
   refractive indices: α, 1.65–1.67; β, 1.66–1.67; γ, 1.67–1.69

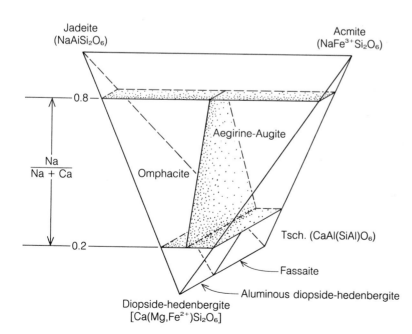

Figure 15-37. Nomenclature for clinopyroxenes of jadeite-acmite-(diopside-hedenbergite)-CaAl$_2$SiO$_6$ (often referred to as Tschermak's molecule) compositions. (By permission, modified after J. R. Clark and J. J. Papike, 1968, "Crystal-chemical characterization of omphacite." *American Mineralogist,* **53,** 840–868.)

Chemistry: As shown in Figure 15-37, Na and Al may be partly re-placed by Ca and Mg, giving *omphacite*, the characteristic clino-pyroxene of eclogites. (Omphacite may be considered alternatively to be a solid solution of jadeite, augite, and acmite with a range in composition of $Jd_{75-25}Ag_{25-75}Ac_{0-25}$.)

Jadeite is a mineral of much significance in petrogenesis since it is intermediate in composition between albite and nepheline; it, however, is rather uncommon, whereas albite and nepheline are common. Analcime, a common zeolite, has the same composition as jadeite with the addition of a molecule of combined water. A significant feature is the high density of jadeite (3.3) as compared with that of the other sodium aluminum silicates (albite, 2.6; neph-eline, 2.6; analcime, 2.3). This feature indicates that its formation is promoted by high pressure. Jadeite cannot be made by melting together its component oxides in the correct proportions; such a melt crystallizes as a mixture of albite and nepheline. For a long time, all efforts at synthesis failed, but jadeite is now made at high pressures and at temperatures between 300°C and 600°C, con-firming the theoretical and geological evidence for its conditions of stability.

Diagnostic features: Resembles nephrite, from which it can be dis-tinguished by a determination of specific gravity.

Occurrence and uses: The mode of occurrence of jadeite was long considered to be somewhat of a mystery, since it was known only as stream-worn boulders brought from Burma and as carved objects from Mayan ruins in Mexico and Guatemala. More recently it has been found *in situ* in several metamorphic rocks associated with serpentine and also as microscopic grains in low-grade metamor-phosed graywackes. It continues as a sought-after mineral for use as a decorative stone.

Spodumene, $LiAlSi_2O_6$

Crystal system, class, and space group: Monoclinic; 2/m or 2; C2/c or C2, respectively.

Cell dimensions and content: $a = 9.45$, $b = 8.39$, $c = 5.22$; $\beta = 110° 20'$; $Z = 4$.

Habit: Crystals common; typically long prismatic crystals and platy masses; extremely large crystals have been recorded—some from the Etta pegmatite, near Keystone, South Dakota, were up to 12 m long and 0.5–2 m wide.

Cleavage: {110} perfect, $(110) \wedge (1\bar{1}0) = 93°$; also a well-developed parting parallel to {100}, which gives a marked splintery fracture.

Hardness: $6\frac{1}{2}$.

Specific gravity: 3.1–3.2.

Color and streak: Typically white or grayish white; some varieties transparent pink and violet (*kunzite*) or green (*hiddenite*). White.

Luster and light transmission: Vitreous. Transparent to translucent.

Optical properties:

anisotropic, biaxial (+): $2V = 54°–69°$

refractive indices: α, 1.648–1.663; β, 1.655–1.670; γ, 1.662–1.679

Chemistry and alteration: Spodumene, unlike the other pyroxenes, is not subject to wide variations in composition as a result of atomic substitutions: lithium, however, may be partly replaced by sodium. Spodumene alters readily to clay minerals and is commonly partly replaced by albite and Li-bearing muscovite, apparently as the result of the action of hydrothermal solutions carrying sodium and potassium.

Diagnostic features: Readily distinguished by its habit and cleavage, its color, and its occurrence.

Occurrence and uses: Spodumene occurs in complex granite pegmatites, such as the one alluded to under *Habit,* generally associated with other lithium aluminosilicates, such as lepidolite, and with beryl and colored tourmalines.

It is mined as a raw material for lithium compounds and for ceramics; the principal producer in the United States is a large pegmatite at at Kings Mountain, North Carolina. Kunzite and hiddenite are cut as gemstones: kunzite is obtained from pegmatites at Pala, California and in Madagascar; the finest hiddenite has come from a small deposit in Alexander County, North Carolina.

PYROXENOID GROUP

Wollastonite, $CaSiO_3$, and some other nonpyroxene minerals with an Si:O ratio of 1:3, such as rhodonite, $(Mn,Fe,Mg,Ca)SiO_3$, pectolite, $NaCa_2Si_3O_8(OH)[= Ca_2NaH(SiO_3)_3]$, and bustamite, $(Mn,Ca)_3(SiO_3)_3$, have long been classed together as constituting a pyroxenoid group. Although all of these minerals are inosilicates, their structures differ so markedly from that of the pyroxenes in the nature and arrangement of the SiO_3 chains that it seems only prudent to abandon the pyroxenoid connotation.

Wollastonite, $CaSiO_3$

Crystal system, class, and space group: Triclinic; $\bar{1}$; $P\bar{1}$.

Cell dimensions and content: $a = 7.94$, $b = 7.32$, $c = 7.07$; $\alpha = 90°$ 02', $\beta = 95° 22'$, $\gamma = 103° 26'$; $Z = 6$.

460

Habit: Crystals rare; typically in cleavable or fibrous masses; also granular and compact.

Cleavage: {100} perfect, (100) $\wedge$ (001) $\cong$ 84°; {$\bar{1}$02} and {001} good, giving a splintery fracture.

Hardness: 5.

Specific gravity: 2.9.

Color and streak: White or grayish. White.

Luster and light transmission: Vitreous, somewhat silky in fibrous varieties. Translucent.

Optical properties:
anisotropic, biaxial (−): 2V = 35°–63°
refractive indices: α, 1.615–1.646; β, 1.627–1.659; γ, 1.629–1.662

Other properties: Some wollastonite is fluorescent, the material from Franklin, New Jersey, being particularly noteworthy in this respect. The fluorescence is dependent on the presence of Mn in solid solution. Unlike the pyroxenes, wollastonite is decomposed by concentrated HCl.

Diagnostic features: Distinguished from other white fibrous silicates, such as tremolite and sillimanite, by its hardness and its decomposition by HCl with the separation of silica.

Occurrence and uses: Wollastonite is formed by the metamorphism of siliceous limestones at temperatures of about 450°C and higher. Thus, it occurs in contact zones around igneous masses and in high-grade regionally metamorphosed rocks.

A large deposit of wollastonite at Willsboro, New York, is being mined for use in ceramics and paints.

Related species: Wollastonite has five recognized polytypes, based on different stacking sequences. The monoclinic polytype, parawollastonite (= wollastonite-2M) occurs in, for example, the Crestmore, California, deposit and in limestone blocks ejected from volcanoes, as at Monte Somma, Italy; its space group is $P2_1/a$. Another noteworthy polytype is *cyclowollastonite* (termed *pseudowollastonite* in much of the literature), which is rare, occurring only in a few high-temperature, thermally metamorphosed rocks; cyclowollastonite is triclinic and pseudo-orthorhombic, and its space group is either *P1* or *P$\bar{1}$*.

Pectolite, $NaCa_2Si_3O_8(OH)$

Crystal system, class, and space group: Triclinic; $\bar{1}$; $P\bar{1}$.

Cell dimensions and content: a = 7.99, b = 7.04, c = 7.02; α = 90°03′, β = 95°17′, γ = 102°28′; Z = 2.

Habit: Single crystals rare; typically occurs as aggregates of needlelike crystals elongated parallel to *b*, commonly radiating and forming spheroidal to globular masses (Figure 15-38). (Specimens should be handled with care because the needlelike crystals readily penetrate the skin!)

Cleavage: {001} and {100} perfect.

Hardness: 5.

Specific gravity: 2.86, higher with increased Mn content.

Color and streak: White.

Luster and light transmission: Vitreous or silky. Translucent.

Optical properties:
 anisotropic, biaxial (+): $2V = 50°–63°$
 refractive indices: α, 1.595–1.610; β, 1.604–1.615; γ, 1.632–1.645

Figure 15-38. Pectolite: radiating, needlelike crystals, Upper New Street Quarry, Paterson, New Jersey. Specimen is $\sim$ 9 cm across. (Photograph courtesy of S. C. Chamberlain.)

Chemistry: Most pectolite has a composition close to the given formula, but, in some specimens, the calcium is extensively replaced by manganese.

Diagnostic features: Habit; decomposed by warm dilute hydrochloric acid with the separation of silica.

Occurrence: Pectolite most commonly occurs associated with zeolites in cavities in basalts, as in northern New Jersey. It also occurs in veins in serpentinites (e.g., in California) and in some metasedimentary rocks (e.g., in Mendocino County, California).

Rhodonite, $(Mn,Fe,Mg,Ca)SiO_3$

Crystal system, class, and space group: Triclinic; $\bar{1}$; $C\bar{1}$.

Cell dimensions and content: a = 10.497, b = 9.797, c = 12.185; $\alpha = 103°01'$, $\beta = 108°33'$, $\gamma = 82°30'$; Z = 20.

Habit: Crystals uncommon, generally tabular parallel to {001}; typically massive, coarse-to-fine granular, or dense aggregates.

Cleavage and fracture: {110}, {1$\bar{1}$0} perfect, {001} good; fine-grained material is very tough and has a conchoidal fracture.

Hardness: 6.

Specific gravity: 3.5–3.75.

Color and streak: Pink to rose-red, commonly veined by black alteration products. White.

Luster and light transmission: Vitreous. Translucent.

Optical properties:
 anisotropic, biaxial (+): $2V = 61°–76°$
 refractive indices: α, 1.711–1.738; β, 1.715–1.741; γ, 1.724–1.751
 pleochroic in oranges and red

Chemistry and alteration: Mn is the chief cation in all rhodonite. In fact, for years, rhodonite was considered to be $MnSiO_3$, and Fe, Mg, and Ca were noted as possible replacement elements for some of the Mn. Some rhodonite contains Zn (up to about 10 percent). The variations in composition are reflected by the density (and therefore the specific gravity values) as well as by optical properties. Rhodonite alters readily by oxidation to black manganese oxides.

Diagnostic features: Rose-red color; rhodonite resembles rhodochrosite, but rhodonite is distinguished by its greater hardness, its lack of effervescence in warm HCl, and its different cleavage.

Occurrence and uses: Rhodonite occurs in veins and irregular masses formed by hydrothermal or metasomatic processes or by metamorphism of sedimentary manganese ores. Excellent crystals have been found at Franklin and Sterling Hill, New Jersey.

Massive rhodonite is sometimes cut and polished as an ornamental stone; a deposit near Sverdlovsk, in the Ural Mountains, has provided a large amount of rhodonite that has been used to make vases, tabletops, and other decorative objects; one of the stations of the Moscow subway is extensively decorated with this rhodonite.

SUBCLASS CYCLOSILICATES

The cyclosilicates are so named because they contain rings of linked SiO_4 tetrahedra. The tetrahedra are linked in the same manner as in the single-chain inosilicates, with each SiO_4 group sharing two oxygens with adjoining tetrahedra on either side, thus giving the same overall formula $(SiO_3^{-2})_n$. However, instead of forming straight chains, the tetrahedra are joined at angles that result in the formation of rings. The minimum number of tetrahedra required to form a ring is, of course, three; a few three-membered ring minerals are known, the best-known being benitoite, $BaTiSi_3O_9$. A few minerals have four-membered rings in their structures (e.g., axinite); several important minerals (e.g., beryl, cordierite, and tourmaline) have six-membered rings. Five-membered rings and single rings with more than six tetrahedra are unknown.

Axinite, $(Ca,Mn,Fe)_3Al_2(BO_3)Si_4O_{12}(OH)$

Crystal system, class, and space group: Triclinic; $\bar{1}$; $P\bar{1}$.

Cell dimensions and content: $a = 7.15$, $b = 9.18$, $c = 8.94$; $\alpha = 91°52'$, $\beta = 98°09'$, $\gamma = 77°19'$; $Z = 2$.

Crystal structure: In most, if not all, other cyclosilicates, the constituent Si_4O_{12} rings have a single orientation. In axinite, adjacent Si_4O_{12}

rings have opposite orientations, as can be readily seen in projections perpendicular to *b*.

Habit: Crystals common; typically wedge-shaped (Figure 15-39); also massive, lamellar, or granular.

Cleavage: {100} good.

Hardness: 7.

Specific gravity: 3.2–3.3.

Color and streak: Typically violet to brown; in some cases yellow, greenish yellow, or pink. White.

Luster and light transmission: Vitreous. Translucent.

Optical properties:
anisotropic, biaxial (−): 2V = 63–90°
refractive indices: α, 1.672–1.693; β, 1.677–1.701; γ, 1.681–1.704
slightly pleochroic

Chemistry: Strictly speaking, axinite is a group name and, depending on different proportions of Ca, Mn, and Fe, individual members include *ferroaxinite, manganaxinite,* and even *magnesio-axinite.* In many specimens, the proportion of Ca to Mn and/or Fe is commonly about 2 : 1.

Diagnostic features: Crystal form and color.

Occurrence and associations: Most axinite has been formed by metasomatic reactions at comparatively high temperatures, the boron being derived from magmatic emanations. Thus, axinite commonly occurs in contact-altered calcareous rocks, in cavities in granite, and in hydrothermal veins. Commonly associated minerals are tourmaline, prehnite, zoisite, and datolite. There is an extensive axinite occurrence in Plumas County, California, where it is the principal rock-forming mineral in some metamorphic rocks along a contact with some ultrabasic igneous rocks. Fine crystals are known from, for example, Riverside, California.

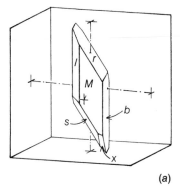

(a)

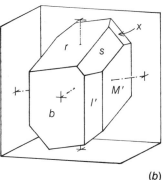

(b)

Figure 15-39. Forms in axinite crystals: $b\{010\}$, $l\{1\bar{2}0\}$, $M\{1\bar{1}0\}$, $s\{\bar{1}21\}$, $r\{011\}$, $x\{\bar{1}11\}$. (a) In the standard presentation used here. (b) Rotated 90° around the *c* axis.

Beryl, $Be_3Al_2Si_6O_{18}$

Crystal system, class, and space group: Hexagonal; 6/m2/m2/m; P6/mcc.

Cell dimensions and content: a = 9.215, c = 9.192; Z = 2.

Habit: Prismatic crystals, combinations of {10$\bar{1}$0} and {0001} (in crystals up to several feet long); also massive.

Cleavage: {0001} distinct.

Hardness: 8.

Specific gravity: 2.65–2.95 (see Figure 15-40).

Color and streak: Commonly pale green, white, or yellow; gem varieties transparent, dark green (*emerald*), pale blue or green (*aquamarine*), yellow (*heliodor*), or pink (*morganite*). White.

Luster and light transmission: Vitreous. Transparent and translucent.

Optical properties:

anisotropic, uniaxial (−)

refractive indices: ω, 1.566–1.602 (Figure 15-40); ϵ, 1.563–1.594

Chemistry: In addition to the elements indicated in the formula, beryl commonly contains appreciable amounts of the alkali elements and relatively large amounts of occluded gases. Until the structure was elucidated, the reason for this was not understood. It is now known that the structure of beryl contains tubular channels defined by the Si_6O_{18} rings. In the mineral, Li can substitute for Al and Al for Be; any charge imbalance resulting from these substitutions is satisfied by the introduction of Na, K, and Cs ions into these tubular channels; the channels also serve to accommodate the occluded gases. Alkali-containing beryls are characterized by higher specific gravities and lower beryllium contents than beryls with compositions corresponding to the ideal formula. The color of emerald is due to Cr^{+3}, of morganite to Mn^{+2}; both cations are present in only trace amounts.

Diagnostic features: Easily recognized as crystals; the only similar mineral is apatite, which is much softer. Massive white beryl, however, may closely resemble feldspar or milky quartz, and differentiation of these can be important in prospecting and mining operations. A simple field test uses acetylene tetrabromide diluted with benzene, in which beryl sinks and both feldspar and quartz float.

Occurrence and uses: The typical mode of occurrence is in granite pegmatites, but beryl has also been found in miarolitic cavities in granite. The fine emeralds from Muso in Colombia occur in cavities in a bituminous limestone. Emeralds have also been found to occur in mica schists constituting wall rock of pegmatite masses.

Beryl is the most common beryllium mineral, and the only

Figure 15-40.   Relationship between omega index of refraction, specific gravity, and BeO content of beryl.

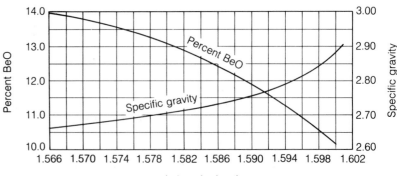

commercial source of this element. It is mined for this use from granite pegmatites in, for example, New Mexico and South Dakota.

Cordierite, $Al_3(Mg,Fe)_2(Si_5AlO_{18})$

Crystal system, class, and space group: Orthorhombic; $2/m2/m2/m$; $Cccm$.

Cell dimensions and content: $a = 17.13$, $b = 9.80$, $c = 9.35$; $Z = 4$.

Crystal structure: The structure of cordierite is similar to that of beryl; three Al atoms are in six-coordination, the fourth substitutes for one Si in the ring structure, giving an $AlSi_5O_{18}$ group. As in beryl there are channels within the ring structure, in which the H_2O molecules and the alkali metals recorded in some analyses are evidently accommodated.

Habit: Crystals uncommon; most are prismatic or pseudo-hexagonal twinned crystals; generally massive or as irregular grains.

Twinning: Twin plane {110}, generally repeated, giving pseudo-hexagonal crystals.

Cleavage: {010} indistinct; parting on {001} develops with the onset of alteration.

Hardness: 7.

Specific gravity: 2.55–2.75, higher with greater Fe content.

Color and streak: Characteristically pale-to-dark blue or violet; also colorless, gray, yellow, or brown; some grains can be seen to be dichroic in violet (or blue) and light yellow with the unaided eye. White.

Luster and light transmission: Vitreous. Transparent.

Optical properties:
anisotropic, biaxial (−): $2V = 40$–$90°$
refractive indices: α, 1.522–1.560; β, 1.524–1.574; γ, 1.527–1.578
pleochroic (slightly)

Alteration: Cordierite alters so readily that it is more commonly found in the altered condition than fresh. The most common type of alteration is to chlorite or muscovite.

Diagnostic features: Easy to recognize when it shows the typical blue or blue-gray color. When it occurs as unaltered colorless or gray grains in a rock, it is very like quartz and can be easily distinguished only by optical or other tests. Dichroism, as noted, is definitive.

Occurrence and uses: Cordierite, formed by the medium- to high-grade metamorphism of aluminum-rich rocks, occurs in schists, gneisses, and hornfelses. It has also been recorded from volcanic rocks, but an investigation of one such occurrence showed it to be a mineral very similar to but distinct from cordierite. This mineral has been called *osumilite*; most, if not all, volcanic "cordierite" is probably this mineral.

Clear blue varieties of cordierite have occasionally been cut as gemstones.

Name: Cordierite is also known as *iolite*, for its violet color, or *dichroite*, for its variation in color with crystallographic direction (dichroism).

Tourmaline, W $X_3Y_6(BO_3)_3(Si_6O_{18})(OH,F)_4$, in which W = Ca, Na, and/or K; X = Al, Fe, Li, Mg, and/or Mn; and Y = Al, Cr, Fe, and/or V

Crystal system, class, and space group: Trigonal; *3m; R3m.*

Cell dimensions and content: a = 15.8–16.0, c = 7.1–7.2; Z = 3.

Habit: Crystals common; typically as prisms, commonly with rounded triangular cross sections due to the predominance of the triangular prism $\{10\bar{1}0\}$ or $\{01\bar{1}0\}$; prism faces commonly vertically striated; doubly terminated crystals show hemimorphic character; also in parallel or radiating groups of columnar to acicular crystals.

Common forms and angles: (Figures 2-52, 15-41).

$(0001) \wedge (10\bar{1}1) = 27°20'$ $(02\bar{2}1) \wedge (\bar{2}021) = 77°00'$

$(0001) \wedge (02\bar{2}1) = 45°57'$ $(32\bar{5}1) \wedge (\bar{3}5\bar{2}1) = 66°01'$

$(10\bar{1}1) \wedge (\bar{1}101) = 46°52'$ $(32\bar{5}1) \wedge (53\bar{2}1) = 42°36'$

Cleavage: $\{11\bar{2}0\}$, $\{10\bar{1}1\}$ indistinct.

Hardness: $7\frac{1}{2}$.

Specific gravity: 3.0–3.2, higher with greater Fe content.

Color and streak: Common tourmaline is black; also brown, dark blue, colorless, pink, green, and blue tourmalines, some of which are zoned parallel and/or perpendicular to c (Figure 15-42). White or slightly tinted by the body color of the specimens.

Luster and light transmission: Vitreous. Transparent to translucent.

Optical properties (schorl):
anisotropic, uniaxial (−)
refractive indices: ω, 1.658–1.698; ϵ, 1.633–1.675
pleochroic

Other properties: Strongly pyroelectric and piezoelectric.

Chemistry: As indicated by the formula, tourmaline group minerals may differ greatly from each other. It also is true, however, that few tourmaline specimens have end-member compositions. Nonetheless, end-member compositions have been named; they are as follows:

Buergerite	$NaFe_3Al_6(BO_3)_3Si_6O_{21}F$
Dravite	$NaMg_3Al_6(BO_3)_3Si_6O_{18}(OH)_4$
Elbaite	$Na(Li,Al)_3Al_6(BO_3)_3Si_6O_{18}(OH)_4$
Ferridravite	$(Na,K)(Mg,Fe)_3Fe_6(BO_3)_3Si_6O_{18}(O, OH)_4$

Figure 15-41.   Tourmaline crystal, black, showing typical striations parallel to c, trigonal outline, and trigonal pyramid $\{10\bar{1}1\}$ termination. Tongafero, Madagascar, ~7 cm wide. (Courtesy of Royal Ontario Museum.)

Liddicoatite	$Ca(Li,Al)_3Al_6(BO_3)_3Si_6O_{18}(OH)_4$
Schorl	$NaFe_3Al_6(BO_3)_3Si_6O_{18}(OH)_4$
Uvite	$(Ca,Na)(Mg,Fe)_3Al_5Mg(BO_3)_3Si_6O_{18}(OH,F)_4$

There are at least partial, and perhaps complete, series between dravite and uvite, between dravite and schorl, between schorl and elbaite, and between elbaite and liddicoatite.

Diagnostic features: Color, triangular cross section, and striations on prism faces.

Occurrence and uses: Schorl is the common black tourmaline most frequently seen in rocks. Both schorl and the colored lithium-bearing types occur in granite pegmatites, commonly in radial clusters; schorl is also widespread in granitic rocks and is common as an accessory mineral in metamorphic rocks, especially schists and gneisses; brown dravite is relatively common in metamorphosed limestones, and it also occurs in some high-temperature metalliferous veins.

The piezoelectric property of tourmaline is utilized in the construction of pressure gauges to measure blast pressures both in air and under water. Gem tourmalines have been mined in the United States from pegmatites in New England and southern California.

Figure 15-42. Tourmaline crystals exhibiting color-banding, commonly described as a manifestation of hemimorphism, parallel to c; specimen is from Tourmaline Queen Mine, San Diego County, California; top of each crystal appears blue, light segment is white, remainder is rose-red. Height of specimen is ~ 25 cm. (Photograph courtesy of John S. White.)

SUBCLASS SOROSILICATES

In the sorosilicates, two SiO_4 tetrahedra are linked by sharing one oxygen, thereby forming discrete $(Si_2O_7)^{-6}$ groups in their structures. A few simple sorosilicates exist as minerals. Structures containing both Si_2O_7 and independent SiO_4 groups also occur, the most important being vesuvianite and minerals of the epidote group.

Lawsonite, $CaAl_2Si_2O_7(OH)_2 \cdot H_2O$

Crystal system, class, and space group: Orthorhombic; $2/m2/m2/m$; *Ccmm.*

Cell dimensions and content: $a = 8.80$, $b = 5.85$, $c = 13.14$; $Z = 4$.

Crystal structure: The composition of lawsonite is similar to that of anorthite, but the structure is quite different, lawsonite being much more closely packed (note its higher density and superior hardness). The lawsonite structure consists of chains of six-coordinated aluminum-oxygen (and hydroxyl) groups, linked sideways by the Si_2O_7 groups; in the framework so formed, there are "holes" occupied by the calcium ions and water molecules.

Habit: Crystals uncommon; prismatic, typically tabular parallel to c; also massive, granular.

Cleavage: {100}, {001} perfect; {110} indistinct.

Hardness: 8.

Specific gravity: 3.09.

Color and streak: White, pale blue, or pale gray. White.

Luster and light transmission: Vitreous to greasy. Transparent to translucent.

Optical properties:
anisotropic, biaxial (+): 2V = 76–87°
refractive indices: α, 1.665; β, 1.672–1.676; γ, 1.684–1.686

Diagnostic features: Hardness, comparatively high specific gravity, and common association with glaucophane.

Occurrence and associations: Lawsonite is a mineral of metamorphic rocks; it typically occurs in glaucophane schists. It is widely distributed in these rocks in California and elsewhere.

Pumpellyite, $Ca_2MgAl_2(SiO_4)(Si_2O_7)(OH)_2 \cdot H_2O$

Crystal system, class, and space group: Monoclinic; 2/m; A2/m.

Cell dimensions and content: a = 8.81, b = 5.94, c = 19.14; β = 97° 36'; Z = 4 (2?).

Habit: Crystals rare; typically acicular, bladed {001}, or fibrous, commonly in subparallel or radiated (rosettes) aggregates.

Twinning: Common on {001} and {100}.

Cleavage: {001} distinct, {100} less distinct.

Hardness: $5\frac{1}{2}$–6.

Specific gravity: 3.18–3.23.

Color and streak: Green to brownish. Nearly white.

Luster and light transmission: Vitreous. Subtranslucent.

Optical properties:
anisotropic, biaxial (+): 2V = 10–85°[may be optically (–)]
refractive indices: α, 1.674–1.748; β, 1.675–1.754; γ, 1.688–1.764
pleochroic in green and yellow

Chemistry: Fe^{+2} commonly substitutes for Mg; Fe^{+3} for Al; minor Na, Mn, and/or Ti^{+4} may be present.

Diagnostic features: Typically deeper colors than epidote group minerals, which it resembles.

Occurrence, associations, and use: Pumpellyite may be more common than reports indicate because of its having been misidentified as epidote. Pumpellyite occurs widely in zeolite, prehnite-pumpellyite, greenschist, and blueschist facies metamorphic rocks, along with zeolites, prehnite, lawsonite and glaucophane, chlorite, epidote, zoisite, albite, etc.; in veins and amygdules in andesites and basalts, along with prehnite, zeolites, datolite, chlorite, epidote, quartz (agate), and native copper; and as an alteration product in, for example, greenstones, apparently replacing plagioclase. It also occurs as a heavy mineral in detrital sediments.

A variety, generally termed *chlorastrolite*, but locally called by the misnomer "greenstone," has been recovered from the Lake Superior basalts of northwestern Michigan and cut and polished into attractive cabachons.

Melilite Group
Gehlenite, $Ca_2Al(AlSiO_7)$
Akermanite, $Ca_2Mg(Si_2O_7)$

Crystal system, class, and space group: Tetragonal; $\bar{4}2m$; $P\bar{4}2_1m$.

Cell dimensions and content: $a = 7.69$ (gehlenite)–7.84 (akermanite); $c = 5.067$ (gehlenite)–5.01 (akermanite); $Z = 2$.

Habit: Crystals uncommon; tabular on {001} or short prismatic; commonly massive, granular.

Cleavage and fracture: {001}, {110} indistinct. Uneven to conchoidal.

Hardness: 5–6.

Specific gravity: 3.04 (gehlenite)–2.94.

Color and streak: Colorless, honey-yellow, gray-green, or brown. White.

Luster and light transmission: Vitreous to resinous. Transparent to translucent.

Optical properties:

    Gehlenite: uniaxial (−)   Akermanite: uniaxial (+)

         ω, 1.651–1.670 ω, 1.632–1.651

         ϵ, 1.651–1.658 ϵ, 1.640–1.651

Chemistry: There is complete solid solution between these end-members, pure end-members being rare in nature. Na and minor K may replace Ca; Fe^{+2} may replace Mg; and Fe^{+3} may replace Al. Rarely, Zn is present (*hardystonite*, $Ca_2ZnSi_2O_7$).

Diagnostic features: Distinguished from apatite by crystal system and tabular habit. Distinguishing the melilites from zoisite and vesuvianite (idocrase) often requires optical determinations.

Occurrence and associations: Melilite occurs in carbonate rocks that have undergone contact metamorphism (associated with such minerals as diopside and wollastonite) and in silica-deficient igneous rocks, such as melilite basalts and melilotites (along with feldspathoids, olivine, titanaugite, acmite, and perovskite). Melilites are also common in furnace slags and ceramics. Melilites are commonly altered to such minerals as calcite, garnet, zeolites, and vesuvianite.

Hemimorphite, $Zn_4Si_2O_7(OH)_2 \cdot H_2O$

Crystal system, class, and space group: Orthorhombic; *mm2*; *Imm2*.

Cell dimensions and content: a = 8.370, b = 10.719, c = 5.120; Z = 2.

Habit: Crystals common; typically thin tabular to bladelike parallel to {101}; also massive, commonly with stalactitic or mammillary forms.

Twinning: not uncommon with {001} twin plane.

Cleavage: {110} perfect.

Hardness: 5.

Specific gravity: 3.4–3.5.

Color and streak: White; in some places stained brown (with iron) or blue or green (with copper). White.

Luster and light transmission: Vitreous. Transparent to Translucent.

Optical properties:
   anisotropic, biaxial (+): 2V = 44–47°
   refractive indices: α, 1.611–1.617; β, 1.614–1.620; γ, 1.632–1.639

Diagnostic features: Soluble in HCl, giving gelatinous silica on partial evaporation. Strongly pyroelectric and piezoelectric. On heating, it decomposes into willemite at about 240°C.

Occurrence, associations, and use: Hemimorphite occurs in the oxidized zone of zinc deposits, generally associated with smithsonite. Fine specimens have been found in, for example, Franklin, New Jersey, and Leadville, Colorado.

   Hemimorphite has constituted a minor ore of zinc at several deposits.

Name: This mineral was originally known as *calamine*, which name, however, has also been used for zinc carbonate. The name *hemimorphite* was proposed in 1853 from the hemimorphic nature of the crystals, and this name has been adopted by international agreement to eliminate the confusion caused by the dual application of calamine.

Vesuvianite $Ca_{10}Mg_2Al_4(SiO_4)_5(Si_2O_7)_2(OH)_4$

Crystal system, class, and space group: Tetragonal; *4/m2/m2/m*; *P4/nnc*.

Cell dimensions and content: a = 15.66, c = 11.85; Z = 4.

Habit: Crystals common; prismatic or pyramidal (Figures 15-43 and 15-44); also massive, granular or compact.

Cleavage: {110} indistinct.

Hardness: 7.

Specific gravity: 3.3–3.6.

Color and streak: Typically brown or green; in some cases yellow or blue (var. *cyprine*). White.

Luster and light transmission: Vitreous to somewhat resinous. Transparent to translucent.

Optical properties:
anisotropic, uniaxial (−)
refractive indices: ω, 1.702–1.752; ϵ, 1.698–1.746

Chemistry: Vesuvianite shows a considerable range in its composition: some Ca may be replaced by Na and rare earths; Mg may be replaced by Fe^{+2} and Mn; Al may be replaced by Fe^{+3} and Ti; also, some varieties contain Be.

Diagnostic features: Easily recognized when in crystals; massive varieties closely resemble some forms of garnet, from which it can usually be distinguished by its lower specific gravity, or jadeite, from which it is readily distinguished by optical properties.

Occurrence and associations: Vesuvianite is formed by the contact metamorphism of impure limestones and as masses enclosed within ultrabasic igneous rocks (e.g., those now identifiable as serpentinites). It is commonly associated with calcite, lime garnet (grossular or andradite), wollastonite, and other calc-silicate minerals. Fine crystals have been found, for example, at Eden Mills, Vermont; near Olmsteadville, New York; and in El Dorado County, California.

Name: The name *idocrase* is also used widely for this mineral.

EPIDOTE GROUP

The epidote group comprises several minerals with the general formula $X_2Y_3Si_3O_{12}(OH)$. X is Ca, which is partly replaced by rare-earth elements (R) in allanite. Y is Al and Fe^{+3} (and may include Ti); it is partly replaced by Mg and Fe^{+2} in allanite and by Mn^{+3} in piemontite. The more common species of this group are:

Zoisite	$Ca_2Al_3Si_3O_{12}(OH)$	Orthorhombic
Clinozoisite	$Ca_2Al_3Si_3O_{12}(OH)$	Monoclinic
Epidote	$Ca_2(Al,Fe)_3Si_3O_{12}(OH)$	Monoclinic
Allanite	$(Ce,Ca,Y)_2(Al,Fe,Mg)_3Si_3O_{12}(OH)$	Monoclinic

That is to say, the simplest composition $Ca_2Al_3Si_3O_{12}(OH)$ is dimorphous, with the orthorhombic zoisite and the monoclinic clinozoisite, whereas the other species in the group are monoclinic.

The structure of minerals of the epidote group is a mixed type, containing both Si_2O_7 and SiO_4 groups; for example, the formula of clinozoisite can be written $Ca_2Al \cdot Al_2O(OH)(Si_2O_7)(SiO_4)$. Strictly speaking, however, the structure contains AlO_4 and AlO_3OH linked in chains

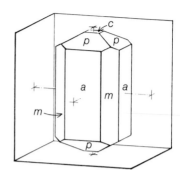

Figure 15-43. Typical forms in vesuvianite crystals: $c\{001\}$, $a\{100\}$, $m\{110\}$, $p\{101\}$.

Figure 15-44. Vesuvianite crystals, greenish gray, showing difference in habit from one locality. Forms: {100}, {110}, {101}, {001}, {211} (one small face), ~ 3.5 cm high. Wilui River, Siberia, the USSR, ~ 3.5 cm high. (Courtesy of Royal Ontario Museum.)

that extend along the *b* direction, which accounts for the typical elongation of crystals in that direction.

Minerals of the epidote group characteristically occur in metamorphic rocks (except for allanite, which occurs in granites and granitic pegmatites). In this connection, it is interesting to note that the composition of zoisite and clinozoisite is similar to that of, for example, anorthite. The specific gravities of the two minerals, however, are different, reflecting the different atomic packing within the different structures (zoisite and clinozoisite, $G = 3.2–3.3$; anorthite, $G = 2.76$). This difference is also reflected in the general absence of anorthite-rich plagioclase in low- and medium-grade schists and gneisses; instead, albite-rich plagioclase and an epidote group mineral are present. Evidently, under the conditions of low- and medium-grade metamorphism, the epidote group minerals are more stable than anorthite.

Zoisite, $Ca_2Al_3Si_3O_{12}(OH)$

Crystal system, class, and space group: Orthorhombic; $2/m2/m2/m$; *Pnmc*.

Cell dimensions and content: $a = 16.24, b = 5.58, c = 10.10; Z = 4$.

Habit: Single crystals uncommon; typically aggregates of long prismatic crystals parallel to *b*, striated along *b*, terminations rare.

Cleavage: {001} perfect.

Hardness: $6\frac{1}{2}$.

Specific gravity: 3.3.

Color and streak: Typically gray; in some cases, pink (var. *thulite*), apple-green, bluish or purple (var. *tanzanite*). White.

Luster and light transmission: Vitreous, pearly on cleavage surfaces. Transparent to translucent.

Optical properties:
anisotropic, biaxial (+): $2V = 0–60°$
refractive indices: α, 1.685–1.707; β, 1.688–1.711; γ, 1.697–1.725

Chemistry: Its composition is generally close to the formula, the only substitution being a small amount (up to about 10 atom percent) of Fe^{+3} for Al.

Diagnostic features: Distinguished from the amphiboles that it resembles in appearance and mode of occurrence by its single perfect cleavage.

Occurrence and use: Zoisite occurs as a constituent of schists and gneisses, but less commonly than its dimorph clinozoisite, and also in quartz veins; most of the materials known as *saussurite* (altered plagioclase in basic igneous rocks) is a mixture of albite and zoisite or clinozoisite.

The variety tanzanite, found in Tanzania, is a prized gem. [It is an essentially iron-free zoisite, the color of which is attributed to the presence of vanadium ($V_2O_3 \cong 0.2$ percent)].

Clinozoisite, $Ca_2Al_3Si_3O_{12}(OH)$
Epidote, $Ca_2(Al,Fe)_3Si_3O_{12}(OH)$

Crystal system, class, and space group: Monoclinic; $2/m$; $P2_1m$.

Cell dimensions and content (epidote): $a \cong 8.93$, $b \cong 5.62$, $c \cong 10.23$; $\beta = 115°24'$; $Z = 2$.

Habit: Crystals common, typically elongated parallel to b (Figure 15-45); also massive, fibrous or granular.

Twinning: {100} uncommon.

Cleavage: {001} perfect.

Hardness: 7.

Specific gravity: 3.3–3.6, higher with greater Fe content.

Color and streak: Pale green or greenish gray for clinozoisite, yellowish to brownish green to black for epidote. White or grayish white.

Luster and light transmission: Vitreous. Transparent to translucent.

Optical properties:

    Clinozoisite: biaxial (+): $2V = 14-90°$

        refractive indices: α, 1.703–1.715; β, 1.707–1.725; γ, 1.709–1.734

    Epidote: biaxial (−): $2V = 64°-90°$

        refractive indices: α, 1.715–1.751; β, 1.725–1.784; γ, 1.734–1.797

Chemistry: Clinozoisite and epidote are in effect a single species, the boundary between them being arbitrarily drawn at a replacement of 10 atom percent Al by Fe; the maximum replacement of Al by Fe is about one-third, equivalent to a formula $Ca_2Al_2FeSi_3O_{12}(OH)$.

Diagnostic features: Most epidote can usually be readily recognized by its hardness and its characteristic yellowish green (pistachio) color, seldom seen in other minerals. Cleavage is characteristic for both epidote and clinozoisite.

Occurrence: Clinozoisite and epidote are important constituents of low- and medium-grade metamorphic rocks, commonly as the principal calcium aluminum silicates in these rocks. Epidote occurs in large amounts in contact-metamorphosed limestones, especially in those associated with iron ores; it is also rather commonly associated with zeolites in amygdules and other cavities in basalts. Very fine crystals of epidote have come from the Untersulzbachtal, in the Austrian Tyrol, and from Prince of Wales Island, Alaska.

Related Species: Piemontite is a member of the epidote group in which some of the Al is replaced by Mn^{+3}; trivalent manganese is a strong chromophore, and piemontite has a violet-red color easily recog-

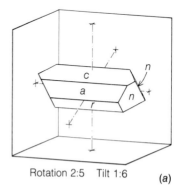

Rotation 2:5   Tilt 1:6   *(a)*

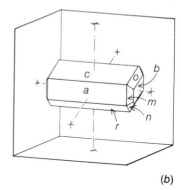

(b)

Figure 15-45. Typical forms in epidote crystals: c{001}, a{100,} r{$\bar{1}$01}, n{$\bar{1}$11}, m{110}, o{011}.

nized even when the mineral is present in very small amounts, as in some schists and quartzites.

Allanite, $(Ce,Ca,Y)_2(Al,Fe,Mg)_3Si_3O_{12}(OH)$

Crystal system, class, and space group: Monoclinic; $2/m$; $P2_1/m$.

Cell dimensions and content: a = 8.92, b = 5.75, c = 10.23; β = 115°; Z = 2.

Crystal structure: Allanite is commonly metamict as a result of radiation damage caused by the radioactive decay of thorium. The wide variation in density is mainly due to this loss of crystallinity and an accompanying absorption of water.

Habit: Crystals uncommon; typically tabular parallel to {100}, or long and slender, elongate parallel to b; also massive and as embedded grains.

Cleavage and fracture: {001} seldom observable. Conchoidal.

Hardness: $6\frac{1}{2}$.

Specific gravity: 2.7–4.0

Color and streak: Typically black; rarely dark brown. Gray-brown.

Luster and light transmission: Vitreous, sometimes pitchy. Transparent to subtranslucent.

Optical properties:

anisotropic, biaxial (±): 2V (−) = 40–90°, 2V(+) = 57–90°

refractive indices: α, 1.690–1.791; β, 1.700–1.815; γ, 1.706–1.828

pleochroism in browns or greens

Chemistry: The composition ranges greatly. In addition to the elements indicated in the formula, most allanite contains some thorium, up to 3 percent having been recorded; beryllium, sodium, and potassium have also been recorded in some analyses.

Diagnostic features: Color, luster, mode of occurrence, and weak radioactivity.

Occurrence: Allanite is not uncommon as an accessory mineral in small grains in granitic rocks. Good crystals and large masses occur in some granitic pegmatites.

Name: The name *orthite* is frequently used for this mineral in Europe, although *allanite* has priority (*allanite* was applied to this mineral in 1811, *orthite* in 1818).

SUBCLASS NESOSILICATES

Nesosilicates are those silicates with isolated SiO_4^{-4} groups in the structure. Under the obsolete silicic acid theory, they were considered to be salts of orthosilicic acid, H_4SiO_4, and called *orthosilicates*. It is noteworthy that in the nesosilicates, Al seldom substitutes for Si, so their formulas generally correspond strictly to (SiO_4) or a multiple thereof. Another feature of the nesosilicates is the absence of compounds with the alkali elements.

The nesosilicates have notably dense atomic packing, and their physical properties reflect this. They are relatively harder and, with exceptions, have a higher density than corresponding compounds of the other silicate structure types. The absence of chains and sheets is reflected in the generally equidimensional nature of the crystals.

Olivine, $(Mg,Fe)_2SiO_4$

Crystal system, class, and space group: Orthorhombic; 2/m2/m2/m; Pmcn.

Cell dimensions and content: a = 4.76, b = 10.20, c = 5.98 (forsterite, Mg_2SiO_4); Z = 4. a = 4.82, b = 10.48, c = 6.11; (fayalite, Fe_2SiO_4).

Habit: Crystals uncommon (see Figure 15-46); typically in granular masses and as rounded grains in igneous rocks.

Cleavage: {010} indistinct.

Hardness: $6\frac{1}{2}$.

Specific gravity: 3.22 (Mg_2SiO_4) to 4.39 (Fe_2SiO_4); common olivine about 3.3–3.4.

Color and streak: Typically olive-green (hence the name); also white (forsterite) and brown to black (fayalite). White or gray.

Luster and light transmission: Vitreous. Transparent to translucent.

Optical properties:
   anisotropic, biaxial (+); 2V = 82°–90° (forsterite)
             biaxial (−); 2V = 46°–90° (other olivines)
   refractive indices: α, 1.635–1.827; β, 1.651–1.869; γ, 1.670–1.879 (see Figure 15-47)
   weak pleochroism

Chemistry and alteration: The olivine series is an example, albeit imperfect, of continuous solid solution of two components, Mg_2SiO_4 and Fe_2SiO_4; three names are used currently: *forsterite* for pure or nearly pure Mg_2SiO_4, *fayalite* for pure or nearly pure Fe_2SiO_4, and *olivine* for the common intermediate varieties. Forsterite and olivine are incompatible with free silica because they react with it to give pyroxene; as a consequence, olivine and quartz cannot crystallize together in a rock. Fayalite, however, does not react in this way, and fayalite occurs in some granites and rhyolites.

The composition of olivine generally corresponds closely to $(Mg,Fe)_2SiO_4$, there being little replacement by other elements. Substitution by calcium is evidently strongly temperature dependent, because only a little of the olivine from plutonic rocks contains more than 0.1 percent CaO, whereas most of the olivine of volcanic rocks contains more than this amount, typically ranging up to a maximum

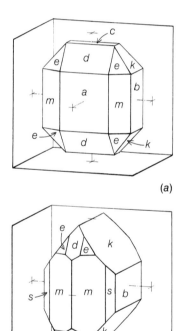

(a)

(b)

Figure 15-46.   Forms in olivine crystals: c{001}, b{010}, a{100}, m{110}, s{120}, d{101}, k{021}, e{111}.

476

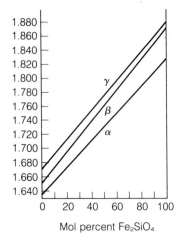

Figure 15-47. Variation of refractive indices with composition in the olivine series.

of about 1 percent CaO. Manganese is present in most olivines and generally correlates positively with Fe content, ranging from about 0.1 percent in forsterites up to 1–2.5 percent in fayalite. Olivines from ultrabasic rocks generally contain some nickel, commonly about 0.3 percent. A noteworthy feature of olivine is the virtual absence of aluminum; evidently replacement of Mg and Si by Al is unacceptable in the olivine structure.

Olivine alters readily. Hydrothermal alteration generally results in the formation of serpentine, whereas surface or near-surface alteration results in oxidation of the iron and removal of the magnesium and silica, commonly leaving a brown or red-brown pseudomorph that consists of goethite or hematite.

Diagnostic features: Distinguished from other rock-forming silicates by color and granular nature. Olivine dissolves easily in hot 1:1 HCl, with the separation of gelatinous silica.

Occurrence and uses: Olivine is typically a mineral of basic and ultrabasic igneous rocks; in some places, it constitutes major rock masses (dunite); some basalts contain nodules of granular olivine, some that are derived from the earth's mantle. Olivine is a common mineral in stony and stony-iron meteorites. Forsterite is formed by the metamorphism of dolomitic limestone.

Fayalite melts at 1205°C, forsterite at 1890°C; thus, magnesium-rich olivine, with a very high melting point, is used in the manufacture of refractory bricks. Transparent olivine of good color has been cut into attractive gemstones (*peridot*).

Related species: There are a number of minerals, all silicates of bivalent elements, that are isomorphous with olivine; they are of rare occurrence except for *monticellite*, $CaMgSiO_4$, which occurs in many metamorphosed limestones.

Willemite, Zn_2SiO_4

Crystal system, class, and space group: Trigonal; $\bar{3}$; $R\bar{3}$.
Cell dimensions and content: a = 13.94, c = 9.31; Z = 18.
Habit: Crystals uncommon; small prismatic or rhombohedral; typically massive, coarse-to-fine granular.
Cleavage: {0001} good, {11$\bar{2}$0} indistinct.
Hardness: $5\frac{1}{2}$–6.
Specific gravity: 3.9–4.2.
Color and streak: White, yellow, green, reddish brown, or black. White.
Luster and light transmission: Vitreous to greasy. Translucent.
Optical properties:
   anisotropic, uniaxial (+)
   refractive indices: ω, 1.691; ϵ, 1.719

Chemistry: A considerable part of the zinc may be replaced by manganese, and a small amount of iron is commonly present.

Other properties: Willemite from Franklin, New Jersey, shows a magnificent green fluorescence in ultraviolet light; this fluorescence is activated by the partial substitution of Mn for Zn; Mn-free willemite is not fluorescent.

Diagnostic properties: Willemite from Franklin is characterized by its associations and fluorescence. Willemite is soluble in hydrochloric acid with the separation of gelatinous silica.

Occurrence, associations, and use: Willemite occurs abundantly in the metamorphic ore deposit at Franklin, New Jersey, where it is a major zinc ore. It occurs there with franklinite and zincite—constituting a green, black, and red mixture—and also as large crystals in the marble host rock. Willemite is not uncommon in the oxidized zone of zinc-bearing deposits, especially in arid regions, but it is generally inconspicuous and easily overlooked.

Zircon, ZrSiO₄

Crystal system, class, and space group: Tetragonal; $4/m2/m2/m$; $I4_1/amd$.

Cell dimensions and content: $a = 6.604$, $c = 5.979$; $Z = 4$.

Habit: Crystals common; most are prisms terminated by pyramids (Figure 15-48).

Twinning: Rarely twinned, twin plane {112}, giving knee-shaped twins.

Common forms and angles: (Figure 15-48; Figure 2-63a):

$(101) \wedge (011) = 56°40'$ $\quad$ $(100) \wedge (101) = 47°51'$
$(301) \wedge (031) = 83°08'$ $\quad$ $(100) \wedge (301) = 20°13'$
$(110) \wedge (211) = 31°43'$ $\quad$ $(211) \wedge (121) = 32°56'$

Cleavage: {110} distinct, {111} indistinct.

Hardness: $7\frac{1}{2}$.

Figure 15-48. (a), (b), (c) Typical forms in zircon crystals: a{100}, m{110}, p{101}, x{211}.

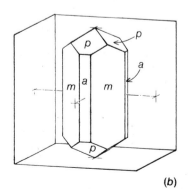

(a)

(b)

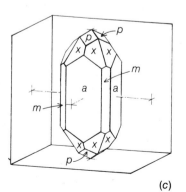

(c)

Specific gravity: 4.6–4.7 for crystalline material; down to 3.9 when metamict.

Color and streak: Typically brown or reddish brown; also colorless, gray, green, or violet. [The transparent blue varieties used as gemstones are produced by heat treatment of natural zircon.] White.

Luster and light transmission: Vitreous to adamantine. Transparent to translucent.

Optical properties:
anisotropic, uniaxial (+)
refractive indices: ω, 1.920–1.960; ϵ, 1.967–2.015

Chemistry: In all zircon, some of the Zr is replaced by Hf (generally about 1 percent, but 4 percent and more has been recorded; where Hf exceeds Zr, the mineral is called *hafnon*). Part of the Zr may also be replaced by rare earths, coupled with the replacement of silicon by phosphorus. Much zircon is radioactive because of the presence of Th and U replacing Zr in the structure; radiation damage from these radioactive elements causes these zircons to be, at least in part, metamict.

Diagnostic features: Habit, hardness, color, and density. Some zircon fluoresces orange in ultraviolet radiation.

Occurrence and uses: Zircon is a common accessory mineral in igneous rocks and pegmatites of the granite, syenite, and nepheline syenite families. The presence of uranium and thorium makes it a useful mineral for age determination of such rocks. Red crystals weighing up to 7 kg have been found in Renfrew County, Ontario. Because it is resistant to mechanical and chemical weathering, zircon occurs as a detrital mineral in river and beach sands.

Zircon is the principal source of zirconium and hafnium for industry. It is extracted from sands, the most important sources being beach deposits on the coast of Queensland in Australia and in Florida and Georgia. Zircon, in both its natural and heat-treated forms, has been cut and polished for use as a gemstone.

Thorite, $ThSiO_4$

Crystal system, class, and space group: Tetragonal; $4/m2/m2/m$; $I4_1amd$.

Cell dimensions and content: $a = 7.117$, $c = 6.295$; $Z = 4$.

Habit: Crystals rare; squat pyramids or square prisms with pyramidal terminations; generally as irregular grains and masses.

Cleavage and fracture: {100} distinct but seldom observed. Conchoidal.

Hardness: 5.

Specific gravity: 6.7 for pure crystalline $ThSiO_4$; generally considerably lower (down to 4 and less) due to different degrees of metamictization and accompanying hydration.

Color and streak: Reddish brown to black; rarely orange (var. *orangite*). Light orange to brown.

Luster and light transmission: Resinous. Translucent.

Optical properties:

anisotropic, uniaxial (+)

refractive indices: ω, 1.78–1.825; ϵ, 1.79–1.840

may appear isotropic

Chemistry and alteration: Some of the thorium may be replaced by uranium (var. *uranothorite*), and some of the $(SiO_4)^{-4}$ may be replaced by $(OH)_4^{-4}$ (var. *thorogummite*). Nearly all thorite is metamict. It is strongly radioactive.

Diagnostic features: Color, luster, habit, and strong radioactivity.

Occurrence, associations, and uses: Thorite is an accessory mineral in granites, syenites, and nepheline syenites, and in pegmatites associated with these rocks. It also occurs as a detrital mineral in sands. Uranothorite is partially responsible for the uranium content of some of the ores in the Bancroft district of Ontario.

Thorite is the most common thorium mineral, and it may become an important raw material when, and if, thorium is used for the production of atomic energy. The present demand for thorium is met by its extraction from monazite.

Garnet Group $X_3Y_2(SiO_4)_3$, in which $X = Ca, Fe^{+2}, Mg$, and/or Mn^{+2}; and $Y = Al, Cr, Fe^{+3}, Mn^{+3}, Ti, V$, and/or Zr.

Crystal system, class, and space group: Isometric; $4/m\bar{3}2/m$; $Ia3d$.

Habit: Crystals common, except for pyrope and uvarovite (for which they are rare); the crystal forms are characteristic: typically dodecahedra {110} or trapezohedra {211}, or combinations of these; rarely combinations of dodecahedron and hexoctahedron {321} (Figures 15-49 and 15-50). Also massive, coarse-to-fine granular.

Cleavage: None.

Hardness: $7–7\frac{1}{2}$.

Specific gravity: 3.6–4.3, the value for each specimen depending on its composition.

Color: Dark red and reddish brown are common. More specifically, almandine is generally red or reddish brown; pyrope, red to nearly black; spessartine, orange to dark red or brown; grossular, white when pure, commonly yellow, pink, green, or brown; andradite, yellow, greenish yellow, greenish brown, or black; and uvarovite a characteristic emerald-green.

Streak: White or pale shade of body color.

Luster and light transmission: Vitreous to resinous. Transparent to subtranslucent.

Optical properties:

isotropic (some appear anisotropic)

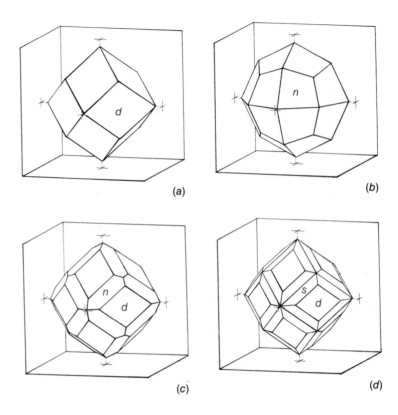

Figure 15-49.   Typical forms in garnet crystals: $d\{110\}$, $n\{211\}$, $s\{321\}$.

refractive index:
> Almandine: 1.830   Grossular: 1.734
> Pyrope: 1.713   Andradite: 1.887
> Spessartine: 1.799   Uvarovite: 1.863

grossular, andradite, and uvarovite may appear biaxial $(-)$

Chemistry: The following names are used for the relatively common garnet species and their components:

Species	Formula	Specific Gravity	Unit Cell Edge (a)
Almandine	$Fe_3Al_2(SiO_4)_3$	4.32	11.53
Pyrope	$Mg_3Al_2(SiO_4)_3$	3.58	11.46
Spessartine	$Mn_3Al_2(SiO_4)_3$	4.19	11.62
Grossular	$Ca_3Al_2(SiO_4)_3$	3.59	11.85
Andradite	$Ca_3Fe_2(SiO_4)_3$	3.86	12.05
Uvarovite	$Ca_3Cr_2(SiO_4)_3$	3.78	11.97

The composition of naturally occurring garnets rarely approaches the formulas first given because of extensive atomic substitution. The specific name applied is that of the component that is

(a)   (b)   (c)   (d)

Figure 15-50. Garnet, dark red and brownish black. (a) Dodecahedron {110}. (b) Dodecahedron beveled by trapezohedron {211}; Fort Wrangel, Alaska. (c) Trapezohedron {211}; Burma. (d) Trapezohedron {211} with minor dodecahedron faces; Burma. In contrast to (a) and (b), (c) and (d) have a_1 turned 45° to the right; (a), (c), and (d) ~ 4 cm and (b) ~ 3 cm along a. (Parts a and b courtesy of Queen's University. Parts c and d courtesy of Royal Ontario Museum.)

present in largest amount. Ferrous iron and magnesium are interchangeable, and a series of intermediate compositions exist between almandine and pyrope; similarly, series of intermediate compositions exist between almandine and spessartine and between grossular and andradite (Figures 15-51a and b). The difference in ionic size between calcium on the one hand and ferrous iron and magnesium on the other, however, results in rather limited substitution and a wide composition gap between grossular-andradite, on the one hand, and almandine-pyrope-spessartine, on the other (Figure 15-51b). There is also a wide composition gap between spessartine and pyrope, which is due, at least in part, to the difference in ionic size between magnesium and manganese, but also possibly due to the different geological environments in which spessartine and pyrope generally form. Grossular commonly contains combined water as a result of the partial substitution of $(OH)_4^{-4}$ for $(SiO_4)^{-4}$ in the structure, and there is apparently a complete series between grossular and *hydrogrossular* $[Ca_3Al_2(SiO_4)_{3-x}(OH)_{4x}]$. In addition, there are series between grossular and uvarovite and the rather rare *schorlomite* $[Ca_3(Fe,Ti)_2(Si,Ti)_3O_{12}]$, and between pyrope and the rather rare *knorringite* $[Mg_3Cr_2(SiO_4)_3]$. In addition, there are other rare garnets such as *kimzeyite* $[Ca_3(Zr,Ti)_2(Si,Al)_3O_{12}]$ and *goldmanite* $[Ca_3(V,Al,Fe)_2(SiO_4)_3]$.

Occurrence and associations: Garnets differ somewhat in their mode of typical occurrence, as follows:

    Almandine: The common garnet of gneisses and schists is almandine. It is also recorded from granites, rhyolites, and pegmatites.

    Pyrope: Less common than the other garnets (except uvarovite), pyrope occurs in ultrabasic igneous rocks and serpentinites derived from them. It also occurs in high-grade, magnesium-rich metamorphic rocks.

    Spessartine: Many garnets from granite pegmatites and in vesicles in rhyolites are spessartine or intermediate between

482

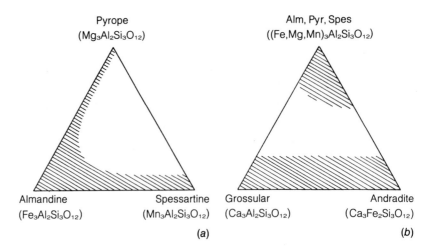

Figure 15-51. Mineral compositions in the garnet group. Shaded areas indicate compositional regions for natural garnets; unshaded areas indicate compositional regions essentially unrepresented by natural garnets.

spessartine and almandine. Spessartine also occurs in metamorphosed manganese-bearing rocks.

Grossular: Grossular is typically formed by contact or regional metamorphism of impure limestones and dolostones and, thus, is associated with calcite, wollastonite, and idocrase.

Andradite: Andradite is formed by the metasomatic alteration of limestones by iron-bearing solutions, and it commonly occurs associated with ore deposits in calcareous rocks.

Uvarovite: Uvarovite, which is rare, occurs in association with chromite and serpentinite.

Garnets, being resistant to both mechanical and chemical breakdown, also occur as detrital grains in sands and sandstones.

Uses: Garnet has some value as an abrasive because it is fairly hard, lacks cleavage, and hence breaks into irregular grains. Although garnet is a common mineral, material suitable for use as an abrasive has seldom been found in workable quantity. The requirements are for large isolated crystals that are crushed to provide the garnet "sand" used to make sandpaper. Several thousand tons of such garnet have been produced annually at Gore Mountain, in the Adirondack Mountains of New York.

Transparent unflawed garnet of good color can be cut into attractive gemstones. Much of the red garnet jewelry consists of pyrope from Czechoslovakia. Uvarovite would make a magnificent gemstone, but it does not occur in sufficiently large pieces. Green garnet gemstones are cut from a variety of andradite known as *demantoid*.

Titanite (Sphene), $CaTiSiO_5$

Crystal system, class, and space group: Monoclinic; *2/m; A2/a.*
Cell dimensions and content: $a = 7.05$, $b = 8.70$, $c = 6.54$; $\beta = 113° 48'$; $Z = 4$.

Habit: Crystals common; typically as wedge-shaped crystals (hence the name, *sphene*) with large {100} and {21$\bar{1}$} faces; also massive or granular.

Twinning: Twin plane {10$\bar{1}$}, giving contact twins or cruciform penetration twins.

Cleavage: {11$\bar{1}$} distinct.

Hardness: 6.

Specific gravity: 3.5.

Color and streak: Typically brown. Also yellow, green, or gray. White.

Luster and light transmission: Adamantine. Transparent to translucent.

Optical properties:

anisotropic, biaxial (+): $2V = 17°–56°$

refractive indices: α, 1.840–1.950; β, 1.870–2.034; γ, 1.943–2.110

slightly pleochroic

Chemistry: The composition ranges somewhat. Calcium may be partly substituted by Na or rare-earth elements; Ti by Al, Fe, Zr, Nb, Ta, Sn, and Sb; and Si by Al.

Diagnostic features: Color, typical wedge-shaped or diamond-shaped cross sections, and adamantine luster.

Occurrence: Titanite is widely distributed as an accessory mineral in acid and intermediate igneous rocks, in their associated pegmatites, and in several metamorphic rocks.

ALUMINUM SILICATE GROUP

The aluminum silicate group of minerals can be considered to include the three polymorphs of Al_2SiO_5 (andalusite, sillimanite, and kyanite), topaz [$Al_2SiO_4(F,OH)_2$], and also staurolite [the fundamental formula of which can be written as $2Al_2SiO_5 \cdot Fe(OH)_2$].

Despite the fact that stability relationships of the minerals of this group have been submitted to experimental investigation by several geochemists, those relationships cannot yet be considered to have been defined absolutely. Nonetheless, it has long been recognized that: kyanite is much denser than the other two, and hence its formation is favored by high pressures; sillimanite is characteristic of high-grade metamorphic rocks, and hence its formation appears to be favored by high temperatures; and andalusite is commonly a product of the moderately high-temperature metamorphism of argillaceous rocks. One of the experimentally established equilibrium diagrams for these polymorphs is shown in Figure 15-52.

On heating to 1300°C, kyanite, sillimanite, andalusite, and topaz decompose into *mullite*, $Al_6Si_2O_{13}$, and a silica-rich glass. Mullite is found in a few localities as a mineral where aluminum-rich rocks have been subjected to extremely high temperatures, such as where fragments

484

Figure 15-52. Stability relations among minerals with the composition Al_2SiO_5. The position of the triple point has been given different values by different investigators. *A* is the triple point for the diagram as given; the circled points represent other noteworthy triple points reported in the literature. The differences apparently depend on the fact that such things as minor impurities greatly affect this system. (Redrawn after S. W. Richardson, M. C. Gilbert, and P. M. Bell, 1969, "Experimental determination of the . . . aluminum silicate triple point." *American Journal of Science,* **267,** 259–272.)

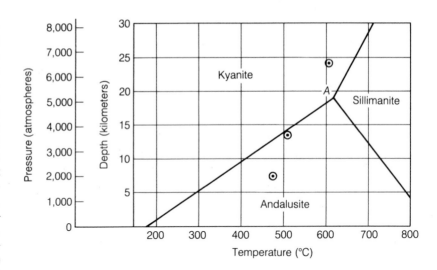

of shale have been picked up by a lava flow. Mullite is used in industry as a refractory, for which purpose it is made by heating kyanite or andalusite.

Andalusite, Al_2SiO_5

Crystal system, class, and space group: Orthorhombic; $2/m2/m2/m$; *Pnnm.*

Cell dimensions: $a = 7.78$, $b = 7.92$, $c = 5.57$; $Z = 4$.

Habit: Crystals uncommon; coarsely prismatic with nearly square cross sections, the prism angle being 89° 12′; also massive. Many crystals have carbonaceous inclusions, arranged in shapes conforming to the symmetry of the mineral (var. *chiastolite,* Figure 15-53).

Cleavage: {110} good.

Hardness: $7\frac{1}{2}$.

Specific gravity: 3.15.

Color and streak: White, gray, rose-red, brown, or green (var. *viridine*); dichroism may be seen in many transparent hand specimens. White.

Luster and light transmission: Vitreous, often dull. Transparent to translucent.

Optical properties:
anisotropic, biaxial (−): 2V = 71°–86°
refractive indices: α, 1.629–1.640; β, 1.634–1.645; γ, 1.638–1.650

Chemistry and alteration: There is a manganian andalusite $[(Al,Mn^{+3})_2SiO_5]$.

Paramorphs of kyanite and sillimanite after andalusite have been recorded. Andalusite also alters readily to sericite.

Diagnostic features: Crystal form and mode of occurrence in metamorphosed shales.

Occurrence and uses: Andalusite occurs mainly in contact-metamorphosed shales; it also occurs in a few regionally metamorphosed rocks and pegmatites.

Andalusite is used in the manufacture of mullite refractories, especially spark plugs; it has been mined in the White Mountains, Mono County, California, and considerable tonnages have also been produced from andalusite-bearing sands in western Transvaal, Union of South Africa.

Sillimanite, Al_2SiO_5

Crystal system, class, and space group: Orthorhombic; $2/m2/m2/m$; Pbnm.

Cell dimensions and content: $a = 7.47$, $b = 7.66$, $c = 5.76$; $Z = 4$.

Habit: Crystals uncommon; typically in finely fibrous or coarse prismatic masses; the fibrous nature is due to the existence of chains of aluminum-oxygen groups parallel to the c axis.

Cleavage: {010} perfect.

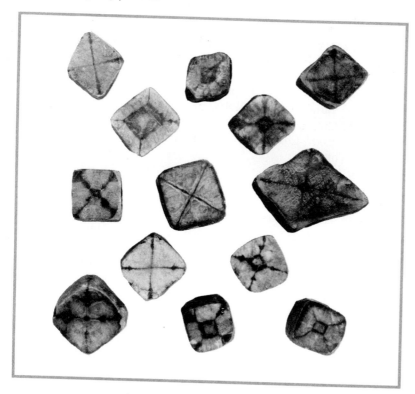

Figure 15-53. Andalusite, var. chiastolite, showing characteristic regular arrangement of inclusions; Madera County, California. Specimens are 2–3 cm across. (Courtesy of the American Museum of Natural History.)

Hardness: 7.
Specific gravity: 3.24.
Color and streak: White; uncommonly brownish or greenish. White.
Luster and light transmission: Vitreous; silky in fibrous material. Transparent to translucent.
Optical properties:
   anisotropic, biaxial (+): $2V = 20°–30°$
   refractive indices: α, 1.653–1.661; β, 1.654–1.670; γ, 1.669–1.684
Diagnostic features: Resembles other fibrous silicates, such as wollastonite and tremolite, but can be distinguished by its single direction of cleavage and its insolubility in acids.
Occurrence and use: Sillimanite, formed in aluminum-rich rocks under conditions of high-grade regional metamorphism, is found in schists and gneisses. Workable deposits of sillimanite are rarer than those of andalusite or kyanite, and hence they have been but little used in the manufacture of mullite.

Kyanite, Al_2SiO_5

Crystal system, class, and space group: Triclinic; $\bar{1}$; $P\bar{1}$.
Cell dimensions and content: a = 7.10, b = 7.74, c = 5.57; $\alpha = 90° 06'$, $\beta = 101° 02'$, $\gamma = 105° 45'$; Z = 4.
Habit: Crystals common; {100} tablets elongate in the c direction, seldom terminated; also as bladed masses. Crystals are distinctly flexible and often give the appearance of having been bent or twisted.
Cleavage and parting: {100} perfect, {010} good; parting on {001}.
Hardness: Variable; on {100} 4–5 parallel to c, 6–7 parallel to b.
Specific gravity: 3.63.
Color and streak: Characteristically patchy blue; also green, white, or gray. White.
Luster and light transmission: Vitreous; some is pearly on cleavage surfaces. Translucent to transparent.
Optical properties:
   anisotropic, biaxial (−): $2V = 82°$
   refractive indices: α, 1.712–1.718; β, 1.720–1.725; γ, 1.727–1.734
Diagnostic features: Variable hardness, color, and habit.
Occurrence and uses: Kyanite is the product of medium-grade regional metamorphism of aluminum-rich rocks; it occurs in gneisses and schists and in quartz veins and pegmatites that transect those rocks.

   Kyanite is important commercially for the manufacture of mullite refractories. There has been some production from the metamorphic belt of Virginia, South Carolina, and Georgia.

Name: Kyanite is sometimes spelled *cyanite*; it is also known as *disthene*, a name derived from its variable hardness.

Staurolite, $(Fe,Mg,Zn)_2Al_9(Si,Al)_4O_{22}(OH)_2$

Crystal system, class, and space group: Monoclinic (pseudo-orthorhombic); $2/m$; $C2/m$.

Cell dimensions and content: $a = 7.83$, $b = 16.62$, $c = 5.65$; $\beta = 90°$; $Z = 4$.

Habit: Crystals common; prismatic with {001} and {010}, commonly twinned (Figures 2-76, 15-54, and 15-55).

Twinning: Cruciform twins are common: twin plane {031} gives a right-angle cross; twin plane {231} gives a 60°/120° cross, commonly called "saw-horse" twins.

Common forms and angles:

  (110) $\wedge$ (1$\bar{1}$0) = 50° 26'   (001) $\wedge$ (201) = 55° 17'

Cleavage: {010} distinct.

Hardness: 7.

Specific gravity: 3.7–3.8.

Color and streak: Brown. Gray.

Luster: Subvitreous to somewhat resinous. Translucent to nearly opaque.

Optical properties:

  anisotropic, biaxial (+): $2V = 80°–90°$

  refractive indices: α, 1.736–1.747; β, 1.740–1.754; γ, 1.745–1.762

  pleochroic

Chemistry: Along with the substitutional possibilities indicated by the formula, some of the ferrous iron may be replaced by manganese and some of the aluminum by ferric iron. In addition, Ti may be

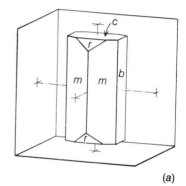

(a)

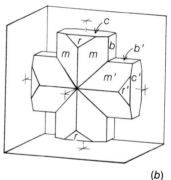

(b)

Figure 15-54. Forms in staurolite crystals. (a) Untwinned. (b) Twinned on {031}. c{001}, b{010}, m{110}, r{201}. (See also Figure 2-76.)

Figure 15-55. Staurolite, grayish brown. Forms: {110}, {010}, {201}. (a) Single crystal {110} parallel to page, c vertical. (b) Twin crystal, {031} twin plane. (c) Twin crystal with {231} twin plane, Blue Ridge, Georgia; greatest dimension—length of (b)—4 cm). (Courtesy of Queen's University.)

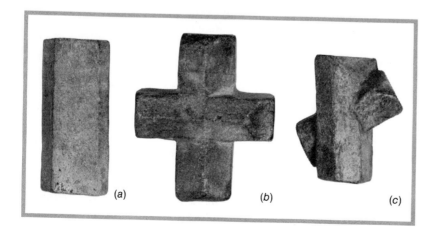

(a)   (b)   (c)

present in minor amounts (0.1–1 percent). Also, cobaltian (to about 8.5 percent CoO) as well as zinc-bearing staurolites have been recorded.

The structure of staurolite, closely related to that of kyanite, consists of layers with the kyanite structure alternating with layers containing the iron atoms and the hydroxyl groups. Individual crystals consisting of kyanite and staurolite in parallel growth are also known.

Diagnostic features: Color and habit.

Occurrence and use: Staurolite is characteristic of medium-grade metamorphic rocks; it occurs in aluminum-rich gneisses and schists.

The crossed twins found in Virginia and elsewhere, sometimes called "fairy stone crosses," are sold as good luck charms. [Imitations made of baked brown clay are also occasionally sold.]

Topaz, $Al_2SiO_4(F,OH)_2$

Crystal system, class, and space group: Orthorhombic; 2/m2/m2/m; Pbnm.

Cell dimensions and content: a = 4.65, b = 8.80, c = 8.40; Z = 4.

Habit: Crystals common; well-developed prisms, some up to several hundred kilograms in weight, typically with wedge-shaped terminations; also massive, coarse-to-fine granular.

Common forms and angles (Figure 15-56):

(010) ∧ (120) = 43°26'	(001) ∧ (011) = 43°41'
(010) ∧ (110) = 62°10'	(001) ∧ (021) = 62°22'
(001) ∧ (121) = 69°12'	(001) ∧ (111) = 63°56'

Cleavage: {001} perfect.

Hardness: 8.

Specific gravity: 3.5–3.6, higher with greater F content.

Color and streak: Colorless or white; also pale blue, yellow, yellow-brown; rarely pink. White.

Luster and light transmission: Vitreous. Transparent.

Optical properties:

anisotropic, biaxial (+): 2V = 48° –68°

refractive indices: α, 1.606–1.630; β, 1.609–1.631; γ, 1.616–1.638

Other properties: Slightly pyroelectric and piezoelectric.

Chemistry: All topaz contains fluorine, the maximum substitution of F by OH being about 30 percent.

Diagnostic features: Readily recognizable in crystals by its habit, hardness, and single perfect cleavage; massive topaz is generally distinguished by its density and hardness.

Occurrence and use: Topaz occurs in pegmatites and high-temperature quartz veins and also in cavities in granites and rhyolites. In spite of its good cleavage, topaz sometimes survives a moderate

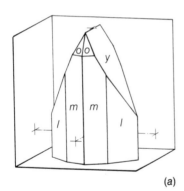

(a)

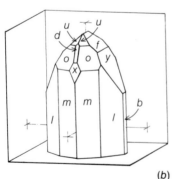

(b)

Figure 15-56.   Typical forms in topaz crystals: c{001}, b{010}, m{110}, l{120}, y{021}, d{101}, x{201}, f{011}, o{111}, u{112}.

degree of natural transportation and occurs as rounded pebbles in alluvial deposits.

Transparent topaz of diverse colors is cut and polished as a gemstone.

Humite Group

	Norbergite	Chondrodite	Humite	Clinohumite
Chemical formula:	$Mg(OH,F)_2 \cdot$ Mg_2SiO_4	$Mg(OH,F)_2 \cdot$ $2Mg_2SiO_4$	$Mg(OH,F)_2 \cdot$ $3Mg_2SiO_4$	$Mg(OH,F)_2 \cdot$ $4Mg_2SiO_4$
Crystal system:	Orthorhombic	Monoclinic	Orthorhombic	Monoclinic
Class and space group	$2/m2/m2/m$; Pmcn	$2/m$; $P2_1/c$	$2/m2/m2/m$; Pmcn	$2/m$; $P2_1/c$
Cell dimensions:	$a = 8.74$ $b = 4.71$ $c = 10.22$	$a = 7.89$ $b = 4.74$ $c = 10.29$ $\beta = 109°02'$	$a = 20.90$ $b = 4.75$ $c = 10.25$	$a = 13.71$ $b = 4.75$ $c = 10.29$ $\beta = 100°50'$
Cell content:	$Z = 4$	$Z = 2$	$Z = 4$	$Z = 2$
Cleavage:	None (?)	{100} indistinct	{100} indistinct	{100} indistinct
Fracture:	Subconchoidal	Subconchoidal	Subconchoidal	Subconchoidal
Hardness:	$6-6\frac{1}{2}$	$6-6\frac{1}{2}$	6	6
Specific gravity:	3.18	3.16–3.26	3.20–3.32	3.17–3.35
Luster:	Vitreous	Vitreous	Vitreous	Vitreous
Light transmission:	Transparent to translucent	Transparent to translucent	Transparent to translucent	Transparent to translucent
Optical properties:	biaxial (+) $2V = 44°-50°$	biaxial (+) $2V = 64°-90°$	biaxial (+) $2V = 65°-84°$	biaxial (+) $2V = 52°-90°$
refractive *indices:*	$\alpha = 1.561-1.567$ $\beta = 1.566-1.579$ $\gamma = 1.587-1.593$	$\alpha = 1.592-1.643$ $\beta = 1.602-1.655$ $\gamma = 1.619-1.675$	$\alpha = 1.607-1.643$ $\beta = 1.619-1.653$ $\gamma = 1.639-1.675$	$\alpha = 1.623-1.702$ $\beta = 1.636-1.709$ $\gamma = 1.651-1.728$

Habit: Rare crystals are of diverse habits; commonly rounded or irregular shaped grains.

Twinning: Common, in chondrodite and clinohumite, on {001}; simple or multiple.

Color and streak: All humite group minerals are typically some shade of yellow or brown; some humite and clinohumite is white, some chondrodite is reddish. White.

Chemistry and alteration: Fluorine generally exceeds OH; less than one-half of Mg may be replaced by Fe^{+2}, Mn, Ca, and/or Zn. All members of the humite group have been found altered to serpentine, chlorite, or limonite.

Diagnostic features: Members of this group are difficult to distinguish from each other and from olivine by macroscopic means.

Occurrence and associations: All members of the humite group typically occur in dolostones and dolomitic limestones that have been subjected to contact metamorphism. Chondrodite has also been found in carbonatites; chondrodite and clinohumite occur rarely in veins; clinohumite occurs in serpentinites and talc schists; and nor-bergite has been found in ore deposits.

Datolite, $CaBSiO_4(OH)$

Crystal system, class, and space group: Monoclinic; $2/m$; $P2_1/a$.
Cell dimensions and content: $a = 9.66$, $b = 7.64$, $c = 4.83$; $\beta = 90°$ 09'; $Z = 4$.
Habit: Crystals uncommon; typically in short prismatic crystals, in many cases with a variety of forms; also as granular or compact porcelainlike masses.
Cleavage: None.
Hardness: $6\frac{1}{2}$.
Specific gravity: 3.0.
Color and streak: Colorless; pale shades of yellow and green; also white and opaque. White.
Luster and light transmission: Vitreous to pearly. Translucent to sub-translucent.
Optical properties:
   anisotropic, biaxial $(-)$: $2V = 72°–75°$
   refractive indices: α, 1.622–1.626; β, 1.649–1.654; γ, 1.666–1.670
Diagnostic features: Color, crystal form, and occurrence.
Occurrence and use: Datolite occurs as a secondary mineral, gener-ally in cavities in basalts where it is associated with zeolites (e.g., in basaltic rocks of New Jersey and Massachusetts). White porcel-lanous masses are associated with native copper in the Lake Su-perior volcanics.

Chloritoid, $(Fe,Mg,Mn)_2(Al,Fe)Al_3O_2(SiO_4)_2(OH)_4$

Crystal system, class, and space group: Monoclinic; $2/m$; $C2/c$. (Also triclinic; $\bar{1}$; $C\bar{1}$.)
Cell dimensions and content: $a \cong 9.5$, $b \cong 5.48$, $c \cong 18.2$; $\beta \cong 101°$ 45'; $Z = 4$. [Triclinic (for $FeAl_2SiO_5(OH)_2$): $a = 9.46$, $b = 5.50$, $c = 9.15$; $\alpha = 97° 03'$, $\beta = 101° 34'$, $\gamma = 90° 06'$; $Z = 4$.]
Habit: Crystals uncommon; typically pseudo-hexagonal, tabular crys-tals; also massive, as thin scales.
Twinning: Common, simple or lamellar on {001}.
Cleavage and parting: {001} perfect, {110} distinct. {010}.
Hardness: $6\frac{1}{2}$.
Specific gravity: 3.61 (monoclinic); 3.58 (triclinic).

Color and streak: Greenish gray to greenish black. Gray.

Luster and light transmission: Pearly (especially on cleavage). Translucent.

Optical properties:

anisotropic, biaxial (−): $2V = 36°–90°$

refractive indices: α, 1.713–1.730; β, 1.719–1.734; γ, 1.723–1.740

pleochroic in greens and gray-blue

Chemistry: Apparently all chloritoids contain Mg. However, the substitution of Mg for Fe^{+2} is thought to be limited to less than 40 atomic percent. In addition, the substitution of Mn for Fe^{+2} appears to be limited to less than 17 percent and that of Fe^{+3} for Al to be less than 10 atomic percent.

Diagnostic features: Hardness; cleavage flakes are brittle; soluble in sulfuric acid.

Occurrence and associations: Chloritoid occurs in phyllites and schists along with such minerals as chlorite, muscovite, garnet, and staurolite; in quartz-calcite veins; and in certain hydrothermally altered rocks.

SILICATES OF UNKNOWN STRUCTURE

Chrysocolla, $(Cu,Al)_2H_2Si_2O_5(OH)_4 \cdot nH_2O$

Crystal system and structure: Chrysocolla has been variously reported as amorphous with Si_4O_{10} layers present in a defect structure (and included with the phyllosilicates); cryptocrystalline, either of unknown or orthorhombic(?) crystal system (if the latter, included with the tectosilicates); monoclinic (with no silicate category indicated); and as a mixture of colloidal *plancheite* $[Cu_8Si_8O_{22}(OH)_4 \cdot H_2O]$ and *shattuckite* $[Cu_5(SiO_3)_4(OH)_2]$.

Habit: Crystals nonexistant (?!); typically very finely fibrous or massive, in some cases earthy.

Cleavage and fracture: None. Conchoidal.

Hardness: 2–4.

Specific gravity: 2.0–2.4

Color and streak: Green or blue, brown to black where discolored with iron or copper oxides. White (when pure).

Luster and light transmission: Vitreous or waxy. Subtranslucent.

Optical properties:

anisotropic, biaxial (−): $2V \cong 24°$

refractive indices: α, 1.575–1.585; β, 1.597; γ, 1.598–1.635

Chemistry: As noted, the published data on chrysocolla are quite diverse, in part contradictory, and thus it is prudent to say that the true nature of this material is not established. In any case, the water content may be variable.

Diagnostic features: Color and habit. Green chrysocolla may resemble malachite, but it does not effervesce with HCl.

Occurrence and uses: Chrysocolla is formed in the oxidized zone of copper deposits where free silica is available. Much chrysocolla was probably formed from a gel flocculated from a colloidal solution.

Chrysocolla is a minor ore of copper, and massive chrysocolla of good color has been cut and polished for ornamental use.

Part III
DETERMINATIONS

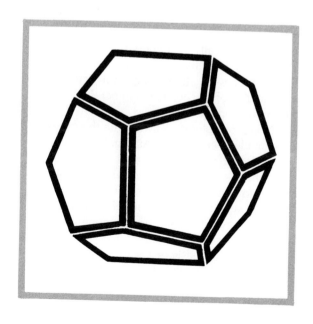

16 Determinative Tables

In the identification of minerals, a set of determinative tables is of great assistance when it is used systematically by a student with keen observation and retentive memory. The whole detective process is stimulating, and determining the identity of an unknown mineral is a most satisfying triumph.

No set of tables does away with the necessity of individual enterprise. Although the student who follows the general precepts and special advice of determinative tables will certainly be helped to avoid the waste of time, there is no foolproof system in the domain of mineral identification. The best key to mineral identification is to "Be smart" and "Remember everything"!

The primary basis of discrimination used in the tables in this book is *luster*—that is, whether the mineral appears metallic or nonmetallic. A few opaque or nearly opaque minerals with a dark brown to black color are sometimes described as having submetallic luster; they are grouped with the minerals with metallic luster, but some of them are also tabulated with the nonmetallic group.

The secondary basis of discrimination is *hardness*. Relative hardness is easily determined with the simplest of tools, and it is one of the few macroscopically determined properties capable of being expressed in numerical terms. The tables are divided into three groups: minerals with hardness less than $2\frac{1}{2}$ (can be scratched with the fingernail); minerals with hardness between $2\frac{1}{2}$ and $5\frac{1}{2}$ (will scratch the fingernail but can be scratched with a knife blade); and minerals with hardness greater than $5\frac{1}{2}$ (will scratch a knife blade). With a little practice, a student should be able to place a figure on the hardness of a mineral in the $2\frac{1}{2}$–$5\frac{1}{2}$ group on the basis of its relative ease of scratching. The hardness given for each mineral is that of a clean smooth surface; minerals in earthy varieties, such as some forms of hematite, appear to be softer than the same minerals in good crystals.

Minerals with nonmetallic luster far exceed in numbers those with metallic luster. Therefore, it is convenient to find a further criterion for subdividing the nonmetallic minerals. In this book, the tables for the nonmetallic minerals are subdivided into two sections based on color: one section includes those minerals that are colorless or white, unless tinted by impurities; the other section includes those minerals that are colored. The latter also have colored streaks whereas the former, even if colored, tend to have white streaks.

By using luster and hardness and dividing the nonmetallic minerals into two sections on the basis of color, we thus obtain a set of tables comprising nine groups, as follows:

Luster	Hardness	Number of Minerals	Page Ref.
Metallic	$H < 2\frac{1}{2}$	4	496
	$H = 2\frac{1}{2}-5\frac{1}{2}$	29	497
	$H > 5\frac{1}{2}$	19	500
Nonmetallic	$H < 2\frac{1}{2}$	8	502
Colorless or	$H = 2\frac{1}{2}-5\frac{1}{2}$	36	503
white	$H > 5\frac{1}{2}$	20	508
Colored	$H < 2\frac{1}{2}$	11	510
	$H = 2\frac{1}{2}-5\frac{1}{2}$	32	511
	$H > 5\frac{1}{2}$	46	515

Within each group, the minerals are described in order of least to greatest specific gravity. Specific gravity is, of course, not so easily estimated as hardness is; however, practice in handling minerals of known density generally develops one's ability to distinguish specific gravities differing by one unit; thus, the student should eventually be able to say whether a specimen has a specific gravity near 3, or 4, or 5, or greater than 5. In this connection, the availability of one or more of the heavy liquids mentioned on page 187 is a useful determinative aid. With a small bottle of tetrabromoethane (acetylene tetrabromide), which has a density of 2.96, it takes only a moment to determine whether a fragment of an unknown mineral sinks or floats—that is, whether it is a heavy or a light mineral.

Having determined luster and hardness and made a reasonable estimate of the specific gravity, you will have narrowed the choice to comparatively few species; thence, the other physical properties—particularly crystal form, cleavage, and color—will often lead to the correct identity. To confirm the identification, you should refer to the full description of the mineral given in Chapters 8 through 15; for most minerals, a section headed *Diagnostic Features* outlines a few distinctive characteristics and/or special tests that are particularly useful. To facilitate such reference, the page references for the full descriptions are given under the mineral names on the tables.

In addition, because optical data that are distinctive for many of the minerals can be determined relatively easily, we have compiled a second group of tables based on those properties for the nonopaque minerals described in Part II. The optical tables are arranged as follows:

	Page
Isotropic	521
Anisotropic	522
Uniaxial (+)	522
Uniaxial (−)	523
Biaxial (+)	526
Biaxial (−)	531

Within the groups, the minerals are described in order of least to greatest index of refraction according to n_ω for uniaxial minerals and n_β for biaxial minerals. Nearly all of the optical data are taken from the compilation of Phillips and Griffen (1981). Again, page references are given to the full descriptions in Chapters 8 through 15.

The following references provide tables that

496 include properties for more minerals than are described here.

Dietrich, R. V. 1969. *Mineral Tables. Hand-Specimen Properties of 1500 Minerals.* New York: McGraw-Hill, 237 pp.

Larsen, E. S., and Berman, Harry. 1934. *The Microscopic Determination of Nonopaque Minerals* (2nd ed.). U.S. Geological Survey, Bulletin 848, 266 pp.

Phillips, W. R., and Griffen, D. T. 1981. *Optical Mineralogy. The Nonopaque Minerals.* San Francisco: W. H. Freeman and Company, 677 pp.

Short, M. N. 1931. *Microscopic Determination of the Ore Minerals.* U.S. Geological Survey, Bulletin 825, 204 pp.

Uytenbogaardt, W., and Burke, E. A. J. 1971. *Tables for Microscopic Identification of Ore Minerals* (2nd ed.). New York: Elsevier, 430 pp.

Winchell, A. N. 1929. *Elements of Optical Mineralogy, Part III. Determinative Tables* (2nd ed.). New York: Wiley, 231 pp.

Luster metallic; $H < 2\frac{1}{2}$

Name, Formula	Color	Streak	SG	H	System and Habit	Cleavage	Observations
Graphite C (p. 248)	Black	Black; black on glazed porcelain	2.09–2.23	1–2	Hexagonal; hexagonal plates or massive, foliated or earthy	{0001} perfect	Greasy feel; plates flexible, inelastic; marks paper; commonly occurs with calcite in metamorphic rocks
Stibnite Sb_2S_3 (p. 270)	Black	Gray	4.6	2	Orthorhombic; prismatic crystals elongated ∥ c; Bladed masses	{010} perfect	Brittle; occurs in quartz veins
Molybdenite MoS_2 (p. 279)	Blue-gray	Blue-gray; greenish black on glazed porcelain	4.62–4.73	1–1½	Hexagonal; hexagonal plates or massive, foliated	{0001} perfect	Like graphite, but heavier and more metallic; marks paper; generally associated with quartz or feldspar
Sylvanite $(Ag,Au)Te_2$ (p. 280)	White to gray	Gray	8.2	1½–2	Monoclinic; prismatic and tabular crystals, also massive bladed	{010} perfect	Occurs in quartz veins with gold on telluride minerals

Luster metallic; $H = 2\frac{1}{2}-5\frac{1}{2}$

Name, Formula	Color	Streak	SG	H	System and Habit	Cleavage	Observations
Goethite $FeO(OH)$ (p. 319)	Brown, black	Yellow to brown	3.3–4.3	$5–5\frac{1}{2}$	Orthorhombic; typically massive, botryoidal or stalactitic	{010} perfect; {100} good	Earthy varieties known as *limonite;* cement and coloring matter in sedimentary rocks; forms pseudomorphs after pyrite
Chalcopyrite $CuFeS_2$ (p. 261)	Brass-yellow (iridescent tarnish)	Greenish, black	4.1–4.3	$3\frac{1}{2}–4$	Tetragonal; sphenoidal crystals; and typically massive, compact	{011} indistinct	Commonly tarnished blue or gray; occurs with pyrite, galena, sphalerite; yellower and softer than pyrite
Manganite $MnO(OH)$ (p. 318)	Black	Brown	4.3	4	Monclinic (pseudo-orthorhombic); prismatic crystals, commonly elongated ∥ c	{010} perfect; {110}, {001} good	Occurs in hydrothermal deposits; alters to pyrolusite, the streak then becoming black
Enargite Cu_3AsS_4 (p. 283)	Black	Black	4.4	3	Orthorhombic; tabular or prismatic crystals; also massive, compact or granular	{110} perfect; {100}, {010} distinct	Widely associated with chalcocite and bornite
Pyrrhotite $Fe_{1-x}S$ (p. 264)	Bronze-yellow	Gray-black	4.58–4.65	$3\frac{1}{2}–4\frac{1}{2}$	Hexagonal; typically massive, granular or compact	None	Magnetic; surface generally tarnished brown
"Psilomelane" Romanechite $BaMn_9O_{16}(OH)_4$ (p. 304)	Black	Black	4.7	5–6	Monoclinic; massive, commonly botryoidal	None visible	Secondary mineral, weathering product of other Mn minerals
Covellite CuS (p. 267)	Indigo-blue to black	Black-gray	4.6–4.76	$1\frac{1}{2}–2$	Hexagonal; hexagonal plates; also massive, foliated or granular	{0001} perfect	Uncommon copper mineral; generally associated with chalcocite or enargite
Pentlandite $(Fe,Ni)_9S_8$ (p. 266)	Bronze-yellow	Brown	4.6–5.0	$3\frac{1}{2}–4$	Isometric; massive	None; parting on {111}	Occurs intermixed with pyrrhotite

Luster metallic; $H = 2\frac{1}{2}-5\frac{1}{2}$ (*Continued*)

Tetrahedrite $Cu_{12}Sb_4S_{13}$ Tennantite $Cu_{12}As_4S_{13}$ (p. 283)	Black	Brown to black	4.6–5.1	$3-4\frac{1}{2}$	Isometric; tetrahedral crystals; also massive, granular or compact	None	Widely distributed in hydrothermal Cu deposits; commonly contains Ag
Bornite Cu_5FeS_4 (p. 253)	Bronze-brown, purple tarnish	Gray-black	5.07	3	Isometric; typically massive, granular or compact	{111} in traces	Widely associated with chalcopyrite or chalcocite; purple tarnish characteristic
Pyrolusite MnO_2 (p. 307)	Gray to black	Black	5.0–5.2	to $5\frac{1}{2}$	Tetragonal; massive, columnar, fibrous, or powdery	{110} perfect	Secondary after other Mn minerals; typically soft and earthy
Millerite NiS (p. 266)	Pale brass-yellow	Greenish black	5.5	$3-3\frac{1}{2}$	Trigonal; typically as hairlike crystals, generally in radiating groups	{10$\bar{1}$1}, {01$\bar{1}$2} perfect	Commonly occurs in cavities associated with calcite or siderite
Chalcocite Cu_2S (p. 251)	Gray to black	Gray to black	5.5–5.8	$2\frac{1}{2}-3$	Orthorhombic; typically massive, compact or granular	{110} indistinct	Sectile; typically has sooty-appearing coating
Arsenic As (p. 240)	Gray-white, tarnishing to dark gray	Gray	5.7	$3\frac{1}{2}$	Hexagonal; typically massive, often botyroidal	{0001} perfect	Volatilizes without melting
Pyrargyrite Ag_3SbS_3 Proustite Ag_3AsS_3 (p. 282)	Deep red to black	Red	5.6–5.9	$2-2\frac{1}{2}$	Trigonal; prismatic crystals; also massive, granular	{10$\bar{1}$1} good	Widely associated with galena in Pb-Ag ores
Bournonite $PbCuSbS_3$ (p. 284)	Gray to black	Gray-black	5.83	$2\frac{1}{2}-3$	Orthorhombic; short prismatic or tabular crystals; also massive, granular or compact	{010} indistinct	Occurs in veins with other sulfides, commonly with galena and tetrahedrite

Luster metallic; $H = 2\frac{1}{2}-5\frac{1}{2}$ (*Continued*)

Boulangerite $Pb_5Sb_4S_{11}$ (p. 285)	Bluish lead-gray	Brownish gray	6.23	$2\frac{1}{2}-3$	Monoclinic; long prismatic crystals, also fibrous and hairlike	{100} good	Occurs in vein deposits associated with other Pb minerals
Antimony Sb (p. 241)	White to gray	Gray	6.7	$3-3\frac{1}{2}$	Hexagonal; massive, lamellar	{0001} perfect	Commonly coated with white antimony oxide
Argentite Ag_2S (p. 251)	Black	Dark gray-black	7.2– 7.4	$2-2\frac{1}{2}$	Isometric; cubes or octahedra; also massive	{001}, {011} indistinct	Sectile; occurs with other silver minerals
Wolframite $(Fe,Mn)WO_4$ (p. 364)	Brown-ish black	Brown-black	7.1– 7.5	$4-4\frac{1}{2}$	Monoclinic; crystals commonly tabular ‖ {100}; also massive, bladed or compact	{010} perfect	Resembles columbite-tantalite, but distinguished by perfect cleavage; occurs in quartz veins
Galena PbS (p. 254)	Lead-gray	Lead-gray	7.6	$2\frac{1}{2}$	Isometric; cubes or cubo-octahedra; generally massive, granular	{001} perfect	The commonest lead mineral; commonly occurs with sphalerite
Iron Fe (p. 239)	Gray	Gray	7.8	4	Isometric; massive	{100} not prominent	Generally as meteorites, which always contain Ni; magnetic
Nickeline NiAs (p. 265)	Copper-red	Brown-black	7.8	$5-5\frac{1}{2}$	Hexagonal; typically massive, compact or granular	None	Generally coated with green annabergite
Copper Cu (p. 236)	Copper-red, typically tarnished	Red; metallic	8.9	$2\frac{1}{2}-3$	Isometric; massive and in dendritic or wirelike forms	None	Commonly in cavities in basic igneous rocks; malleable
Calaverite $AuTe_2$ (p. 280)	Brass-yellow	Green-gray	9.2	$2\frac{1}{2}-3$	Monoclinic; bladed crystals or massive	None	Melts easily giving globules of Au
Bismuth Bi (p. 241)	Silver-white, cream, reddish	Silver-white	9.7– 9.8	$2-2\frac{1}{2}$	Hexagonal; indistinct crystals, commonly hopper-shaped; typically in platy masses	{0001} perfect	Occurs in quartz veins and some pegmatites; generally has brassy tarnish

Luster metallic; $H = 2\frac{1}{2}-5\frac{1}{2}$ (*Continued*)

Silver Ag (p. 235)	Silver-white, commonly tarnished black	Silver	10.1–11.1	$2\frac{1}{2}$–3	Isometric; commonly in dendritic or wirelike forms	None	Occurs widely with barite and calcite in veins; tarnishes gray or black; malleable
Platinum Pt (p. 237)	Steel-gray	Steel-gray	14–19	4–$4\frac{1}{2}$	Isometric; typically as nuggets or grains	None	Typically as grains in alluvial deposits; may be magnetic because it contains iron
Gold Au (p. 232)	Yellow	Yellow	15–19	$2\frac{1}{2}$–3	Isometric; typically in dendritic forms, or as alluvial grains or nuggets	None	Easily recognized by its color, softness, and malleability

Luster metallic; $H > 5\frac{1}{2}$

Name, Formula	Color	Streak	SG	H	System and Habit	Cleavage	Observations
Anatase TiO_2 (p. 310)	Commonly brown	White	3.9	$5\frac{1}{2}$–6	Tetragonal; typically in acute pyramidal crystals	{001}, {101} perfect	Occurs in small crystals in joint planes and veins in schists and gneisses
Brookite TiO_2 (p. 310)	Brown to black	White to gray	4.1	$5\frac{1}{2}$–6	Orthorhombic; tabular and prismatic crystals	{120} indistinct	Occurs in same manner as anatase
Rutile TiO_2 (p. 306)	Red-brown	Yellow to gray	4.2–4.3	6–$6\frac{1}{2}$	Tetragonal; typically in long prismatic crystals; also massive, granular	{110} distinct	Common accessory mineral of igneous and metamorphic rocks
Chromite $FeCr_2O_4$ (p. 293)	Black	Brown	4.5–4.8	$5\frac{1}{2}$–6	Isometric; octahedral crystals rare; typically massive, granular or compact	None; {111} parting	Occurs in ultrabasic rocks and serpentinites
Ilmenite $FeTiO_3$ (p. 301)	Black	Black	4.6–4.7	6	Trigonal; tabular crystals; typically massive, compact or granular	None; parting on {0001}, {01$\bar{1}$2}	Occurs associated with basic igneous rocks and in sands derived therefrom

Luster metallic; $H > 5\frac{1}{2}$ (*Continued*)

Braunite $3Mn_2O_3$ $\cdot MnSiO_3$ (p. 302)	Gray to black	Gray to black	4.7–4.8	$6-6\frac{1}{2}$	Tetragonal; pyramidal crystals or massive granular	{112} perfect		Ore mineral of manganese; occurs in veins and in metamorphic deposits
Hausmannite Mn_3O_4 (p. 295)	Brownish black	Brown	4.84	$5\frac{1}{2}-6$	Tetragonal: pyramidal crystals or massive granular	{001} perfect		Occurs in same way as braunite; distinguished from that mineral by brown streak
Marcasite FeS_2 (p. 276)	Light brass-yellow	Gray-brown	4.89	$6-6\frac{1}{2}$	Orthorhombic; complex twinned crystals; also massive, stalactitic or nodular	{101} distinct		Occurs in low-temperature surface and near-surface deposits
Pyrite FeS_2 (p. 273)	Light brass-yellow	Brown-black	5.0	$6\frac{1}{2}$	Isometric; pyritohedra, cubes, or octahedra; also massive, compact or granular	{001} indistinct		The commonest sulfide mineral; alters to limonite
Franklinite $\sim ZnFe_2O_4$ (p. 295)	Black	Brown	5.1–5.3	$5\frac{1}{2}-6\frac{1}{2}$	Isometric; rounded octahedra or massive granular	None; {111} parting		Occurs in metamorphosed limestone at Franklin, N.J.; weakly magnetic
Magnetite Fe_3O_4 (p. 292)	Black	Black	5.2	$5\frac{1}{2}-6\frac{1}{2}$	Isometric; octahedra or massive granular	None; parting on {111}		Accessory mineral of igneous and metamorphic rock, and in sands; strongly magnetic
Hematite Fe_2O_3 (p. 300)	Red to black	Red-brown	5.26	6	Trigonal; tabular crystals; massive, earthy	None; parting on {0001}, {01$\bar{1}$2}		Common iron mineral of sediments and metamorphosed sediments
Arsenopyrite FeAsS (p. 277)	Silver-white	Gray-black	6.1	6	Monoclinic; prismatic crystals; typically massive, columnar or granular	{101} distinct		Generally occurs in metalliferous veins; gives garliclike odor when scratched or pulverized
Cobaltite CoAsS (p. 275)	Silver-white	Gray-black	6.3	$5\frac{1}{2}-6$	Isometric; pyritohedra, cubes, or massive, granular or compact	{001} perfect		Occurs in vein deposits; in some cases coated with pink erythrite

Luster metallic; $H > 5\frac{1}{2}$ (*Continued*)

Skutterudite (Co,Ni)As$_{2-3}$ (p. 281)	Silver-white	Black	6.5	$5\frac{1}{2}$–6	Isometric; cubes and octahedra, also massive, granular or compact	{100}, {111} distinct	Occurs in vein deposits; in some cases coated with pink erythrite
Columbite-tantalite (Fe,Mn) (Nb,Ta)$_2$O$_6$ (p. 311)	Red-brown to black	Brown to black	5.2–8.0	6–6$\frac{1}{2}$	Orthorhombic; tabular or prismatic crystals, also massive	{010} distinct; {100} less distinct	Occurs in granite pegmatites
Cassiterite SnO$_2$ (p. 309)	Brown to black	White	7.0	6$\frac{1}{2}$	Tetragonal; prismatic crystals and massive, botryoidal or concretionary	{100} distinct; {110} indistinct	Occurs in quartz veins and granite pegmatites; also detrital
Uraninite (pitchblende) UO$_2$ (p. 313)	Black	Brown-black	8–10	6$\frac{1}{2}$	Isometric; cubes; also massive, in some cases botryoidal	None	Occurs in granite pegmatites, hydrothermal veins, and disseminated in sandstones and limestones; yellow alteration products commonly present
Sperrylite PtAs$_2$ (p. 275)	Tin-white	Black	10.6	6–7	Isometric; cubic crystals and rounded grains	{001} indistinct	Occurs in basic igneous rocks and in sands derived therefrom

Luster nonmetallic; typically colorless or white unless tinted by impurities; streak white; $H < 2\frac{1}{2}$

Name, Formula	Color	SG	H	System and Habit	Cleavage	Observations
Epsomite MgSO$_4\cdot$7H$_2$O (p. 359)	White	1.67	2	Orthorhombic; in acicular crystals, elongated $\parallel$ c, or powdery crusts	{010} perfect; {101} distinct	Secondary mineral; soluble in H$_2$O; bitter, salty taste
Nitratite NaNO$_3$ (p. 346)	Colorless, white	2.25	1$\frac{1}{2}$–2	Trigonal; typically in granular masses	{10$\bar{1}$1} perfect	Soluble in H$_2$O; cooling and saline taste; occurs in rainless deserts

Luster nonmetallic; typically colorless or white unless tinted by impurities; streak white; $H < 2\frac{1}{2}$ (Continued)

Gypsum $CaSO_4 \cdot 2H_2O$ (p. 355)	Colorless, white; some is tinted pink, yellow, gray	2.32	2	Monoclinic; tabular prismatic crystals, elongated $\parallel c$, tabular $\parallel \{010\}$; swallow-tail twins; commonly granular, massive or fibrous	{010} perfect; {100}, {011} distinct	Cleavage laminae flexible but not elastic
Montmorillonite $\sim Al_2Si_4O_{10}$ $(OH)_2 \cdot xH_2O$ (p. 424)	White, gray, green-gray	2.0–2.7	2	Monoclinic; clayey masses	{001} perfect, not mega-scopically observable	Clay mineral; absorbs water with swelling; may have soapy feel
Brucite $Mg(OH)_2$ (p. 315)	Colorless, white, pale green	2.4	2	Trigonal; in hexagonal plates or massive compact; in some cases fibrous	{0001} perfect	Occurs in association with dolomite or serpentine or in metamorphosed dolostone
Kaolinite $Al_4Si_4O_{10}(OH)_8$ (p. 422)	White, gray, brown	2.6	2	Triclinic; clayey masses	{001} perfect, not usually observable	Clay mineral; clayey odor when breathed upon; does not swell in water
Talc $Mg_3Si_4O_{10}(OH)_2$ (p. 429)	Green	2.82	1	Monoclinic; in foliated or fine-grained masses	{001} perfect	Greasy feel; occurs in Mg-rich metamorphic rocks
Pyrophyllite $Al_2Si_4O_{10}(OH)_2$ (p. 428)	White, yellow, brown	2.84	1	Monoclinic; radiated fibrous or fine grained	{001} perfect	Greasy feel; occurs in Al-rich metamorphic rocks

Luster nonmetallic; typically colorless or white unless tinted by impurities; streak white; $H = 2\frac{1}{2}$–$5\frac{1}{2}$

Name, Formula	Color	SG	H	System and Habit	Cleavage	Observations
Carnallite $KMgCl_3 \cdot 6H_2O$ (p. 327)	Colorless or white	1.60	$2\frac{1}{2}$	Orthorhombic; typically in coarsely granular masses	None	Bitter taste, occurs in salt deposits; deliquescent

Luster nonmetallic; typically colorless or white unless tinted by impurities; streak white; $H = 2\frac{1}{2}-5\frac{1}{2}$ (Continued)

Borax $Na_2B_4O_7 \cdot 10H_2O$ (p. 347)	White	1.71	$2\frac{1}{2}$	Monoclinic; short prismatic crystals	{100}, {110} perfect	Sweetish alkaline taste; deposited from salt lakes; effloresces in dry air
Sylvite KCl (p. 325)	Colorless, sometimes white, light orangish brown, or gray	1.99	$2\frac{1}{2}$	Isometric; cubic crystals; typically massive, granular	{100} perfect	Salty taste like halite, but somewhat bitter, sectile
Opal $SiO_2 \cdot nH_2O$ (p. 401)	Colorless, white	2.0– 2.2	$5\frac{1}{2}$	Amorphous; massive compact, commonly mammillary	None	Secondary silica deposited at low temperatures
Chabazite $CaAl_2Si_4O_{12}$ $\cdot 6H_2O$ (p. 418)	Colorless, white, pink	2.0– 2.2	4	Trigonal; typically in cubelike rhombohedral crystals	{10$\bar{1}$1} indistinct	Occurs in cavities in basalts; crystal form distinguishes it from other zeolites
Stilbite $NaCa_2Al_5Si_{13}O_{36}$ $\cdot 14H_2O$ (p. 417)	White, pale yellow	2.1– 2.2	4	Monoclinic; sheaflike aggregates of twinned crystals	{010} perfect	Occurs in cavities in basalts; crystal form distinguishes it from other zeolites
Halite NaCl (p. 324)	Colorless, white, red	2.16	$2\frac{1}{2}$	Isometric; cubic crystals; typically massive, granular	{100} perfect	Characteristic salty taste
Heulandite $\sim CaAl_2Si_7O_{18}$ $\cdot 6H_2O$ (p. 416)	White, light brown	2.2	3–4	Monoclinic; trapezoidal crystals tabular $\parallel$ {010}	{010} perfect	Pearly luster on cleavage; occurs in cavities in basalts; crystal form distinguishes it from other zeolites
Natrolite $Na_2Al_2Si_3O_{10}$ $\cdot 2H_2O$ (p. 419)	Colorless, white	2.25	5	Orthorhombic; long prismatic crystals and radiating nodular forms	{110} perfect	Occurs in cavities in basalts; crystal form distinguishes it from other zeolites
Analcime $NaAlSi_2O_6$ $\cdot H_2O$ (p. 418)	Colorless, white	2.26	5	Isometric; trapezohedral crystals; also massive, compact or granular	None	Occurs in cavities in basalts with other zeolites, also in some igneous and sedimentary rocks

Luster nonmetallic; typically colorless or white unless tinted by impurities; streak white; $H = 2\frac{1}{2}-5\frac{1}{2}$ (*Continued*)

Laumontite $CaAl_2Si_4O_{12}$ $\cdot 4H_2O$ (p. 417)	White	2.25– 2.30	4	Monoclinic; small prismatic crystals; also massive, commonly powdery	{010}, {110} perfect	Occurs in cavities in basalts; crystal form distinguishes it from other zeolites
Apophyllite $KCa_4(Si_4O_{10})_2$ $(OH,F)\cdot 8H_2O$ (p. 440)	Colorless, white	2.35	5	Tetragonal; prismatic crystals with basal plane and pyramids	{001} perfect	Occurs in cavities in basalts and in hydrothermal veins, generally with zeolites
Gibbsite $Al(OH)_3$ (p. 316)	White, red, brown	2.3– 2.4	$2\frac{1}{2}-3\frac{1}{2}$	Monoclinic; typically massive, compact, pisolitic, or earthy	{001} perfect	A major component of bauxite
Colemanite $Ca_2B_6O_{11}\cdot 5H_2O$ (p. 348)	Colorless, white	2.42	$4\frac{1}{2}$	Monoclinic; short prismatic crystals, generally with complex terminations; also massive	{010} perfect; {001} distinct	Occurs as geodelike masses in sedimentary rocks
Calcite $CaCO_3$ (p. 331)	Colorless, white	2.71	3	Trigonal; scalenohedral and rhombohedral crystals; also massive, granular or compact	{10$\bar{1}$1} perfect	May be colored brown, gray, or black with impurities; effervesces in cold dilute HCl
Alunite $KAl_3(SO_4)_2(OH)_6$ (p. 361)	White, pink	2.6– 2.9	$3\frac{1}{2}-4$	Trigonal; massive, granular or compact	{0001} distinct	Generally occurs as hydrothermal alteration of volcanic rocks
Muscovite $KAl_2(AlSi_3O_{10})$ $(OH)_2$ (p. 432)	Colorless or pale tints	2.8– 2.9	$2\frac{1}{2}-4$	Monoclinic; typically in irregular platy crystals; also massive	{001} perfect	Large crystals in granite pegmatites; small flakes in many metamorphic and some igneous rocks
Dolomite $CaMg(CO_3)_2$ (p. 338)	White, yellow, gray, pink	2.85	$3\frac{1}{2}-4$	Trigonal; small curved rhombohedral crystals and saddle-shaped groups; also massive, granular	{10$\bar{1}$1} perfect	Effervesces in warm dilute HCl or in cold dilute HCl when freshly powdered
Pectolite $NaCa_2Si_3O_8(OH)$ (p. 460)	White	2.86	5	Triclinic; typically in radiating aggregates of needlelike crystals	{001}, {100} perfect	Typically occurs associated with zeolites in cavities in basalt

506   **Luster nonmetallic; typically colorless or white unless tinted by impurities; streak white; $H = 2\frac{1}{2}-5\frac{1}{2}$ (*Continued*)**

Wollastonite $CaSiO_3$ (p. 459)	White	2.9	5	Triclinic; typically in columnar or fibrous masses	{100} perfect; {1̄02}, {001} good	Occurs in metamorphosed limestones	
Aragonite $CaCO_3$ (p. 341)	White	2.93	$3\frac{1}{2}-4$	Orthorhombic; hexagonal twinned crystals; also massive, coralloid	{010} distinct	Effervesces in cold dilute HCl; much less common than calcite	
Cryolite Na_3AlF_6 (p. 327)	White	2.97	$2\frac{1}{2}$	Monoclinic; massive, coarsely granular	None; parting on {001} and {110}	Wet-snow luster; occurs in large amounts at Ivigtut, Greenland, associated with siderite	
Anhydrite $CaSO_4$ (p. 354)	White, gray, pale blue	2.9–3.0	$3\frac{1}{2}$	Orthorhombic; typically massive, granular	{010} perfect; {100}, {001} good	Forms extensive beds in sedimentary rocks; alters to gypsum	
Magnesite $MgCO_3$ (p. 334)	White	3.0	4	Trigonal; typically massive, compact	{101̄1} perfect	Cleavage not seen in compact forms, which tend to have conchoidal fracture; effervesces in warm dilute HCl	
Boehmite $AlO(OH)$ (p. 317)	White, yellow, brown	3.0–3.1	3	Orthorhombic; massive, compact; commonly pisolitic	{010} not visible	Important constituent of bauxite	
Hemimorphite $Zn_4Si_2O_7(OH)_2$ $\cdot H_2O$ (p. 470)	White	3.4–3.5	5	Orthorhombic; typically massive, commonly in mammillary forms	{110} perfect	Alteration product of sphalerite; generally associated with smithsonite	
Periclase MgO (p. 288)	White	3.58	5	Isometric; typically in small rounded grains	{100} perfect	Occurs as small grains commonly altered to brucite in metamorphosed dolomite	
Strontianite $SrCO_3$ (p. 343)	White	3.78	$3\frac{1}{2}$	Orthorhombic; typically massive, columnar to fibrous	{110} good	Effervesces in cold dilute HCl; colors flame red	

Luster nonmetallic; typically colorless or white unless tinted by impurities; streak white; $H = 2\frac{1}{2}-5\frac{1}{2}$ (*Continued*)

Celestite $SrSO_4$ (p. 352)	White, pale blue	3.9– 4.0	3	Orthorhombic; crystals tabular‖{001}, also massive, often fibrous	{001} perfect; {210} good	Red flame coloration
Witherite $BaCO_3$ (p. 342)	White	4.3	3	Orthorhombic; hexagonal twinned crystals, also massive, columnar or granular	{010} distinct	Effervesces in cold dilute HCl; colors flame green
Smithsonite $ZnCO_3$ (p. 337)	White, yellow, blue, green	4.4	4	Trigonal; typically massive compact, commonly concretionary	{10$\bar{1}$1} good	Effervesces in warm dilute HCl
Barite $BaSO_4$ (p. 351)	Colorless, white, gray	4.5	3	Orthorhombic; crystals typically tabular‖{001}, also massive, granular	{001}, {210} perfect	Green flame coloration
Scheelite $CaWO_4$ (p. 365)	White, yellow, brown	6.1	$4\frac{1}{2}$–5	Tetragonal; typically massive, granular	{101} distinct	Fluoresces white or yellow in ultraviolet light
Anglesite $PbSO_4$ (p. 354)	White, gray	6.2– 6.4	3	Orthorhombic; crystals prismatic or tabular‖{001}; also massive, compact	{001} good; {210} distinct	Commonly has a nucleus of galena
Cerussite $PbCO_3$ (p. 343)	White, gray	6.5– 6.6	3–$3\frac{1}{2}$	Orthorhombic; in latticelike aggregates of twinned crystals; also massive, granular	{110}, {021} distinct	Effervesces weakly in cold dilute HCl, strongly in warm dilute HCl
Mimetite $Pb_5(AsO_4)_3Cl$ (p. 376)	White, yellow	7.2– 7.3	$3\frac{1}{2}$–4	Hexagonal; in small prismatic crystals; also massive, commonly globular and botryoidal	None	Occurs in the oxidized zone of lead deposits

508 Luster nonmetallic, typically colorless or white unless tinted by impurities; streak white; $H > 5\frac{1}{2}$

Name, Formula	Color	SG	H	System and Habit	Cleavage	Observations
Tridymite SiO_2 (p. 399)	Colorless, white	2.26	$6\frac{1}{2}$–7	Hexagonal; small thin hexagonal plates, commonly twinned	None	Occurs in cavities in volcanic rocks
Cristobalite SiO_2 (p. 400)	White	2.32	$6\frac{1}{2}$	Isometric; minute octahedra or small rounded aggregates	None	Occurs in cavities in volcanic rocks
Leucite $KAlSi_2O_6$ (p. 411)	White, gray	2.45	6	Isometric; trapezohedral crystals	{110} indistinct	Occurs as crystals in basic volcanic rocks
Nepheline $NaAlSiO_4$ (p. 412)	White, brown, gray	2.5–2.6	6	Hexagonal; hexagonal prisms, and massive granular	{10$\bar{1}$0} indistinct	Occurs in silica-deficient igneous rocks; gelatinizes in HCl
Orthoclase $KAlSi_3O_8$ (p. 404)	White, flesh to salmon pink	2.56	6	Monoclinic; prismatic crystals, flattened ‖ {010} or elongated ‖ a; also massive, granular	{001} perfect; {010} good	Occurs in many common igneous and metamorphic rocks
Microcline $KAlSi_3O_8$ (p. 406)	White, pink, green	2.56	6	Triclinic; like orthoclase	{001} perfect; {010} good	Like orthoclase, but may show imperfect polysynthetic twinning striations, or grid on {001}
Quartz SiO_2 (p. 395)	Colorless, white, gray, pink, green	2.65	7	Trigonal; prismatic crystals terminated by rhombohedra; also massive, granular or compact (*chalcedony*)	None	Occurs in Si-rich igneous rocks, granite pegmatites, sedimentary and metamorphic rocks
Scapolite $(Na,Ca)_4$ $[(Al,Si)_4O_8]_3$ (Cl,CO_3) (p. 410)	White, yellow, pink, gray	2.56–2.77	6	Tetragonal, prismatic crystals terminated by pyramids; also massive, granular or columnar	{100}, {110} distinct	Occurs in Ca-rich metamorphic rocks
Plagioclase (Na,Ca) $(Al,Si)_4O_8$ (p. 407)	White, gray, brown	2.62–2.76	6	Triclinic; prismatic crystals, flattened ‖ {010}; also massive, granular	{001} perfect; {010} good	Polysynthetic twinning, generally visible on {001} cleavage surfaces

Luster nonmetallic, typically colorless or white unless tinted by impurities; streak white; $H > 5\frac{1}{2}$ (Continued)

Amblygonite (Li,Na)Al(PO$_4$) (F,OH) (p. 371)	White, tints of other colors	2.9–3.1	$5\frac{1}{2}$–6	Triclinic, equant crystals; compact or cleavable masses	{100} perfect; {110} good; {01$\bar{1}$} distinct	Occurs in pegmatites and greisens; commonly has twinning striae on cleavages
Tremolite Ca$_2$Mg$_5$Si$_8$O$_{22}$ (OH)$_2$ (p. 446)	White	3.0	6	Monoclinic; typically as columnar or fibrous aggregates	{110} perfect	Occurs in calcareous and magnesian metamorphic rocks, associated with calcite, dolomite, talc
Datolite CaBSiO$_4$(OH) (p. 490)	Colorless, pale yellow, pale green, pink, white	3.0	$6\frac{1}{2}$	Monoclinic; short prismatic crystals; also massive, granular or compact	None	Occurs in cavities in basic igneous rocks, typically associated with zeolites
Anthophyllite (Mg,Fe)$_7$Si$_8$O$_{22}$ (OH)$_2$ (p. 445)	White, gray, pink	2.9–3.3	6	Orthorhombic; massive, bladed or fibrous aggregates	{210} perfect	Occurs in Mg-rich metamorphic rocks
Lawsonite CaAl$_2$Si$_2$O$_7$ (OH)$_2 \cdot$ H$_2$O (p. 467)	White, pink, gray	3.1	8	Orthorhombic; prismatic and tabular crystals; also massive granular	{100}, {001} perfect; {010} indistinct	Occurs in schists and gneisses with glaucophane
Spodumene LiAlSi$_2$O$_6$ (p. 458)	Colorless, white, pink	3.1–3.2	$6\frac{1}{2}$	Monoclinic; long prismatic crystals, also platy masses	{110} perfect; parting on {100}	Occurs in granite pegmatites; will color a flame red
Andalusite Al$_2$SiO$_5$ (p. 484)	White, gray, pink, brown	3.15	$7\frac{1}{2}$	Orthorhombic; coarse prismatic crystals, square cross section; also massive, columnar or granular	{110} good	Many grains contain inclusions symmetrically arranged (var. chiastolite); typically occurs in hornfelses
Sillimanite Al$_2$SiO$_5$ (p. 485)	White, brown, gray	3.24	7	Orthorhombic; typically as fibrous aggregates	{010} perfect	Occurs in high-grade schists and gneisses
Diaspore AlO(OH) (p. 319)	White, gray, brown	3.3–3.4	$6\frac{1}{2}$	Orthorhombic; massive, bladed, foliated, or compact	{010} perfect	Occurs in Al-rich metamorphic rocks

510   Luster nonmetallic, typically colorless or white unless tinted by impurities; streak white; $H > 5\frac{1}{2}$ (*Continued*)

| Diamond
C
(p. 246) | Colorless,
yellow,
pink,
gray,
black | 3.5 | 10 | Isometric; rounded octahedral crystals, commonly twinned | {111}
perfect | Occurs in ultrabasic igneous rocks or sands derived from them |
| Topaz
$Al_2SiO_4(OH,F)_2$
(p. 488) | Colorless,
white,
yellow,
blue | 3.5–
3.6 | 8 | Orthorhombic; prismatic crystals terminated by pyramids; also massive, granular | {001}
perfect | Occurs in veins and cavities in Si-rich igneous rocks, greisens, and pegmatites |

Luster nonmetallic; typically colored; streak white unless otherwise stated; $H < 2\frac{1}{2}$

Name, Formula	Color	SG	H	System and Habit	Cleavage	Observations
Melanterite $FeSO_4 \cdot 7H_2O$ (p. 359)	Pale green	1.90	2	Monoclinic; typically stalactitic or powdery crusts	{001} perfect, {110} distinct	Secondary mineral after iron sulfides; soluble in H_2O; tastes like ink
Vermiculite ~$Mg_3Si_4O_{10}$ $(OH)_2 \cdot 4H_2O$ (p. 425)	Yellow, brown, green	2.4	$1\frac{1}{2}$	Monoclinic; pseudo-hexagonal plates, pseudomorphs after biotite	{001} perfect	Expands enormously when heated; occurs in Mg-rich rocks
Glauconite ~$K(Fe,Mg,Al)_2$ $Si_4O_{10}(OH)_2$ (p. 435)	Dark green to black	2.5– 2.8	2	Monoclinic; rounded grains	{001} perfect; not mega- scopically observable	Streak pale green; occurs in marine sedimentary rocks as pure beds or as grains in limestone, sandstone, mudstone
Vivianite $Fe_3(PO_4)_2$ $\cdot 8H_2O$ (p. 370)	Blue, blue- green	2.7	$1–1\frac{1}{2}$	Monoclinic; in prismatic crystals, elongated $\parallel c$, or earthy masses	{010} perfect	Color darkens on exposure; streak blue; earthy masses in clay and fossil bones; crystals associated with iron sulfides in ore deposits
Chlorite ~$(Mg,Fe,Al)_{5–6}$ $(Al,Si)_4O_{10}$ $(OH)_8$ (p. 437)	Green	2.6– 3.4	2	Monoclinic; in pseudohexagonal plates or micaceous masses	{001} perfect	Cleavage laminae flexible but not elastic; common in schists

Luster nonmetallic; typically colored; streak white unless otherwise stated; $H < 2\frac{1}{2}$ (Continued)

Erythrite $Co_3(AsO_4)_2$ $\cdot 8H_2O$ (p. 370)	Carmine-red	3.06	$1\frac{1}{2}$–$2\frac{1}{2}$	Monoclinic; prismatic crystals and powdery masses	{010} perfect	Streak pink; occurs as coatings or crusts on cobalt arsenides
Autunite $Ca(UO_2)_2(PO_4)_2$ $\cdot 10$–$12H_2O$ (p. 380)	Lemon-yellow	3.1–3.2	2–$2\frac{1}{2}$	Tetragonal; thin tabular on {001}; scaly aggregates	{001} perfect	Streak yellow; occurs as an oxidation product from veins or pegmatites containing uraninite
Torbernite, Metatorbernite $Cu(UO_2)_2(PO_4)_2$ $\cdot 8$–$12H_2O$ (p. 380, 381)	Emerald-green to apple-green	3.2–3.7	2–$2\frac{1}{2}$	Tetragonal; tabular on {001}; foliated or micaceous	{001} perfect	Streak green; occurs as an oxidation product over veins containing uraninite and copper sulfides
Orpiment As_2S_3 (p. 270)	Yellow	3.49	$1\frac{1}{2}$–2	Orthorhombic; acicular crystals, elongated ∥ c, typically in foliated or granular masses	{010} perfect	Streak yellow; commonly associated with realgar
Realgar AsS (p. 269)	Red	3.56	$1\frac{1}{2}$–2	Monoclinic; short prismatic crystals; typically in granular masses	{010} good	Streak orange; commonly occurs with orpiment
Carnotite $K_2(UO_2)_2(VO_4)_2$ $\cdot 3H_2O$ (p. 381)	Yellow	4–5	2	Monoclinic; typically massive, powdery	{001} perfect, seldom visible	Streak yellow; occurs disseminated in sandstone; like tyuyamunite, but rarer

Luster nonmetallic; typically colored; streak white unless otherwise stated; $H = 2\frac{1}{2}$–$5\frac{1}{2}$

Name, Formula	Color	SG	H	System and Habit	Cleavage	Observations
Sulfur S (p. 242)	Yellow	2.07	$2\frac{1}{2}$	Orthorhombic; pyramidal crystals; typically massive, granular or compact	{001}, {110}, indistinct	Burns, giving choking fumes of SO_2; deposited by fumaroles and occurs in limestones and rock gypsum

Luster nonmetallic; typically colored; streak white unless otherwise stated; $H = 2\frac{1}{2}-5\frac{1}{2}$ (*Continued*)

Chrysocolla $(Cu,Al)_2H_2Si_2O_5$ $(OH)_4 \cdot nH_2O$ (p. 491)	Blue, green	2.0– 2.4	2–4	Orthorhombic (?); massive, compact	None	Secondary mineral in oxidized copper ores
Chalcanthite $CoSO_4 \cdot 5H_2O$ (p. 358)	Blue	2.28	$2\frac{1}{2}$	Triclinic; short prismatic crystals; also massive, granular	$\{1\bar{1}0\}$ indistinct	Streak pale blue; metallic nauseous taste; occurs in oxidized zone of sulfide copper ores
Serpentine $Mg_3Si_4O_{10}(OH)_8$ $\cdot 4H_2O$ (p. 426)	Green, yellow, brown, gray	2.5– 2.6	$2\frac{1}{2}$	Monoclinic; massive, compact or fibrous (asbestos)	None visible	Alteration product of olivine, typically in ultrabasic igneous rocks
Stilpnomelane $\sim K(Fe,Al)_{10}$ $Si_{12}O_{30}(OH)_8$ (p. 439)	Black, greenish black, yellowish to reddish brown	2.59– 2.96	3–4	Monoclinic; foliated masses, thin scales	$\{001\}$ perfect; $\{010\}$ indistinct	Cleavage less perfect than that of biotite, second cleavage perpendicular to first; cleavage folia are brittle; occurs in Fe- and Mn-rich low-grade metamorphic rocks
Lepidolite $\sim KLi_2Al(Si_4O_{10})$ $(OH)_2$ (p. 435)	Colorless, pink, mauve, violet, gray	2.8– 2.9	$2\frac{1}{2}$–3	Monoclinic; massive, coarse-to-fine granular	$\{001\}$ perfect	Pale mauve color often diagnostic; occurs in granite pegmatites
Phlogopite $\sim KMg_3$ $(AlSi_3O_{10})$ $(OH)_2$ (p. 433)	Pale brown	2.8– 2.9	$2\frac{1}{2}$–3	Monoclinic; typically in irregular platy grains	$\{001\}$ perfect	Occurs in ultrabasic rocks and metamorphosed dolostones
Jarosite $KFe_3(SO_4)_2$ $(OH)_6$ (p. 362)	Brown	2.9– 3.3	3	Trigonal; massive, compact	$\{0001\}$ distinct	Resembles limonite, but contains sulfate; yellow streak
Biotite $K(Mg,Fe)_3$ $AlSi_3O_{10}$ $(OH)_2$ (p. 433)	Green, brown, black	2.9– 3.4	$2\frac{1}{2}$	Monoclinic; typically in irregular platy crystals	$\{001\}$ perfect	Occurs in granitic igneous rocks, pegmatites, and metamorphic rocks

Luster nonmetallic; typically colored; streak white unless otherwise stated; $H = 2\frac{1}{2}-5\frac{1}{2}$ 513
(Continued)

Apatite ~$Ca_5(PO_4)_3F$ (p. 373)	Green, blue, brown, yellow, white	3.1– 3.2	5	Hexagonal; prismatic crystals, commonly with "rounded" edges; also massive, granular or compact	{0001} indistinct	Occurs in pegmatites and metamorphosed limestones; also in sedimentary rocks (phosphorite)
Fluorite CaF_2 (p. 325)	Purple, green, yellow, colorless	3.18	4	Isometric; cubic crystals, commonly twinned on (111); also massive, granular	{111} perfect	Common vein mineral, and replacing limestones and dolostones
Kyanite Al_2SiO_5 (p. 486)	Blue, green, white	3.63	4–7	Triclinic; typically as bladed crystals	{100} perfect; {010} good	Occurs in Al-rich schists and gneisses
Rhodochrosite $MnCO_3$ (p. 336)	Pink	3.7	$3\frac{1}{2}$–4	Trigonal; massive, granular	{10$\bar{1}$1} perfect	Effervesces in warm dilute HCl; occurs in ore deposits
Azurite $Cu_3(CO_3)_2(OH)_2$ (p. 346)	Blue	3.77	$3\frac{1}{2}$–4	Monoclinic; crystals of various habit, also massive, spherical aggregates	{011} perfect; {100} distinct	Streak blue; effervesces in cold dilute HCl
Siderite $FeCO_3$ (p. 335)	Yellow, brown	3.8– 4.0	4	Trigonal; rhombohedral aggregates, also massive, granular to compact	{10$\bar{1}$1} perfect	Effervesces in warm dilute HCl; occurs in veins and as replacement of limestone
Antlerite $Cu_3(SO_4)$ $(OH)_4$ (p. 361)	Dark green	3.88	$3\frac{1}{2}$	Orthorhombic; typically in granular aggregates or cross-fiber veinlets	{010} perfect	Streak green; soluble in HCl without effervescence
Malachite $Cu_2(CO_3)(OH)_2$ (p. 345)	Green	3.6– 4.1	$3\frac{1}{2}$–4	Monoclinic; massive compact, commonly stalactitic or botryoidal	{$\bar{2}$01} perfect; {010} distinct	Streak green; effervesces in cold dilute HCl
Goethite (Limonite) $FeO(OH)$ (p. 319)	Yellow, brown	3.3– 4.3	5–$5\frac{1}{2}$	Orthorhombic; massive, compact or earthy, also stalactitic	{010} perfect; {100} good	Streak yellow; dissolves in dilute HCl giving a yellow solution
Brochantite $Cu_4(SO_4)(OH)_6$ (p. 360)	Green	4.0	$3\frac{1}{2}$–4	Monoclinic; prismatic to acicular crystals; also massive granular	{100} perfect	Streak green; dissolves in cold dilute HCl without effervescence

514   Luster nonmetallic; typically colored; streak white unless otherwise stated; $H = 2\frac{1}{2}-5\frac{1}{2}$ (*Continued*)

Sphalerite (Zn,Fe)S (p. 257)	Yellow, brown, black	4.0–4.1	$3\frac{1}{2}-4$	Isometric; typically massive, coarse-to-fine granular	{110} perfect	Dissolves in warm HCl giving off H_2S
Wurtzite ZnS (p. 260)	Brownish black	4.09	$3\frac{1}{2}-4$	Hexagonal; pyramidal or prismatic crystals; typically massive, fibrous, columnar, or as banded crusts	{11$\bar{2}$0} perfect	Streak brown; much less common than sphalerite
Xenotime YPO$_4$ (p. 368)	Yellowish to reddish brown	4.4–5.1	4–5	Tetragonal prismatic with pyramidal faces resembling zircon	{100} perfect	Accessory mineral in acidic igneous rocks, pegmatites, and gneisses
Monazite CePO$_4$ (p. 369)	Yellow, brown	4.6–5.4	$5-5\frac{1}{2}$	Monoclinic; small crystals flattened ‖{100}, also as detrital grains	{100} distinct; parting on {001}	Accessory mineral in granites and pegmatites; as concentrates in sands and sandstones
Pyrochlore–microlite ~NaCa(Nb,Ta)$_2$ O$_6$F (p. 303)	Yellow, brown, black	4.2–6.4	$5-5\frac{1}{2}$	Isometric; octahedral crystals and rounded grains	{111} distinct	Occurs in carbonatites (pyrochlore) and granite pegmatites
Thorite ThSiO$_4$ (p. 478)	Orange, brown, black	4–6.7	5	Tetragonal; pyramidal crystals; also massive, compact	{100} distinct	Strongly radioactive; occurs in pegmatites and as detrital grains
Zincite ZnO (p. 289)	Red	5.7	4	Hexagonal; platy masses or rounded grains	{10$\bar{1}$0} perfect	Streak orange; occurs with franklinite and willemite at Franklin, N. J.
Crocoite PbCrO$_4$ (p. 363)	Orange-red	6.1	$2\frac{1}{2}-3$	Monoclinic; typically as prismatic crystals	{110} distinct	Streak orange-yellow; rare secondary mineral in lead veins
Cuprite Cu$_2$O (p. 287)	Red, often tarnished	6.1	$3\frac{1}{2}-4$	Isometric; cubic crystals, also massive, granular or hairlike	{111} indistinct	Streak red-brown; easily reduced to copper on charcoal
Wulfenite PbMoO$_4$ (p. 366)	Yellow, red, brown	6.5–7.0	3	Tetragonal; typically square, platy crystals, tabular ‖{001}	{101} distinct	Commonly associated with vanadinite

Luster nonmetallic; typically colored; streak white unless otherwise stated; $H = 2\frac{1}{2}$–$5\frac{1}{2}$ 515 (*Continued*)

Name, Formula	Color	SG	H	System and Habit	Cleavage	Observations
Vanadinite $Pb_5(VO_4)_3Cl$ (p. 377)	Yellow, red, brown	6.9	$2\frac{1}{2}$–3	Hexagonal; typically small hexagonal prisms	None	Streak yellow; occurs in the oxidation zone of lead deposits
Pyromorphite $Pb_5(PO_4)_3Cl$ (p. 376)	Green, yellow, brown	7.0–7.1	$3\frac{1}{2}$–4	Hexagonal; in small prismatic crystals; also massive, globular and botryoidal	None	Occurs in the oxidized zone of lead deposits
Cinnabar HgS (p. 267)	Red	8.1	2–$2\frac{1}{2}$	Hexagonal; massive, granular	$\{10\bar{1}0\}$ perfect	Streak red; volatilizes on heating

Luster nonmetallic; typically colored; streak white unless otherwise stated; $H > 5\frac{1}{2}$

Name, Formula	Color	SG	H	System and Habit	Cleavage	Observations
Sodalite $Na_8(AlSiO_4)_6Cl_2$ (p. 413)	Blue, white, pink	2.3	6	Isometric; massive, granular	$\{110\}$ indistinct	Blue color characteristic; occurs in nepheline syenites
Cancrinite $Na_6Ca_2(AlSiO_4)_6$ $(CO_3)_2$ (p. 414)	Yellow, white, pink	2.4–2.5	6	Hexagonal; massive, granular	$\{10\bar{1}0\}$ perfect	Effervesces in warm dilute HCl; yellow color often characteristic
Cordierite $Al_3(Mg,Fe)_2$ $AlSi_5O_{18}$ (p. 465)	Colorless, blue, gray, brown	2.55–2.75	7	Orthorhombic; massive, granular	$\{010\}$ indistinct; parting on $\{001\}$	Occurs in aluminous metamorphic rocks
Turquoise $CuAl_6(PO_4)_4$ $(OH)_8 \cdot 4H_2O$ (p. 378)	Blue, green	2.6–2.8	5–6	Triclinic; massive, compact	$\{001\}$ perfect; $\{010\}$ good but rare	Occurs in veinlets in igneous, metamorphic, and sedimentary rocks
Beryl $Be_3Al_2Si_6O_{18}$ (p. 463)	Colorless, white, green, yellow, blue	2.65–2.85	8	Hexagonal; prismatic crystals or massive granular	$\{0001\}$ distinct	Occurs in granite pegmatites

Luster nonmetallic; typically colored; streak white unless otherwise stated; $H > 5\frac{1}{2}$ (*Continued*)

Prehnite $Ca_2Al_2Si_3O_{10}$ $(OH)_2$ (p. 436)	White, pale to apple-green	2.9	$6\frac{1}{2}$	Orthorhombic; generally massive; encrusting with crystalline surface	{001} good	Occurs in cavities in basic igneous rocks associated with zeolites and in low-grade metamorphic rocks
Melilite $Ca_2Al(AlSiO_7)-$ $Ca_2Mg(Si_2O_7)$ (p. 469)	Nearly colorless, yellow or reddish brown, greenish, gray	2.94–3.05	5–6	Tetragonal; short square prisms	{001}, {110} indistinct	Conchoidal fracture; in metamorphosed carbonate rocks and Si-deficient igneous rocks
Tourmaline Na,Mg,Fe,Al borosilicate (p. 466)	Black, brown, pink, green, blue	3.0–3.2	$7\frac{1}{2}$	Trigonal; prismatic crystals with rounded trigonal cross section; also massive, columnar	{11$\bar{2}$0}, {10$\bar{1}$1} indistinct	Prism faces vertically striated; occurs in granite pegmatites, schists, and gneisses
Glaucophane $Na_2Mg_3Al_2Si_8$ $O_{22}(OH)_2$ (p. 448)	Blue, lavender, blue-black	3.0–3.3	6	Monoclinic; in aggregates of bladed or acicular crystals, or as felt or silky needles	{110} perfect	Occurs in schists and gneisses
Actinolite $Ca_2(Mg,Fe)_5$ $Si_8O_{22}(OH)_2$ (p. 446)	Green	3.0–3.4	6	Monoclinic; commonly in columnar or fibrous aggregates	{110} perfect	Occurs in schists, gneisses, and amphibolites
Hornblende ~$NaCa_2(Mg,Fe,$ $Al)_5(Al,Si)_8O_{22}$ $(OH)_2$ (p. 447)	Green, brown, black	3.0–3.4	6	Monoclinic; long prismatic crystals, also columnar, fibrous or granular	{110} perfect	A common mineral in both igneous and metamorphic rocks
Pumpellyite $Ca_2MgAl_2(SiO_4)$ $(Si_2O_7)(OH)_2$ $\cdot H_2O$ (p. 468)	Bluish green to brown to nearly white	3.18–3.23	$5\frac{1}{2}$–6	Monoclinic; minute fibers or narrow plates	{001} distinct; {100} less distinct	Occurs with prehnite and zeolites in amygdules and low-grade metamorphic rocks

Luster nonmetallic; typically colored; streak white unless otherwise stated; $H > 5\frac{1}{2}$ 517
(Continued)

Humite $(Mg,Fe)_7(SiO_4)_3$ $(F,OH)_2$ (p. 489)	Yellow, honey-yellow, reddish yellow, brownish red	3.15–3.35	$6-6\frac{1}{2}$	Monoclinic (clinohumite and chondrodite); orthorhombic (humite and norbergite); irregular masses	{100} indistinct	Brittle; subconchoidal fracture; occurs in contact metamorphosed dolostones
Axinite Ca,Mn,Fe,Al borosilicate (p. 462)	Brown, yellow	3.2–3.3	7	Triclinic; wedge-shaped crystals; also massive granular	{100} good	Occurs in contact zones of igneous intrusions
Diopside $Ca(Mg,Fe)Si_2O_6$ (p. 454)	White, green	3.25–3.40	6	Monoclinic; prismatic crystals, subspherical grains; massive, granular	{110} good	Occurs in metamorphosed dolostones
Jadeite $NaAlSi_2O_6$ (p. 457)	White, green	3.3	$6-6\frac{1}{2}$	Monoclinic; massive, granular to compact	{110} good	Metamorphic mineral, generally associated with serpentine
Zoisite $Ca_2Al_3Si_3O_{12}$ (OH) (p. 472)	Gray, green, pink	3.3	7	Orthorhombic; massive; columnar or bladed	{001} perfect	Occurs in metamorphic rocks rich in lime and alumina
Allanite $\sim(Ca,Ce)_2(Al,$ $Fe)_3Si_3O_{12}OH$ (p. 474)	Black, dark brown	2.7–4.0	$6\frac{1}{2}$	Monoclinic; platy or columnar crystals; also massive, granular	{001} seldom observable	Streak gray or green-gray; typically somewhat radioactive
Titanite $CaTiSiO_5$ (p. 482)	Brown, yellow, green	3.5	6	Monoclinic; typically in small diamond- or wedge-shaped crystals	{110} distinct	An accessory mineral of igneous and metamorphic rocks
Rhodonite $\sim MnSiO_3$ (p. 461)	Pink	3.5–3.7	6	Triclinic; tabular crystals; massive, granular to compact	{100}, {010} perfect	Commonly veined with black manganese oxides
Hypersthene $(Mg,Fe)SiO_3$ (p. 451)	Brown, black	3.4–4.0	6	Orthorhombic; massive, rarely platy	{210} good	Occurs in basic igneous rocks
Cummingtonite $(Fe,Mg)_7Si_8O_{22}$ $(OH)_2$ (p. 445)	Brown	3.2–3.6	6	Monoclinic; typically massive or as fibrous aggregates	{110} perfect	Occurs in Fe-rich schists and gneisses

Luster nonmetallic; typically colored; streak white unless otherwise stated; $H > 5\frac{1}{2}$
(*Continued*)

Augite (Ca,Na)(Mg,Fe, Al,Ti)(Si,Al)$_2$ O$_6$ (p. 454)	Black, dark green	3.25– 3.55	6	Monoclinic; short prismatic crystals; also massive, granular	{110} good	The common pyrox- ene of igneous rocks
Vesuvianite Ca,Mg,Al silicate (p. 470)	Brown, green, yellow, blue	3.3– 3.5	7	Tetragonal; short pris- matic or low pyram- idal crystals; also massive granular	{110} indistinct	Occurs in metamor- phosed limestones
Riebeckite Na$_2$(Mg,Fe)$_5$ Si$_8$O$_{22}$(OH)$_2$ (p. 448)	Dark blue, black	3.3– 3.6	6	Monoclinic; bladed crystals, also massive granular; in some cases asbestiform	{110} perfect	Blue-gray streak; oc- curs in Na-rich igne- ous rocks
Olivine (Mg,Fe)$_2$SiO$_4$ (p. 475)	Olive- green	3.3– 3.6	$6\frac{1}{2}$	Orthorhombic; granu- lar, massive	{010} indistinct	Occurs in basic and ultrabasic igneous rocks
Clinozoisite, Epidote Ca$_2$(Al,Fe)$_3$ Si$_3$O$_{12}$(OH) (p. 473)	Bilious green to black	3.3– 3.6	7	Monoclinic; prismatic crystals elongated ∥ b; also finely massive, columnar or granular	{001} perfect	Yellow-green color typical; occurs in metamorphic rocks and veins
Hedenbergite Ca(Fe,Mg)Si$_2$O$_6$ (p. 454)	Green to black	3.40– 3.55	6	Monoclinic; massive, granular, or fibrous	{110} good	Occurs in skarn as- sociated with ore deposits
Spinel MgAl$_2$O$_4$ (p. 292)	Black, brown, green, red, blue	3.5– 3.6	$7\frac{1}{2}$–8	Isometric; octahedral crystals, commonly twinned on {111}	None; {111} parting	Occurs in metamor- phic rocks and ultra- basic igneous rocks and in sands derived from them
Chloritoid (Fe,Mg,Mn)$_2$ (Al,Fe)Al$_3$ O$_2$(SiO$_4$)$_2$ (OH)$_4$ (p. 490)	Green (com- monly grassy green)	3.58– 3.61	$6\frac{1}{2}$	Monoclinic; com- monly foliated masses, thin scales, porphyroblasts	{001} perfect; {110} distinct	Cleavage flakes are brittle; occurs in low-to-medium- grade metamorphic Fe- and Al-rich sedi- mentary rocks

Luster nonmetallic; typically colored; streak white unless otherwise stated; $H > 5\frac{1}{2}$ 519
(*Continued*)

Aegirine $NaFeSi_2O_6$ (p. 456)	Dark green, brown, black	3.6	6	Monoclinic; typically in prismatic crystals, often elongated $\parallel c$	{110} perfect	Occurs in Na-rich igneous rocks
Grossular $Ca_3Al_2Si_3O_{12}$ (p. 480)	White, yellow, green, pink, brown	3.6	$7-7\frac{1}{2}$	Isometric; dodecahedra and trapezohedra, and combinations; also massive, granular or compact	None	Occurs in metamorphosed limestones
Pyrope $Mg_3Al_2Si_3O_{12}$ (p. 480)	Red to nearly black	3.6	$7-7\frac{1}{2}$	Isometric; typically in anhedral grains	None	Occurs in high-grade metamorphic rocks
Chrysoberyl $BeAl_2O_4$ (p. 296)	Yellow, green, brown,	3.7–3.8	$8\frac{1}{2}$	Orthorhombic; crystals tabular $\parallel$ {001}, and pseudohexagonal twins	{110} distinct	Occurs in granite pegmatites and in alluvial deposits
Staurolite $(Fe,Mg,Zn)_2Al_9$ $(Si,Al)_4O_{22}$ $(OH)_2$ (p. 487)	Brown	3.7–3.8	7	Orthorhombic; prismatic crystals, commonly twinned in right-angle and diagonal crosses	{010} distinct	Occurs in medium-grade schists and gneisses
Uvarovite $Ca_3Cr_2Si_3O_{12}$ (p. 480)	Green	3.8	$7-7\frac{1}{2}$	Isometric; typically as dodecahedra	None	Generally associated with serpentine and chromite
Andradite $Ca_3Fe_2Si_3O_{12}$ (p. 480)	Yellow, green, brown, black	3.8–3.9	$7-7\frac{1}{2}$	Isometric; dodecahedra and trapezohedra, and combinations; also massive granular	None	Occurs in metamorphosed limestones and associated with ore deposits
Anatase TiO_2 (p. 310)	Brown to colorless	3.9	$5\frac{1}{2}-6$	Tetragonal; typically as acute pyramidal crystals	{001}, {101} perfect	Occurs in small crystals in joint planes and veins in schists and gneisses
Corundum Al_2O_3 (p. 298)	Gray, blue, pink, brown	4.0	9	Trigonal; hexagonal prisms and pyramids; also massive, granular	None; parting on {0001}, {01$\bar{1}$2}	Occurs in metamorphic rocks, and in some Al-rich igneous rocks

Luster nonmetallic; typically colored; streak white unless otherwise stated; $H > 5\frac{1}{2}$ (*Continued*)

Willemite Zn_2SiO_4 (p. 476)	Yellow, green, brown, black	3.9–4.2	$5\frac{1}{2}$–6	Trigonal; massive, granular	{0001} good; {11$\bar{2}$0} indistinct	Much fluoresces green in ultraviolet light
Spessartine $Mn_3Al_2Si_3O_{12}$ (p. 480)	Orange, red, brown	4.0–4.2	7–$7\frac{1}{2}$	Isometric; trapezo-hedra, some in combination with dodeca-hedra; also massive, granular	None	Occurs in granite pegmatites and metamorphosed Mn-rich rocks
Brookite TiO_2 (p. 310)	Brown to black	4.1	$5\frac{1}{2}$–6	Orthorhombic; tabular and prismatic crystals	{120} indistinct	Occurs in joint planes and veins in schists and gneisses
Almandine $Fe_3Al_2Si_3O_{12}$ (p. 480)	Red to black	4.0–4.3	7–$7\frac{1}{2}$	Isometric; dodeca-hedra, trapezohedra, and combinations; also massive, granular	None	Occurs in metamorphic rocks; the most common garnet
Zircon $ZrSiO_4$ (p. 477)	Brown, gray	3.9–4.7	$7\frac{1}{2}$	Tetragonal; prismatic crystals terminated by pyramids, and as rounded grains	{110} distinct; {111} indistinct	An accessory mineral of igneous and metamorphic rocks; a detrital mineral in sands
Rutile TiO_2 (p. 306)	Red-brown to black	4.2–4.5	6–$6\frac{1}{2}$	Tetragonal; commonly as long prismatic crystals or massive	{110} distinct	Common accessory mineral of igneous and metamorphic rocks; also in quartz veins
Cassiterite SnO_2 (p. 309)	Brown to black	7.0	$6\frac{1}{2}$	Tetragonal; prismatic crystals; massive, botryoidal or concretionary	{100} distinct; {110} indistinct	Occurs in quartz veins and granite pegmatites; also detrital

Isotropic Minerals

n Var.	Mineral Name	Color in Thinsection	Page (Part II)
1.433 1.435	Fluorite	Colorless, violet, pale tints (often zoned)	p. 325
1.435[a] 1.460	Opal	Colorless	p. 401
1.483 1.487	Sodalite	Colorless to pale blue	p. 413
1.479 1.493	Analcime	Colorless	p. 418
1.4903	Sylvite	Colorless	p. 325
1.50	Lazurite	Blue	p. 413
1.508[a] 1.511	Leucite	Colorless	p. 411
1.526[a] 1.542	Halloysite	Colorless, opaque	p. 423
1.542	Apophyllite	Colorless	p. 440
1.544	Halite	Colorless	p. 324
1.55[a] 1.59	Antigorite	Colorless, very pale green	p. 426
1.56[a] 1.60	Penninite (a chlorite)	Colorless, pale yellow, pale green	p. 437
1.60[a] 1.65	Prochlorite (a chlorite)	Pale yellow-green, green	p. 437
1.713[b]	Pyrope	Colorless, pale red	p. 480
1.719[b]	Spinel	Colorless, pale colors	p. 292
1.734[b]	Grossular	Colorless	p. 480
1.735 1.745	Periclase	Colorless	p. 288
1.664[b] 1.87	Thorite	Brown, yellow-green	p. 478
1.799[b]	Spessartine	Pink, pale brown	p. 480
1.830[b]	Almandine	Pale pink, light brown	p. 480
1.863[b]	Uvarovite	Pale emerald-green	p. 480
1.887[b]	Andradite	Pale yellow, brownish	p. 480

(continued)

522   Isotropic Minerals (*Continued*)

1.93[a] 2.02	Microlite	Yellow-brown, reddish, green	p. 303	
1.96[a] 2.01	Pyrochlore	Yellow-brown, red-brown	p. 303	
2.0[a] 2.1	Limonite	Yellow to yellow-brown to red-brown	p. 321	
2.071	Cerargyrite (Chlorargyrite)	Colorless	p. 323	
2.4195	Diamond	Colorless	p. 246	
2.37[b] 2.50	Sphalerite	Colorless, pale yellow, pale brown	p. 257	
> 2.72 (Li)	Tennantite	Deep red in very thin splinters	p. 283	
> 2.72 (Li)	Tetrahedrite	Cherry-red in very thin splinters	p. 283	
2.849	Cuprite	Red, orange-yellow	p. 287	

[a] n variation exceeds 0.01
[b] n variation exceeds 0.10

Uniaxial Positive Minerals

n_ω Var.	n_ϵ	Mineral Name	Color in Thinsection	Page (Part II)
1.508 1.511	1.509–1.511	Leucite	Colorless	p. 411
1.531 1.536	1.533–1.538	Apophyllite	Colorless	p. 440
1.526[a] 1.544	1.531–1.553	Chalcedony	Colorless, pale yellow, brown, etc. without pleochroism	p. 397
1.544	1.553	α Quartz	Colorless	p. 395
1.559[a] 1.590	1.580–1.600	Brucite	Colorless	p. 315
1.568[a] 1.585	1.590–1.601	Alunite	Colorless	p. 361
1.651[a] 1.632	1.651–1.640	Akermanite	Colorless	p. 469
1.691	1.719	Willemite	Colorless	p. 476
1.720 1.724	1.816–1.827	Xenotime	O^b = colorless, pale pink, pale yellow, E^c = yellow, gray-brown, yellow-green	p. 368

Uniaxial Positive Minerals (*Continued*)

1.78[a] 1.825	1.79–1.840	Thorite	Shades of green, brown $E > O$	p. 478
1.921	1.9375	Scheelite	Colorless	p. 365
1.920[a] 1.960	1.967–2.015	Zircon	Colorless, patchy brown	p. 477
1.990[a] 2.010	2.091–2.100	Cassiterite	Shades of yellow, red, brown or greenish; $E > O$ (commonly zoned)	p. 309
2.013	2.029	Zincite	Deep red, yellow, not pleochroic	p. 289
2.356 (Na)	2.378	Wurtzite		p. 260
2.506	2.529	Greenockite	Pale yellow, weak pleochroism	p. 261
2.583	2.700	Brookite	Yellow-brown or red-brown to deep brown. $E > O$	p. 310
2.605[a] 2.616	2.890–2.903	Rutile	Red-brown to yellow-brown, $\sim$ opaque, $E > O$	p. 306
2.905	3.256	Cinnabar	Deep red	p. 267

[a] n_ω variation exceeds 0.01
[b] On the tables for uniaxial minerals, O refers to the so-called ordinary ray—that is, the ray vibrating in the plane of the a crystallographic axes.
[c] On the tables for uniaxial minerals, E refers to the so-called extraordinary ray—that is, the ray vibrating parallel to the c crystallographic axis.

Uniaxial Negative Minerals

n_ω Var.	n_ϵ	Mineral Name	Color in Thinsection	Page (Part II)
1.472[a] 1.494	1.470–1.485	Chabazite	Colorless	p. 418
1.487	1.484	Cristobalite	Colorless	p. 400
1.490[a] 1.530	1.488–1.503	Cancrinite	Colorless	p. 414
1.529[a] 1.549	1.526–1.544	Nepheline	Colorless	p. 412
1.537 1.545	1.537–1.544	Apophyllite	Colorless	p. 440
1.538[a] 1.550	1.536–1.541	Marialite	Colorless	p. 410
1.545[a] 1.583	1.525–1.564	Vermiculite	O = yellow-brown, green- brown, E = colorless, pale green	p. 425

524 Uniaxial Negative Minerals (*Continued*)

1.540[a] 1.600	1.535–1.565	Scapolite	Colorless	p. 410
1.577 1.578	1.553–1.555	Autunite	Pale yellow, greenish	p. 380
1.580	1.575	Sodium melilite	Colorless	p. 469
1.56[a] 1.60	1.53–1.57	Hydromuscovite	Colorless	p. 425
1.572[a] 1.592	1.550–1.558	Mizzonite	Colorless	p. 410
1.566[a] 1.602	1.563–1.594	Beryl	Colorless	p. 463
1.587	1.3361	Nitratite	Colorless	p. 346
1.590 1.592	1.581–1.582	Torbernite	O = pale to dark green, E = colorless, pale greenish blue	p. 380
1.557[a] 1.637	1.530–1.590	Phlogopite	O = pale yellow, pale pink-brown, pale green, E = colorless	p. 433
1.592[a] 1.605	1.558–1.564	Meionite	Colorless	p. 410
1.594[a] 1.609	1.564–1.580	Paragonite	Colorless	p. 433
1.58[a] 1.65	1.57–1.62	Delessite (a chlorite)	O = dark green, E = light yellow-green	p. 437
1.61[a] 1.65	1.56–1.61	Glauconite	O = yellow-green, olive-green, blue-green E = pale yellow, green	p. 435
1.633[a] 1.650	1.629–1.646	Fluorapatite	Colorless, pale blue or green, $E > O$	p. 374
1.635[a] 1.658	1.615–1.633	Elbaite	Colorless, pale blue, green, rose, $O > E$	p. 466
1.605[a] 1.696	1.565–1.625	Biotite	O = dark brown, green, red-brown, E = pale yellow, green, brown	p. 433
1.631[a] 1.658	1.610–1.633	Dravite	Colorless, yellow, brown $O \gg E$	p. 466
1.658	1.486	Calcite	Colorless	p. 331
1.650[a] 1.667	1.647–1.665	Chlorapatite	Colorless, pale colors, $E > O$	p. 375
1.576[b] 1.745	1.543–1.634	Stilpnomelane	O = black, dark brown or green, E = colorless, yellow	p. 439

Uniaxial Negative Minerals (*Continued*)

1.670[a] 1.651	1.658–1.651	Gehlenite	Colorless	p. 469
1.658[a] 1.698	1.633–1.675	Schorl	O = black, blue-black, dark green or brown, E = gray, light blue, yellow	p. 467
1.679	1.500	Dolomite	Colorless	p. 338
1.700	1.509	Magnesite	Colorless	p. 334
1.690[a] 1.750	1.510–1.548	Ankerite	Colorless	p. 338
1.702[a] 1.752	1.698–1.746	Vesuvianite	Colorless, light green or brown, $O > E$	p. 470
1.767 1.772	1.759–1.762	Corundum	Colorless, pale colors; red, blue, etc., $O > E$ weak	p. 298
1.816	1.597	Rhodochrosite	O = pale pink, colorless, E = colorless	p. 336
1.815 1.820	1.713–1.715	Jarosite	O = deep yellow-brown, red-brown, E = pale yellow, colorless	p. 362
1.850	1.625	Smithsonite	Colorless	p. 337
1.875	1.633	Siderite	Colorless, pale yellow-brown	p. 335
2.058 (Na)	2.048 (Na)	Pyromorphite	Colorless or pale tints, $E > O$	p. 376
2.124[b] 2.263	2.106–2.239	Mimetite	Colorless, pale yellow, other pale tints, $E > O$	p. 376
2.4053	2.2826	Wulfenite	Yellow to pale orange, $O > E$	p. 366
2.416	2.350	Vanadinite	O = pale tints, E = colorless	p. 377
2.398[b] 2.515	2.260–2.275	Goethite	O = yellow-brown, dark red-orange, E = yellow, colorless	p. 319
2.46 (Li)	2.15 (Li)	Hausmannite	~Opaque, deep red-brown in thin splinters, not pleochroic	p. 295
2.561	2.488	Anatase	Light yellow-brown or red-brown to dark brown, $E > O$	p. 310
~2.7		Ilmenite	Opaque, deep red on thin edges	p. 301
3.084 (Li)	2.881 (Li)	Pyrargyrite	Deep red	p. 282
3.088 (Na)	2.7924 (Na)	Proustite	O = blood-red, E = cochineal-red	p. 282

526   Uniaxial Negative Minerals (*Continued*)

3.15[a] 3.22	2.87–2.94	Hematite	O = deep red-brown, E = yellow-brown	p. 300

[a] n_ω variation exceeds 0.01
[b] n_ω variation exceeds 0.10

Biaxial Positive Minerals

n_β Var.	n_α n_γ	Mineral Name	2V Angle Dispersion	Color in Thinsection	Page (Part II)
1.338	n_α 1.338 n_γ 1.339	Cryolite	~43° $v > r$ weak, horizontal	Colorless	p. 327
1.472 1.475	n_α 1.465–1.467 n_γ 1.494–1.497	Carnallite	~70° $v > r$ weak	Colorless	p. 327
1.469 1.480	n_α 1.469–1.479 n_γ 1.473–1.483	Tridymite	35°–90°	Colorless	p. 399
1.478	n_α 1.471 n_γ 1.486	Melanterite	85° $r > v$ weak, inclined	Colorless, pale green	p. 359
1.476 1.486	n_α 1.473–1.483 n_γ 1.485–1.496	Natrolite	58°–64° $r > v$ weak	Colorless	p. 419
1.487[a] 1.505	n_α 1.487–1.505 n_γ 1.488–1.512	Heulandite	30° $r > v$, crossed	Colorless	p. 416
1.522 1.526	n_α 1.519–1.521 n_γ 1.529–1.531	Gypsum	~58° $r > v$ strong, inclined	Colorless	p. 355
1.526[a] 1.543	n_α 1.520–1.535 n_γ 1.545–1.561	Wavellite	~72° $r > v$ weak	Colorless, pale tints, $X^b > Y^c > Z^d$	
1.531 1.537	n_α 1.527–1.533 n_γ 1.538–1.543	Albite	77°–84° $v > r$ weak	Colorless	p. 403
1.537 1.541	n_α 1.533–1.537 n_γ 1.543–1.547	Oligoclase	84°—90° $v > r$ weak	Colorless	p. 408
1.524[a] 1.574	n_α 1.522–1.560 n_γ 1.527–1.578	Cordierite	90°–75° $r > v$ weak	Colorless, pale blue, $Z > Y > X$	p. 465
1.548 1.558	n_α 1.544–1.555 n_γ 1.551–1.563	Andesine	90°–77° $r > v$ weak	Colorless	p. 408

Biaxial Positive Minerals (*Continued*)

1.558[a] 1.569	n_α 1.555–1.565 n_γ 1.563–1.574	Labradorite	77°–87° $r > v$ weak	Colorless	p. 408
1.569 1.572	n_α 1.565–1.567 n_γ 1.574–1.577	Bytownite	87°–90° $r > v$ weak	Colorless	p. 408
1.566[a] 1.579	n_α 1.561–1.567 n_γ 1.587–1.593	Norbergite	44°–50° $r > v$ weak	Colorless to yellow or brown, $X > Y \sim Z$	p. 489
1.568[a] 1.580	n_α 1.568–1.580 n_γ 1.587–1.600	Gibbsite	0°–40° $r > v$ or $v > r$ strong	Colorless	p. 316
1.576	n_α 1.570 n_γ 1.614	Anhydrite	~43° $v > r$ strong	Colorless	p. 354
1.56[a] 1.60	n_α 1.56–1.59 n_γ 1.57–1.60	Penninite (a chlorite)	Small $v > r$	Colorless, $X \sim Y =$ pale green, $Z =$ pale yellow	p. 437
1.57[a] 1.60	n_α 1.57–1.59 n_γ 1.58–1.61	Clinochlore	15°–45° $v > r$	Colorless, $X \sim Y =$ pale green, $Z =$ pale yellow	p. 438
1.592	n_α 1.586 n_γ 1.614	Colemanite	~55° $v > r$ weak	Colorless	p. 348
1.594 1.600	n_α 1.593–1.599 n_γ 1.597–1.604	Coesite	54°–64°	Colorless	p. 393
1.59[a] 1.62	n_α 1.59–1.60 n_γ 1.61–1.62	Amesite (a chlorite)	10°–20° $v > r$	Colorless	p. 437
1.604[a] 1.615	n_α 1.595–1.610 n_γ 1.632–1.645	Pectolite	50°–63° $r > v$ weak	Colorless	p. 460
1.614 1.620	n_α 1.611–1.617 n_γ 1.632–1.639	Hemimorphite	44°–47° $r > v$ strong	Colorless	p. 470
1.609[a] 1.631	n_α 1.606–1.630 n_γ 1.616–1.638	Topaz	48°–68° $r > v$ moderate	Colorless	p. 488
1.62	n_α 1.61 n_γ 1.65	Turquoise	~40° $v > r$ strong	$X \sim$ colorless, $Z =$ pale blue to pale green	p. 378
1.623 1.624	n_α 1.621–1.622 n_γ 1.630–1.633	Celestite	~50° $v > r$ moderate	Colorless, pale blue, $Z > Y > X$	p. 352
1.602[a] 1.655	n_α 1.592–1.643 n_γ 1.619–1.675	Chondrodite	64°–90° $r > v$ weak	Colorless to yellow or brown, $X > Y \sim Z$	p. 489

Biaxial Positive Minerals (*Continued*)

1.602[a] 1.656	n_α 1.5788–1.616 n_γ 1.6294–1.675	Vivianite	83°–63° $v > r$ weak, horizontal	X = blue, Y = pale yellow-green or blue-green, Z = pale yellow-green	p. 370
1.615[a] 1.647	n_α 1.610–1.637 n_γ 1.632–1.673	Prehnite	60°–70° $r > v$ or $v > r$ weak to strong	Colorless	p. 436
1.61[a] 1.65	n_α 1.60–1.65 n_γ 1.61–1.66	Prochlorite (a chlorite)	20°–50° $v > r$	$X \sim Y$ = pale yellow-green, Z = pale to dark green	p. 437
1.602[a] 1.672	n_α 1.588–1.663 n_γ 1.613–1.683	Anthophyllite	79°–90° $r > v$ or $v > r$ weak to moderate	Colorless to pale brown or green, weak to moderate pleochroism, $Z > Y = X$	p. 445
1.619[a] 1.653	n_α 1.607–1.643 n_γ 1.639–1.675	Humite	65°–84° $r > v$ weak	Colorless to yellow or brown, $X > Y \sim Z$	p. 489
1.636 1.639	n_α 1.634–1.637 n_γ 1.646–1.649	Barite	36°–40° $v > r$ weak	Colorless, pale tints, $Z > Y > X$	p. 351
1.62 1.66	n_α 1.61–1.66 n_γ 1.63–1.67	Pargasite	70°–90° $r > v$ weak	X = colorless, Y = blue-green, Z = deep blue-green	p. 443
1.622[a] 1.676	n_α 1.610–1.667 n_γ 1.632–1.684	Gedrite	70°–90° $r > v$ or $v > r$ weak to moderate	Colorless, pale brown to green, weak to moderate pleochroism, $Z > Y = X$	p. 445
1.65 1.66	n_α 1.64–1.65 n_γ 1.65–1.67	Boehmite	~80°	Colorless	p. 317
1.654[a] 1.670	n_α 1.653–1.661 n_γ 1.669–1.684	Sillimanite	20°–30° $r > v$ strong	Colorless, light brown or yellow $Z > Y > X$	p. 485
1.640[a] 1.676	n_α 1.628–1.658 n_γ 1.652–1.692	Cummingtonite	70°–90° $v > r$ weak	Colorless to pale green or brown, weak pleochroism in yellow, brown, green, $Z > Y \geq X$	p. 445
1.658 1.663	n_α 1.622–1.629 n_γ 1.681–1.701	Erythrite	~90° $r > v$	X = pale pink, Y = pale violet, Z = red	p. 370

Biaxial Positive Minerals (*Continued*)

1.651[a] 1.673	n_α 1.635–1.653 n_γ 1.670–1.690	Forsterite	82°–90° $v > r$ weak	Colorless to pale green, $X = Z$ = pale yellow-green, Y = yellow-orange	p. 475
1.655[a] 1.669	n_α 1.654–1.664 n_γ 1.665–1.675	Enstatite	35°–90° $r > v$ weak	Colorless	p. 451
1.655[a] 1.670	n_α 1.648–1.663 n_γ 1.662–1.679	Spodumene	54°–69° $v > r$ weak	Colorless	p. 458
1.659[a] 1.674	n_α 1.654–1.665 n_γ 1.667–1.688	Jadeite	70°–75° $v > r$ moderate	Colorless	p. 457
1.618[a] 1.714	n_α 1.610–1.700 n_γ 1.630–1.730	Hornblende	85°–90° $v > r$ or $r > v$ moderate	Green, blue-green, brown, strong pleochroism	p. 447
1.636[a] 1.709	n_α 1.623–1.702 n_γ 1.651–1.728	Clinohumite	52°–90° $r > v$ weak	Colorless to yellow or brown $X > Y \sim Z$	p. 489
1.672 1.676	n_α 1.665 n_γ 1.684–1.686	Lawsonite	76°–87° $r > v$ strong	Colorless, rarely pale blue-green $X > Y > Z$	p. 467
1.672 1.681	n_α 1.664–1.672 n_γ 1.694–1.702	Diopside	58°–57° $r > v$ weak	Colorless, pale green Weak pleochroism	p. 454
1.677[a] 1.689	n_α 1.670–1.689 n_γ 1.684–1.696	Lithiophilite	63°–0° $v > r$ strong	Colorless, X = deep pink, Y = pale greenish yellow, Z = pale buff	
1.670[a] 1.700	n_α 1.662–1.691 n_γ 1.688–1.718	Omphacite	58°–83° $r > v$ moderate	X = colorless, $Y = Z$ = very pale green	p. 458
1.688[a] 1.711	n_α 1.685–1.707 n_γ 1.697–1.725	Zoisite	0°–60° $v > r$ (α) or $r > v$ (β) strong	Colorless	p. 472
1.690[a] 1.712	n_α 1.690–1.702 n_γ 1.702–1.719	Riebeckite	0°–90° $v > r$ strong	X = dark blue, Y = dark gray-blue, Z = yellow brown	p. 448
1.672[a] 1.741	n_α 1.671–1.735 n_γ 1.703–1.761	Augite	25°–60° $r > v$ weak, inclined	Colorless, neutral gray, pale green, brown-violet	p. 454
1.705[a] 1.725	n_α 1.682–1.706 n_γ 1.730–1.752	Diaspore	84°–86° $v > r$ weak	Colorless, rarely pale tints	p. 319

530   Biaxial Positive Minerals (*Continued*)

1.675[a] 1.754	n_α 1.674–1.748 n_γ 1.688–1.764	Pumpellyite	10°–85° (rarely −) $v > r$, rarely $r > v$	Blue-green to yellow, red-brown $Y > Z > X$	p. 468
1.707[a] 1.725	n_α 1.703–1.715 n_γ 1.709–1.734	Clinozoisite	14°–90° $v > r$ strong	Colorless	p. 473
1.742[a] 1.710	n_α 1.722–1.700 n_γ 1.758–1.730	Aegirine-augite	90°–70° $r > v$ moderate-strong	X = bright green, Y = yellow-green, Z = pale yellow	p. 456
1.715[a] 1.741	n_α 1.711–1.738 n_γ 1.724–1.751	Rhodonite	61°–76° $v > r$ weak, inclined	Colorless to pale pink, X = orange, Y = rose pink, Z = light yellow-orange	p. 461
1.73	n_α 1.73 n_γ 1.74	Ottrelite (a chloritoid)	Moderate $r > v$ strong	X = olive green, Y = blue, Z = yellow-green	p. 490
1.730 1.736	n_α 1.722–1.728 n_γ 1.750–1.757	Hedenbergite	62°–63° $r > v$ strong	X = dark green, blue-green, Y = brown-green, blue-green Z = yellow-green	p. 454
1.738	n_α 1.726 n_γ 1.789	Antlerite	53° $v > r$ very strong	X = yellow-green, Y = blue-green, Z = green	p. 361
1.734[a] 1.749	n_α 1.732–1.747 n_γ 1.741–1.758	Chrysoberyl	70°–10° $r > v$ weak	Colorless X = pale rose, Y = greenish yellow, Z = pale green	p. 296
1.740[a] 1.754	n_α 1.736–1.747 n_γ 1.745–1.762	Staurolite	80°–90° $r > v$ weak-moderate	X = colorless, light yellow, Y = yellow, yellow-brown, Z = golden yellow, red-brown	p. 487
1.700[e] 1.815	n_α 1.690–1.791 n_γ 1.706–1.828	Allanite	90°–57° $r > v$ strong (also $v > r$)	Light to dark brown, yellow, green, $Z > Y > X$	p. 474
1.730[a] 1.807	n_α 1.725–1.794 n_γ 1.750–1.832	Piemontite	50°–86° $r > v$ strong ($v > r$ rare)	X = pale yellow, pink, Y = pale violet, red-violet, Z = dark red-brown	p. 473

Biaxial Positive Minerals (*Continued*)

1.754	n_α 1.730	Azurite	~68	X = clear-blue,	p. 346
1.758	n_γ 1.835–1.838		$r > v$ distinct	Y = azure-blue, Z = deep violet-blue	
2.038	n_α 1.958	Sulfur	~69°	Pale yellow, weakly	p. 242
	n_γ 2.245		$v > r$ weak	pleochroic	
2.22	n_α 2.17–2.20	Huebnerite	~73°	Yellow, orange, red-	p. 364
	n_γ 2.30–2.32			brown, green, red, $Z > Y > X$	
2.17[a]	n_α 2.15–2.20	Tantalite	90°-moderate	Red-brown,	p. 311
2.30	n_γ 2.25–2.35		$v > r$ moderate	$Z > Y = X$, ~opaque	
2.25	n_α 2.25	Manganite	Small	Red-brown,	p. 318
	n_γ 2.35		$r > v$ very strong	~opaque $Z > Y = X$	
2.22	n_α 2.17–2.26	Wolframite	73°–79°	X = Yellow, orange,	p. 364
2.32	n_γ 2.32–2.42			Y = greenish yel-low, red-brown Z = olive, dark red-brown	
2.36	n_α 2.29	Crocoite	57°	$X = Y$ = orange-red,	p. 363
	n_γ 2.66		$r > v$ strong, inclined	Z = blood-red	
2.584	n_α 2.583	Brookite	0°–30°	Yellow-brown or	p. 310
	n_γ 2.700		Crossed very strong	red-brown to deep brown, $Z > Y > X$	

[a] n_β variation exceeds 0.01
[b] On the tables for biaxial minerals, X is the direction of vibration of the fast ray.
[c] On the tables for biaxial minerals, Y is the direction of vibration of the intermediate ray.
[d] On the tables for biaxial minerals, Z is the direction of vibration of the slow ray.
[e] n_β variation exceeds 0.10

Biaxial Negative Minerals

n_β Var.	n_α n_γ	Mineral Name	2V Angle Dispersion	Color in Thinsection	Page (Part II)
1.452	n_α 1.430–1.440	Epsomite	~50°	Colorless	p. 359
1.462	n_γ 1.457–1.469		$v > r$ weak		
1.469	n_α 1.447	Borax	~40°	Colorless	p. 347
	n_γ 1.472		$r > v$ strong, crossed		
1.472	n_α 1.454	Kernite	80°	Colorless	p. 348
	n_γ 1.488		$r > v$ distinct		
1.491[a]	n_α 1.482–1.500	Stilbite	30°–49°	Colorless	p. 417
1.507	n_γ 1.493–1.513		$v > r$, inclined		

Biaxial Negative Minerals (*Continued*)

1.512 1.522	n_α 1.502–1.514 n_γ 1.514–1.525	Laumontite	25°–47° $v > r$ strong, inclined	Colorless	p. 417
1.518 1.519	n_α 1.514–1.516 n_γ 1.521–1.522	Microcline	66°–68° $r > v$	Colorless	p. 406
1.522 1.524	n_α 1.518–1.520 n_γ 1.522–1.525	Orthoclase	35°–60° $r > v$ distinct, horizontal	Colorless	p. 404
1.522 1.529	n_α 1.518–1.524 n_γ 1.522–1.530	Sanidine	18°–42° $r > v$ weak, horizontal	Colorless	p. 404
1.529 1.532	n_α 1.524–1.526 n_γ 1.530–1.534	Anorthoclase	42°–52° $r > v$ weak	Colorless	p. 403
1.539	n_α 1.516 n_γ 1.546	Chalcanthite	56° $v > r$	Pale blue	p. 358
1.541 1.548	n_α 1.537–1.544 n_γ 1.547–1.551	Oligoclase	90°–86° $r > v$ weak	Colorless	p. 408
1.50[a] 1.59	n_α 1.48–1.57 n_γ 1.50–1.60	Montmorillonite	0°–30°	Colorless, tints of yellow, brown or pink, $Z = Y > X$	p. 424
1.530[a] 1.564	n_α 1.529–1.559 n_γ 1.537–1.567	Chrysotile	30°–50°	$X \sim Y$ = colorless, pale yellow-green, Z = yellow, pale green	p. 426
1.524[a] 1.574	n_α 1.522–1.560 n_γ 1.527–1.578	Cordierite	40°–90° $v > r$ weak	Colorless, pale blue, $Z > Y > X$	p. 465
1.545[a] 1.583	n_α 1.525–1.564 n_γ 1.545–1.583	Vermiculite	0°–8° $v > r$ weak	X = colorless, pale green, $Y = Z$ = yellow- brown, green- brown	p. 425
1.559 1.569	n_α 1.553–1.565 n_γ 1.560–1.570	Kaolinite	23°–60° $r > v$ weak	Colorless, ~opaque	p. 422
1.551[a] 1.585	n_α 1.525–1.548 n_γ 1.554–1.587	Lepidolite	0°–58° $r > v$ weak	Colorless	p. 435
1.55[a] 1.59	n_α 1.54–1.58 n_γ 1.55–1.59	Antigorite	~20° $r > v$	Colorless, very pale green	p. 426
1.572 1.580	n_α 1.567–1.573 n_γ 1.577–1.585	Bytownite	90°–76° $v > r$ weak	Colorless	p. 408
1.56[a] 1.60	n_α 1.53–1.57 n_γ 1.56–1.61	Hydromuscovite	Small (0°–5°)	Colorless	p. 425
1.580 1.585	n_α 1.573–1.577 n_γ 1.585–1.590	Anorthite	77°–78° $v > r$ weak	Colorless	p. 403

Biaxial Negative Minerals (*Continued*)

1.575[a] 1.594	n_α 1.538–1.550 n_γ 1.565–1.600	Talc	0°–30° $r > v$	Colorless	p. 429
1.586 1.589	n_α 1.534–1.556 n_γ 1.596–1.601	Pyrophyllite	53°–62° $r > v$ weak	Colorless	p. 428
1.57[a] 1.61	n_α 1.54–1.57 n_γ 1.57–1.61	Illite	0°–30°	Colorless	p. 425
1.56[a] 1.63	n_α 1.53–1.61 n_γ 1.56–1.64	Nontronite	25°–70°	X = yellow-green, $Y = Z$ = bright green, brownish green	p. 424
1.582[a] 1.610	n_α 1.552–1.574 n_γ 1.587–1.610	Muscovite	30°–47° $r > v$ distinct, horizontal	Colorless	p. 432
1.557[a] 1.637	n_α 1.530–1.590 n_γ 1.558–1.637	Phlogopite	0°–15° $v > r$	X = colorless, $Y \sim Z$ = pale yel- low, pale pink- brown, pale green	p. 433
1.597	n_α 1.575–1.585 n_γ 1.598–1.635	Chrysocolla	~24°	Colorless, pale blu- ish green $X > Z$	p. 491
1.587[a] 1.610	n_α 1.575–1.595 n_γ 1.590–1.622	Amblygonite	50°–90° $r > v$ weak, crossed-inclined	Colorless	p. 371
1.594[a] 1.609	n_α 1.564–1.580 n_γ 1.600–1.609	Paragonite	0°–40° $r > v$ moderate	Colorless	p. 433
1.612[a] 1.629	n_α 1.600–1.616 n_γ 1.626–1.640	Tremolite	88°–84° $v > r$ weak	Colorless	p. 441
1.626	n_α 1.605 n_γ 1.633	Jimthompsonite	62° $r > v$ weak	Colorless	p. 384
1.61[a] 1.65	n_α 1.56–1.61 n_γ 1.61–1.65	Glauconite	0°–20° $r > v$, inclined	X = pale yellow, green, $Y = Z$ = yellow- green, olive-green, blue-green	p. 435
1.632	n_α 1.617 n_γ 1.640	Chesterite	71° $r > v$ weak	Colorless	p. 384
1.602[a] 1.672	n_α 1.588–1.663 n_γ 1.613–1.683	Anthophyllite	78°–90° $r > v$ or $v > r$ weak to moderate	Colorless to pale brown or green; weak to moderate pleochroism, $Z = Y > X$	p. 445
1.612[a] 1.663	n_α 1.594–1.647 n_γ 1.618–1.663	Glaucophane	50°–0° $v > r$ weak	X = yellow, Y = violet, Z = blue	p. 448

534   Biaxial Negative Minerals (*Continued*)

1.633[a] 1.644	n_α 1.629–1.640 n_γ 1.638–1.650	Andalusite	71°–86° $v > r$ weak, rarely $r > v$	Colorless, X = pink, yellow, Y = colorless, light green, Z = colorless, olive	p. 484
1.627[a] 1.659	n_α 1.615–1.646 n_γ 1.629–1.662	Wollastonite	35°–63° $r > v$ weak	Colorless	p. 459
1.622[a] 1.676	n_α 1.610–1.667 n_γ 1.632–1.684	Gedrite	75°–90° $r > v$ or $v > r$ weak to moderate	Colorless, pale brown to green, weak to moderate pleochroism, $Z = Y > X$	p. 445
1.605[a] 1.696	n_α 1.565–1.625 n_γ 1.605–1.696	Biotite	0°–25° $v > r$	X = pale yellow, pale green, pale brown, $Y \sim Z$ = dark brown, dark green, dark red- brown	p. 433
1.649 1.654	n_α 1.622–1.626 n_γ 1.666–1.670	Datolite	72°–75° $r > v$ weak	Colorless	p. 490
1.629[a] 1.682	n_α 1.616–1.669 n_γ 1.640–1.686	Actinolite	84°–75° $v > r$ weak	Pale yellow-green, pleochroic in yel- low and green, $Z > Y \geq X$	p. 446
1.576[b] 1.745	n_α 1.543–1.634 n_γ 1.576–1.745	Stilpnomelane	~0°	X = colorless, yellow, $Y = Z$ = black, dark brown, dark green	p. 439
1.658 1.663	n_α 1.622–1.629 n_γ 1.681–1.701	Erythrite	~90° $r > v$	X = pale pink, Y = pale violet, Z = red	p. 370
1.664	n_α 1.516 n_γ 1.666	Strontianite	~7° $v > r$ weak	Colorless	p. 343
1.618[a] 1.714	n_α 1.610–1.700 n_γ 1.630–1.730	Hornblende	35°–90° $v > r$ or $r > v$ moderate	Green, blue-green, brown, strong pleochroism, $Z \geq Y > X$	p. 447
1.676	n_α 1.529 n_γ 1.677	Witherite	16° $v > r$ very weak	Colorless	p. 342

Biaxial Negative Minerals (*Continued*)

1.680	n_α 1.530 n_γ 1.685	Aragonite	18° $v > r$ weak	Colorless	p. 341
1.676 1.684	n_α 1.658–1.664 n_γ 1.692–1.700	Cummingtonite	90°–85° $v > r$ weak	Colorless to pale green or brown; weak pleochroism in yellow brown, green, $Z > Y \geq X$	p. 445
1.669[a] 1.695	n_α 1.664–1.686 n_γ 1.675–1.699	Bronzite	90°–64°	X = pale orange, Y = pale yellow, Z = pale green	p. 451
1.677 1.701	n_α 1.672–1.693 n_γ 1.681–1.704	Axinite	63°–90° $v > r$ strong	Pale purple, brown, yellow; weak pleochroism, $Y > X > Z$	p. 462
1.677[a] 1.710	n_α 1.672–1.700 n_γ 1.684–1.715	Arfvedsonite	0°–70° $v > r$ very strong	Pleochroic in yellow, green to indigo, $X > Y > Z$	p. 448
1.66[a] 1.73	n_α 1.65–1.70 n_γ 1.67–1.73	Hastingsite-Ferrohastingsite	90°–10° $v > r$ moderate	X = yellow, Y = green, Z = dark green	p. 443
1.700 1.696	n_α 1.691–1.685 n_γ 1.707–1.701	Barkevikite (-Ferro-hornblende)	40°–50° $r > v$ weak to strong	X = yellow-brown, Y = red-brown, Z = dark brown	p. 443
1.690[a] 1.712	n_α 1.690–1.702 n_γ 1.702–1.719	Riebeckite	90°–50° $r > v$ strong	X = dark blue, Y = dark gray-blue, Z = yellow-brown	p. 448
1.695[a] 1.722	n_α 1.686–1.710 n_γ 1.699–1.725	Hypersthene	64°–53°	X = salmon-pink, Y = pale brown, yellow, Z = pale bluish green	p. 451
1.720 1.725	n_α 1.712–1.718 n_γ 1.727–1.734	Kyanite	82° $r > v$ weak	Colorless to pale blue; $Z > Y > X$	p. 486
1.719[a] 1.734	n_α 1.713–1.730 n_γ 1.723–1.740	Chloritoid	36°–90° $r > v$ strong	X = gray-green, olive-green, Y = slate-blue, indigo, Z = colorless, light green	p. 490

536   Biaxial Negative Minerals (*Continued*)

1.725[a] 1.784	n_α 1.715–1.751 n_γ 1.734–1.797	Epidote	90°–64° $r > v$ strong	Colorless, pale yellow-green, $Y > Z > X$	p. 473
1.700[b] 1.815	n_α 1.690–1.791 n_γ 1.706–1.828	Allanite	40°–90° $r > v$ strong	Light to dark brown, yellow, green, $Z > Y > X$	p. 474
1.771	n_α 1.728 n_γ 1.800	Brochantite	~77° $v > r$ distinct	Bluish green, weak pleochroism, $Z > Y = X$	p. 360
1.760[a] 1.804	n_α 1.730–1.768 n_γ 1.774–1.816	Hortonolite (an olivine)	72°–62° $r > v$ weak	Colorless to pale yellow-green, $Y > X = Z$	p. 476
1.742[a] 1.820	n_α 1.776–1.722 n_γ 1.836–1.758	Aegirine	58°–90° $r > v$ moderate-strong	X = bright green, Y = yellow-green, Z = yellow	p. 456
1.81	n_α 1.77 n_γ 1.82	Tephroite (an olivine)	70°–64° $r > v$ weak	X = reddish brown, Y = pink, Z = blue-green	p. 476
1.848[a] 1.869	n_α 1.806–1.827 n_γ 1.858–1.879	Fayalite	51°–46° $r > v$ weak	Colorless to pale yellow, red-brown, $X = Z$ = pale yellow, Y = yellow-orange	p. 475
1.861	n_α 1.831 n_γ 1.880	Atacamite	75° $v > r$ strong	X = pale green, Y = yellow-green, Z = green	p. 323
1.875	n_α 1.655 n_γ 1.909	Malachite	~43° $v > r$ distinct, inclined weak	X = colorless, pale green, Y = yellow-green, Z = dark green	p. 345
1.870[a] 1.93	n_α 1.670–1.77 n_γ 1.895–1.97	Tyuyamunite	40°–55° $v > r$ weak	X = colorless, Y = very pale yellow, Z = pale yellow	p. 381
1.901[b] 2.06	n_α 1.750–1.78 n_γ 1.92–2.08	Carnotite	43°–60°	X = colorless, $Y = Z$ = pale yellow	p. 381
2.074	n_α 1.803 n_γ 2.076	Cerussite	~9° $r > v$ strong	Colorless	p. 343

Biaxial Negative Minerals (*Continued*)

2.20	n_α 1.94 n_γ 2.51	Lepidocrocite	~83° $r > v$ weak	X = yellow to colorless, Y = yellow-brown, red-orange, Z = yellow-orange, dark red	p. 318
2.30[b] 2.45	n_α 2.20–2.30 n_γ 2.35–2.45	Columbite	Moderate-90° $v > r$ moderate	Red-brown, $Z > Y = X$, ~opaque	p. 311
2.393[a] 2.409	n_α 2.260–2.275 n_γ 2.398–2.515	Goethite	0°–27° $r > v$ extreme	X = yellow to colorless, Y = yellow-brown, red-orange, Z = yellow-orange, dark red	p. 319
2.684	n_α 2.538 n_γ 2.704	Realgar	39° $r > v$ strong, inclined	X ~ colorless, Y = pale yellow, Z = pale yellow	p. 269
2.81	n_α 2.4 (Li) n_γ 3.02	Orpiment	76° $r > v$ strong	Y = pale yellow, Z = greenish yellow	p. 270

[a] n_α variation exceeds 0.01
[b] n_γ variation exceeds 0.10

Part IV
APPENDIXES

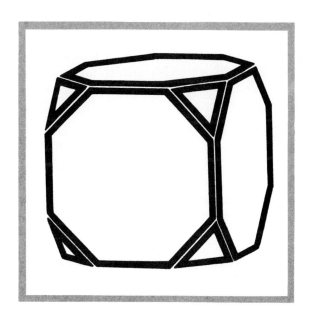

Appendix A
Natural Glasses
and Macerals

Minerals are sometimes confused with natural glasses and macerals, especially during macroscopic examination of rocks. Both of these kinds of nonminerals are widely, though sporadically, distributed as components of many diverse rocks.

NATURAL GLASSES

Glasses are produced when molten silica-bearing mineral matter is quenched. Natural glasses have been formed from melts that have originated as a consequence of the fusion of country rock adjacent to magma (these glasses are called *buchites*); the collision of, for example, meteorites and Earth (*impact melts* and *tektites*); the fusion of rocks by frictional heat in response to faulting (*hyalomylonites*); the striking of sand or rock by lightning (*fulgurites*); and the heat produced by naturally initiated combustion of plant matter (*ash glass*); as well as by the chilling of intrusive or extrusive magma (*glassy igneous rocks*).

The composition of a glass is determined by chemical analyses. The general composition of most glasses—frequently expressed on the basis of SiO_2 content—can be estimated from the glass's index of refraction, dispersion, or specific gravity, particularly if its mode of occurrence and/or the identity of its associated minerals or rocks is also known.

Properties of representative natural glasses are given in Table A1-1.

Occurrences

Obsidian is the name generally applied to massive volcanic glasses other than those derived from basaltic magmas. Although most obsidians are of rhyolitic (i.e., granitic) composition, some

Table A1-1. *Properties of representative natural glasses*

Name:	Lechatelierite[a]	Obsidian[b]	Tachylyte[c]	Impact melt[d]
SiO_2 content				
(an example)	99.0	74.37	49.74	92.88
Structure:	amorphorus	amorphous	amorphous	amorphous
Cleavage:	none	none	none	none
Fracture:	conchoidal	conchoidal	subconchoidal	conchoidal
Hardness:	$5–5\frac{1}{2}$	$5\frac{1}{2}$	5	$5\frac{1}{2}$
Specific gravity:	1.15 ± 0.10	2.363	2.85	2.10
Tenacity:	extremely brittle	very brittle	brittle	brittle
Color:	light green	dark gray	black	white
Streak:	dull white to yellowish brown	white	gray	white
Luster:	vitreous	vitreous	greasy	vitreous
Light transmission:	transparent	translucent	opaque	transparent
Optical properties:				
isotropism	isotropic	isotropic	isotropic	isotropic
index of				
refraction (n)	1.462	1.490	1.605	1.468
Other properties:	has overall shape of a hollow cylinder	transparent in thin slivers	dissolves readily in acid	cellular (pumice-like)

[a] Fulgurite found near South Amboy, New Jersey. (Data from W. M. Myers and A. B. Peck, 1925, *American Mineralogist*, **10**, 152–155.)
[b] Forgia Vecchia, Lipari. (SiO_2 content from H. S. Washington, 1903, *U. S. Geological Survey Professional Paper 14*; specific gravity and *n* from C. E. Tilley, 1922, *Mineralogical Magazine*, **19**, 275–294.)
[c] "Lava dipped from Halemaumau, July 23, 1911," (Kilauea), Hawaii. (SiO_2 content and *n* from Day, A. L., and Shepherd, E. S. 1913, *Smithsonian Institution, Annual Report*, 275–305; specific gravity from Tilley, op. cit.).
[d] Wabar, Rub 'al Kali, Saudi Arabia. (Data from Spencer, L. J. 1933, *Mineralogical Magazine*, **23**, 387–404).

are of dacitic, andesitic, trachytic, or phonolitic compositions. Some of the typical silica contents (weight percent SiO_2) and approximate refractive index values (*n*) for these glasses are:

rhyolitic obsidian:	73% SiO_2,	n = 1.49
dacitic obsidian:	65% SiO_2,	n = 1.52
andesitic obsidian:	59% SiO_2,	n = 1.545
trachytic obsidian:	61% SiO_2,	n = 1.54
phonolitic obsidian:	57% SiO_2,	n = 1.54

As is evident, refractive index appears to correlate rather well with SiO_2 content (see also Figure A1-1). It is especially noteworthy, however, that Huber and Rinehart (1966) found that "the refractive index of glass beads fused from volcanic rocks shows that the relationships may vary appreciably from one petrologic suite to another." This fact should be kept in mind! And, along with silica content, some other general controls obtain; for example, increased amounts of water of hydration tend to increase index of refraction, and increased alkali (Na_2O and K_2O) contents tend to decrease index of refraction.

Tachylyte, sometimes referred to as *sider-*

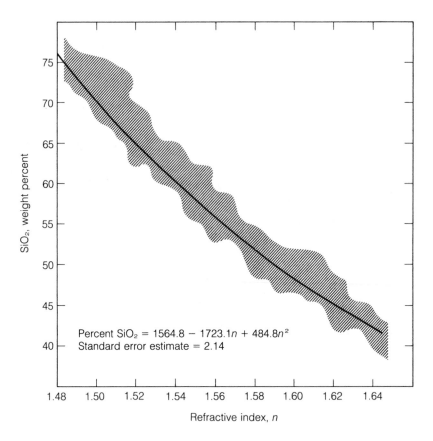

Figure A1-1. Correlation of SiO₂ content and index of refraction for glass beads made by fusing igneous rocks. Although it might differ because fusion alters the original chemical composition of the rock as a consequence of the fact that different constituents have different volatabilities, a curve for natural glasses would probably have a similar form and values. This average curve diagram is modified after Huber and Rinehart (1966), whose original curve was based on 178 points representing fused glass beads from 8 volcanic rock suites. The authors noted that "the standard error of estimate for this curve at the 68 percent confidence level is slightly over 2.1 and at the 95 percent confidence level is nearly 4.3"; the lined area outlines the general limits of data scatter.

In the figure:

$$\text{Percent SiO}_2 = 1564.8 - 1723.1n + 484.8n^2$$

Standard error estimate = 2.14

omelane, is glass of basaltic composition (SiO₂ content ~46–54 weight percent, index of refraction ~1.57–1.61).

Obsidians are typically associated with rhyolites or other consanguineous rocks in volcanic piles or, less commonly, in border zones of dikes and other igneous masses intruded into near-surface, relatively cool country rocks. Obsidian-composition glass also occurs as a constitutent of vitrophyres and of some aphanophyres and tuffs. Tachylyte is rather rare, occurring chiefly as the rinds of some pillow basalts.

Additional glassy rocks of igneous origin include: *pumice,* essentially frothy obsidian, most of which constitutes crusts on lavas or fragments ejected from explosive volcanic centers; *pele's hair,* sometimes called *basaltic pumice,* which consists of brown threadlike fibers of basaltic glass that occur as sporadic matted masses in some volcanic areas; *perlite,* which is apparently partially hydrated obsidian, typically of rhyolitic composition; *pitchstone,* which is partially devitrified and hydrated obsidian; and *palagonite,* which is partially devitrified and hydrated tachylyte.

Buchites have been found to occur in contact zones between basaltic masses and their adjacent country rocks (e.g., shale) and also in xenoliths. They are apparently rather rare. As you might expect, buchites tend to have compositions that reflect the fact that, among rocks, granitic-composition materials have relatively low melting temperatures.

Hyalomylonites are glassy rocks formed when the heat caused by friction associated with faulting is great enough to cause fusion of the rocks involved. These glasses have been recorded here and there from a few fault zones that transect rocks such as granite and arkose. Deep-seated zones where the rocks were already rather hot prior to faulting appear to be prerequisite to the formation of these glasses.

Impact melts and *tektites* may be of essentially any composition, in that any rock material could be involved in the causative high-speed collisions. Hence, the preceding tabulated information should be considered strictly as an example. In most cases, glasses formed from impact melts can be identified on the basis of their occurrence in or adjacent to an impact crater and/or their associated minerals (e.g., native iron-nickel or coesite). Tektites are generally recognizable on the basis of their shapes (e.g., globular or dumbell-like) and the fact that they occur in strewn groups generally consisting of thousands of individual stones. Terrestrial tektites are now believed to represent windblown and deposited splash formed as a result of meteoric (or perhaps comet) collisions with Earth. Many tektites have been given geographically based names (e.g., indochinites, philippinites, and australites).

Fulgurite is the name given to silica glass formations produced where lightning has caused melting and fusion of rock materials and the melt has then been quenched. There are two kinds: sand fulgurites and rock-crust fulgurites. Sand fulgurites, which result from lightning striking unconsolidated sand, are typically long, hollow cylinders, the glass of which is bubble-bearing and has attached sand grains. Rock-crust fulgurites are similarly constituted glass that coats rock exposures that have been hit by lightning. Most sand fulgurites and some rock-crust fulgurites are essentially silica glass, *lechatelierite*.

Ash glass is the name applied to scoriaceous and slaglike glasses, typically greenish to dark gray in color, that have been produced by the heat accompanying combustion of vegetal materials. Several glass stones, up to nearly 15 kg in weight, have been recorded as having been formed as a result of haystack and straw (grain) fires. Although some gramineous species secrete free silica (e.g., epidermal cells of oats may yield opal), much of the silica of at least some of these glasses is apparently derived from subjacent mineral matter. Most ash glass that has been reported has been characterized by a relatively high alkali content (e.g., 16–23 percent total alkalis and up to nearly 14 percent K_2O) and also the presence of carbonaceous matter.

543

MACERALS

Macerals are rock-forming organic units derived by maceration of pieces and products of vegetation. The designation was first suggested by Marie Stopes (1935), who also stated "The word 'macerals' will, I hope, be accepted as a pleasantly sounding parallel to the word 'minerals'. . . ." Subsequently, the term *maceral* has been accepted widely by coal petrographers and, more recently, it has been extended to include the same and similar nonmineral constituents of rocks other than the coals.

Stopes (op. cit.) also suggested that individual macerals should be given distinctive names, each with the suffix -*inite*. This proposal has been followed by most workers in the field.

There is, nonetheless, some confusion relating to how some of the names can and should be applied. Most of the confusion relates to (1) the tendency of coal petrographers, organic geochemists, and micropaleontologists to apply the terms somewhat differently, apparently because of disciplinarian predilections, and (2) the assumption by some workers that they can distinguish and thus name some macerals on the basis of macroscopic examination, whereas other workers steadfastly maintain that the names were given and should be applied only on the basis of microscopic examination (plus or minus a few simple measurements or tests).

General properties of the three main maceral groups, modified after a summary by Crelling and Dutcher (1980), are given in Table A1-2.

Tables A1-3 and A1-4 and the following definitions serve as a useful introduction to the terms for students and perhaps as a reference for nonexperts who deal only occasionally with the pertinent literature. Both macerals and terms for a few related materials are given. We have attempted to make the definitions consistent with most of the recent literature and general usage.

544   Table A1-2. *General properties of the three main maceral groups*

Name of group:	Vitrinite macerals	Liptinite macerals[a]	Inertinite macerals
Derivation:	cell-wall material or woody tissue	waxy and resinous materials (e.g., spores and resins)	altered/degraded matter (e.g., fossil charcoal)
Abundance, in North American coals:	50–90 percent	5–15 percent	< 5–40 percent
Density, g ml^{-1}:	1.3–1.8	1.18–1.28	~1.3–1.9
Coking properties:	constitute bulk of coke mass	contribute to coke; also go to gas and tar	become physically incorporated in coke
Chemistry:	C and H content is intermediate to that of other groups	highest H and lowest C contents of groups	highest C and lowest H contents of groups
Toughness, upon polishing:	give negative relief	give positive relief	tend to give strong positive relief
Reflectance:	intermediate between those of other groups	lowest of three groups	highest of three groups
Fluorescence:	under ultraviolet, some give weak brownish fluorescence	under ultraviolet, all fluoresce	under ultraviolet, no fluorescence

[a] Sometimes referred to as exinite macerals.

Alginite: Liptinite derived from algal remains; most alginites have flowerlike appearances; they fluoresce a brilliant yellow in ultraviolet radiation; they are characteristic of boghead coals.

Anthraxylon: Name formerly applied rather widely to all bright, lustrous components of coal.

Attritus: Name formerly applied rather widely to dull components of coal. (Attritus is typically interlaminated with anthraxylon.)

Bitumen: General designation given flammable hydrocarbons; they are similar to kerogen (see below), but soluble in organic solvents (e.g., carbon disulfide).

Clarain: Coal type, recognizable macroscopically, that consists of intercalated dull and glossy layers; clarain tends to break with smooth fractures that are nearly perpendicular to the bedding (i.e., the intercalations). (Clarain is intermediate between vitrain and durain.)

Coal: Combustible rock consisting of more than 50 weight percent and more than 70 volume percent of carbonaceous material.

Collinite: Featureless vitrinite that fills spaces surrounded by cell walls, which are also vitrinite; it is thought to be derived from jellified vegetal materials.

Cutinite: Liptinite derived from leaf cuticles; cutinite typically occurs as long stringers.

Durain: Coal type, macroscopically recognizable, that ranges from essentially homogeneous to poorly banded and is characterized by a lead-gray to brownish black color and a dull, mattelike luster; it breaks to give granular surfaces.

Exinite: Term used variously as a synonym of liptinite or as a term including only those liptinites that are derived from spores and pollen. (In current, common usage, exinite includes sporinite, cutinite, alginite, and resinite.)

Fusain: Coal type, macroscopically recognizable, that resembles and is sometimes called *mineral charcoal* because of its cellular-porous nature, its powderlike to fibrous appearance, and its silky to satiny luster; most fusain is dirty to the touch.

Fusinite: Cellular inertinite, commonly fragmented, that has a charcoal-like appearance; much fusinite may have been formed from vegetal materials subjected to forest fires.

Huminite: Group name for macerals that exhibit average reflectance and a brownish yellow to reddish brown color in transmitted light, are the precursors of vitrinites, and occur chiefly in low-rank brown coals.

Inertinite: Group name given to macerals that have high reflectance, are C-rich, and are essentially nonreactive (inert) during coking processes (i.e., it includes micrinite, semifusinite, fusinite, and sclerotinite); inertinites are derived from aromatic and/or oxidized components of plant materials.

Inertodetrinite: Clastic fragments of diverse inertinites.

Kerogen: Structureless, plasticlike organic material of marine or terrestrial origin that occurs disseminated in sedimentary rocks (e.g., shales) and can be converted to petroleum by distillation; it is insoluble in nonoxidizing acids, bases, and organic solvents; kerogens are the chief precursors of petroleum.

Liptinite: Name applied to low-reflectance macerals (e.g., alginite, resinite, sporinite, and cutinite) that have retained their original organic forms; kerogens in, for example, oil shales are thought to be derived from liptinites.

Liptodetrinite: Clastic fragments of diverse liptinites.

Macrinite: Inertinite that occurs as structureless ovoid masses with reflectance similar to that of fusinite.

Micrinite: An opaque inertinite that occurs as fine granular structureless particles with high reflectance; particles of micrinite that are greater than approximately 10 microns across are termed macrinites.

Table A1-3. Coal types and constituent macerals*

Rock Types (Overall name given hand specimens)	Macerals (Petrographic units commonly requiring microscopic identification)
Vitrain	Vitrinite
Fusain	Fusinite
Clarain	Vitrinite Xylinite Exinite etc.[a]
Durain	Micrinite Exinite Cutinite etc.[a]

* Modified after Cady (1945).
[a] See Table A1-2 and definitions.

Table A1-4. The common macerals*

Maceral Group	Maceral Individual
Vitrinite	Telinite Collinite Vitrodetrinite
Liptinite (≅ exinite)	Sporinite Cutinite Resinite Alginite Liptodetrinite
Inertinite	Micrinite Macrinite Fusinite Semi-fusinite Sclerotinite Inertodetrinite

* There are also submacerals, maceral varieties, and cryptomacerals.

546

Phyteral: Any recognizable plant fossil preserved in coal.

Resinite: Liptinite derived from resin; resinites commonly occur as small ovoidal or spindle-shaped masses that are translucent orange or greenish; resinites occur in many Cretaceous and younger coals, commonly as cleat and/or joint fillings.

Sclerotinite: Inertinite derived from fungal spores or fruiting bodies; sclerotinites typically occur as spherical or ovoid masses with cell structure.

Semi-fusinite: An inertinite that resembles fusinite but shows a lower reflectance; some semi-fusinite has woody structures preserved.

Sporinite: Liptinite derived from spore exines, which tend to be compressed parallel to the bedding of the enclosing rock; sporinites are typically black in reflected white light and yellow in transmitted light.

Telinite: The cell walls in vitrinite that exhibit cell structure.

Vitrain: Coal type recognizable macroscopically, that is characterized by a uniform brilliant luster, black color, and conchoidal fracture.

Vitrinite: Name applied to uniform gray-to-black anisotropic macerals that exhibit moderate reflectance and constitute the matrix throughout which liptinites and inertinites are disseminated; vitrinites are generally thought to have been derived largely from the humic acid fraction of woody tissues; vitrinites are the major component of vitrain. *Vitrinite reflectance* is a frequently used standard reference value because it varies directly with degree of organic metamorphism.

Vitrodetrinite: Clastic fragments of diverse vitrinites.

Xylinite: A maceral that consists of lignified tissue or xylem; xylinites are constituents of some clarains and appear to be precursors of some vitrinites.

SELECTED REFERENCES

Cady, G. H. 1945. "Coal petrography" *in* Lowry, H. H. (chairman), *Chemistry of Coal Utilization* (Vol. I). New York: Wiley, pp. 86–131.

Crelling, J. C. and Dutcher, R. R. 1980. Principles and Applications of Coal Petrology: *Society Economic Paleontologists and Mineralogists, Short Course Notes*. No. 8, pp. 14–32.

Frondel, Clifford. 1962. *The System of Mineralogy of James Dwight Dana and Edward Salisbury Dana*, Yale University 1837–1892. (7th ed.), *Volume III, Silica Minerals*. New York: Wiley, 334 pp.

Huber, N. K. and Rinehart, C. D. 1966. "Some relationships between the refractive index of fused glass beads and the petrologic affinity of volcanic rock suites." *Geological Society of America Bulletin*, **77**, 101–110.

Spackman, William. 1958. "The maceral concept and the study of modern environments as a means of understanding the nature of coal." *Transactions of the New York Academy of Science*, Ser. II, **20**, 411–423.

Stach, E., Mackowsky, M.-Th., Teichmuller, M., Taylor, G. H., Chandra, D., and Teichmuller, R. 1975. *Stach's Textbook of Coal Petrology* (2nd ed.). Berlin, West Germany: Gebruder Borntraeger, 428 pp.

Stopes, M. C. 1919. "On the four visible ingredients in banded bituminous coal." *Proceedings of the Royal Society*, **90B**, 470–487.

Stopes, M. C. 1935. "On the petrology of banded bituminous coals." *Fuel*, **14**, 4–13.

Tissot, B. P. and Welte, D. H. 1978. *Petroleum Formation and Occurrence, A New Approach to Oil and Gas Exploration*. New York: Springer-Verlag, 538 pp.

Appendix B
Periodic Tables: Atomic Numbers, Atomic Weights, and Ionic Radii

The Periodic Table of the Elements

Number in upper left corner of each box is atomic number. Boldface letter or letters constitute symbol used for element in formulae, etc. Number below symbol is atomic weight based on $^{12}C = 12.000$; numbers in parentheses, given for radioactive elements, give the mass of the most abundant or most stable isotope.

IA	IIA	IIIB	IVB	VB	VIB	VIIB	$\longrightarrow$
1 **H** 1.008							
3 **Li** 6.939	4 **Be** 9.012						
11 **Na** 22.99	12 **Mg** 24.31						
19 **K** 39.102	20 **Ca** 40.08	21 **Sc** 44.956	22 **Ti** 47.90	23 **V** 50.942	24 **Cr** 51.996	25 **Mn** 54.938	26 **Fe** 55.847
37 **Rb** 85.47	38 **Sr** 87.62	39 **Y** 88.906	40 **Zr** 91.22	41 **Nb** 92.906	42 **Mo** 95.94	43 **Tc** 98.906	44 **Ru** 101.07
55 **Cs** 132.905	56 **Ba** 137.33	57 **La** 138.91	72 **Hf** 178.49	73 **Ta** 180.948	74 **W** 183.85	75 **Re** 186.2	76 **Os** 190.2
87 **Fr** (223)	88 **Ra** 226	89 **Ac** (227)					

58 **Ce** 140.12	59 **Pr** 140.908	60 **Nd** 144.24	61 **Pm** (145)	62 **Sm** 150.35	63 **Eu** 151.96
90 **Th** 232.038	91 **Pa** 231.036	92 **U** 238.03	93 **Np** 237.05	94 **Pu** (244)	95 **Am** (243)

								VIIIA
								2 **He** 4.003
				IIIA	IVA	VA	VIA	VIIA
				5 **B** 10.811	6 **C** 12.011	7 **N** 14.007	8 **O** 15.999	9 **F** 18.998	10 **Ne** 20.18

GROUP
−VIII——

IB	IIB	13 **Al** 26.982	14 **Si** 28.086	15 **P** 30.974	16 **S** 32.07	17 **Cl** 35.453	18 **Ar** 39.948

27 **Co** 58.933	28 **Ni** 58.71	29 **Cu** 63.546	30 **Zn** 65.38	31 **Ga** 69.735	32 **Ge** 72.59	33 **As** 74.922	34 **Se** 78.96	35 **Br** 79.904	36 **Kr** 83.80
45 **Rh** 102.905	46 **Pd** 106.4	47 **Ag** 107.868	48 **Cd** 112.41	49 **In** 114.82	50 **Sn** 118.69	51 **Sb** 121.75	52 **Te** 127.60	53 **I** 126.904	54 **Xe** 131.30
77 **Ir** 192.2	78 **Pt** 195.09	79 **Au** 196.967	80 **Hg** 200.59	81 **Tl** 204.37	82 **Pb** 207.19	83 **Bi** 208.980	84 **Po** (~210)	85 **At** (210)	86 **Rn** (222)

64 **Gd** 157.25	65 **Tb** 158.925	66 **Dy** 162.50	67 **Ho** 164.930	68 **Er** 167.26	69 **Tm** 168.934	70 **Yb** 173.04	71 **Lu** 174.97
96 **Cm** (247)	97 **Bk** (247)	98 **Cf** (251)	99 **Es** (254)	100 **Fm** (257)	101 **Md** (258)	102 **No** (259)	103 **Lr** (260)

APPENDIX B.2. Effective Ionic Radii

Values are based on $r\,(^{VI}O^{2}) = 1.40$ Å. Data from R. D. Shannon and C. T. Prewitt, 1969, *Acta Crystallographica*, 925–946, and R. D Shannon, 1976, *Acta Crystallographica*, 751–767. Values are in angstrom units: those set in boldface type are the same in both references; those set in italics are values revised from original Shannon–Prewitt values; those set in regular type are additions given only in the 1976 reference.

Suffixes: *A*, from Ahrens, L. H. 1952. "The use of ionization potentials—Part I. Ionic radii of the elements." *Geochimica Cosmochimica Acta*, **2,** 155–169; *P*, from Pauling, Linus. 1960. *The Nature of the Chemical Bond*. Ithaca, New York: Cornell University Press, 644 pp.

Li — 1 IV **0.590**; VI *0.76*; VIII *0.92*

Be — 2 III *0.16*; IV **0.27**; VI *0.45*

Na — 1 IV **0.99**; V **1.00**; VI **1.02**; VII **1.12**; VIII **1.18**; IX **1.24**; XII *1.39*

Mg — 2 IV *0.57*; V *0.66*; VI **0.720**; VIII **0.89**

K — 1 IV *1.37*; VI **1.38**; VII **1.46**; VIII **1.51**; IX **1.55**; X *1.59*; XII *1.64*

Ca — 2 VI **1.00**; VII *1.06*; VIII **1.12**; IX **1.18**; X *1.23*; XII *1.34*

Sc — 3 VI *0.745*; VIII **0.870**

Ti — 2 VI **0.86**; 3 VI **0.670**; 4 IV *0.42*; V *0.51*; VI **0.605**; VIII *0.74*

V — 2 VI **0.79**; 3 VI **0.640**; 4 V *0.53*; VI *0.58*; VIII *0.72*; 5 IV **0.355**; V **0.46**; VI **0.54**

Cr — 2 VI L **0.73**; H *0.80*; 3 VI **0.615**; 4 IV *0.41*; VI **0.55**; 5 IV *0.345*; VI *0.49*; VIII **0.57**; 6 IV *0.26*; VI *0.44*

Mn — 2 IV *0.66*; V *0.75*; VI L **0.67**; H *0.830*; VII *0.90*; VIII *0.96*; 3 V *0.58*; VI L **0.58**; H *0.645*; 4 IV *0.39*; VI *0.530*; 5 IV *0.33*; 6 IV *0.255*; 7 IV *0.25*; VI *0.46A*

Fe — 2 IV H **0.63**; VI L *0.64*; H *0.780*; VIII *0.92*; 3 IV H **0.49**; V *0.58*; VI L **0.55**; H *0.645*; VIII *0.78*; 4 VI *0.585*; 6 IV *0.25*

Rb — 1 Vi *1.52*; VII **1.56**; VIII *1.61*; IX *1.63*; X *1.66*; XI *1.69*; XII *1.72*; XIV *1.83*

Sr — 2 VI *1.18*; VII **1.21**; VIII *1.26*; IX *1.36*; X **1.44**; XII *1.31*

Y — 3 VI *0.900*; VII *0.96*; VIII *1.019*; IX *1.075*

Zr — 4 IV 0.59; V 0.66; VI **0.72**; VII *0.78*; VIII **0.84**; IX 0.89

Nb — 3 VI **0.72**; 4 Vi *0.68*; VIII *0.79*; 5 IV *0.48*; VI **0.64**; VII *0.69*; VIII *0.74*

Mo — 3 VI *0.69*; 4 VI **0.650**; 5 IV *0.46*; VI *0.61*; 6 IV *0.41*; V **0.50**; VI *0.59*; VII *0.73*

Tc — 4 VI *0.645*; 5 VI 0.60; 7 IV 0.37; VI *0.56A*

Ru — 3 VI **0.68**; 4 VI **0.620**; 5 VI 0.565; 7 IV 0.38; 8 IV 0.36

Cs — 1 VI *1.67*; VIII **1.74**; IX **1.78**; X **1.81**; XI **1.85**; XII **1.88**

Ba — 2 VI *1.35*; VII *1.38*; VIII *1.42*; IX *1.47*; X *1.52*; XI *1.57*; XII *1.61*

La–Lu

Hf — 4 IV 0.58; VI **0.71**; VII 0.76; VIII **0.83**

Ta — 3 VI *0.72*; 4 VI *0.68*; 5 VI **0.64**; VII 0.69; VIII *0.74*

W — 4 VI *0.66*; 5 VI 0.62; 6 IV *0.42*; V 0.51; VI 0.60

Re — 4 VI **0.63**; 5 VI 0.58; 6 VI 0.55; 7 IV *0.38*; VI *0.53*

Os — 4 VI **0.630**; 5 VI 0.575; 6 V 0.49; VI 0.545; 7 VI 0.525; 8 IV 0.39

Fr — 1 VI 1.80A

Ra — 2 VIII 1.48; XII 1.70

Ac–Lw

La — 3 VI *1.032*; VII **1.10**; VIII *1.160*; IX *1.216*; X *1.27*; XII *1.36*

Ce — 3 VI *1.08*; VII *1.07*; VIII *1.143*; IX *1.196*; X *1.25*; XII *1.34*; 4 VI *0.87*; VIII **0.97**; X *1.07*

Pr — 3 VI *0.99*; VIII *1.126*; IX *1.179*; 4 VI *0.85*; VIII *0.96*

Nd — 2 VIII *1.29*; IX *1.35*; 3 VI *0.983*; VIII *1.109*; IX *1.163*; XII *1.27*

Pm — 3 VI *0.97*; VIII *1.093*; IX *1.144*

Sm — 2 VII *1.22*; VIII *1.27*; IX *1.32*; 3 VI *0.958*; VII *1.02*; VIII *1.079*; IX *1.132*; XII *1.24*

Eu — 2 VI **1.17**; VII *1.20*; VIII **1.25**; IX *1.30*; X *1.35*; 3 VI *0.947*; VII *1.01*; VIII *1.066*; IX *1.120*

Ac — 3 VI **1.12**

Th — 4 VI *0.94*; VIII *1.05*; IX **1.09**; X *1.13*; XI *1.18*; XII *1.21*

Pa — 3 VI *1.04*; 4 VI *0.90*; VIII *1.01*; 5 VI *0.78*; VIII *0.91*; IX **0.95**

U — 3 VI *1.025*; 4 VI *0.89*; VII *0.95*; VIII *1.00*; IX *1.05*; XII *1.17*; 5 VI **0.76**; VII *0.84*; 6 II **0.45**; IV *0.52*; VI *0.73*; VII *0.81*; VIII *0.86*

Np — 2 VI **1.10**; 3 VI *1.01*; 4 VI *0.87*; VIII **0.98**; 5 VI *0.75*; 6 VI *0.72*; 7 VI *0.71A*

Pu — 3 VI **1.00**; 4 VI *0.86*; VIII **0.96**; 5 VI *0.74*; 6 VI *0.71*

Am — 2 VII *1.21*; VIII *1.26*; IX *1.31*; 3 VI *0.975*; VIII *1.09*; 4 VI *0.85*; VIII **0.95**

B

3	III		*0.01*
	IV		*0.11*
	VI		0.27

C

4	III		**0.08**
	IV		0.15*P*
	VI		0.16*A*

N (with bar)

3	IV		1.46
	VI		0.16*A*
5	III		*0.104*
	VI		0.13*A*

O (with bar)

2	II		**1.35**
	III		**1.36**
	IV		**1.38**
	VI		**1.40**
	VIII		**1.42**

F (with bar)

1	II		**1.285**
	III		**1.30**
	IV		**1.31**
	VI		**1.33**
7	VI		0.08*A*

Al

3	IV		**0.39**
	V		**0.48**
	VI		*0.535*

Si

4	IV		**0.26**
	VI		**0.400**

P

3	VI		0.44*A*
5	IV		**0.17**
	V		0.29
	VI		0.38

S

2	VI		1.84*P*
4	VI		0.37*A*
6	IV		**0.12**
	VI		0.29

Cl

1	IV		1.81*P*
5	III		**0.12**
7	IV		*0.08*
	VI		0.27*A*

Co

2	IV	H	0.58
	V		0.67
	VI	L	**0.65**
		H	0.745
	VIII		0.90
3	VI	L	*0.545*
		H	0.61
4	IV		0.40
	VI		0.53

Ni

2	IV		0.55
	V		0.63
	VI		*0.690*
3	VI	L	**0.56**
		H	**0.60**
4	VI		0.48

Cu

1	II		**0.46**
	IV		0.60
	VI		0.77
2	IV		0.57
	V		**0.65**
	VI		**0.73**
3	VI		0.54

Zn

2	IV		**0.60**
	V		**0.68**
	VI		**0.740**
	VIII		0.90

Ga

3	IV		**0.47**
	V		**0.55**
	VI		**0.620**

Ge

2	VI		0.73*A*
4	IV		*0.390*
	VI		*0.530*

As

3	VI		0.58*A*
5	IV		**0.335**
	VI		0.46

Se (with bar)

2	VI		1.98*P*
4	VI		0.50*A*
6	IV		*0.28*
	VI		0.42

Br (with bar)

1	VI		1.96*P*
3	IV		0.59
5	III		0.31
7	IV		0.25
	VI		0.39*A*

Rh

3	VI		**0.665**
4	VI		*0.60*
5	VI		0.55

Pd

1	II		**0.59**
2	IV		**0.64**
	VI		**0.86**
3	VI		**0.76**
4	VI		*0.615*

Ag

1	II		**0.67**
	IV		*1.00*
	V		*1.09*
	VI		**1.15**
	VII		*1.22*
	VIII		*1.28*
2	IV		*0.79*
	VI		*0.94*

Cd

2	IV		*0.78*
	V		**0.87**
	VII		*1.03*
	VIII		*1.10*
	XII		**1.31**

In

3	IV		0.62
	VI		*0.800*
	VIII		*0.92*

Sn

4	IV		0.55
	V		0.62
	VI		**0.690**
	VII		0.75
	VIII		0.81

Sb

3	IV		*0.76A*
	V		**0.80**
	VI		*0.76*
5	VI		*0.60*

Te (with bar)

2	VI		2.21*P*
4	III		**0.52**
	IV		0.66
	VI		0.97
6	IV		0.43
	VI		0.56*A*

I (with bar)

1	VI		2.20*A*
5	III		0.44
	VI		**0.95**
7	IV		0.42
	VI		0.53

Ir

3	VI		*0.68*
4	VI		*0.625*
5	VI		0.57

Pt

2	IV		0.60
	VI		0.80*A*
4	VI		0.625
5	VI		0.57

Au

1	VI		1.37*A*
3	IV		0.68
	VI		0.85*A*
5	VI		0.57

Hg

1	III		**0.97**
	IV		1.19
2	II		**0.69**
	IV		**0.96**
	VI		**1.02**
	VIII		**1.14**

Tl

1	VI		**1.50**
	VIII		1.59
	XII		1.70
3	IV		0.885
	VI		0.98
	VIII		0.75

Pb

2	IV		*0.98*
	VI		*1.19*
	VII		1.23
	VIII		**1.29**
	IX		*1.35*
	X		1.40
	XI		*1.45*
	XII		**1.49**
4	IV		0.65
	V		0.73
	VI		**0.775**
	VIII		**0.94**

Bi

3	V		0.96
	VI		*1.03*
	VIII		*1.17*
5	VI		0.76

Po

4	VI		0.94
	VIII		1.08
6	VI		0.67*A*

At

7	VI		0.62*A*

Gd

3	VI		**0.938**
	VII		1.00
	VIII		*1.053*
	IX		1.107

Tb

3	VI		**0.923**
	VII		0.98
	VIII		**1.040**
	IX		1.095
4	VI		**0.76**
	VIII		**0.88**

Dy

2	VI		1.07
	VII		1.13
	VIII		1.19
3	VI		*0.912*
	VII		0.97
	VIII		*1.027*
	IX		1.083

Ho

3	VI		*0.901*
	VIII		*1.015*
	IX		1.072
	X		1.12

Er

3	VI		*0.890*
	VII		0.945
	VIII		*1.004*
	IX		1.062

Tm

2	VI		1.03
	VII		1.09
3	VI		*0.880*
	VIII		*0.994*
	IX		1.052

Yb

3	VI		*0.868*
	VII		0.925
	VIII		0.985
	IX		1.042

Lu

3	VI		*0.861*
	VIII		*0.977*
	IX		1.032

Cm

3	VI		*0.97*
4	VI		0.85
	VIII		**0.95**

Bk

3	VI		**0.96**
4	VI		0.83
	VIII		**0.93**

Cf

3	VI		**0.95**
4	VI		0.821
	VIII		0.92

Es

Fm

Md

No

Lw

Index

Mineral names in boldface type are those described in detail; page numbers in boldface type refer to the principal description of the mineral. Mineral occurrence and use data and authors of Selected Readings are not indexed.